Biophysics of Nucleic Acids Celebrating the 75th Birthday of Professor Kenneth J. Breslauer

Biophysics of Nucleic Acids Celebrating the 75th Birthday of Professor Kenneth J. Breslauer

Editors

Tigran Chalikian
Jens Völker

MDPI • Basel • Beijing • Wuhan • Barcelona • Belgrade • Manchester • Tokyo • Cluj • Tianjin

Editors
Tigran Chalikian
Leslie Dan Faculty
of Pharmacy
University of Toronto
Toronto
Canada

Jens Völker
Chemistry and Chemical
Biology
Rutgers, The State University
of New Jersey
Piscataway
United States

Editorial Office
MDPI
St. Alban-Anlage 66
4052 Basel, Switzerland

This is a reprint of articles from the Special Issue published online in the open access journal *Life* (ISSN 2075-1729) (available at: www.mdpi.com/journal/life/special_issues/75th_Birthday).

For citation purposes, cite each article independently as indicated on the article page online and as indicated below:

LastName, A.A.; LastName, B.B.; LastName, C.C. Article Title. *Journal Name* **Year**, *Volume Number*, Page Range.

ISBN 978-3-0365-6098-4 (Hbk)
ISBN 978-3-0365-6097-7 (PDF)

Cover image courtesy of Tigran Chalikian

Contents

About the Editors

Tigran Chalikian

Dr. Tigran V. Chalikian is a Professor in the Leslie Dan Faculty of Pharmacy at the University of Toronto. He received his Ph.D. in 1990 from the Institute of Biological Physics, Acad. Sci. USSR. His research interests are the biophysical characterization of molecular forces involved in the stabilization of proteins, nucleic acids, and their complexes.

Jens Völker

Dr. Jens Völker is an Associate Professor in the Department of Chemistry and Chemical Biology at Rutgers, The State University of New Jersey. He received his Ph.D. in 1994 from the University of Cape Town. His research interests lie in the areas of thermodynamics of nucleic acid structure, interactions, and function.

Preface to "Biophysics of Nucleic Acids Celebrating the 75th Birthday of Professor Kenneth J. Breslauer"

This Special Issue of Life celebrates the pivotal contributions of Professor Kenneth J. Breslauer to the field of DNA biophysics. The Special Issue represents a collection of 15 papers (4 dedicated reviews and 11 primary research articles) that deal with different aspects of nucleic acid structure, stability, and interactions with ligands and ions. Taken together, this Special Issue presents a timely overview of recent trends in nucleic acid biophysical research from a theoretical and experimental perspective. Although not exhaustive, the issue should represent a useful guide for scientists who wish to familiarize themselves with current topics in nucleic acid biophysics research. It should also be of interest to those who wish to better understand the physical and chemical principles underlying and governing biological functions of nucleic acids. A brief summary of the range of topics covered in this Special Issue is given below.

The first single crystal X-ray structure of a B-DNA oligonucleotide was published by Dickerson and coworkers back in 1980. Since then, the number of published X-ray and NMR DNA structures that have been deposited in the nucleic acid database repository (NDB) has increased dramatically. Berman, Lawson, and Schneider review recent developments and future plans for the NDB as a publicly accessible and centrally important community resource driving research into nucleic acid properties.

Although nearly 70 years have passed since the double helical nature of DNA was first proposed by Watson and Crick, details of the physical properties of the double helix are still actively debated and investigated. In their review of mechanical unzipping experiments of nucleic acids, Rissone and Ritort focus on equilibrium and non-equilibrium stretching behavior of DNA and RNA molecules and evaluate nearest neighbor thermodynamic parameters from a perspective of mechanical unfolding experiments (as opposed to the more conventional temperature unfolding thermodynamic analysis). Young and colleagues conduct an extensive reevaluation of nucleic acid nearest neighbor effects on local DNA conformation by making use of the greatly expanded number of newly deposited single crystal X-ray and NMR structure entrees to the NDB database. The extended database allows these authors a more rigorous and statistically more sound evaluation of such effects than prior efforts.

Sequence and site-specific ligand binding to desirable drug targets with minimal nonspecific binding to off-targets is the holy grail of drug development. Minetti and Remeta review the thermodynamic principles underlying ligand binding that can aid in drug development. In a subsequent research paper, Minetti and coworkers apply these principles to optimize the design of aurintricarboxylic acid (ATA) as viral inhibitor for targeting, *inter alia*, SARS-CoV-2 Viral Targets. In a related article, Paul and coworkers explore the use of minimalist pyridine ligands derived from well-established DNA minor groove binders to gain better insights into the forces that control molecular recognition and binding thermodynamics of minor groove complexes. Whereas the Wilson group (above) systematically varies the ligand to modulate DNA recognition, Scott and Chalikian turn this process on its head, by using known classical minor groove binders to discriminate between different DNA targets. They aim to use minor groove binders as a means to specifically bind and stabilize a subset of DNA quadruplexes containing hairpin like loop domains, while excluding or minimizing binding to other G quadruplex DNA forms. The structural similarity of the core quadruplex that is independent of details of the surrounding sequences make identifying ligands that bind only to specific G quadruplexes sequences to the exclusion of all others a challenging prospect, making targeting features of the loop a more promising option. G-quadruplex-forming sequences are often able to adopt several conformationally distinct, but closely related, four-stranded structures in response to variations in environmental conditions. Takahashi and coworkers highlight how different ligands targeted specifically to the core G quadruplex domain derived from a telomer sequence can preferentially interact and stabilize different conformational isomers of the G-quadruplex and alter DNA replication outcomes in an in vitro replication assay. These observations suggest that targeting G-quadruplexes holds promise for therapeutic interventions if ligands can be made to target specific G-quadruplexes to the exclusion of all or most others.

As indicated above, DNA G-quadruplexes represent particularly interesting noncanonical DNA structures, not only because of their multitude of closely related and often dynamically interchanging conformations, but also because sequences that are able to adopt G quadruplex structures are often closely associated with biologically relevant sequence domains and have been reported to play crucial roles in the control of gene expression and in telomer stability, for example. The article by Prislan, Urbic and Poklar-Ulrih reflects the dynamically interchanging nature of the d(G4T4G3) G-quadruplex by promoting an unconventional kinetic model to analyze heat capacity data at variable heating and cooling rates for this construct. Tariq and coworkers use differences in the pressure dependence of the TM for the intermolecular parallel stranded G-quadruplex formed from d(TGGGGT)4 in comparison to the structurally related intramolecular parallel-stranded G-quadruplex formed by the Pu22-T12T13 variant of the VEGF transcription initiation site to argue for the importance of loop bases in influencing quadruplex hydration.

The success of DNA modeling efforts often depends on the relevance and accuracy of the force field chosen to model DNA interactions. Schneider et al. report a self-consistent set of parameters for DNA residues for application with the AMBER force field that will allow improved modeling of dynamics in novel DNA polymers, an area that is of increasing importance going forward.

The polyanionic nature of nucleic acids makes their physical properties highly depend on salt conditions. The article by Knop and colleagues investigates the properties of nucleic acids at extremes of salt and pressure such as encountered on Mars as one example. The study represents an effort to assess if conditions on Mars are supportive of DNA-based life forms as we know them.

The polyanionic nature of DNA also represents a repulsive force between adjacent DNA duplex chains. Nevertheless, attractive interactions between neighboring DNA chains have been observed, and they play a critical role in DNA packaging and in homologous strand exchange events in nature. They are also important for various DNA surface depositing protocols for biotechnological applications. The article by Chen and Pettitt investigates the potential of mean force (PMF) between neighboring DNA chains to understand how DNA flexibility impacts attractive interactions between neighboring DNA chains.

Besides nonspecific ionic interactions typical for alkali metal ions, DNA can form specific DNA metal complexes with transition metal ions. The article by Ihara, Kitamura, and Katsuda reviews some of their recent work on transition metal nucleic acid interactions that were designed to stabilize specific DNA structures (e.g., DNA triplexes), assist in dynamic DNA splicing, and/or amplify metal DNA signals (e.g., lanthanide luminescence) for biotechnological applications. The article demonstrates how judicious application of insights into DNA structure formation together with knowledge of specific transition metal nucleic acid interactions can be combined to develop clever concepts for nucleic acid-based biotechnology and sensing applications.

Last but by no means least, the famous experiments by Stanley Miller and Harold Urey in the 1950s established that the basic building blocks for the development of life could form in the early earth environment, but how does one get from such basic building blocks to the complexity that is necessary for life to develop? In an intriguing article, Farquharson, Agozzino, and Dill attempt to address this fundamental question with a simple non-equilibrium thermodynamic model that allows for spontaneous polymerization and interactions between two kinds of molecules (information and functional molecules, loosely associated with nucleic acids and proteins, respectively) confined within a membrane system that can grow to a limited size only before dividing. Time evolution of such a simple system, coupled with an application of only a very few physical laws, quickly generates complexities not unlike those seen in real life. Such model studies show that life in all its complexities could indeed have evolved from simple precursors.

To summarize, this Special Issue covers topics ranging from the structure and thermodynamics of canonical and non-canonical DNAs and their complexes to efforts to understand interchain interactions between double helices and to develop nucleic acid-based sensors. The authors employ a gamut of biophysical techniques ranging from the classical spectroscopic, calorimetric and, pressure perturbation tools to the more modern single molecule force extension methods as well as a range of computational approaches for their studies. As such, it is hoped that the articles in this Special Issue will be informative and helpful to an interested reader in highlighting recent developments in the field of nucleic acid biophysics.

Tigran Chalikian and Jens Völker
Editors

life

MDPI

Editorial

Tribute to Kenneth J. Breslauer

Tigran V. Chalikian

Department of Pharmaceutical Sciences, University of Toronto, Toronto, ON M5S 3M2, Canada; t.chalikian@utoronto.ca

Citation: Chalikian, T.V. Tribute to Kenneth J. Breslauer. *Life* **2022**, *12*, 1325. https://doi.org/10.3390/life12091325

Received: 29 July 2022
Accepted: 24 August 2022
Published: 27 August 2022

It is an exciting experience to serve as guest editor for a Special Issue celebrating the 75th birthday of Professor Kenneth J. Breslauer. For more than five decades, Ken Breslauer has been one of the giants shaping the field of biopolymer thermodynamics. After doctoral studies in the laboratory of Prof. Julian Sturtevant and postdoctoral training in the laboratory of Prof. Ignacio Tinoco, Ken joined the Department of Chemistry of Rutgers University. That was in 1974, and he has remained loyal both to Chemistry and to Rutgers ever since.

Ken's lasting impact on molecular biophysics has been especially prominent in his fundamental quest to elucidate the energetic determinants of DNA polymorphism, drug–DNA interactions, DNA repair, and, more recently, the molecular mechanisms of genetic diseases associated with uncontrolled triplet expansion. In retrospect, Ken's pivotal and pioneering studies, particularly into the nearest-neighbor-based thermodynamics of DNA stability and the differential energetics of drug binding to compositionally similar but structurally and hydrationally distinct DNA duplexes, emerge as high points in the progress in our understanding of the physical principles governing the structure and function of DNA.

Ground-breaking work in the Breslauer lab convincingly showed that the association of the quintessential minor groove binder netropsin with the poly(dA)poly(dT) and poly(dAdT)poly(dAdT) duplexes, while being characterized by nearly identical binding free energies, has dramatically different enthalpic and entropic origins [1]. Ken's seminal paper on the nearest-neighbor thermodynamics of DNA, published in *PNAS* in 1986, inspired wide-spread endeavors that continue to this day [2]. Those efforts have resulted in the development of nearest-neighbor databases for computing the full thermodynamic profiles of the duplex-to-single-strand transitions of any duplex DNA, including canonical B-DNA, mismatched DNA, mutagenic DNA lesions, and hairpins, under different environmental conditions (for summary, see ref. [3]). More recently, his studies have provided novel thermodynamic insights into the evolution of the genetic code [4] and resulted in the discovery of "rollamers" [5]. The latter represent dynamic polymorphic DNA structures that may explain the inception and progression of triplet expansion-based genetic diseases and suggest possible avenues for their treatment.

For decades, the Breslauer lab has been a conceptual powerhouse in the development of thermodynamic insights into the functioning of DNA. As such, it has been a hub for collaboration among, and a destination for visits from, highly accomplished scientists from around the world. I was fortunate to spend five years between 1992 and 1997 in this vibrant atmosphere. Ken's mentorship was unobtrusive and unforced, giving you all the freedom you needed to test your ideas. At the same time, he was generously supportive should you need his guidance to overcome a seemingly insurmountable obstacle. More than anything, he was a constant and wonderful example of a superb scholar and effective principal investigator. I learned a great deal from him. I learned the importance of looking broadly into the scientific problem at hand and of thinking outside the limits of a specific experimental technique, employing to that end a combination of techniques, each offering a unique insight into the problem.

I gratefully acknowledge that, throughout my subsequent career as an independent researcher, I felt Ken's gentle support and guiding attention. Recently, helping my son with his biology class, I recounted the Hershey–Chase experiment in which DNA, and not the protein, was shown to carry the infectiousness of the phage. It occurred to me that I knew first-hand that DNA is infectious. Before joining Ken's lab, proteins had been the center of my research interests. Ken infected me with his enthusiasm and passion for DNA. Twenty-five years later, I still retain that infection.

This Special Issue is a tribute to Ken's influential presence in the field of DNA biophysics. It comprises 15 papers that broadly encompass the current state-of-the-art in the field. They include experimental and theoretical studies into the stability and interactions of canonical and noncanonical DNA structures. In some respects, these studies parallel, complement, and extend concepts Ken Breslauer articulated decades ago and has been developing ever since.

Below, we present, in alphabetical order, the testimonials and personal accounts from people who have known Ken at different stages of his career as a mentor, a colleague, or a friend.

1. Robert Boikess, Professor

When I was Ken's department chair, and even afterwards, looking after his best interests was always one of my major concerns. When he was up for promotion and tenure, some senior faculty were concerned that he didn't have enough funding. I assured them that a lot more would be coming (something of an understatement). Over the years, his rare combination of skills as a research scientist and as a scientific administrator attracted the attention of other universities. Fortunately, I was able to convince the Rutgers administration to do what was necessary to keep him at Rutgers. These actions were something that neither I, nor anyone else at Rutgers, ever regretted.

2. Walter A. Dickerhof, Lab Technician

My employment under Ken at Wright-Riemann Laboratories was the first full-time position in my field of study after having graduated college. Despite his impressive stature at the University, Ken proved to be remarkably approachable, as well as extremely welcoming to those new to his group thereby facilitating a seamless integration into same. Though the essence of our affiliation was implicitly professional, Ken possessed a genuine sincerity underpinning this professionalism and serving to foster a relationship at a more personal level while at the same time preserving his authority.

Years after I had left the University, I contacted Ken regarding a reference for potential employment and was pleasantly surprised as to how enthusiastically receptive he was after so many years; this moment being void of any awkwardness and having a feel as if it were days, not years, which had passed since our last conversation. That communication, as well as those following during the brief period thereafter, had this old-friend kind of vibe to them. For this, I cannot find words sufficient but can only state, simply, that it speaks to his character, to wit: typically, it is the employer who expects loyalty from their subordinates; nevertheless, nearly two decades later, still I had Ken's support. This was an exceptional act of loyalty, one by which I felt flattered and, even more so, humbled.

Inarguably intrinsic to Ken's nature, the afore-described attributes instilled in me a visceral sense of belonging as well as a strong feeling of pride to be a member of his research group; sentiments which have persisted with me until this very day.

Thank you, Ken, for your kindness, your generosity and all that you had done for me then and what you have done for me since.

Happy Birthday, my friend.

3. Ann Doeffinger, Administrative Assistant

Simply having the opportunity to wish "75th Happy Birthday" to—and help celebrate the life of—such a Distinguished Professor and a gentleman as Dr. Kenneth Breslauer is in itself a gift.

It has been an immense honor and pleasure to work in the Chemistry Department alongside one of its most pre-eminent professors, as he is truly a remarkable person in every respect.

> "I have come to believe that a great professor is a great artist and that there are few as there are any other great artists. Teaching might even be the greatest of arts since the medium is the human mind and spirit" —*John Steinbeck*

4. Dick Foley, Dean, Professor Emeritus

As Dean of Arts and Science at Rutgers for a decade, I worked with Ken and saw first-hand what a remarkable academic administrator he is. "Remarkable" understates the case. His record of successes at Rutgers is unprecedented. It was during that decade that he oversaw a reorganization of the life sciences and established the Division of Life Sciences, but it was anything but an ordinary reorganization. Every university I have been associated with has an exaggerated sense of itself as uniquely complex, but at Rutgers in those years, my love for the place notwithstanding, this was no exaggeration. Its unusual history and administrative structures had produced a confusing university-wide tangle of schools, departments, and institutes that duplicated each other's research and teaching portfolios in the life sciences and did so, frankly, with notably different degrees of quality. The task of unsorting and then reassembling this jumble demanded equal measures of academic vision, political skill, and sheer tenacity. It was a herculean task which Ken pulled off in ways that no one else could have approximated, the eventual result being a well-integrated collection of prospering departments, centers, institutes, state-of-the-art buildings, core facilities, and the list goes on and on. Impressive as all this is, the most important of Ken's successes over the years are the result of his eye for intellectual talent and ability to attract it. Time after time he competed successfully against far better endowed universities to recruit and retain top faculty. He did so with his characteristic persistence but also ingenious arrangements, all done with a fraction of the resources of the competition. With apologies to Churchill, never has so much been accomplished with so little. It was exciting to be Ken's partner and co-conspirator during that decade, but it was also just great fun. Little in life is better than working on undeniably valuable projects with someone you enormously admire but also hugely enjoy.

5. Barbara Gaffney, Lecturer

I was a Lecturer in the Douglass College Department of Chemistry when, in 1974, our esteemed chair, Stan Mandeles, was fortunate in recruiting Ken Breslauer as a young Assistant Professor. Stan had himself been recruited to build a cutting-edge research department at Douglass that was to be focused at the interface of chemistry and life sciences (this was well before the 1981 reorganization that combined the various college chemistry departments into a single unit at Rutgers Wright Labs). Our small Douglass faculty were friendly and collegial, and Ken fit right in, with his great charm and wit. He was a natural teacher, so he was exceedingly popular with the outstanding Douglass undergraduates. Research funding was scarce, so Stan's new hires were all encouraged to seek external funding, at which Ken showed exceptional skill. With great care, Ken set up his biophysical lab with good instrumentation: his first spectrophotometer and calorimeter. He mentored several graduate students but depended primarily on post-doctoral fellows. After the 1981 reorganization of the college chemistry departments, Ken would soon move his operation to Wright Labs, where it flourished. At Douglass, he had overseen the medical technology program, so perhaps it was natural that at Busch he would again choose to combine research with administration. In 1985, he became the Associate Dean of Life Sciences, and the rest, as they say, is history.

6. Craig A. Gelfand, CEO

It was a tremendous privilege to have been part of the Breslauer lab in the late 1990s, and only later could I look back with full appreciation upon the spectrum of scientific and career learnings. It is impossible to distill 4 years into a few paragraphs. To my non-scientist friends, I often tell silly one-liners like the fact that a lot of us developed an unhealthy addiction to caffeine to offset the slow pace at which a DSC generates data. I will leave deeply scientific topics to my esteemed colleagues, the amazing people that continued to work and thrive in the thermodynamics community. While my career path moved me away from thermodynamics as a main research topic, the lessons I learned in the Breslauer lab, much like the science of thermodynamics itself, always seem apply to the projects that I am dealing with. So, I would like to take this opportunity to thank Dr. Ken Breslauer for his patience in teaching an excited kid that had a lot to learn and to describe a few of the key universal lessons that I was privileged to have learned from a wonderful mentor.

One lesson that applied over-and-over during my career was that hypothesis-driven experiment planning needed to be balanced with flexibility to let resulting data tell their own story—in essence, do not let your preconceived notions interfere with thorough and thoughtful interpretation of your data. A few aspects of my own work as a post-doctoral fellow and research associate in Ken's lab are perhaps a small example of this, but the body of work coming from the Breslauer lab at large serves as tribute to the breadth of inventiveness that this approach enables: the data themselves are the main storytellers; successful researchers listen to those stories, communicate them by publication or presentation, and then help write the next chapter of the story by devising new experiments based on the prior data. I write this knowing that to many this will seem superficial (e. g., "that's how science is supposed to be done"), but throughout the years of my own career that followed, I observed others (and occasionally caught myself) making the mistake of preconceiving desired outcomes, often leading to derailed projects; so, it's a true career benefit that I received this learning during my post-doc years. Productive scientists succeed by being clever, flexible, and patient observers and good stewards of data. Among the achievements that I take most pride in from my own work are the few times when I was asked to solve 'project-killing messes' merely by letting each problem tell itself to me first through the existing data, then navigating to successful solutions.

A correlated lesson is the ability to tell those stories. Ken is a gifted communicator, able to convert the complicated into the understandable, capable of presenting, seemingly effortlessly, to a wide spectrum of people from deeply technical to lay audiences and, even more uniquely, with flair and enthusiasm. There is an intrinsic career barrier that I think is a particularly acute burden in complex fields such as thermodynamics: a communication 'activation energy barrier' that needs to be overcome. The work of any scientist can be rendered essentially irrelevant if we cannot communicate it, sometimes with the paradox that the more unusual or novel the story that needs to be told, the higher the burden for precise and concise communication. Ken is a master, and it is clear that he took extra effort to mentor us lab folk to be better communicators, in parallel with teaching us the science. I clearly recall being exposed to the equally important yet somewhat different skills required to prepare effective oral presentations (through hours of editing and re-working of slides—especially in the prior era when physical transparencies and slides needed to be finished well ahead of time) and written works (I recall at least a few times that Ken somehow managed to write more by hand in red ink in the margins and on backs of pages than I had typed in preparing a draft). These are lessons that are more easily appreciated in retrospect, and skills that I am certain benefitted my career. My Ph. D. advisor, Dr. Joyce Jentoft, had a similar passion for teaching the critical skill of communication of science, so I was doubly fortunate of having two mentors drive this topic. I hope that I have honored the importance of these lessons by passing them on to scientists that I trained and worked with, helping support their careers in the way that my mentors helped me.

I look forward to the privilege of continuing to learn from Ken and from my continued association with his lab.

7. Vera Gindikin, Research Associate

I am grateful for this opportunity to express the appreciation I have had for many years now for Dr. Breslauer and to convey how lucky I feel to have been working in his lab.

Dr. Breslauer, first and foremost, is an incredible, brilliant, and highly devoted scientist, who never ceased to amaze me with his extraordinary mind, vast knowledge, and unique perspective on things. These qualities are well known and appreciated by many who have met and collaborated with Prof. Breslauer, quite a few of whom are contributing to this publication. However, I am also fortunate to have had a personal perspective of Dr. Breslauer by having been a part of the Breslauer lab for the past three decades.

I met Dr. Breslauer for the first time a few years before I joined Rutgers University, when I visited the US while still being an undergraduate student majoring in Biochemistry and Molecular Biology. I was planning to apply to graduate schools and wanted to talk to faculty at Rutgers chemistry department. Dr. Breslauer not only agreed to meet with me but spent a long time of what I am sure was a very busy day talking to me and then introducing me to some of the other faculty members. I also remember how incredible comfortable he made me feel; it was all very surprising and unusual. I had never met people like him before, and now, more than 30 years later, I can honestly say that I have never met people like him since.

I consider myself very fortunate for having joined Ken's lab as a graduate student in 1992 and to be able to continue working in the lab after graduation. Not only did I have the luxury of scientific research flexibility and the input and care of a wonderful scientific advisor, but I also always had incredible personal support from Dr. Breslauer during some difficult times in my life. Dr. Breslauer always found kind words and great solutions to help me, and it made me feel stronger because I felt that support. I think it is an incredible gift to feel such support from one's advisor, which only a very few lucky people get to experience. I am among these few lucky ones, and I am eternally grateful.

Happy Birthday from one very grateful and appreciative "last graduate student"!

8. Arthur P. Grollman, MD, Professor

The Shakespearian theme "What's past is prologue", coupled with the quest of "How small molecules do great things" that guides research in the Zickler Laboratory of Chemical Biology (LCB), epitomizes my 30-year collaboration with Ken Breslauer and Francis Johnson, Director of the Division of Medicinal Chemistry at Stony Brook. Our collaboration continues apace on Ken's 75th birthday as it did when the three of us were considered young Turks.

In 1993, funding was obtained from the National Cancer Institute to support a Program Project grant (PPG), "Exocyclic DNA Adducts and Oxidative DNA Damage." The central theme of this multidisciplinary program involved relationships between molecular structure, energetics, and biological activity. The integration of these factors proved to be central to understanding the molecular mechanism(s) of DNA replication, DNA repair, and mutagenesis. At that time, the mapping of energy landscapes that link structures with biological functions were entirely lacking. Breslauer's research and databases proved critical in bridging this experimental and conceptual gap. His group's seminal research was a unique feature of our PPG, contributing to enthusiastic reviews and continuous NIH funding for this research program for more than 20 years.

The site-specific introduction of single lesions into DNA, using innovative synthetic methods developed in the Johnson lab at Stony Brook, provided modified DNAs with which Breslauer and his talented research associates at Rutgers generated novel energetic insights and concepts, designed to complement structure-based interpretation of biological processes [3].

As a physician-scientist, I was acutely aware of the paucity of drugs considered safe and effective in treating human viral disease. Accordingly, when I initiated a program on rational antiviral drug design, supported by private philanthropy, it was clear that Breslauer and his group were among the few who could meet the exacting goals of such a program. Ken serves on the three-person Executive Committee that guides our blue-ribbon consortium. Our first report (on the active principle of ATA) appears in this Festschrift. As orally effective drugs are urgently needed to combat SARS-CoV-2, we have chosen to repurpose emetine as an antiviral, using low doses to minimize the well-known side effects of this established amebicide. Ken's extensive experience with drug development, together with administrative skills honed over decades, which are summarized in this Editorial, are playing an important role in overcoming the bureaucratic roadblocks associated with our drug development research.

As a personal friend and colleague, Ken has an upbeat attitude toward others diplomatically helping where he can. This is supported by an imaginative capability to solve problems through useful suggestions, always with a light touch and a wonderful sense of humor. He is a scientist for all reasons in all seasons.

9. Martha Haviland, Professor

I first met Ken Breslauer when he was the Dean, Division of Life Sciences, Linus C. Pauling Distinguished Professor, and Vice President, Health Science Partnerships, and I was the Director of Undergraduate Advising for the Division of Life Sciences. As I understand the history, it is Ken's vision and his ability to work with and motivate others that led to the development of the Division and, ultimately, the undergraduate program in life sciences at Rutgers. But it was not until I moved into the position of Director of the Undergraduate Instruction for the Division and reported directly to Ken that I truly understood his dedication to undergraduate education. As just one example, under his leadership and guidance, we revamped our introductory General Biology curriculum, received funding from NSF and the State of New Jersey for our efforts, and leveraged this external funding to secure internal funds to renovate undergraduate classroom laboratory spaces as well as increase the number of teaching assistant positions assigned to the Division. You might expect that a person with his responsibilities might not prioritize individual undergraduate students, but Ken ensured that every e-mail from a student was answered. Some of my fondest memories are when Ken attended undergraduate celebrations, whether they be research symposia, graduation events, or Rutgers Day. His smile is engaging, he always has his hand out to shake yours, and he made each student he spoke with, and their parents, know that they were heard and respected. Finally, on a personal note, Ken encouraged me and supported my growth as an academic and an administrator. He provided me with the autonomy and authority to make decisions and move forward, but I always knew he had my back and would step in to provide support if asked.

10. Roger Jones, Professor Emeritus

Ken has always been a visionary.

I first met Ken in 1977, upon my hire into the Chemistry Department of Douglass College, which is part of Rutgers University. Ken was a few years ahead of me in the Assistant Professor ranks, but he was already influential in the department. Ken was universally liked by both faculty and students. He never missed an opportunity to assist, support, and encourage his students to do their best in addition to being a loyal and professional colleague. We had both been hired as part of the plan of the Chair, Stan Mandeles, to add nucleic acids research to the department, and we quickly developed a lifelong friendship. Nucleic acids research was not common in chemistry departments at the time and was not represented in the much larger Rutgers College department. It happened that the two departments were merged in 1981, and it was my fate to come up for tenure in 1982. Likely, it was Ken's eloquence that convinced the newly combined faculty that my synthetic nucleic acid program was worth promoting. He was always ready to help a colleague.

Ken's long-standing goal was to grow life sciences at Rutgers, which he did with great success! From his time at Douglass College, and continuing through his university administrative positions, Ken proved as adept at raising Rutgers funds to support his life science goals as he was at raising NIH funds to support his research program. In my 42 years at Rutgers, Ken is unique in his ability to sustain a high-quality research program while simultaneously a highly successful series of substantial administrative positions.

> "The qualities of a great man are vision, integrity, courage, understanding, the power of articulation, and profundity of character" —*Dwight Eisenhower*

11. Mike Kiledjian, Professor

Happiest of birthdays Ken!

Thank you for all your tireless dedication and support of the Life Sciences at Rutgers University. Your vision to coalesce life science research within the School of Arts and Sciences into a single division and build a first-rate research program was visionary, and we are all better now because of your forward thinking.

Ken and I first met when I was an undergraduate student at Rutgers working in a lab within the Department of Chemistry. He and my research advisor were colleagues, and Ken was a valuable source of knowledge and guidance. A decade later in 1995, Ken was instrumental in recruiting me back to Rutgers as a junior faculty member in what is now the Department of Cell Biology and Neuroscience (CBN). I have had the pleasure of working closely with Ken in his role as Dean of the Division of Life Science. Ken's tireless efforts to establish, foster, and lead the Division of Life Sciences at Rutgers is a testament to his commitment to scientific excellence.

On a personal level, I am grateful to Ken for his support and mentorship during my appointment as Chair of CBN in 2012. The CBN faculty and I are forever indebted to Ken for his unwavering support throughout the years that enabled us to recruit outstanding junior faculty and establish a superb foundation for the growth of the department. This momentum continues to this day and was initiated in large part to Ken's support.

Cheers to 75 amazing years and to many, many more to come!

12. Elsa Klump, Artist, Widow of Professor Horst Klump

What do I recollect of the Ken Breslauer stories, with which my beloved Horst Klump entertained me? Of course, it was not academic. I am a sculptor, a listener, and an enthusiastic identifier with the lives, thinking, and products of others. This account is therefore second-hand creative hearsay.

How did Horst and Ken connect? To me they recognized each other as brilliant entrepreneurs and scientists, streetwise survivors who could compete, work, and play hard—both top academics with talents in many more directions, bordering on genius.

Horst was an Assistant Professor in the Dept of Physical Chemistry 2, and at times Acting-Head, at the University of Freiburg, Germany, when Ken Breslauer arrived in 1981/82 on a Humboldt Fellowship for post-doctoral research. There were complications though. Both the families of Ken's father and mother, Breslauer and Schaeffer, had fled Germany during the Nazi regime, ahead of the Hitler horrors. Their home was on the East-Coast USA now, and Ken's father had misgivings about his son's return to Germany. So, the parents accompanied their very adult son—Ken Breslauer—back to Germany to assure his well-being, comfort, and safety.

Horst received the Breslauers and welcomed Ken into his family life with two young sons. For Horst it was an honor, and he formed a strong connection with Ken. They were both top researchers, breaking new ground; they understood and appreciated each other's projects, even when one was more than a decade senior to the other. They were both keen sportsmen. Horst was a yachtsman, a tennis and soccer player, and skied in winter. Ken could have become a professional sportsman/baseball player. Together they had fun. As I see it, Ken is a team sports person, whereas Horst was into individual sports. Somehow, through Ken's personality, Horst was roped in. The parents could leave their son behind in Horst's trustworthy hands. In turn, Horst could trust Ken and eventually was even willing to entrust his lifetime research to him.

When Ken returned to academic life in the USA, they stayed in contact. Horst would visit the Breslauer family when on sabbatical or attending conferences in USA, first at Ken's parental home in Queens and later near Rutgers, Ken now having two sons of his own. At first, Ken focused on research, but he quickly extended his talents into managing, developing, and expanding the status of life sciences at Rutgers, The State University of New Jersey, USA. He kept his research lab, entrusted to capable researchers, but most of his time was taken up running Rutgers.

In 2000, when Ken was a Linus Pauling Professor of Chemistry and the Dean and Director of Life Sciences at Rutgers, Horst (and I) came to Ken's lab, initially for the first 3 months of Horst's sabbatical year from UCT. Horst found this setup at Rutgers so conducive to his own research that (although he also planned periods at West Coast Universities) he preferred to stay for most of that year and, from 2002 on, returned annually for the allowed three-month stints for another 14 times. Horst, as a brilliant original researcher and talented lateral thinker, found it difficult when others would take his ideas to appropriate the work for themselves. However, Horst was willing to entrust Ken with his unpublished research accomplishments—"ploughing with other's oxen". Ken could comprehend the thrust of a thought direction, a research paper, facilitate possibilities and even repackage the presentation of a publication to make the work more accessible and marketable, but he would always acknowledge the creator/initiator/pioneer. The Breslauer family CAN write. So could Horst, but better in his mother tongue, German.

I got the impression, as an outsider in science, that Ken appointed scientists he trusted and then allowed them a measure of freedom to develop science projects that suited their own talents and personalities best. Jens Voelker, a former PhD student of Horst's from UCT, now part of Ken's group, put it this way: "This (research) could not be done without Ken's support and his willingness to let us play".

Nancy Ludowicki, PA, was Ken's backbone and was hands on with Ken's many Rutgers University projects. Nancy could achieve directives on many levels. I think Nancy's capabilities were wide. Nancy equipped and installed us, for example, into our living quarters at Treetops Apartments at the edge of Rutgers' "Ecological Forest", where wildlife abounded. Apparently, the day before she died, when Ken visited Nancy in hospital, Nancy still gave Ken the where's and whatall's that she felt Ken needed to know about the tasks that she was leaving behind.

During Horst's sabbaticals, Ken arranged for me to do sculpture at the Rutgers arts school lab on Livingston campus. What a privilege! I had interesting discussions on art with Ken Breslauer, and I was amazed at the depth of Ken's understanding, insights, and advice upon explaining my sculptural art.

I worked in clay and wood and received all the support needed from Rutgers staff. How grateful I am. Thanks to this arrangement enabled by Ken I was able to exhibit with the art lecturers at the Gallery of Mason Gross School of Art and to connect with so many creatively different post-graduate students. I would like to believe I contributed to the development of their thought processes.

13. Ajar Kochar, MD, MHS

I had the privilege of first meeting Dr. Breslauer circa 2005–2006 as a young, highly impressionable Rutgers undergraduate student. At the time, stem cell research was very much in vogue as a potential treatment modality for a panoply of medical pathologies. However, the ethical dilemmas surrounding the use of embryonic stem cells was extremely politically charged. Rutgers, and New Jersey more broadly, were being considered as a central hub for a major stem cell research center. Rutgers was and remains a driving force for academic scholarship and community engagement. Rutgers is able to play this role in large part due to the inimitable leadership of icons like Dr. Breslauer. Dr. Breslauer, in partnership with another college mentor of mine—Dr. Wise Young—helped organize a series of community-based events focused on stem cell research.

I initially got to know Dr. Breslauer in helping to plan these stem-cell-research-based conferences. I vividly recall the first-time meeting with Dr. Breslauer in his office. I was simply awed by his inimitable presence as he radiated a unique combination of academic brilliance, administrative nous, and genuine compassion. Looking back, I had no business being anywhere near those conferences. I had at best a paltry understanding of the science let alone the complex interplay between the science and politics of high-level scientific investigation. Yet, Dr. Breslauer never let me or my classmates (including the incredibly talented Dr. Nakul Raykar, who is now a rock-star trauma surgeon/clinical researcher at Brigham and Women's Hospital/Harvard Medical School) ever feel like our opinions were immaterial. In fact, he encouraged us drive the ship while skillfully shepherding us through the process. Under Dr. Breslauer's mentorship, we were able to help execute a very well attended and insightful series of programs focused on stem cell research.

Much to my delight, after the completion of these stem cell research events, Dr. Breslauer continued to remain a key college mentor and role model. The transition from college to "what's next" is always an incredibly stressful period. Dr. Breslauer played an integral role with my application to medical school, even going so far as to make phone calls to advocate on my behalf. I am confident that without his support I would not have ultimately landed at Brown Medical School, where I had a tremendous experience in medical training. Not only did Dr. Breslauer go above and beyond to support me and many others, but he was incredibly encouraging and nurturing during the process. Moreover, he provided a fantastic example on how to pursue a career in science employing steadfast dedication, passion, and a pinch of humor. Admittedly, I am still at the beginning phases of my own academic career—but I often think about Dr. Breslauer as a North Star role model to emulate.

I would be remiss not to mention that some of my most enjoyable experiences with Dr. Breslauer have been beyond the realm of academics. I was a senior when Rutgers beat Louisville in 2006. I recall celebrating with Dr. Breslauer a few days later and Dr. Breslauer explaining the downstream positive impacts of such a victory on the entire University community. More recently, I have relished celebrating the incredible accomplishments of our Men's Basketball team successfully qualifying for the NCAA tournaments in back-back (really should be a third back . . .) years. Dr. Breslauer's passion beyond science speaks to the Renaissance man that he is. I only hope I will continue to have the opportunity to learn from his incredible expertise and will conclude by simply saying: thank you for a lifetime of memories and mentorship.

14. Ernie Lepore, Board of Governors Professor

I have known Ken for decades now, and I have always thought of him as something like a big brother, probably because he always helps me get out of various messes I created; twice he bailed out two of my graduate students with summer funding when I forgot to secure the funding myself. One of them is at NYU now; the other is at Princeton. Thank you, Ken, from them both. Several times he's allowed me to use the life sciences building for conferences I organized. This was especially helpful during the hurricanes we suffered. And the list goes on and on. Oddly, I even trust his judgment more than I trust the judgment of my own internist and his internist, too. His storehouse of knowledge is something to marvel—ranging from chemistry to life sciences to sports to academia to campus politics and, yes, even to medicine. To envy his long list of accomplishments would not only be presumptuous but would also deprive you of enjoying them. I have been pretty successful in moving Rutgers forward: building an internationally recognized department, hiring National Academy of Sciences members. But then there's Ken—one building, two buildings, three buildings. Incredible. I sometimes wonder whether Rutgers has enough space to house all the buildings that Ken's planning. But it would be remiss not to acknowledge the magic of his voice. Whenever I hear it, it brightens my day.

15. Charles Martin, Professor Emeritus

Ken's extraordinary science will be clear from others who write here. Like many others, I followed his research and benefitted from his scientific achievements. I can also testify about as his skills as an administrator and program builder. In my opinion, much of expansion of the life sciences at Rutgers and the reunification of the medical school with Rutgers can be directly attributed to Ken's efforts.

In 1996, Ken was appointed as the Dean of the new Division of Life Sciences within the School of Arts and Sciences. Ken asked me to assist him and got me appointed as the Director of the Bureau of Biological Research—it was at this point that I was able to closely watch and marvel at his administrative, social, and political abilities. Ken has a remarkable gift for bringing people together, extracting money and effort from them, and getting them to do big things. His first efforts were to raise money for recruiting and building. And what a job he did! Through his well-founded connections built during the first reorganization in the 1980s, he managed to acquire internal funds from various sources within the University. He also recognized that he would need to hire "rainmakers"—well-funded and nationally recognized scientists to build and head new departments and research centers. His first effort was to hire Wise Young to establish the Center for Collaborative Research in Neuroscience. Wise was a prominent NYU Neurobiologist who specialized in spinal cord injury research. Ken recruited Wise and supported his fund-raising efforts from sources ranging from the Keck Foundation (a USD 2.1 million grant), Wall Street bankers, pharmaceutical firms, you name it. We had many meetings, and I got to watch up close (with some "I didn't know you could do that", jaw-dropping amazement) as he and Wise organized and ran fund-raising events on Wall Street and at the Lincoln Center that were used to build the Center in a large space connected to Nelson Laboratories. This effort paid off very well in the early days of the Division and brought a lot of good press to the University and the Division of Life Sciences.

Ken's second big effort was recruiting Jay Tischfield to establish the Department of Genetics and the Human Genetics Institute. Jay was a well-known human geneticist at the University of Indiana Medical School who had established a well-funded human cell repository. When Jay moved to Rutgers, he greatly expanded the repository and its funding base, which was then used to recruit top faculty and establish the impressive Department of Genetics that we see today. Ken worked in many ways to support this new department, including raising funds to build the Life Sciences building to house it. Another of Ken's efforts was in support of the Molecular Biosciences Graduate Program, formed from a number of graduate programs in the life sciences at Rutgers and UMDNJ (New Brunswick). The programs consolidated their first-year fellowships and acquired additional ones from the Rutgers Graduate School; these were then used to recruit students into a first-class first year program involving an advanced curriculum and laboratory rotations. This program still exists today and is a model for graduate programs at universities across the country.

16. Patricia Morton, Associate Professor

Dr. Kenneth Breslauer: leader, colleague, mentor, and most of all, treasured friend. He always has time, always has good advice, and always has your back. Ken is a visionary: he sees beyond what others see, then builds a path which he enables everyone to traverse. He is a unique balance between dreamer and clear-eyed practitioner.

I came to Rutgers from outside academia, and thus into a world of different processes and practices. Dr. Breslauer took me under his wing and, more than once, guided me onto a safe and effective route, thus avoiding many invisible landmines.

He immediately understood the vision and commitment of our new W. M. Keck Center and was amazingly supportive as we tried new ways to raise funds, such as the very successful 'CURE' events in New York City; initiated programs across the normal lines of operations like the Presidential Lecture Series; and absolutely radically welcomed people with spinal cord injuries into our Research Center. His appearance at events was always very meaningful to people in the community, as they were honored that someone in his position would take time to speak to and be with them.

There are many stories I could tell but will sum things up by saying: The world would be a better, more creative, and genuinely caring place if there were more people like Ken Breslauer.

I am deeply grateful that Dr. Kenneth Breslauer has been, and is, part of my life.

17. G. Eric Plum, Lecturer

For more than 30 years, I have benefitted from Ken Breslauer's mentorship and friendship. We all celebrate his stellar career as a scientist and university administrator. In addition to his intellect, Ken exhibits a rare gift for personal engagement and empathy. Everyone who worked with him appreciated his sensitivity about curveballs—although it was my understanding that he could not hit one.

The characteristic of Ken's that I still find most remarkable is his enthusiasm about science. Despite his many conflicting responsibilities, he was always eager to discuss my latest ideas and experimental results. Early in my time in his laboratory, Ken and I made a trip to Long Island to confer with our collaborators at SUNY Stonybrook. Unlike the graduate students, who all had pretty nice cars, Ken drove an old Volvo with at least one wheel in the junkyard. The weather was bad and the Volvo's wipers barely cleared the windshield of rain. I was quite unnerved. Throughout the trip, Ken exuberantly discussed the project with me while occasionally looking at the road. At that point, I believed that "I could do some really interesting science with this guy"—so long as we make it back to New Jersey.

Congratulations Ken on reaching this milestone, and thanks for the "interesting science" and your friendship over the years.

18. Jamshid Rabii, Professor Emeritus

My 33 years at Rutgers University have left me with many fond memories. Like many of my colleagues, a number of these memories stem from my experiences in research, teaching, as well as mentoring undergraduate and graduate students. In my case, however, I was fortunate enough to be encouraged by Ken Breslauer to join his team when he undertook development of the newly created Division of Life Sciences. That Ken was extremely successful in creating a world-class division engaged in research and teaching is a well-established fact. What may be less obvious is his invaluable influence on the professional lives of his team members. I recall accepting Ken's offer of the position of Director of Undergraduate Affairs in the Division of Life Sciences with a degree of apprehension. Although at that time I knew Ken casually from being on several committees with him, I was not familiar with either his leadership qualities or his work ethics. It did not take very long for me to appreciate his impressive leadership ability and management style. It quickly became evident to me that Ken took genuine interest in all aspects of the Division's development and performance and, when needed, offered his support and advice to every member of his team. Adding an ever-present composure and a good sense of humor to his other qualities made Ken Breslauer the ideal "boss." There were many challenges along the way as I strived to manage the undergraduate affairs of the Division up to the high standards that were expected by Ken. It would have been near impossible to get through such challenges had it not been for Ken's guidance and encouragement. Ken Breslauer's support and mentorship, as well as his friendship, was an integral part of my career advancement at Rutgers University. I would be remiss if I did not mention that Ken's support of his team members went beyond their professional activities within the Divison of Life Sciences. On more than one occasion, when frustrated in dealing with the local medical community for a health-related issue for myself or my family, all I needed was to reach out to Ken and he invariably employed his vast range of contacts to remedy the situation.

I am proud to have had Ken Breslauer as a boss, a mentor, and, especially, as a good friend.

19. Glen Ramsay, CEO

When Jack Aviv first introduced to me to Ken Breslauer, I was at that awkward stage of my professional life: transitioning from a postdoctoral fellow to my career. Career choices are a matter of self-preparation but also of opportunities created by others. I was, and still am, a bit of a duck: not the best flyer, nor swimmer, nor walker, but capable of all and of spanning the interfaces. Not being the best can stymie a career, but Ken and Jack saw an opportunity in me. Ken's lab needed not only the best people, but also top-notch instrumentation. Jack used his companies to keep Ken well stocked with the necessary hardware. It was into this mutually beneficial relationship that I landed.

The job offer made to me, which I readily accepted, was a joint position in both of their institutions. The position was uniquely created for me, which I'm sure was an administrative feat. My work involved developing instrumentation that could benefit both men's aspirations. The point I wish to make is that it was these men's faith in myself (and others) that is supreme among their attributes. They always recognized that the collective advancement of others would ultimately be beneficial to all. This "faith in others", I believe, is a chief contributor to their own successes. The result has been the launching of numerous careers, a huge number of publications, and, ultimately, improvement in our society.

This journal's many articles and kind words are proof of Ken's scope of influence. But let my experience demonstrate the depth of his generosity and commitment.

20. Armen Sarvazyan, CSO

Throughout his career, Ken Breslauer has been a leading figure in the field of thermodynamics of nucleic acids and their complexes. As a true visionary, Ken was quick to recognize the benefits of combining traditional calorimetric investigations with volumetric studies based on ultrasonic velocimetric and densimetric techniques. This combination was the foundation of an extremely fruitful collaboration between Ken, my former graduate student Tigran Chalikian, and myself. The collaboration is still ongoing, resulting in dozens of papers that provide unique insights into the physical nature of inter- and intramolecular interactions governing the biological function of proteins and nucleic acids. I want to use this occasion to wish Ken good health and continued scientific leadership for many years to come.

21. Jay A. Tischfield, Ph.D., FFACMG, Professor and Founding CEO

Ken Breslauer is an internationally renowned scientist and academic administrator, but perhaps most importantly, he is a visionary and builder of colleagues and programs. Ken and I had an extraordinarily productive partnership at Rutgers over a 20-year period, during which time I learned of and often appropriated the unorthodox ways in which he conducted his decadal role in the Division of Life Sciences and later as University Vice President for Health Science Partnerships. He taught me to act first and apologize later when we had a good idea that others ignored, frequently suborning co-conspirators whose interests coincided with ours. In retrospect, some of these individuals had divergent goals but they cooperated based on our pledge to support their interests in the future. We could do this because Ken's word and handshake were viewed as his bond, no paper necessary in most instances, a characteristic that I seek to emulate.

Ken was the key individual responsible for recruiting me to Rutgers as the first permanent chair of the Department of Genetics. Initially, I did not want to take the job because I had spent the previous 25 years in various positions at schools of medicine and undergraduate programs were too intimidating. But then Ken did the most unusual thing: he telephoned my childhood friends who were still living in the New York metropolitan area, imploring them to convince me to come back to the region where I spent the first 21 years of my life. I was so impressed that anyone would have the imagination and chutzpah to do this, which clearly required a good spiel and the force of a persevering personality, that I came to Rutgers in 1998. I thought to build faculty research laboratories in Nelson Labs, but it proved to be wholly inadequate for the plans that Ken and I incubated. With unerring technique, Ken convinced the University administrative hierarchy to lend me the money to renovate space for what became the Rutgers Cell and DNA Repository (RUCDR). The project took off, and I was able to quickly repay the money. At every opportunity, Ken touted my success, knowing full well that we could use it as a launch pad for our greater goals. Next, he convinced the administration to build the Life Sciences Building at a time when there was very little construction on campus. In doing this, Ken demonstrated deftness and agility at maneuvering within the Rutgers political system. Almost everybody in the administration succumbed to the promises emanating from his velvet tongue. This was followed by a series of highly original agreements that Ken brokered between the University and RUCDR, creating the wealth and infrastructure (e.g., more buildings and equipment) that allowed RUCDR to deliver hundreds of millions of dollars of federal and industry grants and contracts, providing employment for well over a hundred people and recent privatization at a significant profit to Rutgers.

I stand in awe of Ken's fundamental scientific contributions. His publications describe advances in our understanding of the thermodynamics of nucleic acids, a subject that I can only approach at 30,000 feet. My understanding of some of the biophysical principles that Ken describes in his papers is a tribute to their fundamental importance and his clear and highly articulate writing. I am honored that he has occasionally solicited my input or opinion, especially as it relates to medical implications.

Ken and I were both raised in New York City, though we occasionally regard our home boroughs as different planets. Oddly, we were both at Yale for graduate studies at the same time, though we didn't know each other because he was in the Department of Biophysics with the smart people. In fact, I almost flunked the graduate-level physical chemistry course offered by Ken's advisor. After Yale, we both gravitated to the University of California at Berkeley for postdoctoral studies, but we never met until he led the Rutgers recruitment dance.

I pay homage to Ken Breslauer on his 75th birthday, though I've been a fan and acolyte since we first met. I thank him for being a great friend and remarkable human being and for teaching me to navigate the Rutgers system. I regret that our paths did not cross before I joined Rutgers, especially in view of our prior geographic proximity.

22. Douglas H. Turner, Professor Emeritus

Ken and I were postdocs together in Nacho Tinoco's lab for about 20 months in 1973–1974. We were both Yankee fans from New York, so we bonded immediately. He has been a good friend ever since. At Berkeley, he reminded me of Mickey Mantle because he was the best softball center fielder I ever saw, especially his breaks on fly balls. We also bonded on watching Monday Night Football games. Along with Eric Weitz (now at Northwestern University), we were the greatest chefs of TV dinners on those nights. After Ken was set up at Rutgers, he kindly let me, and Sue Freier (now at IONIS Pharmaceuticals), come to his lab for a series of calorimetry experiments on poly-C. Together with Luis Marky, the four of us pioneered high throughput calorimetry by perfecting sleeping on the lab floor while experiments were running. Ken's calorimetry has had a major effect on my career due to his 1975 JMB publication with Sturtevant and Tinoco establishing that optical melting experiments best agree with calorimetry when linear lower baselines are subtracted from the UV absorption curves. My group has applied this crucial insight in at least 60 papers.

From later visits, it became clear that Rutgers was lucky that Ken expanded his interests to include academic administration. Our original bonding with the Yankees was also important for those visits. When Ken gave me directions for driving to his house, he said, "When you come to a fork in the road, take it." Turns out Ken lived near Yogi Berra, and Yogi famously gave the same directions, but expanded them to include living life. That made them easy for me to remember.

23. Jens Völker, Associate Research Professor

It is a great honor and privilege, and no easy task, to be involved in editing a Festschrift/Special Issue on the occasion of Professor Kenneth J. Breslauer's milestone birthday and to come up, in a few words, with something to say about Ken. Where does one start with someone as multifaceted as Ken Breslauer, and how does one keep it to just a few words?

Perhaps it is best to start where it all began for me, as my first introduction to Ken speaks volumes about the kind of person Ken is, beyond his numerous scientific and administrative achievements. I first met Ken when I attended, as a lowly and very much overawed PhD student from a far corner of the earth, the Gordon Conference on Biopolymers Ken co-chaired in 1992. I met him by literally bumping into him and a group of invited speakers and other luminaries in the elevator of the conference center. Ken immediately turned around and introduced himself by saying, "I thought I knew most people at the conference. I don't know you. I am Ken Breslauer." Upon hearing I was the odd PhD student from South Africa (we had communicated by fax so that I could get an invite to the conference), Ken proceeded to ask what my PhD was all about (thermodynamics of DNA triple helices), and if I had a poster at the conference. Upon hearing that indeed I had, he turned to his colleagues and said, "Carry on without me, I need to see this" and proceeded to spend the next hour and then some talking with me about my PhD project and invited me to visit his lab, if I got the chance. In the process of talking to me, whether intentionally or not I do not know, he managed to greatly alleviate my considerable anxieties of how my research done in isolation at the far end of the world compared with world-class research done in the US. It is this ability to talk to people from all walks of life and to make them feel comfortable I have since come to value and appreciate as one of Ken's greatest assets, a view that is also reflected in the numerous personal reminiscences listed in this editorial and the wide range of authors willing to contribute to this Special Issue.

Two years later, I accepted a postdoc position in Ken's lab, and I have been collaborating with Ken ever since. Having worked with Ken (often in collaboration with my former PhD advisor Horst Klump until his untimely passing last year), I have since come to appreciate many of Ken's other qualities, such as his willingness and encouragement to let me pursue my data to wherever they may lead (however outlandish it initially might have seemed); his patience with my struggles to make sense of my data and write them up in a semi-coherent manner (and whose importance he seemed to grasp within minutes when I finally felt comfortable to share them); his ability to convert my semi-coherent texts into something even a non-specialist can follow; and, not in the least, his willingness to find time to deal with research results despite his full administrative plate as dean and vice president at Rutgers. (Even if it sometimes took a while for him to get around to read/edit the many manuscripts produced in the lab—a standing joke in the lab for many years was that KJB (Ken's Journal of Biology) was a most exclusive journal with a readership of ONE, containing the hottest results unavailable to anyone else.) However, despite all these (and many other) academic/professional achievements, it is my opinion that Ken's ability to relate to people, to take their concerns seriously, and, when needed, his unstinting willingness to help out when life throws a curveball (as several other contributors have commented and that I can attest to from personal experience) that represent Ken's greatest gifts and that make it such an honor and privilege to have been involved in this Special Issue and to write this testimonial.

Happy Birthday Ken, and I hope we will find many an interesting and fruitful research topic to pursue in the years to come.

24. Peter von Hippel, Professor

I send my enthusiastic best wishes on the occasion of this celebration of your 75th birthday by your many friends, colleagues, and admirers. Achieving this three-quarter-century mark represents a momentous milestone in your busy and productive personal and scientific life and provides a point at which one could legitimately consider 'hanging up one's boots'. However, you show no signs of slowing down, and therefore, this celebration and Festschrift could be considered instead as a 'mid-career assessment' opportunity for you, an opportunity for your friends and colleagues to celebrate all that you have accomplished, and an opportunity to help you think about what you might still want to attempt in the 'second-half' of your scientific career.

I had hoped to mark the occasion by providing an overview of some recent and (thanks to the new methods developed by my research collaborators) perhaps 'ground-breaking' single molecule and 2D fluorescence spectroscopy work we have been conducting on the structural characterization and rates of inter-conversion of single-stranded (ss)DNA conformations on microsecond time scales, which has also helped us to more directly approach issues involving how various DNA conformations actually interact with ssDNA-binding proteins in real time. Clearly, this work builds on the major calorimetric and other biophysical chemical studies of DNA (and RNA) structures and sequences and their interactions with binding ligands that you and your lab have been engaged in for so many years. In our recent studies, we have been able to use some new methodologies to actually begin to monitor the appearance and disappearance of elements of the secondary structure of ssDNA on microsecond time scales, rather than just inferring (through monitoring the slower protein rearrangements that follow) what the DNA must be doing at rates that have been too fast to observe directly. We have also been studying the fluctuations in and the balance of forces that control the structure, stability, and inter-conversion rates of double-stranded DNA, and the fork and ss–dsDNA junctions that connect ds- and ssDNA segments in the DNA scaffolds of replication and transcription complexes. These latter loci serve as the positions at which the protein components of these systems actually function. This is, of course, a field to which you and your collaborators have contributed much, and thus such a paper seemed particularly appropriate for your Festschrift. However, now that the deadline is actually upon us, I find that I have, unfortunately (and not for the first time), over-estimated the rate at which this research could proceed before a review could legitimately be written. Therefore, our contribution will have to be limited to this shorter congratulatory message.

I close with a few more personal thoughts. We have been friends and have interacted, both scientifically and personally, for many years. In that capacity, I have much enjoyed reading and thinking about (and on occasion, serving as editor of or reviewer for) the seminal scientific work that you and your collaborators have produced. Those studies have mostly used physical biochemical techniques to characterize the structures and interactions of nucleic acid components with one another and with the proteins that drive the various processes of genome expression at the cellular and evolutionary levels. Your group's work has resulted in the production of unique and invaluable databases that have helped to put nucleic acid studies on a sound thermodynamic footing and have provided the necessary parameters to facilitate the discovery of ligands and drugs that have both increased our basic knowledge and have helped to devise rational approaches to the treatment of various diseases. You have also served as an insightful and constructive editor and reviewer for our papers. I can't now remember exactly when we first met, but our paths have crossed in multiple contexts, including formal and informal occasions connected with basic and applied research issues at meetings and seminars and visits and correspondence over at least the last 40 years. I look forward to the next 40! My colleagues and I are grateful for your many scientific and administrative and personal contributions to our field and to our lives. Thank you!

25. Gabrielle Wilders, Senior Executive Associate

I have worked with Dr. Breslauer for several years at the School of Arts and Sciences. He frequently goes out of his way get to know the staff, tell a joke, or share a story.

When I was diagnosed with a serious medical illness, I was in a state of shock and unsure of next steps. I reached out to Dr. Breslauer for help. I knew he appreciated the urgency of my situation; I knew he would know the top doctors in the area, and most importantly, I knew that helping his friends and colleagues in these situations was a priority to him.

Without hesitation, he paved the way for me to see leading experts in the field until I found one that I was completely comfortable with. I would not have been able to navigate my way to the excellent care I received without Dr. Breslauer's skilled guidance and compassion.

I am eternally grateful for the kindness he has shown me and the concerted effort he put forth to help ensure that I regained my health.

Funding: This research received no external funding.

Conflicts of Interest: The author declares no conflict of interest.

References

1. Marky, L.A.; Breslauer, K.J. Origins of netropsin binding affinity and specificity: Correlations of thermodynamic and structural data. *Proc. Natl. Acad. Sci. USA* **1987**, *84*, 4359–4363. [CrossRef] [PubMed]
2. Breslauer, K.J.; Frank, R.; Blocker, H.; Marky, L.A. Predicting DNA duplex stability from the base sequence. *Proc. Natl. Acad. Sci. USA* **1986**, *83*, 3746–3750. [CrossRef] [PubMed]
3. Breslauer, K.J. The shaping of a molecular linguist: How a career studying DNA energetics revealed the language of molecular communication. *J. Biol. Chem.* **2021**, *296*, 100522. [CrossRef] [PubMed]
4. Klump, H.H.; Volker, J.; Breslauer, K.J. Energy mapping of the genetic code and genomic domains: Implications for code evolution and molecular Darwinism. *Q. Rev. Biophys.* **2020**, *53*, e11. [CrossRef] [PubMed]
5. Volker, J.; Gindikin, V.; Klump, H.H.; Plum, G.E.; Breslauer, K.J. Energy landscapes of dynamic ensembles of rolling triplet repeat bulge loops: Implications for DNA expansion associated with disease states. *J. Am. Chem. Soc.* **2012**, *134*, 6033–6044. [CrossRef] [PubMed]

life

MDPI

Article

Characterization of Aurintricarboxylic Acid (ATA) Interactions with Plasma Transporter Protein and SARS-CoV-2 Viral Targets: Correlation of Functional Activity and Binding Energetics

Conceição A. Minetti [1,*,†], David P. Remeta [1,*,†], Keiji Hashimoto [2], Radha Bonala [2], Rajesh Chennamshetti [2], Xingyu Yin [2], Miguel Garcia-Diaz [2], Arthur P. Grollman [2,3], Francis Johnson [2,4,*] and Viktoriya S. Sidorenko [2,*]

1 Department of Chemistry and Chemical Biology, Rutgers—The State University of New Jersey, Piscataway, NJ 08854, USA

2 Department of Pharmacological Sciences, State University of New York at Stony Brook, Stony Brook, NY 11794, USA; keiji.hashimoto@stonybrook.edu (K.H.); radha.bonala@stonybrook.edu (R.B.); rajesh.chennamshetti@stonybrook.edu (R.C.); xingyu.yin@stonybrook.edu (X.Y.); miguel.garcia-diaz@stonybrook.edu (M.G.-D.); arthur.grollman@stonybrook.edu (A.P.G.)

3 Department of Medicine, State University of New York at Stony Brook, Stony Brook, NY 11794, USA

4 Department of Chemistry, State University of New York at Stony Brook, Stony Brook, NY 11794, USA

* Correspondence: cminetti@chem.rutgers.edu (C.A.M.); dpremeta@chem.rutgers.edu (D.P.R.); francis.johnson@stonybrook.edu (F.J.); victoriya.sidorenko@stonybrook.edu (V.S.S.)

† These authors contributed equally to this work.

Citation: Minetti, C.A.; Remeta, D.P.; Hashimoto, K.; Bonala, R.; Chennamshetti, R.; Yin, X.; Garcia-Diaz, M.; Grollman, A.P.; Johnson, F.; Sidorenko, V.S. Characterization of Aurintricarboxylic Acid (ATA) Interactions with Plasma Transporter Protein and SARS-CoV-2 Viral Targets: Correlation of Functional Activity and Binding Energetics. *Life* **2022**, *12*, 872. https://doi.org/10.3390/life12060872

Academic Editors: Tigran Chalikian and Jens Völker

Received: 25 April 2022
Accepted: 2 June 2022
Published: 10 June 2022

Abstract: In an effort to identify functional-energetic correlations leading to the development of efficient anti-SARS-CoV-2 therapeutic agents, we have designed synthetic analogs of aurintricarboxylic acid (ATA), a heterogeneous polymeric mixture of structurally related linear homologs known to exhibit a host of biological properties, including antiviral activity. These derivatives are evaluated for their ability to interact with a plasma transporter protein (human serum albumin), eukaryotic (yeast) ribosomes, and a SARS-CoV-2 target, the RNA-dependent RNA polymerase (RdRp). The resultant data are critical for characterizing drug distribution, bioavailability, and effective inhibition of host and viral targets. Promising lead compounds are selected on the basis of their binding energetics which have been characterized and correlated with functional activities as assessed by inhibition of RNA replication and protein synthesis. Our results reveal that the activity of heterogeneous ATA is mimicked by linear compounds of defined molecular weight, with a dichlorohexamer salicylic-acid derivative exhibiting the highest potency. These findings are instrumental for optimizing the design of structurally defined ATA analogs that fulfill the requirements of an antiviral drug with respect to bioavailability, homogeneity, and potency, thereby expanding the arsenal of therapeutic regimens that are currently available to address the urgent need for effective SARS-CoV-2 treatment strategies.

Keywords: aurintricarboxylic acid (ATA); salicylic acid polymers; methylene-salicylic acid; SARS-CoV-2; RNA-dependent RNA polymerase (RdRp); yeast ribosomes; human serum albumin (HSA); molecular recognition; biological function and viral disease; inhibitor binding; thermodynamics

1. Introduction

The emergence of highly infectious life-threatening diseases, as evidenced by the SARS-CoV-2 global pandemic, necessitates effective treatment strategies, including the development and evaluation of novel pharmacological compounds as prospective antiviral therapeutics. Over the past several decades, the Breslauer, Grollman, and Johnson research laboratories have pursued a collaborative synergistic program to elucidate specific structure–function–energetic correlations in systems of biomedical relevance. This comprehensive multiparametric strategy has enabled us to characterize the overall impact of carcinogenic and mutagenic DNA lesions [1–9] on nucleic acid recognition [10], replication [11], and repair [10,12] (as reviewed in [13]). Applying the methodology developed and refined during this timeframe towards identifying effective treatment protocols to combat infectious diseases, our laboratories are currently exploring investigational compounds and repurposed molecules to assess their overall therapeutic potential as anti-SARS-CoV-2 inhibitors. The latter includes aurintricarboxylic acid (ATA), a salicylic acid polymer that exhibits a broad range of biological activities, including its efficacy as an antiviral agent [14,15].

The synthesis of ATA traces its origin to 1892 as the primary product of a condensation reaction between salicylic acid and formaldehyde in the presence of sulfuric acid and nitrite ion [16]. Multiple applications have been reported since its discovery, including seminal studies conducted during the 1960s and 1970s [17] that explored ATA as an inhibitor of protein synthesis in eukaryotes while elucidating specific mechanisms of several antiviral and antibiotic compounds at the ribosomal level [18–26]. Harnessing insights gained from parallel investigations on emetine [17–19,24,27], Grollman and colleagues focused particular attention on the initiation of viral protein synthesis as a target for anti-viral drug design. These researchers demonstrated that triphenylmethane dyes such as ATA inhibit initiation of protein synthesis [21] and prevent the attachment of viral RNA to ribosomes isolated from rabbit reticulocytes [20,23]. Subsequent investigations identified nucleic acid binding proteins as an alternate target for ATA [28] via mechanisms suggested to involve inhibition of nucleic acid interactions with the template binding sites [29,30]. Significantly, the groundbreaking studies conducted by Grollman and associates have inspired the publication of over 750 articles on ATA describing its biomedical applications. A long-standing challenge that remains is to characterize the active principle constituent(s) in the ATA mixture and develop lead compounds for further optimization.

Our approach to antiviral drug design is consistent with these observations as we explore initiation of protein synthesis at the ribosomal level and inhibition of RNA-dependent RNA polymerase (RdRp). Moreover, compounds that are structurally related to ATA and share its mode of action might lead to the design of novel antiviral agents [20,31]. The polymeric heterogeneous nature of ATA mixtures imposes a significant challenge for medicinal chemistry investigations of such compounds. Nevertheless, the ability to inhibit an early event in viral replication suggests that the active principle(s) of this heterogeneous mixture might represent a novel chemical entity. In view of its effective utility as both a protein and nucleic acid synthesis inhibitor, heterogeneous ATA mixtures have been demonstrated to exhibit antiviral activity by inhibiting replication of influenza [25], *coxsackievirus* [32], and SARS-CoV [33]. Recently, screenings of prospective anti-SARS-CoV-2 inhibitors have identified ATA as exhibiting antiviral activity via inhibition of viral cell attachment and invasion by blocking spike protein interactions with ACE2 [34].

Considering the myriad of biological and antiviral properties associated with ATA, identification and resolution of the active component(s) represents a critical hurdle in developing these as lead compounds for further optimization to achieve both a higher potency and selectivity. A feasible and practical alternative is to design synthetic protocols that afford generation of homogeneous lead compounds which mimic and/or exceed the polymeric mixture activity. As part of our ongoing efforts to correlate biological and biophysical properties, we are pursuing leads generated via the design of defined synthetic ATA derivatives that optimize inhibitory activity, antiviral selectivity, bioavailability, and the ability to penetrate cells. The current synthetic scheme to produce ATA involves treatment of salicylic acid with formaldehyde, sulfuric acid, and sodium nitrite resulting in a heterogeneous polymeric mixture [35]. The degree of product heterogeneity is dependent on synthetic conditions such as reactant concentrations, reaction time, and temperature [36,37]. The heterogeneity of ATA preparations has been acknowledged by several investigators [36,38] who have provided detailed information on optimal synthetic routes and product characterization [39,40].

Although specific efforts have been undertaken to isolate/segregate the active fraction(s) and identify their respective biological properties [41], attempts to purify and characterize a homogeneous ATA species have not succeeded. These observations inevitably lead to the proposition that aurintricarboxylic acid (i.e., monomeric ATA) as depicted in Scheme 1A is in fact an inactive compound [36,41] and its functional activities are conceivably the consequence of "impurities" including formaurindicarboxylic acid and higher molecular weight species [36]. Despite its predominantly polymeric nature, ATA is often regarded as a monomeric triphenylmethane dye that is presumed to represent the active ligand state in molecular docking studies. Consequently, biological properties attributed to ATA are interpreted via computational and prediction methods as arising from specific interactions between the monomer and molecular target(s). The body of experimental evidence accumulated to date supports the notion that ATA biological properties are primarily attributed to compounds of average molecular weight $\geq$ 2000 [39,42] with a predominant species proposed as illustrated in Scheme 1B.

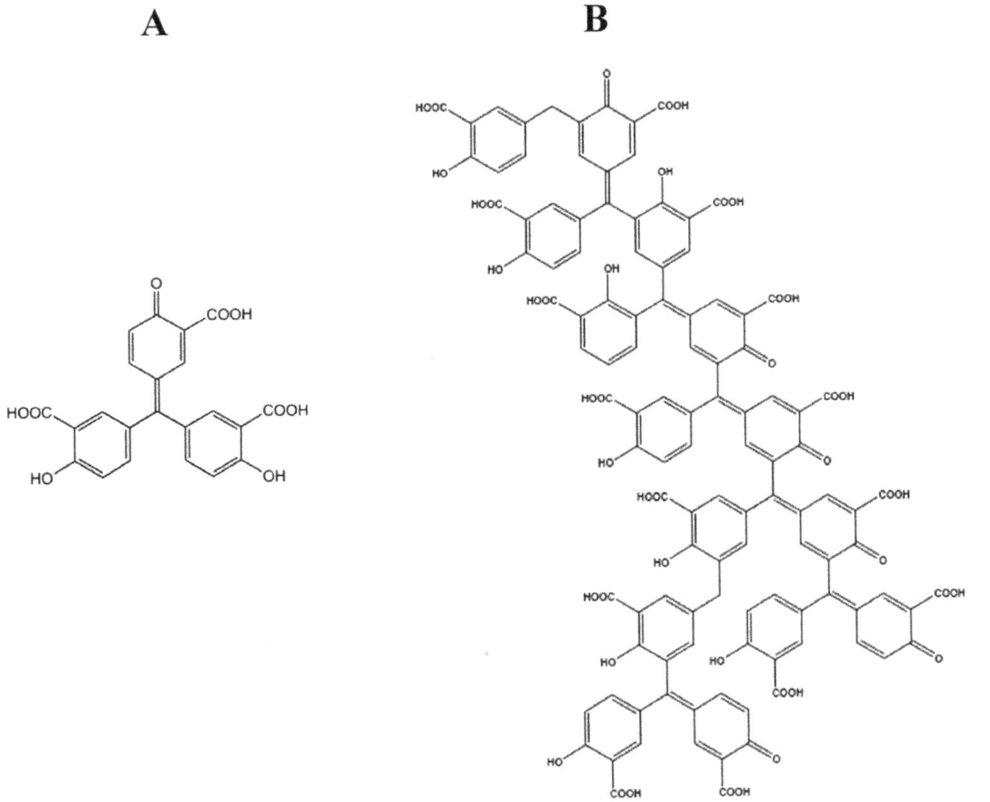

Scheme 1. Chemical Structures of Aurintricarboxylic Acid (ATA). (**A**) Monomer (triphenylmethane unit). (**B**) Idealized representative ATA polymer within the heterogeneous mixture (based on [35]).

Since the discovery of ATA antiviral properties [20,21], there have been extensive investigations regarding the components responsible for biological activity. Perhaps the most significant studies conducted by Cushman and associates examined both low [39,43] and high [35] molecular weight components of this dyestuff for antiviral activity. These investigators concluded that amongst the former, compounds **2** and **3** (Scheme 2) are the most active against HIV-1 in cultured mammalian cells. Significantly, neither of the methyl esters is active thereby indicating the important role of carboxylic anions. Moreover, the low molecular weight compounds containing a quinomethine residue are inactive. The greatest activity appears in the 7000–12,000 MW range with weight average (M_w) and number average (M_n) molecular weights of approximately 2937 and 2547, respectively [35].

Basic matrix–like structures **1a**, **1b** and **4** (Scheme 2) have been proposed [35,39,43] as repeating units in the higher MW polymers yet scant evidence exists to support such chemical structures. Nevertheless, a prominent feature in both series of compounds is the quinomethine group, which accounts for the orange–red color observed in most substances.

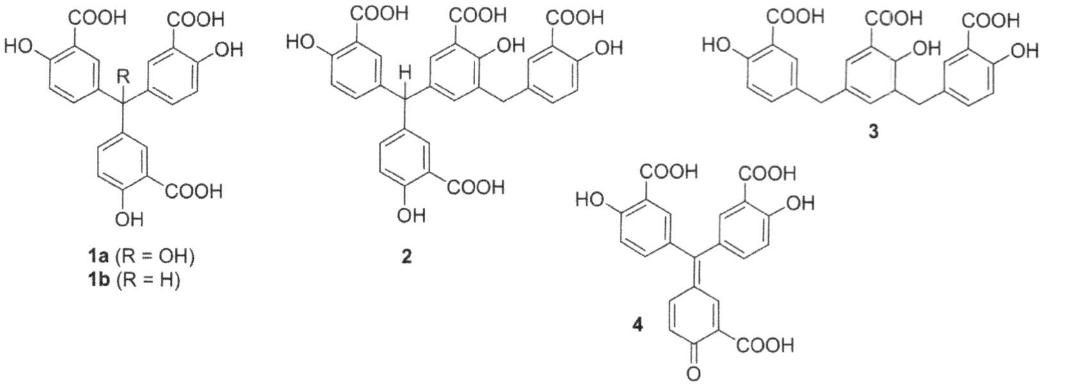

Scheme 2. Chemical Structures of ATA Species Isolated from a Fractionated Preparation (based on [35]).

Considering the heterogeneity of polymeric ATA mixtures, it is desirable to identify the active component(s) responsible for biological and selective antiviral activities. Our experimental strategy involves the synthesis of defined chemical structures that mimic specific components comprising the heterogeneous mixture. This is accomplished by synthesizing homogeneous linear polymers (i.e., dimers, tetramers, hexamers, octamers, etc.) *in lieu of* triphenylmethane units. It is important to note that when a commercial sample of ATA is reduced by zinc dust in acetic acid to convert the quinomethine groups to salicylic acid residues, antiviral activity is retained in the colorless product. This finding in conjunction with the studies of Cushman and colleagues [35,38–40,42,43] infers that biological activity is due primarily to a run of methylene-salicylic residues in the ATA polymers and suggests that the quinomethine groups represented in **4** (Scheme 2) do not contribute to this activity. These observations set the scene for establishing a synthetic strategy in which oligomers comprised solely of methylene-salicylic acid units are evaluated for the purpose of determining whether un-oxidized regions are the primary source of biological activity.

Our systematic multidisciplinary protocol focuses on identifying a prospective lead compound of defined chemical structure and molecular weight that exhibits functional activity comparable to or exceeding the heterogeneous ATA mixture. This study describes the synthesis of a dichlorohexamer that mimics polymeric ATA in terms of its biological activities toward initiation of protein synthesis and SARS-CoV-2 RdRp, yet retains the characteristic features of a lead compound with respect to homogeneity and ADMET qualities. We employ a complementary array of binding assays that utilize a combination of calorimetric and optical techniques to characterize the properties of lead compounds that inhibit RdRp and mammalian protein synthesis in cell-free systems. Parallel experiments profile the binding of heterogeneous polymeric ATA to human serum albumin (HSA), a universal drug-transporter protein in the bloodstream that possesses a unique ability to interact with a myriad of compounds harboring a broad range of hydrophobicities and molecular weights. An accurate assessment of HSA binding energetics is essential, as transporter proteins can diminish the overall bioavailability and/or bioactivity of lead compounds by competing for drug binding to viral or infected host cell receptor sites. Acquisition of the relevant biological and biophysical properties for ATA interactions with antiviral targets facilitates function–energetic correlations. The resultant data serve as a baseline for investigations on small molecule interactions with specific viral targets, including SARS-CoV-2 RdRp and the eukaryotic ribosome.

2. Results

2.1. Experimental Strategy

This study employs a complementary array of biophysical binding assays that utilize a combination of calorimetric and optical techniques to identify and characterize the properties of aurintricarboxylic acid (ATA) (refer to Scheme 1) as a prospective lead compound. ATA is available commercially or as a component of LOPAC screening libraries (refer to Materials and Methods) and comprises a heterogeneous mixture of oligomeric/polymeric species that exhibit antiviral activity [35,44]. We initially assess the binding properties of ATA to human serum albumin (HSA) as characterization of such interactions provides a measure of plasma bioavailability which can be integrated within ADMET properties to evaluate lead compound pharmacokinetics. Whereas a low binding affinity suggests immediate availability of the administered compound, a high binding affinity infers that HSA may function as a reservoir maintaining homogeneous distribution to the tissues thereby increasing biological lifetimes [45]. Considering its role on overall drug pharmacokinetics [46], HSA may limit and/or control toxicity while modulating metabolic inactivation and elimination through excretory pathways. A secondary yet equally relevant objective underlying initial assessment of HSA binding properties is to identify predominant molecular species in the compound mixture and determine an average molecular weight. The resultant compilation of biophysical properties sets the stage for investigations on ATA derivative interactions with specific host targets including SARS-CoV-2 RdRp which is crucial for viral replication and the 40s ribosomal subunit where initiation of viral protein synthesis occurs.

2.2. Biophysical Properties and Binding Energetics

2.2.1. Optical Characterization of ATA–HSA Interactions: Binding Is Accompanied by Fluorescence Quenching and Energy Transfer

We have characterized the biophysical and physicochemical properties of ATA by exploring specific optical (i.e., UV/Vis, Fluorescence) profiles. Analysis of the resultant UV/Vis spectrum acquired over the wavelength range of 200–800 nm reveals major and minor absorbance peaks centered at ~310 nm and 528 nm, respectively. The latter is only detected at sufficiently high concentrations (Figure S1) and reflects the presence of higher molecular species which exhibit more extended conjugation relative to the lower molecular weight compounds [35]. Concentration-dependent absorbance (Figure 1A) and fluorescence (Figure 1B) profiles suggest that ATA polymers do not undergo further intermolecular self-association in solution as evidenced by a linear increase of intensity (Figure 1A,B insets). The absorbance peak wavelength (i.e., 310 nm) in the UV/Vis spectrum (Figure 1A) has been selected for excitation of ATA and the resultant emission spectra recorded over 270–600 nm (Figure 1B) with a characteristic maximum emission intensity observed at 425 nm.

A detailed analysis of ATA–HSA interactions monitored via fluorescence spectroscopy reveals that ATA promotes a reduction of the HSA aromatic residue intensity and the resultant quenching is accompanied by fluorescence resonance energy transfer (FRET) that is simultaneously detected at 425 nm. Specifically, when ATA is titrated into HSA and the aromatic residues are excited at 280 nm, one observes a binding-induced quenching of HSA emission intensity at 330 nm and a concomitant increase in ATA fluorescence emission intensity at 425 nm as depicted in Figure 2A. This enhancement is only evident in the presence of HSA, as isolated ATA standards excited at 280 nm do not exhibit an increased intensity at 425 nm (Figure S2). A secondary optical probe of the drug–protein interaction involves monitoring titration of HSA into an ATA solution excited at 310 nm which results in a concentration-dependent enhancement of ATA fluorescence at 425 nm (Figure S3).

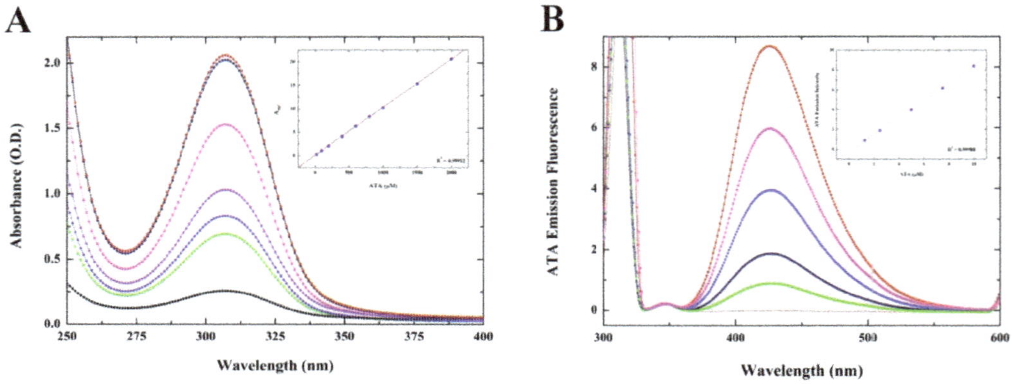

Figure 1. UV/Vis Absorbance and Fluorescence Spectra of the ATA Polymer Mixture. (**A**) Concentration-dependent absorbance spectra and corresponding linear correlation (inset); (**B**) Fluorescence emission profiles exhibiting a maximum intensity at 425 nm that are linearly correlated with increasing ATA concentration (inset).

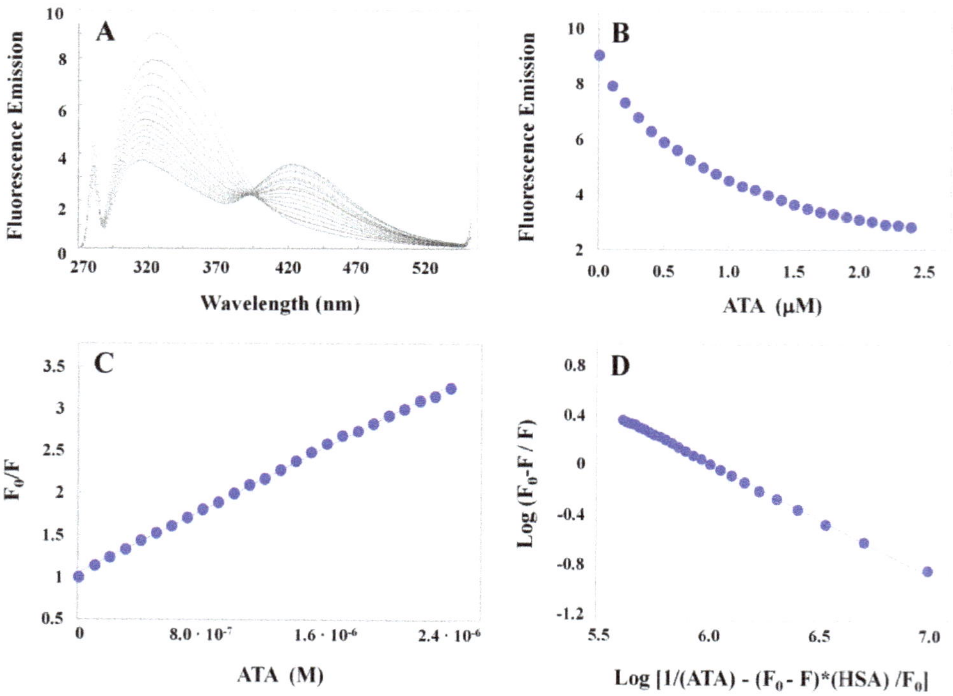

Figure 2. (**A**) Family of HSA Fluorescence Profiles Monitoring ATA-Induced Quenching. The quenching of HSA fluorescence monitored at 325 nm is accompanied by a concomitant increase in ATA emission intensity at 425 nm. Each spectral profile is corrected by subtracting an appropriate reference standard to eliminate the intrinsic fluorescence of ATA upon excitation at 280 nm. (**B**) Emission intensities monitored at 325 nm as a function of ATA concentration. (**C**) Stern–Volmer plot revealing a K_{sv} of 9.2×10^5 M^{-1} ($R^2 = 0.998$). (**D**) Double logarithmic plot of the ATA concentration-dependent quenching yielding a single set of binding sites ($n = 0.9$) and an affinity (K_a) of 5.98×10^5 M^{-1}.

The measurable increase in fluorescence intensity exhibited by ATA when excited at 280 nm only occurs in the presence of HSA and presumably arises as a consequence of energy transfer which can be exploited to assess ATA interactions at specific protein binding sites. This observation infers that upon excitation of HSA Tyr and Trp residues, the proximity of ATA to these aromatic residue(s) at the ligand-protein binding site(s) facilitates detection and quantization of the interaction via FRET that can occur from an excited fluorescent donor (HSA) to a suitable acceptor (ATA). Significantly, one observes a characteristic isoemissive wavelength at ~395 nm, which suggests the presence of two distinct emitting species. Exploitation of these unique fluorescent properties represents a useful probe of ligand–receptor interactions and validation of this protocol is a powerful tool for evaluating the interaction of antiviral molecules with SARS-CoV-2 targets.

The HSA fluorescence quenching data monitored at 325 nm as a function of increasing ATA concentrations (Figure 2B) are recast in the form of a Stern–Volmer plot (Figure 2C). A linear Stern–Volmer plot is consistent with static quenching experienced by a single class of fluorophores that is equally accessible to the quencher. The ATA–HSA binding profile has also been analyzed via a double-reciprocal logarithmic plot (Figure 2D). Inspection of the data reveals that this interaction involves a single set of sites (n = 0.9) and is characterized by a binding constant (K_b) of 6.0×10^5 M^{-1} (i.e., K_d = 1.7 μM).

2.2.2. Elucidating ATA–HSA Binding Energetics via ITC: Protein-Ligand Complex Formation Is Enthalpy-Driven with Favorable Entropic Contributions

Isothermal Titration Calorimetry (ITC) facilitates characterization of ligand–receptor binding energetics and thereby, furnishes a complete thermodynamic description of the ATA–HSA association process. The standard experimental protocol consists of titrating a concentrated ATA stock solution (500 μM) via thirty successive 10 μL aliquots into an HSA standard (10 μM) employing a 300 s integration period between injections. A representative ITC thermogram reflecting the exothermic reaction accompanying binding of ATA to HSA is depicted in the upper panel of Figure 3. The resultant heats are integrated to yield a binding profile (bottom panel) that is fit via nonlinear least squares analysis to a single site model (red line). Formation of the ATA–HSA complex is characterized by a Gibbs Free Energy (ΔG) of -7.6 kcal·mol^{-1} that is primarily enthalpy driven (ΔH = -6.4 kcal·mol^{-1}) with a favorable entropic contribution (TΔS = 1.2 kcal·mol^{-1}) and an overall binding affinity (K_a) of 5.0×10^5 M^{-1}. The K_a values derived from fluorescence and ITC measurements (i.e., ~5–6 $\times 10^5$ M^{-1}) are in excellent agreement. The enthalpic nature of these interactions is consistent with the polar character of ATA and entropically favorable contributions may be attributed to binding-induced desolvation of the HSA binding site.

Recasting the binding energetics in terms of an ATA monomer (M_w = 422.3) as the modular/structural unit (refer to Scheme 1A), the resultant stoichiometric ratio (n ~ 4.5) reveals that polymeric ATA preparations comprise an average molecular weight species approximating 2000. These findings suggest that the predominant species are on the order of 4–5 ATA monomeric units. Systematic studies employing well-defined molecular weight compounds (i.e., monomer, dimer, trimer, etc.) should assist in evaluating our working hypothesis regarding the stoichiometric ratio observed. Determining the potential self-assembly of small molecules is pivotal for evaluating their spatial distribution in a cellular environment [47] and a primary factor in assessing the IC_{50} towards a specific target. These non-covalent interactions can be sufficiently ordered to form large assemblies or nanostructures yet may also be responsible for disordered aggregation and precipitation. A control experiment in which the ATA stock solution is diluted into dialysate appears in Figure S4 and reveals negligible reaction heats, thereby confirming that the ATA polymers do not undergo non-covalent intermolecular self-association. Collectively, our results suggest that ATA interacts with one of the HSA subdomains in a single site binding mode with the ligand averaging a molecular weight approximating 2000, which is equivalent to 4–5 ATA monomeric species.

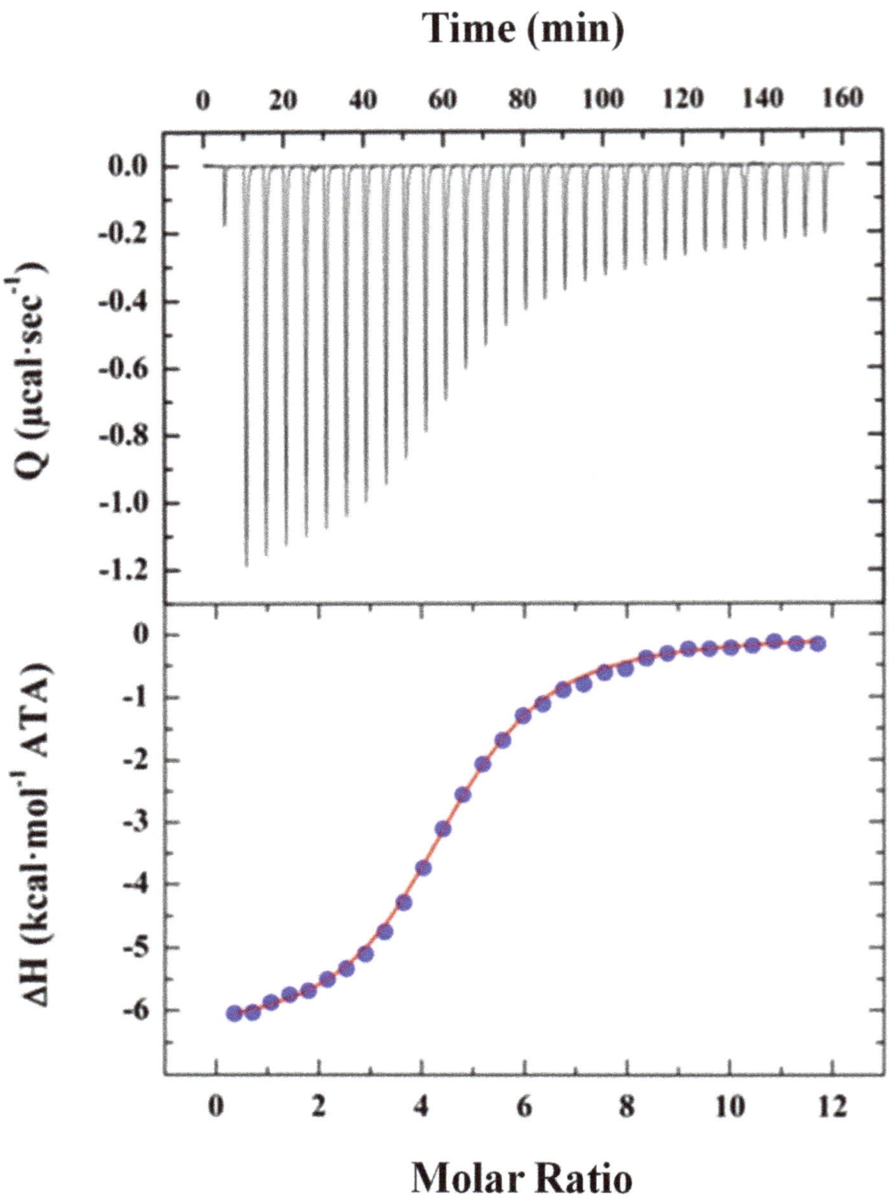

Figure 3. ITC Profile Monitoring ATA–HSA Association. The binding isotherms are acquired at 25.0 °C in 10 mM sodium phosphate buffer (pH 7.4). Titration of HSA (10 µM) with ATA (500 µM) is reflected in the thermogram (top panel), integrated peak areas (blue circles), and corresponding fit (red line) to a single site binding model (bottom panel).

2.2.3. Characterizing HSA Binding of a Dichlorohexamer ATA Analog via ITC: Protein-Ligand Complex Formation Is Enthalpy-Driven with Unfavorable Entropic Contributions

Comparable to its interaction with heterogeneous polymeric ATA, the calorimetric profile of HSA binding to a chlorinated hexamer designated as compound **11** (Refer to Scheme 3) is characterized by an enthalpy-driven ($\Delta H = -15.1$ kcal·mol^{-1}) process as illustrated in Figure 4. In contrast with the heterogeneous polymeric mixture that is expressed in terms of triphenylmethane units (yielding a stoichiometric ratio of ~4.5), compound **11** is expressed as a single hexameric unit (M_w = 956.11). Accordingly, inspection of the resultant ITC profile reveals formation of a 1:1 protein–ligand complex with a single dichlorohexamer occupying an HSA subdomain. Another distinguishing characteristic of the interaction between compound **11** and HSA is its unfavorable entropic nature ($T\Delta S = -7.1$ kcal·mol^{-1}) which results in a binding free energy (ΔG) of -8.0 kcal·mol^{-1}. While HSA association with the chlorinated hexamer and heterogeneous polymeric ATA are characterized by comparable Gibbs Free Energies ($\Delta\Delta G = 0.4$ kcal·mol^{-1}), their thermodynamic binding signatures are remarkably distinct. These results can be rationalized in terms of binding-induced motion restriction within the dichlorohexamer that incurs an entropic penalty. Table 1 presents a summary of thermodynamic binding parameters deduced via analysis of ITC profiles characterizing HSA-ligand interactions for polymeric ATA and the dichlorohexamer.

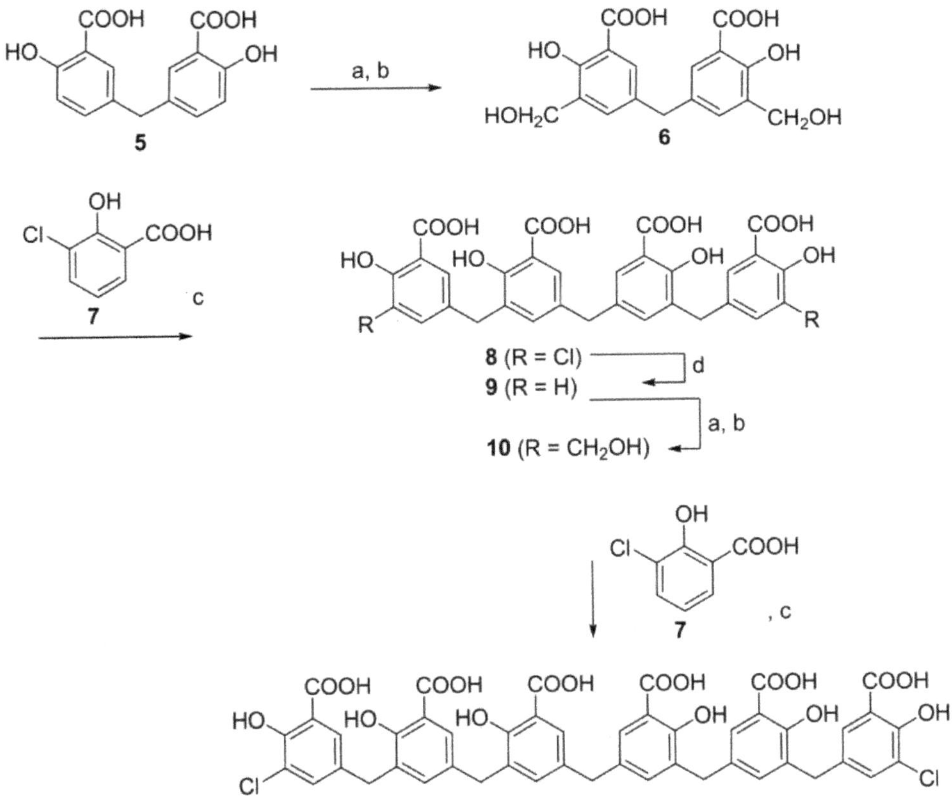

Scheme 3. Synthesis of a Chlorinated Polysalicylic Acid Hexamer (Compound **11**) Evaluated as a Prospective Antiviral Therapeutic. Chemical Reagents: (a) hexamethylene tetramine/triethylamine; (b) $LiBH_4$; (c) sulfuric acid; (d) 10% PdO, H2.

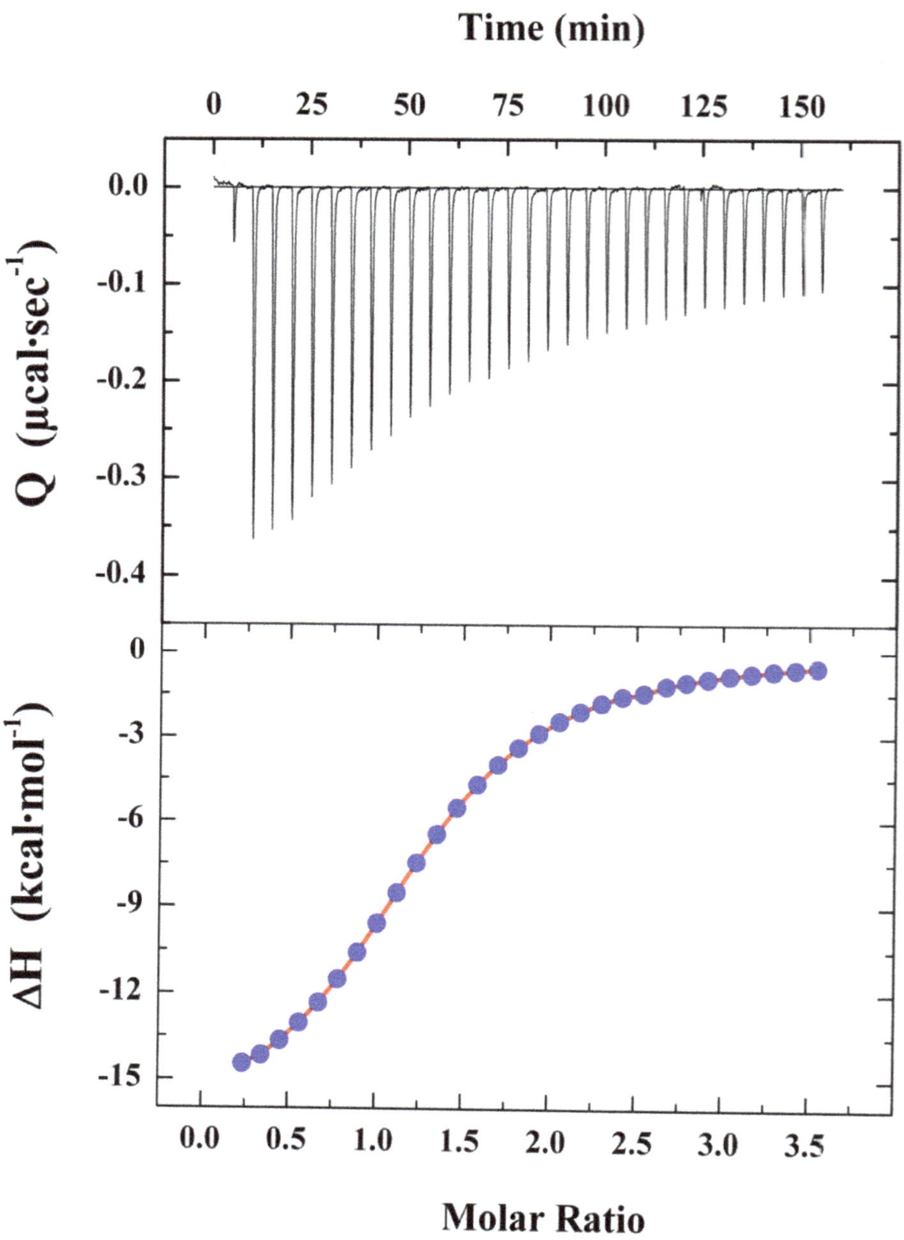

Figure 4. ITC Profile Monitoring Association of the Dichlorohexamer and HSA. The binding isotherm is acquired at 25.0 °C in 10 mM sodium phosphate buffer (pH 7.4). Titration of HSA (5 µM) with the dichlorohexamer (70 µM) is reflected in the thermogram (top panel), integrated peak areas (blue circles), and corresponding fit (red line) to a single site binding model (bottom panel).

Table 1. Thermodynamic Binding Parameters for ATA–HSA Interactions.

Compound	$K_a \times 10^5$ (M^{-1})	K_d (μM)	n [a]	ΔG (kcal·mol^{-1})	ΔH (kcal·mol^{-1})	TΔS (kcal·mol^{-1})
ATA	5.0 ± 0.1	6.7 ± 0.1	4.5 ± 0.5	−7.6 ± 0.1	−6.4 ± 0.2	1.2 ± 0.1
Dichlorohexamer	9.9 ± 0.5	1.0 ± 0.1	1.0 ±0.0	−8.0 ± 0.2	−15.1± 0.2	−7.1 ± 0.2

[a] The stoichiometry (n) is calculated based on the molecular weight for a monomeric unit of ATA (M_w 422.34) and the dichlorohexamer (M_w 956).

2.2.4. Interaction of ATA-Derivatives with Host and/or Viral SARS-CoV-2 Targets: A Synthetic Chlorinated Hexamer Binds Ribosomal Particles with Moderate Affinity

ATA has been reported to prevent the attachment of bacteriophage messenger RNA to ribosomes [21]. The initiation of protein synthesis in cell-free extracts from *E. coli* or rabbit reticulocytes is inhibited at ATA concentrations (<100 nM) which do not prevent chain extension [23]. We have evaluated the ability of ATA derivatives to interact specifically with yeast 80S ribosomes via fluorescence anisotropy. In this experiment, aliquots of purified ribosome stock are titrated into the ligand solution equilibrated in a quartz cuvette under continuous stirring at 25 °C. Fluorescence anisotropy (r) values monitoring ribosome–ligand binding are measured at excitation/emission wavelengths of 305 nm and 420 nm, respectively. The resultant G-factor-corrected r values are employed to calculate the binding affinities as illustrated in Figure 5. Following analysis of the data via linearized plots as described in Materials and Methods, we measure an affinity of 4.8×10^6 M^{-1} (K_d = 0.21 ± 0.02 μM) for the synthetic dichlorohexamer.

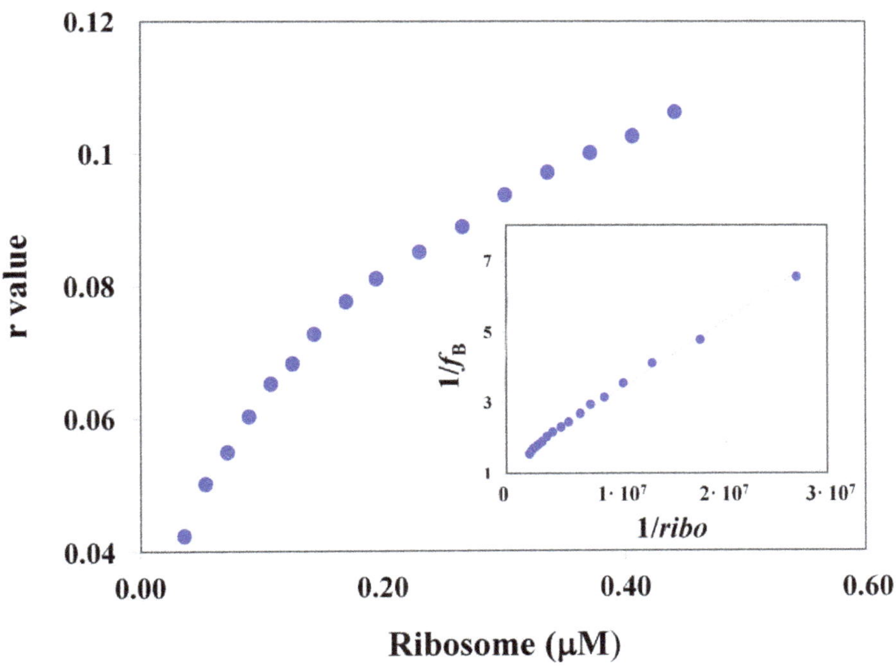

Figure 5. Dichlorohexamer Binding to Yeast Ribosomes Monitored via Fluorescence Anisotropy. Aliquots of purified ribosome stock (27 μM) are titrated into the ligand solution (1 μM) under continuous stirring at 25 °C. Fluorescence anisotropy (r) values monitoring ribosome–ligand binding are measured at excitation/emission wavelengths of 305 nm and 420 nm, respectively. The double reciprocal plot yields an affinity of 4.8×10^6 M^{-1} (K_d = 0.21 ± 0.02 μM) for the synthetic chlorinated hexamer **11**.

2.3. Biological Properties and Functional Activity

2.3.1. Heterogeneous ATA and a Synthetic Chlorinated Hexamer Inhibit Protein Synthesis in Rabbit Reticulocyte Lysates

Since ATA has been originally described as an inhibitor of mammalian protein synthesis, we evaluated commercial heterogeneous ATA and synthetic oligomers of various lengths for their potential to affect translation in lysates of rabbit reticulocytes using mRNA coding for firefly luciferase. The dose–response change in chemoluminescense monitored in the presence of various concentrations of investigational drugs following addition of luciferin allows comparison across compounds. Cycloheximide, a known inhibitor of protein synthesis elongation, has been used as a positive control with an EC_{50} (50% inhibitory concentration) of 0.106 µM (Figure 6A). Employing this assay, we analyzed heterogeneous ATA and several intermediates in the synthetic pathway (Refer to Scheme 3). Whereas polymeric ATA exhibits an EC_{50} of 17.6 µM, small molecular weight analogs represented by chlorinated dimer **4a** (a derivative of compound **5** in Scheme 3), chlorinated tetramer (**8**), and dechlorinated tetramer (**9**) did not impact protein synthesis over a concentration range extending to 50 µM. Significantly, the chlorinated hexameric compound (**11**) inhibits protein synthesis with an efficiency (i.e., EC_{50} of 28.13 µM) similar to that of the heterogeneous ATA preparation (Figure 6B). It is worth noting that at 100 µM concentrations, both polymeric ATA and the dichlorohexamer completely inhibited translation, exhibiting a signal comparable to background values (i.e., in the absence of mRNA substrate). These results suggest that a hexamer is the minimum length of ATA-derived species required to inhibit protein synthesis activity.

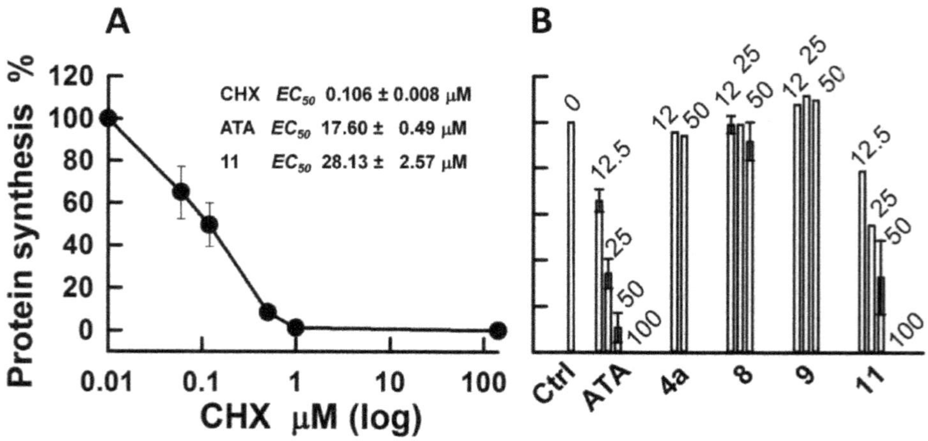

Figure 6. Inhibition of Protein Synthesis by CHX and ATA Derivatives in Rabbit Reticulocyte Lysates. (**A**) Dose-dependent Inhibitory activity of CHX. (**B**) Inhibitory activities of ATA and synthetic derivatives. Concentrations of ATA species are indicated above each bar. Test compounds are incubated with lysates of rabbit reticulocytes in the presence of mRNA coding for luciferase. The chemiluminescence of reaction mixtures is measured following addition of luciferin. Samples incubated in the absence of investigational drugs are used as 100% reference. Results are presented as mean values and standard deviations for three independent measurements; otherwise, data are presented as average values of two independent experiments with 10–15% agreement. EC_{50} values are calculated and displayed in Panel A.

2.3.2. Heterogeneous ATA and Synthetic Chlorinated Hexamer Inhibit SARS-CoV-2 RNA Dependent RNA Polymerase

The mechanisms underlying ATA-mediated anti-viral activities consist of early stages in the viral life cycle and conceivably comprise inhibition of viral RNA-dependent RNA polymerase (RdRp). In this study, we assessed the binding and inhibition activities for several host and viral targets including SARS-CoV-2 RdRp. The heterogeneous ATA preparation efficiently inhibits RdRp-mediated RNA replication (IC_{50} = 56 nM), confirming previous assumptions that ATA interacts with and inhibits RdRp. Moreover, amongst the various molecular weight lead compounds evaluated, we find that only the dichlorohexamer retains an ability to inhibit RdRp with a potency comparable to the ATA mixture (IC_{50} = 108 nM). Our data on RdRp polymerase activity reveal that the compounds studied herein rank according to the following order of inhibitory activity: Polymeric ATA (IC_{50} = 56 nM) ~ Dichlorohexamer **11** (IC_{50} = 108 nM) >>> Tetramer **8** (IC_{50} = 163 μM) > Tetramer **9** (IC_{50} = 514 μM). A representative inhibition profile for these compounds is presented in Figure 7. Inspection of the data indicates that a hexamer is the minimum length required for inhibitory activity which is consistent with previous reports that the bioactive species in ATA preparations comprise molecular weights ≥ 1000. These findings suggest that the synthetic dichlorohexamer exhibits an optimal footprint to interact with and inhibit the RdRp polymerase complex at nanomolar potency.

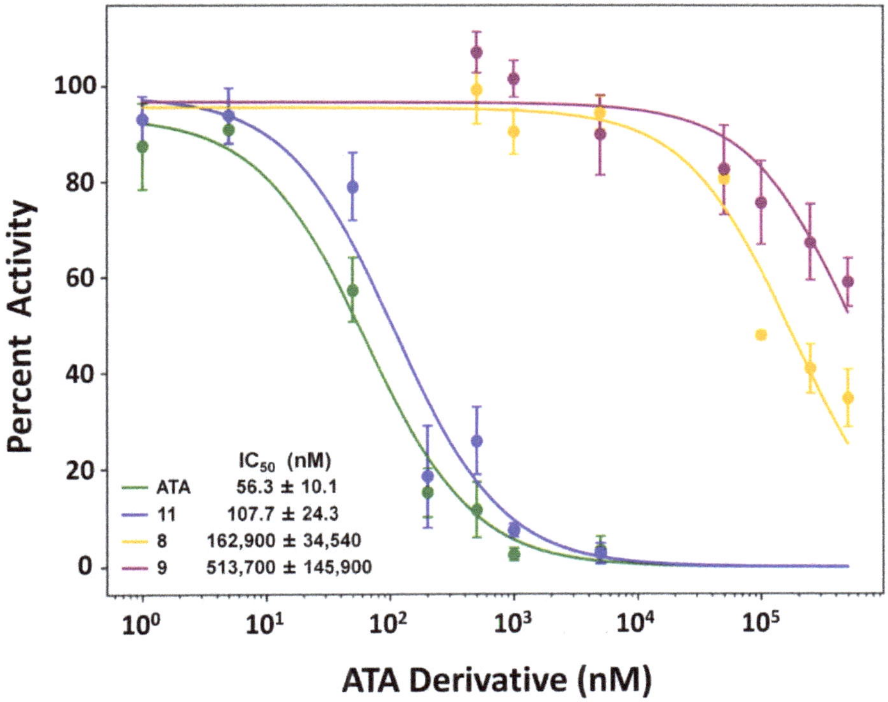

Figure 7. ATA and the Dichlorohexamer Inhibit SARS-CoV-2 RdRp. ATA derivatives are incubated with nsp12/7/8, an active complex of RdRp, in the presence of hybridized RNA substrate and ribonucleoside triphosphates. Cy3-labelled products are resolved via gel electrophoresis and inhibitory activity (IC_{50}) of each compound evaluated by fitting the inhibitory curves to a four-parameter logistic regression. Each dose-response curve is generated from three independent experiments. Mean and standard deviation values are plotted for each concentration of test compounds. The full activity (100%) of RdRp in the absence of inhibitors ranged from 33–72 pmol of product depending on the reaction date.

2.3.3. Heterogeneous ATA Is Non-Toxic and Exhibits Limited Activity as a Protein Synthesis Inhibitor in Cultured Vero E6 Cells

In an effort to establish the effects of ATA species in cultured cells, we have studied cell growth and viability of a Vero E6 cell line in response to heterogeneous ATA, tetramer **8**, and tetramer **9**. In view of our long-term goal to generate antiviral drug candidates, we selected the Vero E6 cell line that is amenable to infections with various coronaviruses including hCoV-OC43 and SARS-CoV-2. The SRB assay is employed for in vitro cytotoxicity screening and measures the absorbance of total protein proportional to cell biomass. Application of this assay reveals that none of the three compounds affect cell growth and viability at concentrations up to 400 µM over 48 h (Figure 8A). The lack of host cell toxicity is a desirable property for investigational drugs, which usually suggests that there are no mammalian targets for a given compound and antiviral mechanism(s) is/are selective toward the virus. When puromycin labeling is applied to establish whether polymeric ATA inhibits protein synthesis in Vero E6 cells, we observe only a 30–40% response between 150–300 µM ATA, in contrast with CHX, which inhibits protein synthesis at concentrations of 1 µM. Representative data illustrating the differential impacts of ATA and CHX on protein synthesis inhibition are depicted in Figure 8B. The inability of ATA to inhibit protein synthesis in Vero E6 cells might be attributed to the fact that such cells are of renal origin which express P-glycoprotein, a transporter responsible for drug efflux [48]. Another possibility is the charged nature of ATA derivatives, which might prevent passive diffusion into the intracellular space. Since compounds with similar charge exhibit no effects in human lung A549 cells (data not shown) and the latter do not overly express P-glycoprotein, our second hypothesis appears more plausible. Accordingly, other groups have demonstrated that while ATA might inhibit protein synthesis in cell free systems, it may lack this activity in cell culture [49,50]. Our results indicate that further modifications of the carboxyl groups in synthetic ATA species might be required to observe antiviral effects in vitro and in vivo.

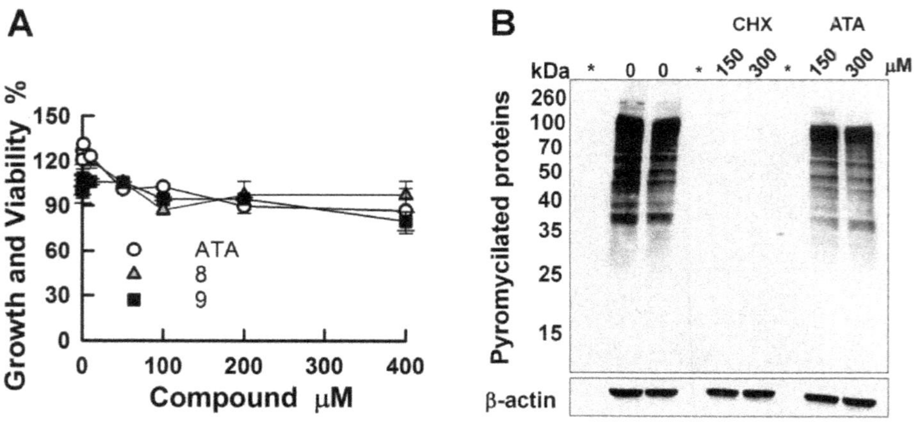

Figure 8. Impact of Heterogeneous ATA and Synthetic Derivatives on Vero E6 Cells in Culture. (**A**) SRB assay following 48 h exposure in culture with ATA, tetramer **8**, and tetramer **9**. (**B**) Representative immunoblottings for puromycilated proteins and ß-actin following 1 h exposure to ATA and cycloheximide (CHX) in culture and puromycin labelling. An asterisk (*) indicates wells that include commercial protein standard. Refer to Figure S5 and Table S1 for additional details.

3. Discussion

Seminal studies on the remarkable antiviral properties of triphenylmethane dyes discovered over four decades ago [21–23,25,26] have prompted investigations of their potential use as lead compounds in targeting SARS-CoV-2. In the search for novel activities via HTS assays utilizing the LOPAC arsenal of compounds, ATA is often identified among positive *hits* in drug discovery campaigns, including those in the quest for SARS-CoV-2 therapies [34]. Considering the potential utility of ATA in terms of its multitarget effects, a thorough characterization of biophysical and molecular features including drug-like properties is severely lacking. In view of the intrinsic heterogeneity observed amongst polymeric ATA preparations [35–37], there is an urgent need to isolate and/or identify active components for the express purpose of developing and synthesizing lead compounds that retain the characteristic molecular properties. The latter must be achieved while simultaneously ensuring the requisite homogeneity and purity that enables such compounds to be considered as viable drug candidates.

3.1. Biophysical Studies Assist Synthetic Efforts in Selecting a Representative Oligomeric ATA Analog

ATA is an anionic polymer that has been demonstrated to bind a variety of viral protein targets including gp120 of HIV-1 and HIV-2 and reported to prevent SARS-CoV replication in cultured cells [33]. A complete understanding of the biological mechanisms underlying antiviral activity necessarily requires a comprehensive multidisciplinary approach to characterize the physicochemical and structural properties of ATA as lead candidate(s). Considering the heterogeneous nature of such compounds, our experimental strategy aims at identifying and evaluating the distribution of oligomeric species/states to assist chemists in the overwhelming task of synthesizing compounds that mimic the bioactive properties of these mixtures yet exhibit well-defined structures and molecular weights.

3.2. Characterization of ATA Intrinsic Biophysical Properties

As a prerequisite for studying ATA binding to specific cellular targets involved in SARS-CoV-2 infectivity, we have characterized the optical and calorimetric properties of ATA–HSA interactions. Knowledge of the HSA binding affinity for lead compounds is required to assess the bioavailability of these drug candidates in vivo. Moreover, such studies are critical in terms of advancing/evaluating ADMET properties that are required in drug development. The resultant data provide unique insights into specific binding modes, stoichiometry, and driving forces underlying ATA–HSA interactions. Our biophysical approach for characterizing ATA binding properties to the ubiquitous protein carrier/receptor HSA has proven particularly insightful.

Employing a combination of optical and calorimetric techniques, we have defined the primary ATA polymeric binding species and determined the requisite binding affinity, stoichiometry, and thermodynamic signatures to dissect the enthalpic and entropic components driving protein–ligand association. Characterization of the ATA–HSA thermodynamic profiles serves as a reference for elucidating specific interactions with other competing cellular targets, including SARS-CoV-2 RdRp and/or the ribosomal site(s) where viral protein synthesis initiation occurs. Collectively, these studies map the binding profiles and thermodynamic signatures of drug–target and off-target interactions that are related to their antiviral biological activities and thereby provide an additional layer of information for rational drug design/optimization, decision-making protocols during screenings, and structure–energetic activities in antiviral drug discovery.

This study provides the basis for synthesis of a lead compound resembling the polymeric ATA mixture that has been characterized in terms of its structural identity and homogeneity via mass spectrometric and NMR analysis in conjunction with concentration-dependent optical and calorimetric studies. Our data corroborate a wealth of published studies in which the ATA bioactive species is comprised of a polymeric mixture *in lieu of* well-defined monomeric structures. Following confirmation of the identity, purity, and molecular weight, we have characterized the binding energetics of this bioactive compound with an off-target transporter protein (HSA) and its respective viral target(s). Specifically, we have identified structural features of ATA-derived compounds that drive the observed antiviral activity.

3.3. Binding Profiles of Polymeric ATA versus the Synthetic Dichlorohexamer

ATA is suggested to adopt polymeric structures that are able to interact with and inhibit the helicase from hepatitis C NS3 virus [51]. However, unlike the branched structure proposed by Gonzalez et al. [36], the authors contend that ATA adopts a more linear polymeric arrangement which mimics nucleic acid structures [51,52]. Inspection of the ITC profile for

ATA binding to HSA (Figure 3) reveals an average stoichiometry of ~5 ATA monomeric units per protein, a finding that infers the predominant molecular weight species is on the order of 2000 Da, which is consistent with fractionation studies conducted previously [36]. Considering the idealized structure proposed by Cushman et al. [35] and our observation that the active ATA component is comprised of approximately five monomeric units, we have designed an experimental strategy to synthesize linear oligomeric compounds with the goal of identifying an optimal molecular weight and length that exhibits maximal activity.

Pursuing this synthetic strategy, we have isolated a dichlorohexamer that interacts with HSA via a single site binding mode and exhibits a stoichiometry of one hexamer per HSA molecule (Figure 4). Comparison of the respective HSA binding profiles for polymeric ATA and the dichlorohexamer reveals comparable Gibbs free energies yet their thermodynamic signatures are quite distinct (Table 1) as the ATA analog exhibits both a higher enthalpy and unfavorable entropy. While ITC studies on the viral targets are deferred to a subsequent study, we can infer that the dichlorohexamer retains biophysical properties commensurate with desirable drug-like qualities such as the hydrophilicity required for solubility and bioavailability. Moreover, as a lead compound that exhibits a high enthalpic efficiency, the ATA analog may outperform prospective competitors that are predominantly entropy-driven with concomitant insolubility and loss of target selectivity [53].

Enthalpic efficiency is considered a unique advantage in drug discovery [53,54] as this metric is proposed to confer higher target selectivity while optimizing ADMET properties (as reviewed in [55]). The enthalpic nature of dichlorohexamer–HSA interactions is consistent with the polarity of prospective lead compounds and suggests that this ATA analog may exhibit the requisite specificity of an antiviral drug provided interactions with specific targets are comparable to those observed for HSA binding. Studies are currently in progress to evaluate this hypothesis by characterizing the thermodynamics of ATA derivative interactions with various host and viral targets. The chlorinated hexamer has been compared with polymeric ATA in terms of its biological and biomolecular properties, revealing similar or superior potency relative to commercially available ATA preparations as described in the following sections.

3.4. ATA Derivatives Bind and Inhibit SARS-CoV-2 Targets

Antiviral drugs exhibit a broad range of mechanisms to target one or more steps in the virus life cycle within mammalian cells. These include blocking cell entry, inhibiting viral proteases required for polypeptide cleavage yielding active viral proteins, and preventing interactions between the virus and host cell that are crucial for viral reproduction. Specific drugs inhibit various components that are exploited by the virus for protein synthesis (e.g., the mammalian ribosome) and replication of its genetic material (i.e., viral RNA-dependent RNA polymerase). This study focuses on host ribosomal protein synthesis machinery and SARS-CoV-2 RdRp. Recent findings suggest that ATA exhibits antiviral activities against SARS-CoV [33] and SARS-CoV-2 [34,56] which reinforces our strategy to identify the antiviral bioactive species in ATA preparations and characterize their respective target(s).

3.5. ATA Derivatives Bind Host Ribosomes and Inhibit Protein Synthesis Initiation

Early studies [21,31,57] reported on ATA inhibition of messenger RNA attachment to ribosomes with negligible effect on aminoacyl transfer to peptide. The initiation of protein synthesis in rabbit reticulocytes is impacted by ATA at concentrations (<100 nM) which do not prevent chain extension [21,58]. The remarkable properties of ATA inhibiting ribonucleic acid attachment to the ribosome and halting protein synthesis initiation has sparked broader interest in exploring its prospective role as an effective antiviral agent and prompted us to investigate the binding properties and inhibition mechanisms of ATA derivatives towards eukaryotic ribosomes. Inspection of fluorescence anisotropy binding profiles (Figure 5) reveal that the dichlorohexamer exhibits a preferential affinity for purified yeast ribosomes (i.e., K_d~200 nM) nearly ten-fold higher than its association with the non-target HSA.

In an effort to explore the underlying mechanisms triggered by specific interactions with purified ribosomes, we have assessed the impact of various ATA-derived compounds on protein synthesis by reticulocyte lysate ribosomes (Figure 6). Analysis of the resultant data reveals that these compounds inhibit protein synthesis in reticulocytes thereby corroborating prior studies. In contrast with equilibrium binding assays, the IC_{50}s for protein synthesis inhibition in reticulocytes are substantially higher (on the order of 20–30 μM). At concentrations of 100 μM, both ATA and the dichlorohexamer inhibit translation by nearly 100%. While the origin of these results requires further exploration, the data suggest that there are a number of concurrent events (i.e., on target/off target) in the reticulocyte lysate ribosome samples. These might recruit ATA and its analog which compete with protein synthesis inhibitory mechanisms and thereby require higher concentrations for effective protein synthesis inhibition. Our ongoing studies are aimed at identifying the origins of these intriguing results.

3.6. ATA Derivatives Bind and Inhibit SARS-CoV-2 RdRp

Direct ATA-target binding interactions have been reported for a number of nucleic acid processing enzymes including DNase I (K_d = 9 μM), RNase A (K_d = 2.3 μM), reverse transcriptase (K_d = 0.25 μM), and Taq polymerase (K_d = 82 μM) as measured via fluorescence studies [59]. The overall mechanisms ascribed to ATA derivatives reside in the ability to inhibit crucial macromolecular processes via substrate/cofactor displacement and/or competition mechanisms in addition to direct interaction with the target enzymes. Examples of ATA-mediated inhibition processes that likely displace their natural substrate(s) include several polymerases, helicases, and other viral enzyme targets [29,52,60–62]. The fact that ATA is efficacious against a number of viruses [26] including coronaviruses and a potential mechanism of action is viral replication inhibition [25], a logical strategy is to assess the impact of such compounds on SARS-CoV-2 RdRp.

The current study reports direct inhibition of SARS-CoV-2 RdRp by polymeric ATA and synthetic oligomers with nanomolar potencies. Amongst several derivatives studied, the dichlorohexamer is nearly equipotent (IC_{50} = 108 nM) to heterogeneous ATA (IC_{50} = 58 nM) in inhibiting the RdRp-dependent RNA replication complex (Figure 7). Comparing these results with data on other nucleic acid enzymes (i.e., K_d/IC_{50} ranging from 0.25 to 82 μM) [59], and assuming the interchangeability of K_d and IC_{50} (an approximation that is not always warranted), the dichlorohexamer represents a high affinity candidate with an IC_{50} ranking among some of the highest potency inhibitors. In fact, recent studies report RdRp inhibitors with IC_{50}s in the low micromolar range, values that are comparable to those observed for remdesivir as a control compound [63].

3.7. ATA Analog Bioactive Conformation as RdRp Inhibitor

Despite the wealth of experimental evidence accumulated to date that confirms the heterogeneous polymeric nature of ATA preparations, computational models with docking poses suggest that the ATA monomer (Scheme 1) binds SARS-CoV RdRp at the target site(s). Our experimental strategy employing a combination of synthetic, biological, and biophysical protocols has confirmed a minimal length for effective inhibitor activity. Specifically, the dichlorohexamer retains comparable activity to the polymeric ATA mixture, whereas tetrameric compounds **8** and **9** exhibit minimal or no RdRp inhibition activity (i.e., IC_{50} ~ 163 and 514 μM, respectively) as illustrated in Figure 7. Efforts towards identifying suitable lead compounds with proven antiviral activity and preferably targeting RdRp should provide the basis for developing effective therapeutics against SARS-CoV-2 and facilitating the design of broad-spectrum antivirals that may be useful in combating future coronavirus outbreaks. This study describes the characterization of ATA-derived compounds selected on the basis of their homogeneity, binding, inhibition, and antiviral activities, essential characteristics that serve as a lead for the generation of novel compounds with a clear therapeutic efficacy.

3.8. *ATA and Derivatives in the Cellular Context: Permeability and Toxicity*

Although antiviral activity has been demonstrated for ATA against SARS-CoV in the absence of toxicity to host cells [33], the latter is achieved only at significantly high concentrations (i.e., 200–500 μM). Recently, it has been reported that ATA inhibits propagation of SARS-CoV-2 in VERO cells [56] albeit at concentrations as high as 100 μM to achieve full antiviral activity. Our findings are generally consistent, as we observe modest activity when assessing the antiviral properties of ATA and dichlorohexamer at concentrations ≥200 μM (data not shown) in hCoV-OC43 and VERO E6 cells. Given the inhibitory activity of ATA towards mammalian protein synthesis in cell-free systems and the lack of toxicity in host cells, we hypothesize that ATA antiviral activity generally occurs at high doses and is elicited via inhibition of viral entry. In fact, investigators have identified another mechanism by which ATA may exert antiviral activity that involves inhibition of cell-surface receptor (i.e., acetyl cholinesterase 2) binding by the RBD (receptor binding domain) of the SARS-CoV-2 spike protein [34]. While a recent study claims that ATA is membrane-permeable in a rat brain PC12 cell line, the concentrations required to monitor its accumulation within cells is on the order of 300 μM [64]. The unusual choice of host cells selected for these measurements warrants caution in interpreting their results.

In view of these disparate findings and our experimental observations, we contend that neither ATA nor the isolated synthetic derivatives retain an ability to readily penetrate cultured cells. Moreover, the ATA species that affect viral entry are conceivably distinct from those inhibiting viral RdRp and/or protein synthesis, which require resolution of ATA components within heterogeneous preparations. The most plausible explanation for a lack of cell permeability is that the charged nature of ATA compounds precludes passive diffusion into the intracellular space, thereby exhibiting minimal or no activity in cultured cells. In an effort to address this deficiency, our ongoing studies are aimed at optimizing ATA-derived lead compounds in order to preserve potency while simultaneously improving their ability to enter mammalian cells. As an alternate approach, specific ATA analog delivery systems will be evaluated in terms of the ability to facilitate cell entry/permeability and thereby exert their full potential as antiviral molecules.

3.9. *Structure-Activity Considerations regarding ATA Derived Lead Compounds*

Although officially reported as triphenylmethane (CID: 2259; M_W = 422.34 g/mol), ATA in fact comprises a heterogeneous polymeric population of various molecular weight species. Investigations utilizing ATA in screenings generally assume with rare exception that this compound exists in the monomeric state which is employed for in silico docking. Unlike numerous studies that persist in perpetuating the "active" ATA species as a monomer, a recent study suggested that ATA indeed adopts polymeric structures and the latter are effective helicase inhibitors [51]. These authors have correctly suggested that ATA adopts a more linear polymeric arrangement resembling nucleic acid structures similar to the proposal of Cushman and colleagues [52]. Our combined biological/biochemical/biophysical data characterizing the ATA-elicited RdRp inhibition properties suggest that a minimum optimal oligomeric size is required for viral activity. Considering the proposed idealized structure [35] and controlled synthesis of linear oligomeric compounds presented herein, we envision specific interactions of the ATA-derived polymers to occur via a polysalicylic stretch linearly distributed as illustrated in Figure 9, which may resemble nucleic acid substrates. While the dichlorohexamer mimics or even exceeds the biological properties of heterogeneous ATA, we are expanding our synthetic repertoire to generate higher molecular weight structures (e.g., an octamer) that are the focus of a forthcoming study.

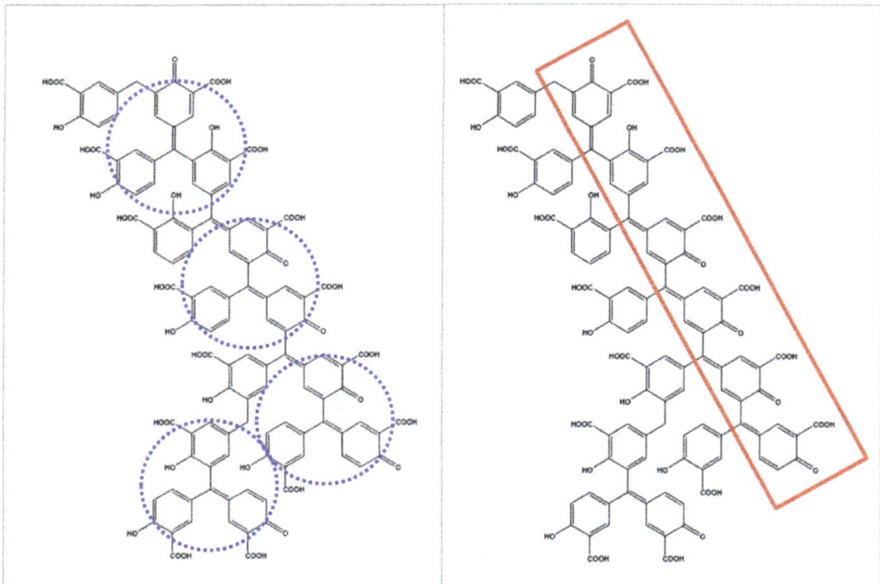

Figure 9. An Idealized Binding Mode Based on Biological and Biophysical Properties. Considering the structure proposed by Cushman et al. [35] depicting triphenylmethane units in blue (**left** panel) and molecular weight deduced from the ITC stoichiometric ratio in conjunction with inhibition of RdRp exhibited by the dichlorohexamer, we envision that the polysalicylic acid stretch outlined in red (**right** panel) corresponds to a minimal optimal enzyme footprint for ATA activity. Reduction of the quinomethine groups yielding methylsalicylic acid polymers such as the dichlorohexamer does not abolish the activity of these ATA analogs.

4. Concluding Remarks

Given the recent emergence of SARS-CoV-2 as a global pandemic, coupled with the need to maintain vigilance in a post-pandemic/endemic era, the identification of potent antiviral therapies represents an urgent priority with the surge of novel variants exhibiting resistance to current vaccine regimens. Recent studies screening thousands of novel and repurposed compounds have identified ATA as a prospective lead candidate in SARS-CoV-2 treatment strategies [34,56,65–67]. ATA preparations currently employed for screening purposes comprise a heterogeneous mixture of active/inactive components that preclude drug development and lead optimization. The current study represents an inaugural attempt to alleviate such shortcomings by synthesizing potent homogeneous ATA derivatives harboring specific therapeutic properties. The latter must meet or exceed that of the active heterogeneous ATA preparations, thereby serving the purpose of identifying effective anti-SARS-CoV-2 agents with potential broad spectrum therapeutic applications in the drug discovery arena. The finding that a hexameric salicylic acid derivative exhibits optimal SARS-CoV-2 RdRp inhibition activity underscores the potential viability of these compounds as lead candidates in antiviral treatment therapies and remains the focus of our ongoing studies.

5. Materials and Methods

5.1. Materials

Aurintricarboxylic Acid (ATA) was procured from Millipore Sigma (St. Louis, MO, USA) as catalog number A1895 with a designated M_w of 422.34. ATA was suspended without further purification/treatment in 10 mM sodium phosphate buffer (pH 7.4) or dialysate for calorimetric characterization of protein-ligand binding interactions. Defatted

Human Serum Albumin (HSA) was obtained from Millipore Sigma (St. Louis, MO, USA) as catalog number A1887. HSA stock solutions (100 μM) were prepared by suspending the lyophilized powder in 10 mM phosphate buffer containing 1 mM mercaptoethanol (BME) to ensure disruption of Cys-34 crosslinked dimers. The HSA solution was dialyzed extensively against three 1.0 L exchanges of 10 mM sodium phosphate buffer (pH 7.4) to remove excess BME and the resultant monomeric mercaptoalbumin used for biophysical measurements. The HSA standard concentration was determined spectrophotometrically employing an extinction coefficient of 35,000 M^{-1} cm^{-1} at 280 nm. Cycloheximide (CHX), TWEEN20, sodium dodecyl sulfate (SDS), trypsin-EDTA, TRIS-buffered saline, Dulbecco's phosphate-buffered saline (PBS), methanol, puromycin, and film for chemoluminescence detection were obtained from Millipore Sigma (St. Louis, MO, USA). Unless indicated otherwise, reagents, materials and equipment for electrophoresis and immunobloting were purchased from Bio-Rad Laboratories (Hercules, CA, USA). The sources of other relevant reagents, biological kits, and materials are noted in the protocols described below.

5.2. *UV/Vis Absorbance and Fluorescence Spectroscopy*

5.2.1. Characterization of HSA and ATA Optical Properties

The optical properties of HSA, heterogeneous polymeric ATA, and oligomeric derivatives have been characterized by monitoring their respective absorbance profiles on an AVIV Biomedical Model 14 UV/Vis Spectrophotometer (Lakewood, NJ, USA). Absorbance spectra have been acquired at 1.0 nm intervals over the wavelength range of 200–800 nm employing an averaging time of 5 s and slit width of 1 nm. The profiles are characterized by absorbance peaks centered at 280 nm and 310 nm for HSA and ATA, respectively. In view of the fact that drugs tend to aggregate as a consequence of their intrinsic hydrophobicities, it is important to assess the state of these ligands in terms of biophysical and structural properties. Concentration-dependent measurements of polymeric ATA standards spanning the 10 to 2500 μM range have been conducted to determine its propensity for intramolecular aggregation. Expanding these measurements to the nanomolar range, concentration-dependent studies on the fluorimetric properties of polymeric ATA standards ranging from 10 to 1000 nM have been evaluated. Analysis of the respective absorbance and fluorescence spectra reveals a strictly linear dependence of optical density and emission intensity over this expanded concentration range. These findings suggest that polymeric ATA does not exhibit appreciable intermolecular aggregation and/or form supramolecular assemblies.

5.2.2. Characterization of HSA and ATA Fluorescence Properties

The fluorescent properties of HSA, heterogeneous polymeric ATA, and oligomeric derivatives have been characterized by monitoring their respective profiles on an AVIV Biomedical Model 107 Differential Fluorescence Spectrophotometer (Lakewood, NJ, USA). Fluorescence emission spectra of HSA and ATA analogs have been acquired in a 10 mm quartz cuvette at 1.0 nm intervals over the wavelength range of 250–600 nm employing an averaging time of 5 s and excitation/emission slits of 10 nm. Excitation of an HSA standard at 280 nm yields an emission spectrum with a maximum intensity at 320 nm, whereas excitation of a polymeric ATA standard at 310 nm yields an emission spectrum with a maximum intensity at 425 nm.

5.2.3. Characterization of ATA–HSA Binding via Fluorescence Quenching

Fluorescence quenching measurements of protein–ligand binding interactions monitor the emission intensity of aromatic chromophores including Tyr and Trp residues at 280 and 295 nm, respectively. The impact of polymeric ATA on HSA fluorescent properties has been evaluated via measurement of the protein emission intensity in the presence of increasing ligand. The ATA–HSA interaction has been characterized by titrating aliquots of ligand stock into the protein standard in 10 mM phosphate buffer (pH 7.4) and monitoring the resultant fluorescence emission at 325 nm employing an excitation wavelength of 280 nm. The HSA fluorescence quenching data monitored at 325 nm as a function of increasing

ATA concentrations (Figure 2) are recast in the form of a Stern-Volmer plot and analyzed in accordance with the following relation: $F_0/F = 1 + K_{SV}\ [ATA]$. The ATA–HSA binding profile has also been analyzed via a double-reciprocal logarithmic plot according to the relation: $log\ [F_0 - F/F] = n\ log\ K_b - n\ log\ [1/[Q] - (F_0 - F) \times [P_t]/F_0]$ where: Q = ATA concentration; P_t = total HSA concentration; F_0 and F are HSA fluorescence in the absence and presence of ATA. Employing a reverse titration mode monitoring the addition of HSA into ATA, one observes a significant increase in ATA emission intensity at 425 nm upon excitation at 310 nm, suggesting the existence of protein-induced enhancement of ligand fluorescence. This phenomenon is interpreted as the result of enhanced ligand fluorescence upon migrating from an aqueous to protein site environment and corroborates the existence of a specific binding interaction.

5.2.4. Characterization of Ligand–Ribosome Binding via Fluorescence Anisotropy

Fluorescence anisotropy (r) of ribosome–ligand binding interactions are performed by exciting the ligand fluorophore (ATA and/or derivatives) with polarized light (310 nm) and measuring the fluorescence intensities (420 nm) that are both parallel (I_{VV}) and perpendicular (I_{VH}) to the excitation polarization. The resultant values of I_{VV}, I_{VH}, and grating factor (G) are introduced into the following relation to calculate fluorescence anisotropy: $r = I_{VV} - G{\cdot}I_{VH}/I_{VV} + 2G \times I_{VH}$. The G value corrects for efficiency differences in the instrument optics [68] and is determined by measuring the intensity ratio of vertical and horizontal components when the free fluorescent ligand solution is excited with horizontally polarized light. Fluorescence anisotropy measurements are performed in 1.0 cm quartz cuvettes under continuous stirring at 25 °C in the presence of increasing ribosome concentrations until saturation is achieved. The interaction of ATA derivatives with ribosomes is characterized by titrating 1–2 μL aliquots of ribosome stock (27 μM) into the ligand solutions in 20 mM PIPES (pH 7.4), 5 mM $MgCl_2$, 30 mM KCl, and measuring the r values for calculation of fraction bound (f_B). The latter is determined via the following relation: $f_B = r - r_F/R \times (r_B - r) + r - r_F$, where r is the anisotropy measured at a given ribosome concentration with r_F and r_B corresponding to the anisotropy of free and fully bound ligand, respectively. Since the ligand fluorescence intensity increases in the presence of a targets as observed for ATA–HSA interactions (refer to Figures S1 and S2), a correction factor (R) is introduced in the equation to account for these changes [69]. Parallel fluorescence intensity measurements are performed to determine the R value which is calculated as follows: $R = F_B/F_F$, where F_B and F_F represent ligand fluorescence intensities in the bound and free states, respectively. Recasting f_B as a function of ribosome concentration via a double-reciprocal plot yields the binding constant (K_b) according to the relation: $1/f_B = 1/K_b \times 1/[ribosome] + 1$.

5.3. Characterization of Binding Energetics via Isothermal Titration Calorimetry

Thermodynamic binding parameters for the association of polymeric ATA and oligomeric derivatives with Human Serum Albumin (HSA) were determined calorimetrically employing a MicroCal VP-ITC (Malvern Panalytical, Northampton, MA, USA). The protein was prepared as described in Section 5.1 and dialyzed exhaustively against a buffer comprised of 10 mM sodium phosphate adjusted to pH 7.4. The ATA polymer and dichlorohexamer were dissolved directly in final dialysate at the respective standard concentrations. Each ITC experiment consisted of thirty consecutive 10.0 μL injections during which the reaction heat is monitored and integrated over a 5.0 min period under continuous stirring. The experimental protocol has been designed to ensure that there is a tenfold excess of ligand in the titration syringe relative to standard protein concentration in the sample cell. The resultant binding isotherms are generated by recording the integrated reaction heats normalized for ligand concentration versus the HSA:ligand ratio. A nonlinear least squares fit of the resultant profile to a single site binding model yields thermodynamic parameters for the HSA–ligand complex including the affinity (K_a), Gibbs free energy (ΔG), enthalpy (ΔH), and entropy (ΔS).

5.4. Biochemical and Biological Assays

The following solutions of ATA analogs were prepared for biological studies. Stock solutions of synthetic ATA species (compounds **8**, **9**, and **11**) were prepared in DMSO at 50–100 mM and stored at −80 °C. Commercial heterogeneous polymeric ATA (catalog number A1895) was obtained from Millipore Sigma (St. Louis, MO, USA). A stock solution containing 15.8 mg/mL was prepared in DMSO and labeled 37 mM ATA employing the molecular weight of 422.3 for a monomeric ATA unit.

5.4.1. Inhibition of SARS-CoV-2 RNA Dependent RNA Polymerase Active Complex

Non-structural proteins 12, 7, and 8 (Nsp12, 7, and 8) of SARS-CoV-2 were purified as described previously [70]. The RNA extension assay was performed in a reaction buffer containing 50 mM KCl, 100 mM Tris-HCl (pH 8.0) and 1 mM DTT. A fluorescent-labeled RNA primer (5′-Cy3-Gr Ur Cr Ar Ur Ur Cr Ur Cr Cr Ur Ar Ar Gr Ar Ar Gr Cr Ur Ar-3′) was annealed to a 40 nt RNA template (5′-Cr Ur Ar Ur Cr Cr Cr Cr Ar Ur Gr Ur Gr Ar Ur Ur Ur Ur Ar Ar Ur Ar Gr Cr Ur Ur Cr Ur Ur Ar Gr Gr Ar Gr Ar Ar Ur Gr Ar Cr-3′) by heating the reaction to 75 °C and gradually cooling to 4 °C which generated a double-strand RNA substrate. SARS-CoV-2 RdRp (nsp12) was incubated with its co-factors nsp7 and nsp8 (1:2:2 molar ratio) on ice for 20 min prior to polymerase extension. The reactions (20 µL) contained 15 mM $MgCl_2$, 500 nM RNA substrate, 500 µM NTPs, varying concentrations of ATA species and were initiated by the addition of 1 µM pre-incubated RdRp complex at 37 °C. The reactions were terminated after 60 min by adding 10 µL of stop solution comprised of 95% formamide in 20 mM EDTA. A 5 µL aliquot of reaction product was loaded onto a 16% denaturing polyacrylamide gel and the Cy3-labelled RNA products visualized using a Typhoon Imager. Experiments on each ATA analog were repeated independently at least three times.

5.4.2. Inhibition of Protein Synthesis in Rabbit Reticulocyte Lysates

The TnT® Quick Coupled Transcription/Translation System from Promega (Madison, WI, USA; L1170) was used to evaluate the inhibitory effects of ATA derivatives and cycloheximide on protein synthesis. Assay conditions were employed as recommended by the manufacturer. In summary, reactions (20 µL) contained 80% TNT Quick master mix, 20 µM methionine, and 40 ng/µL luciferase RNA (Promega; L4561) at the indicated concentration of test chemicals. Samples were incubated at 30 °C for 90 min and diluted ten-fold with PBS. An aliquot (2 µL) was mixed with 20 µL of luciferase assay reagent (Promega; E1500) and luminescence was measured with a Turner Designs Model TD-20/20 Luminometer. Depending on the compound evaluated, experiments were repeated two-four times with good agreement between the runs. For compounds that exhibited minimal or no activity, two independent experiments were conducted to confirm the findings. The results are reported as average values including the mean and standard deviation wherever applicable.

5.4.3. Cell Culture

African green monkey kidney cells (Vero E6) were purchased from the American Type Culture Collection (Manassas, VA, USA; Cat. No. VERO C1008 (Vero 76, clone E6, Vero E6) CRL-1586™). Complete details regarding this cell line may be accessed at the following link: https://www.atcc.org/products/crl-1586#detailed-product-information (accessed on 24 April 2022). Prior to commercial availability, the cell lines are tested by the vendor for mycoplasma contamination and authenticated via short tandem repeat profiling. The Vero E6 cell line was handled and maintained according to manufacturer's instructions in Eagles's Minimum Essential Medium (EMEM) supplemented with 10% fetal bovine serum (10% FBS-EMEM) in a humidified incubator in 5% CO_2 at 37 °C.

5.4.4. Protein Synthesis in Cultured Cells: Puromycin Pulse Labeling

Prior to the experiment, Vero E6 cells were seeded at 30,000 cells per cm^2 in 6-well plates in 5% FBS-EMEM. Two days following plating, the medium was replaced with 5%

FBS–EMEM containing polymeric ATA at concentrations spanning the range of 1–300 μM and the plates incubated for one hour followed by addition of puromycin (0.01 mg/mL final concentration). Control experiments were conducted in the absence and presence of cycloheximide (1–150 μM) or DMSO. Following incubation with puromycin for 10 min, the medium was quickly removed and cells washed with warm PBS and allowed to recover in drug-free 5% FBS-EMEM for 30 min. Cell pellets collected by trypsinization and centrifugation were stored at −80 °C while awaiting protein preparation and immunoblotting as described below.

5.4.5. Protein Preparation and Immunoblotting to Detect Puromycilated Proteins

Cell lysates were prepared in RIPA (radio-immunoprecipitation) buffer (Sigma; R0278) containing protease inhibitors (Sigma; P8340). Protein concentrations were determined via the Pierce™ bicinconinic acid (BCA) assay (Thermo Fisher Scientific, Waltham, MA, USA) using bovine serum albumin as a standard. All procedures were performed in accordance with protocols recommended by the manufacturers. For immunoblotting, 8 μg of protein was combined with Laemmli buffer containing 10% β-mercaptoethanol, heated at 95 °C for 5 min and loaded on 8–16% Mini-PROTEAN® TGX™ precast protein gels. Samples were resolved via electrophoresis in Tris-Gly-SDS buffer and transferred to nitrocellulose membranes in Tris-Gly buffer containing 20% methanol at 4–6 V/cm^2 for one hour. Membranes were blocked in 10 mM sodium phosphate buffer containing 150 mM NaCl (pH 7.8), 5% milk and 0.1% TWEEN20 (PBST) at room temperature for at least one hour, then incubated overnight with primary anti-puromycin antibodies (TFS; MABE343 mouse monoclonal antibodies clone 12D10; 1:25,000) at 4 °C. Membranes were washed three times for five minutes with PBST and incubated with secondary goat anti-mouse horseradish peroxidase (HRP)-conjugated antibodies (TFS, Invitrogen; G21040; 1:25,000) for 90 min. Following three additional rounds of washing as described above, membranes were incubated with an HRP substrate (TFS; 34,077) for one minute and chemoluminescence visualized by exposing membranes to film for various times.

For β-actin staining, the same membranes used for puromycin immunostaining were washed four times for five min with 20 mM Tris-buffered saline containing 500 mM sodium chloride (pH 7.5) and 0.1% TWEEN20 (TBST) and stripped by two rounds of 20- and 10-min washes at 50 °C in glycine buffer (pH 2.2) containing 0.1% SDS (*w/v*) and 1% TWEEN20. Membranes were then washed at room temperature with PBS (3 × 10 min) and PBST (2 × 5 min). Blocking, primary (1:4000), and secondary (1:5000) antibody staining were conducted as described above. Primary rabbit anti-β-actin antibodies (4967S, lot 9) and secondary HRP-conjugated goat-anti-rabbit antibodies (7074S, lot 27) were obtained from Cell Signaling. The results were quantified with Image J 1.52a software using β-actin staining for each sample as an internal reference. A second set of puromycin labeling experiments was conducted to confirm the findings and results are displayed as a representative immunoblotting.

5.4.6. Evaluation of Growth Inhibition and Toxicity in Vero E6 Cells

A day before exposure, Vero E6 cells were plated on 24-well plates (500 μL per well) at 6250 cells per cm^2 in 10% FBS-EMEM. Test compounds were dissolved in EMEM at two-fold concentrations from their designated amounts in culture and added to cells in equal volume (500 μL) to achieve concentrations of 1–400 μM in 5% FBS-EMEM. The cells were exposed under standard culture conditions for 48 h. Controls containing varying amounts of DMSO and sans DMSO were included as well. A sulforhodamine B (SRB) assay was conducted to evaluate the impact of polymeric ATA and oligomeric analogs on cell growth and survival. All procedures were conducted according to protocols employed by the National Cancer Institute that are available online at https://dtp.cancer.gov/discovery_development/nci-60/methodology.htm (accessed on 24 April 2022). In summary, the exposed cells were fixed by addition of trichloroacetic acid (TCA, 10% *v/v*) and incubated for one hour at 4 °C after which the plates were washed five times with tap water and air dried. For color development, 500 μL of 0.4% SRB reagent prepared in 1% acetic acid was added to fixed cells and the samples were incubated at room temperature for 15 min under gentle rocking. Unbound dye was removed via five washes in 1% acetic acid and the plates air dried. Trizma base (10 mM) was used to dissolve the stain and absorbances recorded at 515 nm on a UV/Vis Model Ultrospec2000 Spectrophotometer (PharmaciaBiotech, Piscataway, NJ, USA). Since DMSO did not affect cells at the concentrations employed in our assay, the absorbance values of cells incubated in media with or without DMSO for 48 h were combined and established as 100% growth/survival. We also conducted an SRB assay on separate wells of cells the same day of exposure (day 0) taking advantage of the ability to store fixed plates for parallel processing with samples following exposure. All experiments were repeated two/three times independently, two/three wells per each condition within an experiment. Dose-dependent changes in survival and cell growth were analyzed via the Sigma Plot Program v.13.0 (Systat Software, Inc., San Jose, CA, USA) and displayed as mean values with standard deviations.

5.4.7. Isolation of 80S Yeast Ribosomes

Saccharomyces cerevisiae YPL 154C strain (MATa leu2Δ0 met15Δ0 ura3Δ0 pep4Δ::kan^r, Horizon Discovery, Lafayette, CO, USA) was grown in YPD medium and disrupted in liquid nitrogen by SPEX 6870 freezer mill. Yeast cell extracts were prepared in a buffer system comprised of 50 mM Hepes-KOH (pH 7.4), 11 mM magnesium acetate, 1 mM dithiothreitol, 1 mM phenylmethylsulfonyl fluoride, cOmplete Protease Inhibitor Cocktail (1 tablet/50 mL, Roche 11697498001), and 60 U/mL Protector RNase Inhibitor (Roche 03335402001). The supernatant was obtained via ultracentrifugation in a SW32Ti rotor (Beckman, Brea, CA, USA) at 13,000 rpm (max 30,000× *g*) for 45 m at 4 °C. The supernatant was loaded onto 37.65% (*w/v*) sucrose, 20 mM Bis-Tris (pH 6.5), 11 mM magnesium acetate, 0.2 M ammonium acetate, 0.3 M potassium chloride and 5 mM dithiothreitol. Ultracentrifugation was performed in a SW32Ti rotor (Beckman) at 28,000 rpm (max 140,000× *g*) for 22 h at 4 °C. The supernatant was decanted and the pellet washed twice with a buffer composed of 20 mM Hepes-KOH (pH 7.4), 11 mM magnesium acetate, 30 mM potassium chloride, 2 mM dithiothreitol, cOmplete Protease Inhibitor Cocktail (1 tablet/50 mL, Roche 11697498001), and 40 U/mL Protector Rnase Inhibitor (Roche 03335402001). The pellet was suspended in identical buffer, transferred to a 1.5 mL tube, and centrifuged at 20,817× *g* for 10 m at 4 °C. The supernatant was transferred to a fresh tube and frozen in liquid nitrogen. The ribosome preparation obtained from 90 g of wet yeast pellet was 122 mg/mL (27,120 pmol/mL). This procedure has been described in detail elsewhere [71]. As a quality control measure, 8000 pmol of the sample was analyzed by gradient centrifugation using 15–45% sucrose (Supplementary Material; Figure S6 and Table S2).

5.5. Synthesis of Defined ATA-Derived Analogs

Several ATA derivatives of defined length and molecular weight were synthesized in our laboratory. The quality and purity of samples was verified by high performance liquid chromatography (HPLC), nuclear magnetic resonance (H^1-NMR) and liquid chromatography tandem mass spectrometry (LC-MS/MS). The synthetic route of ATA-derivative **11** is delineated in Scheme 3 and may be summarized as follows. Bis-formylation of **5** prepared according to Cushman et al. [39] (a) with hexamethylene tetramine/triethylamine followed by reduction with LiBH4 (b) led to **6,** which in the presence of sulfuric acid could be coupled with two equivalents of **7** (c) to produce tetramer **8**. Upon catalytic reduction with 10% PdO and H2 (d), tetramer **8** yielded **9**. The latter, when subjected to the same series of reactions used to convert **8** to **10** via **9**, yielded the desired dichlorohexamer **11**. This oligomer exhibited excellent activity when tested in reticulocytes and as an inhibitor of SARS-CoV-2 RdRp. The synthesis and purification of each ATA derivative is currently in progress and specific details will be reported elsewhere.

Supplementary Materials: The following supporting information can be downloaded at: https://www.mdpi.com/article/10.3390/life12060872/s1, Figure S1: UV/Vis Absorbance Spectrum of the ATA Polymer Standard (2.0 mM) Revealing Major and Minor Peaks at 310 and 530 nm, respectively. Figure S2: Impact of has on ATA Emission Fluorescence; Figure S3: Fluorescence Intensity of ATA in the Presence of Increasing HSA Concentrations; Figure S4: ITC Profile of ATA Dilution into Dialysate; Figure S5: Full Size Immunoblotting Corresponding to Figure 8; Figure S6: Fractions of Yeast 80S Ribosomes Obtained via Sucrose Gradient Centrifugation. Table S1: Densitometry Results for Figure 8 and Supplementary Figure S6; Table S2: Optical Density of Yeast 80S Ribosome Fractions Obtained via Sucrose Gradient Centrifugation.

Author Contributions: Conceptualization, A.P.G. and F.J.; Methodology, V.S.S., F.J., C.A.M., D.P.R., M.G.-D., K.H., R.B., R.C. and X.Y.; Data analysis, C.A.M., D.P.R., V.S.S. and M.G.-D.; Writing—original draft preparation, C.A.M., D.P.R., V.S.S. and F.J.; Writing—Editing and Review, C.A.M., D.P.R., V.S.S. and A.P.G.; Illustrations, C.A.M., D.P.R., V.S.S. and M.G.-D.; Funding acquisition, A.P.G., V.S.S. and M.G.-D. All authors have read and agreed to the published version of the manuscript.

Funding: Support for his research was provided by the Laufer Family Foundation (V.S.S., K.H., R.B., R.C., F.J., A.P.G.), the Zickler Family Foundation (V.S.S., K.H.), and a COVID Seed Grant from Stony Brook University OVPR (M.G.-D., X.Y.).

Institutional Review Board Statement: Not applicable.

Informed Consent Statement: Not applicable.

Data Availability Statement: Not applicable.

Conflicts of Interest: The authors declare no conflict of interest.

Abbreviations

ATA: aurintricarboxylic acid (commercial heterogeneous preparation); CHX, cycloheximide; COVID, coronavirus disease; HTS, high throughput screening; PBS, phosphate buffered saline; RdRp, SARS-CoV-2 RNA dependent RNA polymerase (complex comprising SARS-CoV-2 nsp7, nsp8 and nsp12); SARS-CoV-2, severe acute respiratory syndrome coronavirus 2; SDS, sodium dodecyl sulphate.

References

1. Vesnaver, G.; Chang, C.N.; Eisenberg, M.; Grollman, A.P.; Breslauer, K.J. Influence of Abasic and Anucleosidic Sites on the Stability, Conformation, and Melting Behavior of a DNA Duplex—Correlations of Thermodynamic and Structural Data. *Proc. Natl. Acad. Sci. USA* **1989**, *86*, 3614–3618. [CrossRef]
2. Plum, G.E.; Grollman, A.P.; Johnson, F.; Breslauer, K.J. Influence of the oxidatively damaged adduct 8-oxodeoxyguanosine on the conformation, energetics, and thermodynamic stability of a DNA duplex. *Biochemistry* **1995**, *34*, 16148–16160. [CrossRef] [PubMed]
3. Gelfand, C.A.; Plum, G.E.; Grollman, A.P.; Johnson, F.; Breslauer, K.J. The impact of an exocyclic cytosine adduct on DNA duplex properties: Significant thermodynamic consequences despite modest lesion-induced structural alterations. *Biochemistry* **1998**, *37*, 12507–12512. [CrossRef] [PubMed]
4. Gelfand, C.A.; Plum, G.E.; Mielewczyk, S.; Remeta, D.P.; Breslauer, K.J. A quantitative method for evaluating the stabilities of nucleic acids. *Proc. Natl. Acad. Sci. USA* **1999**, *96*, 6113–6118. [CrossRef] [PubMed]
5. Minetti, C.A.S.A.; Remeta, D.P.; Dickstein, R.; Breslauer, K.J. Energetic signatures of single base bulges: Thermodynamic consequences and biological implications. *Nucleic Acids Res.* **2010**, *38*, 97–116. [CrossRef]
6. Minetti, C.A.S.A.; Remeta, D.P.; Johnson, F.; Iden, C.R.; Breslauer, K.J. Impact of alpha-Hydroxy-Propanodeoxyguanine Adducts on DNA Duplex Energetics: Opposite Base Modulation and Implications for Mutagenicity and Genotoxicity. *Biopolymers* **2010**, *93*, 370–382.
7. Lukin, M.; Minetti, C.A.S.A.; Remeta, D.P.; Attaluri, S.; Johnson, F.; Breslauer, K.J.; de los Santos, C. Novel post-synthetic generation, isomeric resolution, and characterization of Fapy-dG within oligodeoxynucleotides: Differential anomeric impacts on DNA duplex properties. *Nucleic Acids Res.* **2011**, *39*, 5776–5789. [CrossRef]
8. Minetti, C.A.S.A.; Remeta, D.P.; Iden, C.R.; Johnson, F.; Grollman, A.P.; Breslauer, K.J. Impact of thymine glycol damage on DNA duplex energetics: Correlations with lesion-induced biochemical and structural consequences. *Biopolymers* **2015**, *103*, 491–508. [CrossRef]

9. Minetti, C.A.; Sun, J.Y.; Jacobs, D.P.; Kang, I.; Remeta, D.P.; Breslauer, K.J. Impact of bistrand abasic sites and proximate orientation on DNA global structure and duplex energetics. *Biopolymers* **2018**, *109*, e23098. [CrossRef]
10. Minetti, C.A.S.A.; Remeta, D.P.; Zharkov, D.O.; Plum, G.E.; Johnson, F.; Grollman, A.P.; Breslauer, K.J. Energetics of lesion recognition by a DNA repair protein: Thermodynamic characterization of formamidopyrimidine-glycosylase (Fpg) interactions with damaged DNA duplexes. *J. Mol. Biol.* **2003**, *328*, 1047–1060. [CrossRef]
11. Minetti, C.A.S.A.; Remeta, D.P.; Miller, H.; Gelfand, C.A.; Plum, G.E.; Grollman, A.P.; Breslauer, K.J. The thermodynamics of template-directed DNA synthesis: Base insertion and extension enthalpies. *Proc. Natl. Acad. Sci. USA* **2003**, *100*, 14719–14724. [CrossRef] [PubMed]
12. Minetti, C.A.S.A.; Remeta, D.P.; Breslauer, K.J. A continuous hyperchromicity assay to characterize the kinetics and thermodynamics of DNA lesion recognition and base excision. *Proc. Natl. Acad. Sci. USA* **2008**, *105*, 70–75. [CrossRef] [PubMed]
13. Breslauer, K.J. The shaping of a molecular linguist: How a career studying DNA energetics revealed the language of molecular communication. *J. Biol. Chem.* **2021**, *296*, 100522. [CrossRef] [PubMed]
14. De Clercq, E. The next ten stories on antiviral drug discovery (part E): Advents, advances, and adventures. *Med. Res. Rev.* **2011**, *31*, 118–160. [CrossRef] [PubMed]
15. De Clercq, E. Potential antivirals and antiviral strategies against SARS coronavirus infections. *Expert Rev. Anti. Infect. Ther.* **2006**, *4*, 291–302. [CrossRef]
16. Caro, N. Über Oxyaurin und Oxyaurincarbonsäuren (About oxyaurine and oxyaurine carboxylic acids). *Ber. Dtsch. Chem. Ges.* **1892**, *25*, 939–949. [CrossRef]
17. Grollman, A.P. Emetine: New Uses for an Old Drug. *Ohio State Med. J.* **1970**, *66*, 257–259.
18. Grollman, A.P. Structural Basis for Inhibition of Protein Synthesis by Emetine and Cycloheximide Based on an Analogy between Ipecac Alkaloids and Glutarimide Antibiotics. *Proc. Natl. Acad. Sci. USA* **1966**, *56*, 1867–1874. [CrossRef]
19. Grollman, A.P. Inhibitors of Protein Biosynthesis.V. Effects of Emetine on Protein and Nucleic Acid Biosynthesis in Hela Cells. *J. Biol. Chem.* **1968**, *243*, 4089–4094. [CrossRef]
20. Grollman, A.P.; Stewart, M.L. Inhibition of Attachment of Messenger Ribonucleic Acid to Ribosomes. *Proc. Natl. Acad. Sci. USA* **1968**, *61*, 719–725. [CrossRef]
21. Stewart, M.L.; Grollman, A.P.; Huang, M.T. Aurintricarboxylic acid: Inhibitor of initiation of protein synthesis. *Proc. Natl. Acad. Sci. USA* **1971**, *68*, 97–101. [CrossRef] [PubMed]
22. Grollman, A.P.; Horwitz, S.B. Rational Design of Antiviral Agents. In *Molecular Pharmacology—Drug Design*; Ariens, E.F.E., Ed.; Academic Press: Cambridge, MA, USA, 1971; pp. 261–276.
23. Huang, M.; Grollman, A.P. Effects of Aurintricarboxylic Acid on Ribosomes and Biosynthesis of Globin in Rabbit Reticulocytes. *Mol. Pharmacol.* **1972**, *8*, 111–127. [PubMed]
24. Grollman, A.P.; Jarkovsky, Z. Emetine and Related Alkaloides. In *Mechansim of Action of Antimicrobial and Antitumor Agents Antibiotics*; Springer: Berlin/Heidelberg, Germany, 1975; pp. 420–435.
25. Liao, L.L.; Horwitz, S.B.; Huang, M.T.; Grollman, A.P.; Steward, D.; Martin, J. Triphenylmethane Dyes as Inhibitors of Reverse-Transcriptase, Ribonucleic-Acid Polymerase, and Protein-Synthesis-Structure-Activity-Relationships. *J. Med. Chem.* **1975**, *18*, 117–120. [CrossRef] [PubMed]
26. Steward, D.L.; Martin, J.; Grollman, A.P. Inhibition of influenza virus by triphenylmethane compounds. *Ann. N. Y. Acad. Sci.* **1977**, *284*, 638–649. [CrossRef]
27. Grollman, A.P. Mechanism of Action of Emetine-Demonstration of Its Inhibitory Action on Protein Synthesis. *J. Clin. Investig.* **1966**, *45*, 1018.
28. Blumenthal, T.; Landers, T.A. The inhibition of nucleic acid-binding proteins by aurintricarboxylic acid. *Biochem. Biophys. Res. Commun.* **1973**, *55*, 680–688. [CrossRef]
29. Givens, J.F.; Manly, K.F. Inhibition of RNA-directed DNA polymerase by aurintricarboxylic acid. *Nucleic Acids. Res.* **1976**, *3*, 405–418. [CrossRef]
30. Akiyama, S.; Kuwano, M.; Yamamoto, M.; Endo, H. Effect of Aurintricarboxylic Acid on RNA-Polymerase from Rat-Liver. *J. Biochem.* **1977**, *81*, 135–141. [CrossRef]
31. Grollman, A. Inhibition of messenger ribonucleic acid attachment to ribosomes. II. Proposed mechanisms for the design of novel antiviral agents. *Antimicrob. Agents Chemother.* **1968**, *8*, 36.
32. Polatnick, J.; Bachrach, H.L. Effect of zinc and other chemical agents on foot-and-mouth-disease virus replication. *Antimicrob. Agents Chemother.* **1978**, *13*, 731–734. [CrossRef]
33. He, R.; Adonov, A.; Traykova-Adonova, M.; Cao, J.; Cutts, T.; Grudesky, E.; Deschambaul, Y.; Berry, J.; Drebot, M.; Li, X. Potent and selective inhibition of SARS coronavirus replication by aurintricarboxylic acid. *Biochem. Biophys. Res. Commun.* **2004**, *320*, 1199–1203. [CrossRef] [PubMed]
34. Ben David, A.; Diamant, E.; Dor, E.; Barnea, A.; Natan, N.; Levin, L.; Chapman, S.; Mimran, L.C.; Epstein, E.; Zichel, R.; et al. Identification of SARS-CoV-2 Receptor Binding Inhibitors by In Vitro Screening of Drug Libraries. *Molecules* **2021**, *26*, 3213. [CrossRef] [PubMed]
35. Cushman, M.; Wang, P.L.; Stowell, J.G.; Schols, D.; De Clercq, E. Structural Investigation and Anti-Hiv Activities of High-Molecular-Weight ATA Polymers. *J. Org. Chem.* **1992**, *57*, 7241–7248. [CrossRef]

36. Gonzalez, R.G.; Blackburn, B.J.; Schleich, T. Fractionation and Structural Elucidation of the Active Components of Aurintricarboxylic Acid, a Potent Inhibitor of Protein Nucleic-Acid Interactions. *Biochim. Biophys. Acta* **1979**, *562*, 534–545. [CrossRef]
37. Gonzalez, R.G.; Haxo, R.S.; Schleich, T. Mechanism of action of polymeric aurintricarboxylic acid, a potent inhibitor of protein-nucleic acid interactions. *Biochemistry* **1980**, *19*, 4299–4303. [CrossRef]
38. Cushman, M.; Kanamathareddy, S. Synthesis of the Covalent Hydrate of the Incorrectly Assumed Structure of Aurintricarboxylic Acid (Ata). *Tetrahedron* **1990**, *46*, 1491–1498. [CrossRef]
39. Cushman, M.; Kanamathareddy, S.; De Clercq, E.; Schols, D.; Goldman, M.E.; Bowen, J.A. Synthesis and anti-HIV activities of low molecular weight aurintricarboxylic acid fragments and related compounds. *J. Med. Chem.* **1991**, *34*, 337–342. [CrossRef]
40. Cushman, M.; Wang, P.L.; Chang, S.H.; Wild, C.; De Clercq, E.; Schols, D.; Goldman, M.E.; Bowen, J.A. Preparation and anti-HIV activities of aurintricarboxylic acid fractions and analogues: Direct correlation of antiviral potency with molecular weight. *J. Med. Chem.* **1991**, *34*, 329–337. [CrossRef]
41. Tsutsui, K.; Seki, S.; Tsutsui, K.; Oda, T. Fractionation of Aurintricarboxylic Acid and Effects of Its Components on Nuclear Swelling and Nucleic-Acid Synthesis. *Biochim. Biophys. Acta* **1978**, *517*, 14–23. [CrossRef]
42. Cushman, M.; Kanamathareddy, S. Synthesis and Evaluation of a Triphenylcarbinol Related to the Incorrectly Assumed Structure of Aurintricarboxylic Acid. *Ann. N. Y. Acad. Sci.* **1990**, *616*, 499–502. [CrossRef]
43. Wang, P.; Kozlowski, J.; Cushman, M. Isolation and Structure Elucidation of Low-Molecular-Weight Components of Aurintricarboxylic Acid (Ata). *J. Org. Chem.* **1992**, *57*, 3861–3866. [CrossRef]
44. Smith, T.J. Aurintricarboxylic Acid-Derived Polysalicylates as Platforms for Drug Development: A Mini-Review. *IOSR J. Pharm. Biol. Sci.* **2018**, *13*, 44–47.
45. Zhang, Q.B.; Qian, M.X.; Wu, Y.; Wang, Y.P.; Shangguan, W.W.; Lu, J.G.; Zhao, W.J.; Feng, J. Design and biological evaluation of novel long-acting adalimumab Fab conjugated with the albumin binding domain. *Eur. J. Pharmacol.* **2021**, *904*, 174152. [CrossRef] [PubMed]
46. Tayyab, S.; Feroz, S.R. Serum albumin: Clinical significance of drug binding and development as drug delivery vehicle. *Adv. Protein Chem. Struct. Biol.* **2021**, *123*, 193–218. [PubMed]
47. Gao, Y.; Kuang, Y.; Du, X.W.; Zhou, J.; Chandran, P.; Horkay, F.; Xu, B. Imaging Self-Assembly Dependent Spatial Distribution of Small Molecules in a Cellular Environment. *Langmuir* **2013**, *29*, 15191–15200. [CrossRef]
48. Owen, D.R.; Allerton, C.M.N.; Anderson, A.S.; Aschenbrenner, L.; Avery, M.; Berritt, S.; Boras, B.; Cardin, R.D.; Carlo, A.; Coffman, K.J.; et al. An oral SARS-CoV-2 M-pro inhibitor clinical candidate for the treatment of COVID-19. *Science* **2021**, *374*, 1586–1593. [CrossRef]
49. Liang, F.B.; Huang, Z.H.; Lee, S.Y.; Liang, J.; Ivanov, M.I.; Alonso, A.; Bliska, J.B.; Lawrence, D.S.; Mustelin, T.; Zhang, Z.Y. Aurintricarboxylic acid blocks in vitro and in vivo activity of YopH, an essential virulent factor of Yersinia pestis, the agent of plague. *J. Biol. Chem.* **2003**, *278*, 41734–41741. [CrossRef]
50. Balzarini, J.; Mitsuya, H.; De Clercq, E.; Broder, S. Aurintricarboxylic acid and Evans Blue represent two different classes of anionic compounds which selectively inhibit the cytopathogenicity of human T-cell lymphotropic virus type III/lymphadenopathy-associated virus. *Biochem. Biophys. Res. Commun.* **1986**, *136*, 64–71. [CrossRef]
51. Shadrick, W.R.; Mukherjee, S.; Hanson, A.M.; Sweeney, N.L.; Frick, D.N. Aurintricarboxylic acid modulates the affinity of hepatitis C virus NS3 helicase for both nucleic acid and ATP. *Biochemistry* **2013**, *52*, 6151–6159. [CrossRef]
52. Cushman, M.; Sherman, P. Inhibition of HIV-1 integration protein by aurintricarboxylic acid monomers, monomer analogs, and polymer fractions. *Biochem. Biophys. Res. Commun.* **1992**, *185*, 85–90. [CrossRef]
53. Freire, E. Do enthalpy and entropy distinguish first in class from best in class? *Drug Discov. Today* **2008**, *13*, 869–874. [CrossRef] [PubMed]
54. Velazquez-Campoy, A.; Todd, M.J.; Freire, E. HIV-1 protease inhibitors: Enthalpic versus entropic optimization of the binding affinity. *Biochemistry* **2000**, *39*, 2201–2207. [CrossRef] [PubMed]
55. Tarcsay, A.; Keseru, G.M. Contributions of Molecular Properties to Drug Promiscuity Miniperspective. *J. Med. Chem.* **2013**, *56*, 1789–1795. [CrossRef] [PubMed]
56. Canal, B.; McClure, A.W.; Curran, J.F.; Wu, M.; Ulferts, R.; Weissmann, F.; Zeng, J.K.; Bertolin, A.P.; Milligan, J.C.; Basu, S.; et al. Identifying SARS-CoV-2 antiviral compounds by screening for small molecule inhibitors of nsp14/nsp10 exoribonuclease. *Biochem. J.* **2021**, *478*, 2445–2464. [CrossRef]
57. Marcus, A.; Bewley, J.D.; Weeks, D.P. Aurintricarboxylic acid and initiation factors of wheat embryo. *Science* **1970**, *167*, 1735–1736. [CrossRef]
58. Siegelman, F.; Apirion, D. Aurintricarboxylic acid, a preferential inhibitor of initiation of protein synthesis. *J. Bacteriol.* **1971**, *105*, 902–907. [CrossRef]
59. Ghosh, U.; Giri, K.; Bhattacharyya, N.P. Interaction of aurintricarboxylic acid (ATA) with four nucleic acid binding proteins DNase I, RNase A, reverse transcriptase and Taq polymerase. *Spectrochim. Acta A* **2009**, *74*, 1145–1151. [CrossRef]
60. Hallick, R.B.; Chelm, B.K.; Gray, P.W.; Orozco, E.M., Jr. Use of aurintricarboxylic acid as an inhibitor of nucleases during nucleic acid isolation. *Nucleic Acids. Res.* **1977**, *4*, 3055–3064. [CrossRef]
61. Nakane, H.; Balzarini, J.; De Clercq, E.; Ono, K. Differential inhibition of various deoxyribonucleic acid polymerases by Evans blue and aurintricarboxylic acid. *Eur. J. Biochem.* **1988**, *177*, 91–96. [CrossRef]

62. Myskiw, C.; Deschambault, Y.; Jefferies, K.; He, R.; Cao, J. Aurintricarboxylic acid inhibits the early stage of vaccinia virus replication by targeting both cellular and viral factors. *J. Virol.* **2007**, *81*, 3027–3032. [CrossRef]
63. Zhang, G.N.; Zhao, J.Y.; Li, Q.J.; Wang, M.H.; Zhu, M.; Wang, J.X.; Cen, S.; Wang, Y.C. Discovery and optimization of 2-((1H-indol-3-yl)thio)-N-benzyl-acetamides as novel SARS-CoV-2 RdRp inhibitors. *Eur. J. Med. Chem.* **2021**, *223*, 113622. [CrossRef] [PubMed]
64. Cho, H.; Lee, D.Y.; Shrestha, S.; Shim, Y.S.; Kim, K.C.; Kim, M.-K.; Lee, K.-H.; Won, J.; Kang, J.-S. Aurintricarboxylic acid translocates across the plasma membrane, inhibits protein tyrosine phosphatase and prevents apoptosis in PC12 cells. *Mol. Cells* **2004**, *18*, 46–52. [PubMed]
65. Zhou, H.; Fang, Y.; Xu, T.; Ni, W.J.; Shen, A.Z.; Meng, X.M. Potential therapeutic targets and promising drugs for combating SARS-CoV-2. *Br. J. Pharmacol.* **2020**, *177*, 3147–3161. [CrossRef] [PubMed]
66. Khursheed, A.; Jain, V.; Rasool, A.; Rather, M.A.; Malik, N.A.; Shalla, A.H. Molecular scaffolds from mother nature as possible lead compounds in drug design and discovery against coronaviruses: A landscape analysis of published literature and molecular docking studies. *Microb Pathog.* **2021**, *157*, 104933. [CrossRef] [PubMed]
67. Tun, M.M.N.; Morita, K.; Ishikawa, T.; Urata, S. The Antiviral Effect of the Chemical Compounds Targeting DED/EDh Motifs of the Viral Proteins on Lymphocytic Choriomeningitis Virus and SARS-CoV-2. *Viruses* **2021**, *13*, 1220.
68. Lakowicz, J.R. *Principles of Fluorescence Spectroscopy*, 1st ed.; Plenum Press: New York, NY, USA, 1983.
69. Ingersoll, C.M.; Strollo, C.M. Steady-state fluorescence anisotropy to investigate flavonoids binding to proteins. *J. Chem. Educ.* **2007**, *84*, 1313–1315. [CrossRef]
70. Hillen, H.S.; Kokic, G.; Farnung, L.; Dienemann, C.; Tegunov, D.; Cramer, P. Structure of replicating SARS-CoV-2 polymerase. *Nature* **2020**, *584*, 154–156. [CrossRef]
71. Lomakin, I.B.; Steitz, T.A. The initiation of mammalian protein synthesis and mRNA scanning mechanism. *Nature* **2013**, *500*, 307. [CrossRef]

life

MDPI

Article

Thermally Induced Transitions of d($G_4T_4G_3$) Quadruplexes Can Be Described as Kinetically Driven Processes

Iztok Prislan [1,*], Tomaz Urbic [2] and Natasa Poklar Ulrih [1,*]

1 Biotechnical Faculty, University of Ljubljana, Jamnikarjeva 101, 1000 Ljubljana, Slovenia
2 Faculty of Chemistry and Chemical Technology, University of Ljubljana, Vecna Pot 113, 1000 Ljubljana, Slovenia; tomaz.urbic@fkkt.uni-lj.si
* Correspondence: iztok.prislan@bf.uni-lj.si (I.P.); natasa.poklar@bf.uni-lj.si (N.P.U.)

Abstract: DNA sequences that are rich in guanines and can form four-stranded structures are called G-quadruplexes. Due to the growing evidence that they may play an important role in several key biological processes, the G-quadruplexes have captured the interest of several researchers. G-quadruplexes may form in the presence of different metal cations as polymorphic structures formed in kinetically governed processes. Here we investigate a complex polymorphism of d($G_4T_4G_3$) quadruplexes at different K^+ concentrations. We show that population size of different d($G_4T_4G_3$) quadruplex conformations can be manipulated by cooling rate and/or K^+ concentration. We use a kinetic model to describe data obtained from DSC, CD and UV spectroscopy and PAGE experiments. Our model is able to describe the observed thermally induced conformational transitions of d($G_4T_4G_3$) quadruplexes at different K^+ concentrations.

Keywords: G-quadruplexes; potassium; calorimetry; polymorphism; kinetics; folding; kinetic models; fitting

Citation: Prislan, I.; Urbic, T.; Poklar Ulrih, N. Thermally Induced Transitions of d($G_4T_4G_3$) Quadruplexes Can Be Described as Kinetically Driven Processes. *Life* **2022**, *12*, 825. https://doi.org/10.3390/life12060825

Academic Editors: Tigran Chalikian, Jens Völker and Paolo Mariani

Received: 6 April 2022
Accepted: 27 May 2022
Published: 1 June 2022

1. Introduction

Understanding the relationship between the structure of biological macromolecules and the thermodynamic or kinetic properties that dictate stability and binding with other molecules remains one of the most important problems in biochemistry and biotechnology [1]. Such knowledge is crucial to understanding biological processes and to design more efficient pharmaceutical ligands. The laws of thermodynamics and kinetics do not tell us directly about molecular structures or mechanisms, although we can use the results of thermodynamic property measurements to help us interpret molecular mechanisms of a system [2].

Over the past couple of decades, DNA sequences that are both rich in guanines and can form four-stranded structures, called G-quadruplexes, have captured the interest of several major laboratories all over the world. This is mainly due to the growing evidence that G-quadruplexes may play an important role in several key biological processes [3,4]. Even though much progress has been made in understanding kinetic and thermodynamic factors responsible for the structural interconversion of several quadruplex structures, there are still many factors that govern the formation of G-quadruplexes and their physico-chemical properties that are poorly understood.

Four guanine bases linked by Hoogsteen type hydrogen bonds form a cyclic coplanar G-tetrads which can stack on top of another tetrad to form G-quadruplex structures [5,6]. These structures are additionally stabilized by metal cations that are selectively bound in the central cavity between the G-quartets. Cation coordination is essential for the stabilization of G-quadruplexes, and monovalent and divalent ions have been shown to influence the structure and stability of G-quadruplexes. By compiling a number of studies, one can estimate that G-quadruplex stabilization follows the general trend: $Sr^{2+} > K^+ > Ca^{2+} > NH^{4+}, Na^+, Rb^+ > Mg^{2+} > Li^+ \geq Cs^+$ [7].

Changing the type or concentration of cations can induce the formation of new long-lived G-quadruplex conformations with identical nucleotide sequence. For instance, oligonucleotide d(G_2AG_2AG) forms a bimolecular G-quadruplex at low Na^+ concentration and a tetramolecular G-quadruplex at higher Na^+ concentration [8]. Furthermore, d($G_3T_4G_3$) forms only hairpin dimer G-quadruplex structures in the presence of Na^+ ions, while in the presence of K^+ ions both hairpin dimer and linear, four-stranded G-quadruplex structures were observed [9]. This phenomenon is known as G4 polymorphism [10–12] and can even lead to concurrent folding isomers in heterogenous ensembles [13–15]. The polymorphism of GQ structures is determined by strand orientation (i.e., parallel vs. antiparallel), the conformation of the glycosidic bonds, and the loop topology (e.g., lateral loops vs. diagonal loops). These variations in the structure might be due to several reasons, such as the cationic coordination, π-stacking interactions, hydrogen bonding and hydrophobic effects [16]. Different mechanisms have been proposed to describe folding/unfolding pathways of G-quadruplexes. Some of them are sequential and do not consider side-reactions [17] while others consist of several parallel pathways [18,19].

The behavior of G-quadruplexes in the presence of potassium ions is of particular interest due to the high intracellular concentrations. Chaires et al. have suggested, based on NMR, fluorescence, CD experiments, and molecular modeling, that model telomeric human DNA, 5′-AGGGTTAGGGTTAGGGTTAGGG-3′ (Tel22) in K^+ solutions appears as an equilibrium mixture of two (3 + 1) hybrid-type G-quadruplex structures, Hybrid-1 and Hybrid-2 [20]. This was later confirmed by Bončina et al., who successfully described the unfolding/folding mechanism of Tel22 and showed that each conformational transition depends on K^+ concentration [11]. Grün et al. were studying the cMYC gene-promoter region and used a kinetic model to describe a novel, noncanonical folding isomerism of G-quadruplexes with more than four G-rich tracts [21]. Grey et al. utilized a variety of spectroscopies to follow the kinetics of human telomeric DNA sequences folding and unfolding, and identified previously unreported slow kinetic steps [17]. Kinetics of potassium binding to telomeric (hybrid) and c-myc (parallel) G-quadruplexes have been investigated with electrospray mass spectrometry where authors have reported that misfolded states can linger for several minutes [18]. Different folding timescales of DNA and RNA G-quadruplexes have been reported by Zhang and Balasubramanian, who discussed their findings in relation to a biological timescale [22]. The folding rate of quadruplexes and folding timescale were also investigated by Nguyen et al., who showed that a stem loop accelerates the folding of quadruplexes significantly as compared to a non-structured loop [23]. One can conclude that the formation of G-quadruplexes takes place through different mechanisms on various timescales, making the folding rate of a G-quadruplex and the effect of the environment and loop properties very important parameters to investigate.

Our laboratory has used kinetic models to describe the polymorphism of d($G_4T_4G_3$) quadruplexes in Na^+ and K^+ solutions [24,25]. To test whether this approach can be used under different environmental conditions, we followed in this work the thermally induced folding/unfolding transitions d($G_4T_4G_3$) quadruplexes at different K^+ concentrations, and tried to interpret them in terms of a kinetically governed coexistence of several folded structures and their unfolded forms. We used a variety of experimental techniques (UV and CD spectroscopy, DSC, gel electrophoresis) in order to improve the significance of the proposed model and calculated parameters. We demonstrate that the d($G_4T_4G_3$) sequence can fold into different quadruplex structures and that the mechanism of thermally induced folding/unfolding depends on the cooling/heating rate and potassium concentration.

2. Materials and Methods

2.1. Sample Preparation and UV Spectroscopy

The d($G_4T_4G_3$) oligonucleotide was obtained HPLC pure from Invitrogen Co., (Carlsbad, CA, USA) and Midland Co. (Midland, CO, USA)., and used without further purification. Its concentrations in buffer solutions were determined at 25 °C spectrophotometrically using for the extinction coefficient of its single stranded form at 25 °C the value of

ε_{260} = 105,100 $M^{-1}cm^{-1}$ estimated from the nearest-neighbor data of Cantor et al. [26]. The quadruplex concentration used in this study were 0.82, 0.60 and 0.52 mM in single strands in 25, 35 and 50 mM K^+ solutions, respectively. The buffer used in all experiments consisted of 10 mM K-cacodylic buffer, 1 mM EDTA and different concentrations of KCl (15, 25 and 45 mM) at pH = 6.9. The starting samples were first unfolded into single-stranded form by heating in an outer thermostat at 95 °C for 5 min, cooled down to 4 °C at the cooling rates of 0.05 or 1.0 °C/min to form quadruplex structures and then used in the UV, CD, DSC and PAGE experiments.

2.2. CD Spectroscopy Melting Experiments

CD spectra of d($G_4T_4G_3$) quadruplexes in K^+ solutions were measured as a function of temperature in an Jasco J-1500 CD Spectropolarimeter. Ellipticity, Θ, was measured between 5 and 95 °C in the temperature intervals of 3 °C at the average heating rate of 0.5 °C/min. CD spectra of samples (0.82, 0.60 and 0.52 mM in single strands) prepared at cooling rates of 0.05 or 1.0 °C/min, corrected for the corresponding buffer contribution, were collected between 215 and 320 nm in a 0.25 mm cuvette at 60 nm/min, with a signal averaging time of 2 s and a 5 nm bandwidth. Melting curves were obtained by plotting ellipticity at λ = 290 nm vs. temperature.

2.3. Gel Electrophoresis

G-quadruplex structures formed upon the cooling of single-stranded DNA at different rates of 0.05 or 1.0 °C/min and at different potassium concentrations were studied by non-denaturing PAGE performed on 20% polyacrylamide gels supplemented with 25, 35 and 55 mM KCl. G-quadruplex samples were loaded on gels and the electrophoreses were run at 5 °C (5 h), 20 °C (3.5 h) and 35 °C (1.7 h), all at 10 V/cm (I = 300 mA). Bands in the gels were followed by UV shadowing at λ = 254 nm. To facilitate comparisons between the bands observed with different samples, the single-stranded d(5′-AGAAGAAAAGA-3′) and d(5′-TCTTTTCTTCT-3′) and double-stranded (5′-AGAAGAAAAGA-3 mixed with 5′-TCTTTTCTTCT-3′) 11-mere control oligonucleotides were used.

After UV shadowing, experiment gels were submerged in ethidium bromide solution and bands were recorded in fluorescent mode at an excitation wavelength of 290 nm.

2.4. Differential Scanning Calorimetry (DSC)

DSC experiments were performed using the Nano DSC III instrument (Calorimetry Sciences Corp., Lindon, UT, USA) and the Nano DSC instrument (TA Instruments, New Castle, DE, USA) on samples prepared by cooling in an outer thermostat. Samples were first heated to 95 °C and cooled down to four °C at the selected rate (0.05 or 1.0 °C/min). Samples were then loaded to DSC and the first DSC melting scans were recorded at the selected heating rate (0.5, 1.0 and 2.0 °C/min). Next, an annealing scan was recorded at a cooling rate of 1.0 °C, followed by another melting scan at the selected heating rate. The corresponding baseline (buffer-buffer) scans were subtracted from the unfolding/folding scans prior to their normalization and analysis. The total enthalpy of unfolding or folding, ΔH_{tot}, was obtained from the measured DSC thermograms as the area under the $\Delta c_p = \bar{c}_{p,2} - \bar{c}_{p,\text{S}}$ versus T curve, where Δc_p is the measured heat capacity $\bar{c}_{p,2}$ corrected for the baseline and normalized to 1 mole of quadruplex in single strands, and $\bar{c}_{p,\text{S}}$ is the corresponding partial molar heat capacity of the unfolded single stranded state extrapolated from high temperatures over the whole measured temperature interval. For samples prepared either at the cooling rate of 0.05 or 1.0 °C/min, the measured DSC heating and cooling curves were highly reproducible at all measured heating and cooling rates above ~10 °C.

2.5. Model Analysis of Structural Transitions

We attempted to interpret the observed DSC data in terms of the already described complex model (Figure 1) [24], which considers the thermally induced folding/unfolding transitions of d($G_4T_4G_3$) quadruplexes in the presence of K^+ ions as a global kinetic transition process that involves one parallel tetramolecular quadruplex structure (T_4), three bimolecular quadruplexes that possibly exhibit characteristics of parallel, anti-parallel and/or hybrid structures (A_2, B_2 and C_2), and the corresponding single strands (S).

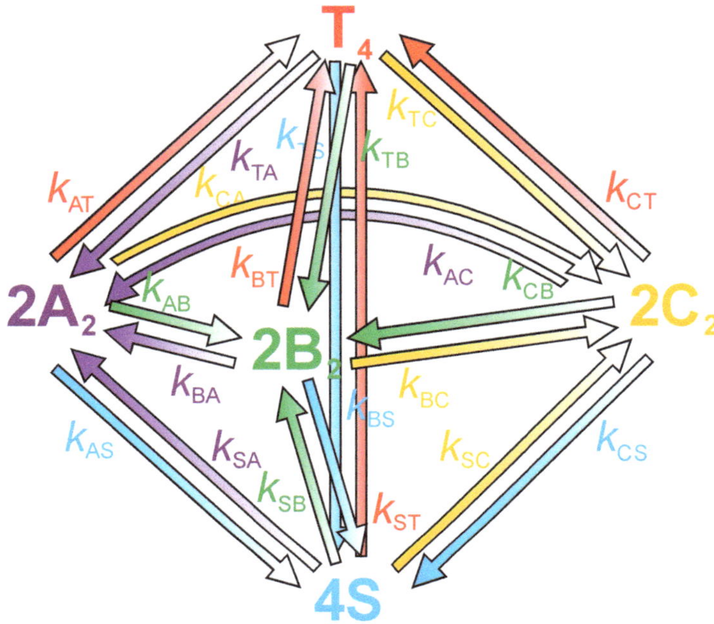

Figure 1. Proposed kinetic model for the thermally induced folding/unfolding transitions of d($G_4T_4G_3$) in K^+ solutions which assumes the coexistence and interconversions of one tetramolecular quadruplex T_4, three bimolecular quadruplexes A_2, B_2 and C_2 and the unfolded single strands S. k_{ij} are rate constants for each i→j transition step.

In order to describe the DSC data, we used the previously reported model function [24]:

$$\Delta c_p = \bar{c}_{p,2} - \bar{c}_{p,S} = -\frac{d\alpha_A}{dT}\Delta H_A - \frac{d\alpha_B}{dT}\Delta H_B - \frac{d\alpha_C}{dT}\Delta H_C - \frac{d\alpha_T}{dT}\Delta H_T \quad (1)$$

in which at any T in the measured temperature interval $\bar{c}_{p,2}$ is the measured partial molar heat capacity of the solute (DNA) expressed per moles of single strands, $\bar{c}_{p,S}$ is the corresponding measured heat capacity of the unfolded quadruplex form occurring at high temperatures extrapolated to T, and α_A, α_B, α_C and α_T are the corresponding fractions of quadruplex species present in the solution. The $\frac{d\alpha}{dT}$ terms needed to calculate Δc_p are obtained by taking into account the rates of reactions predicted by the model and the heating or cooling rate, $r = \frac{dT}{dt}$, at which the DSC experiment is performed. The rates of reactions can be expressed using the Arrhenius relation $k_{ij} = A_{ij}e^{(-E_{ij}/RT)} = e^{A'_{ij}}e^{(-E_{ij}/RT)} = e^{(A'_{ij}-E_{ij}/RT)}$, where k_{ij} represents rate constant, $A_{ij} = e^{A'_{ij}}$ frequency factor and E_{ij} activation energy for corresponding ij step. Unit of $e^{A'_{ij}}$ is s^{-1} for unfolding and s^{-1} M^{-1} for folding transitions of A_2, B_2 and C_2 (s^{-1} M^{-3} for folding transition of T_4 from S, i.e. $e^{A'_{ST}}$) and unit of E_{ij} is cal/mol.

For a given total concentration of DNA in the single-stranded form, c_{tot} (in unit mol/L), expressed as $c_{\text{tot}} = c_S + 2c_A + 2c_B + 2c_C + 4c_T$, the $d\alpha/dT$ terms needed to calculate Δc_p are obtained by taking into account the rates of reactions predicted by the model (Figure 1):

$$-\frac{d\alpha_A}{dT} = \frac{1}{r}\Big[(k_{AB} + k_{AS} + k_{AC})\alpha_A + C_{\text{tot}}k_{AT}\alpha_A^2 - k_{BA}\alpha_B - k_{TA}\alpha_T - k_{CA}\alpha_C - 2k_{SA}C_{\text{tot}}(1 - \alpha_A - \alpha_B - \alpha_C - \alpha_T)^2\Big]$$

$$-\frac{d\alpha_B}{dT} = \frac{1}{r}\Big[(k_{BS} + k_{BC} + k_{BA})\alpha_B + C_{\text{tot}}k_{BT}\alpha_B^2 - k_{AB}\alpha_A - k_{TB}\alpha_T - k_{CB}\alpha_C - 2k_{SB}C_{\text{tot}}(1 - \alpha_A - \alpha_B - \alpha_C - \alpha_T)^2\Big]$$

$$-\frac{d\alpha_C}{dT} = \frac{1}{r}\Big[(k_{CB} + k_{CS} + k_{CA})\alpha_C + C_{\text{tot}}k_{CT}\alpha_C^2 - k_{BC}\alpha_B - k_{TC}\alpha_T - k_{AC}\alpha_A - 2k_{SC}C_{\text{tot}}(1 - \alpha_A - \alpha_B - \alpha_C - \alpha_T)^2\Big]$$

$$-\frac{d\alpha_T}{dT} = \frac{1}{r}\Big[(k_{TA} + k_{TB} + k_{TC} + k_{TS})\alpha_T - C_{\text{tot}}k_{AT}\alpha_A^2 - C_{\text{tot}}k_{BT}\alpha_B^2 - C_{\text{tot}}k_{CT}\alpha_C^2 - 4k_{ST}C_{\text{tot}}^3(1 - \alpha_A - \alpha_B - \alpha_C - \alpha_T)^4\Big]$$

When fitting the kinetic model (Figure 1) to our experimental data, all parameters were allowed to be used. We found out that not all of them were necessary to obtain an acceptable fit of model functions to experimental data. We neglected these parameters (set k_{ij} to 0), thus simplifying the system of equations and increasing the significance of the remaining parameters:

$$-\frac{d\alpha_A}{dT} = \frac{1}{r}[(k_{AB} + k_{AS})\alpha_A - k_{BA}\alpha_B]$$

$$-\frac{d\alpha_B}{dT} = \frac{1}{r}\Big[(k_{BS} + k_{BC} + k_{BA})\alpha_B + C_{\text{tot}}k_{BT}\alpha_B^2 - k_{AB}\alpha_A - 2k_{SB}C_{\text{tot}}(1 - \alpha_A - \alpha_B - \alpha_C - \alpha_T)^2\Big]$$

$$-\frac{d\alpha_C}{dT} = \frac{1}{r}[k_{CS}\alpha_C - k_{BC}\alpha_B]$$

$$-\frac{d\alpha_T}{dT} = \frac{1}{r}\Big[k_{TS}\alpha_T - C_{\text{tot}}k_{BT}\alpha_B^2\Big]$$

To solve this system of differential equations for a given set of adjustable parameters at each measured heating and cooling rate, the Cash-Karp adaptive step-size controlled Runge-Kutta method was employed [27], and the obtained solutions (α_i and $\frac{d\alpha_i}{dT}$) were used to calculate the corresponding model function. The calculation started at a high temperature and a set cooling rate. The initial composition was $\alpha_S = 1$. When the low temperature was reached, the obtained composition was a starting point for heating. The "best fit" adjustable parameters were calculated from global fitting model functions to the experimental DSC curves using the non-linear minimization of the corresponding χ^2 function. The values of adjustable parameters at the global minimum of χ^2 are considered to be the best descriptors of the experimental Δc_p vs. T curve, and therefore are used to characterize the kinetics and thermodynamics for all steps in the suggested model mechanism.

3. Results and Discussion

3.1. Observing Polymorphism of d($G_4T_4G_3$) Quadruplexes in K^+ Solution

We measured the CD spectra of d($G_4T_4G_3$) samples prepared in 25, 35 and 50 mM K^+ solutions at the cooling rate of 0.05 °C/min (Figure 2a), and at the cooling rate of 1.0 °C/min (Figure 2b). The CD spectra is often used to interpret the structure of G-quadruplexes. CD-spectra with positive peak at ~290 nm and negative peak at ~263 nm are characteristic of the anti-parallel type quadruplex structures. On the other hand, the parallel type structures exhibit CD spectra with a positive peak at ~263 nm and negative peak at ~240 nm [28,29]. Hybrid quadruplex structures may also be involved in thermally induced folding/unfolding transitions [11,30,31]. The comparison of three spectra in Figure 2a shows that the peak at 263 nm increases, and the peak at 290 nm decreases with increasing K^+ concentration. The same behavior can be observed when samples are prepared at the cooling rate of 1.0 °C/min (Figure 2b). Thus, we can conclude that increasing K^+ concentration favors the formation of a structure with spectral properties similar to parallel type quadruplexes. Interestingly, when comparing the CD spectra of samples prepared at different cooling rates, we can observe that a slower cooling rate results in increased peak at 263 nm and a decreased peak at 290 nm. These results show that besides changing K^+ concentration, different cooling rates can be used to manipulate the structure of $d(G_4T_4G_3)$ quadruplexes during thermally induced folding. The cooling rate dependence of structural properties of $d(G_4T_4G_3)$ quadruplexes point to the conclusion that thermally induced folding/unfolding transitions of $d(G_4T_4G_3)$ quadruplexes are a kinetically driven process. Furthermore, because of the increased peak at 263 nm after a slow cooling rate, we can assume that a slower cooling rate also favors the formation of a structure with spectral properties similar to parallel-type quadruplexes.

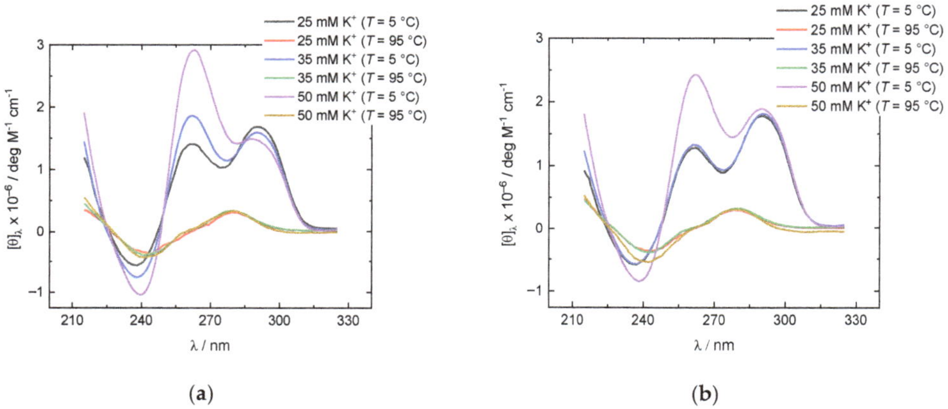

Figure 2. Temperature dependence of CD spectra of $d(G_4T_4G_3)$ quadruplexes in K^+ solutions (25, 35 and 50 mM): (**a**) $d(G_4T_4G_3)$ quadruplexes prepared at the cooling rate of 0.05 °C/min; (**b**) $d(G_4T_4G_3)$ quadruplexes prepared at the cooling rate of 1 °C/min. The measured ellipticity was normalized to 1 M single strand concentration and 1 cm light path length.

The CD spectra suggest that at least two structures with antiparallel and parallel spectral properties (for reasons of clarity we will call them "antiparallel" and "parallel" structures) coexist in solution after cooling, and that the ratio between antiparallel and parallel structures depends on the potassium concentration and a cooling rate. To confirm these results, we performed PAGE experiments where slow and fast cooled $d(G_4T_4G_3)$ samples in 25 mM and 50 mM K^+ solutions were run on the gels at 5 °C, the same temperature at which CD-spectra were recorded (Figure 3). Control oligonucleotide markers (single-stranded d(5′-TCTTTTCTTCT-3′), single-stranded d(5′-AGAAGAAAAGA-3′) and double-stranded (5′-AGAAGAAAAGA-3′ mixed with 5′-TCTTTTCTTCT-3′)) were used to ascribe the size of the molecules to the positions of the bands, and to compare different gels. The gel experiments show that all samples exhibit at least three bands where two fast migrating bands are very close to each other, indicating the presence of at least two structures similar in size and charge. Furthermore, the samples prepared at the cooling rate of 0.05 °C/min (2nd lane) exhibit a more pronounced band, positioned just behind the fastest migrating band, which in accordance with CD-spectra probably represents a parallel structure. This observation can be further confirmed by comparing gels at different K^+ concentrations since the slower of the two fast migrating bands is more pronounced at 50 mM K^+ concentration. Besides the two fast migrating bands, one slow migrating band can be observed at both K^+ concentrations and cooling rates, but the intensity is less pronounced in samples prepared at cooling rate of 1.0 °C/min (1st lane). Interestingly, staining the gels with ethidium bromide results in high intensity fluorescence signals for the slowest migration band and double stranded marker, whereas very weak signals can be observed for the fastest migrating bands. Based on the fluorescence intensity, we can conclude that ethidium bromide interacts with the structure in the slowest migrating band in a similar way to the double stranded marker, and that interaction with quadruplexes is weaker compared to duplex DNA as already observed before [32]. The position and difference in fluorescence intensity between fast and slow migrating bands indicates the significant differences of complexes, and we believe that the slower migrating band corresponds to some highly ordered structure, for example the parallel tetramolecular $d(G_4T_4G_3)_4$ quadruplex.

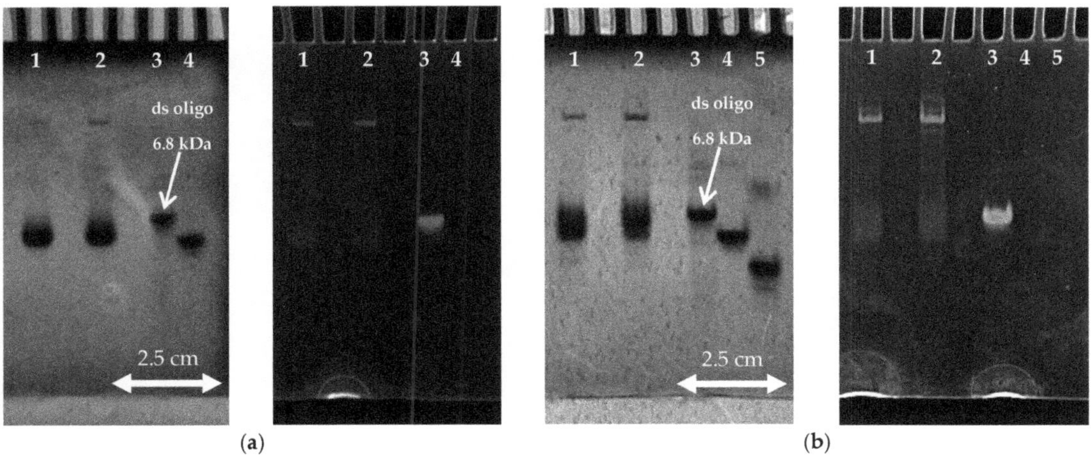

Figure 3. Non denaturing PAGE (20%) of $d(G_4T_4G_3)$ quadruplexes prepared at the cooling rate of either 1 °C/min (1st lane) or 0.05 °C/min (2nd lane) and followed by UV shadowing (λ = 254 nm) and Fluorescence (λ = 590 nm): (**a**) $d(G_4T_4G_3)$ quadruplexes prepared in 25 mM K^+ solution and measured at a constant temperature of 5 °C. As a control oligonucleotide double stranded d(5′-AGAAGAAAAGA-3′; 5′-TCTTTTCTTCT-3′) (3rd lane) and 11-mere single stranded d(5′-TCTTTTCTTCT-3′) (4th lane) markers were used;): (**b**) $d(G_4T_4G_3)$ quadruplexes prepared in 50 mM K^+ solution and measured at a constant temperature of 5 °C. As control oligonucleotide, double stranded d(5′-AGAAGAAAAGA-3′; 5′-TCTTTTCTTCT-3′) (3rd lane), single stranded d(5′-TCTTTTCTTCT-3′) (4th lane) and single stranded d(5′-AGAAGAAAAGA-3′) (5th lane) markers were used.

3.2. Thermally Induced Structural Transitions and Model Analysis

To observe the thermal behavior of $d(G_4T_4G_3)$ starting samples, we have conducted a series of DSC and CD experiments. DSC heating thermograms of $d(G_4T_4G_3)$ samples measured in 25, 35, and 50 mM K^+ solutions and prepared at the cooling rate of either 0.05 or 1.0 °C/min exhibit at least three peaks suggesting that the melting processes consist of several conformational transitions (Figure 4). Samples prepared at the cooling rate of 0.05 °C/min exhibit peaks at ~20 °C, ~45 °C and ~65 °C, suggesting the coexistence of at least three quadruplex structures, which is in line with PAGE experiments. Increasing the K^+ concentration decreases the intensity of the second peak and increases the intensity of the third. Together with the data from CD experiments, this suggests that the third peak corresponds to the unfolding of a parallel structure and that the parallel structure is thermally more stable than the antiparallel. This can be further proved by PAGE experiments of samples prepared at a slow cooling rate at 50 mM K^+ concentration (Figure 3b) where the increased intensity of the slower of the two fast migrating bands can be observed. Samples prepared at the cooling rate of 1.0 °C/min also exhibit three peaks, but the third peak is not as expressed as in the case of a slower cooling rate. These results, together with the CD

(Figure 2) and PAGE (Figure 3) experiments, strongly suggest that the thermal unfolding transition of d($G_4T_4G_3$) in the presence of K^+ ions may be considered as a combination of several kinetically governed steps.

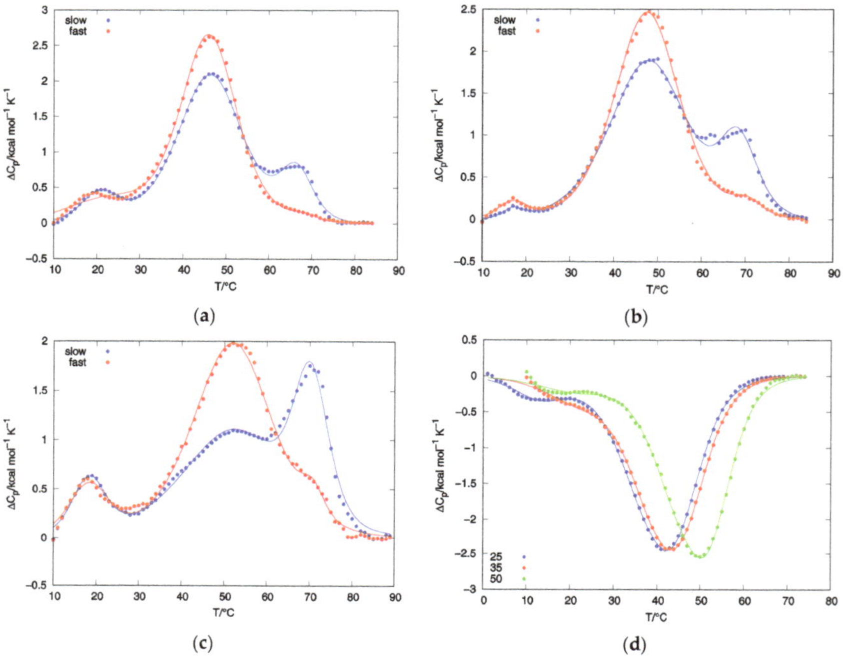

Figure 4. Fitting the model function to the heating and cooling DSC thermograms of d($G_4T_4G_3$) quadruplexes in the presence of 25, 35 and 50 mM K^+ ions: (**a**) The Δc_p vs. T curves measured at the heating rate of 1 °C/min in the presence of 25 mM K^+ ions for samples prepared by controlled cooling in an outer thermostat at the rate of 0.05 °C/min (red line) or 1.0 °C/min (blue line); (**b**) The Δc_p vs. T curves measured at the heating rate of 1 °C/min in the presence of 35 mM K^+ ions for samples prepared by controlled cooling in an outer thermostat at the rate of 0.05 °C/min (red line) or 1.0 °C/min (blue line); (**c**) The Δc_p vs. T curves measured at the heating rate of 1 °C/min in the presence of 50 mM K^+ ions for samples prepared by controlled cooling in an outer thermostat at the rate of 0.05 °C/min (red line) or 1.0 °C/min (blue line); (**d**) The Δc_p vs. T curves measured at the cooling rate of 1 °C/min in the presence of 25 (blue line), 35 (red line) and 50 (green line) mM K^+ ions. In all panels full lines represent model–based Δc_p vs. T curves calculated from (1).

The total enthalpy of unfolding, ΔH_{tot}, was calculated from the experimentally obtained DSC thermograms (total area under the ΔC_P versus T curve). Analysis of the total area under measured heating DSC thermograms resulted in very similar overall enthalpies of unfolding, $\Delta H_{tot} \approx 56$ kcal/mol of double stranded quadruplex, regardless of the cooling rate or K^+ concentration used. From these ΔH_{tot} values, the enthalpy of quadruplex unfolding is estimated to be about 19 kcal/mol of G-quartets, which agrees well with the literature values reported for G-quartet formation in the presence of K^+ ions [33–35]. After the first melting scan was recorded at the selected heating rate (0.5, 1.0 and 2.0 °C/min), the sample was left in DSC and a cooling scan was recorded at the rate of 1.0 °C/min, followed by another melting scan at the selected heating rate. Almost no differences were observed when comparing the second melting scans to the melting scan at corresponding heating rates obtained from samples prepared in an outer thermostat at the cooling rate of 1 °C/min. This confirms the kinetic nature of the folding/unfolding transition of d($G_4T_4G_3$) in the presence of K^+ ions.

Interestingly, DSC thermograms show transition at low temperatures. Even though the reproducibility and reliability of the measured DSC peaks between 4 °C and ~20 °C is rather poor, PAGE experiments confirmed that there is an additional structure present at low temperatures that disappears at temperatures over 20 °C (Figure S2). The position and difference in fluorescence intensity between fast and slow migrating bands (Figure 3) indicates significant differences in structures of complexes, and we believe that a slower migrating band corresponds to some highly ordered structure, for example, the parallel tetramolecular $d(G_4T_4G_3)_4$ quadruplex.

The results of DSC, PAGE and CD experiments performed on $d(G_4T_4G_3)$ quadruplexes in K^+ solutions reveal that their thermally induced folding/unfolding transitions are kinetically governed and include the participation of at least three structures. A model function (1) was used to describe thermally induced folding/unfolding transitions of $d(G_4T_4G_3)$ quadruplexes in the presence of K^+ ions at different concentrations. Figure 4 shows the best fit of model function to the DSC data. To further test the validity of our model, we have collected and fitted data at several heating rates (0.5, 1.0 and 2.0 °C/min) for DSC experiments. By increasing the heating rate, we can observe shifting of the DSC peaks to higher temperatures (Figure 5), which is another indication that conformational transitions of $d(G_4T_4G_3)$ in the presence of K^+ ions are kinetically driven processes. The model function (1) with the same set of parameters (Table 1) was used to describe experimental data at all heating rates at selected K^+ concentration.

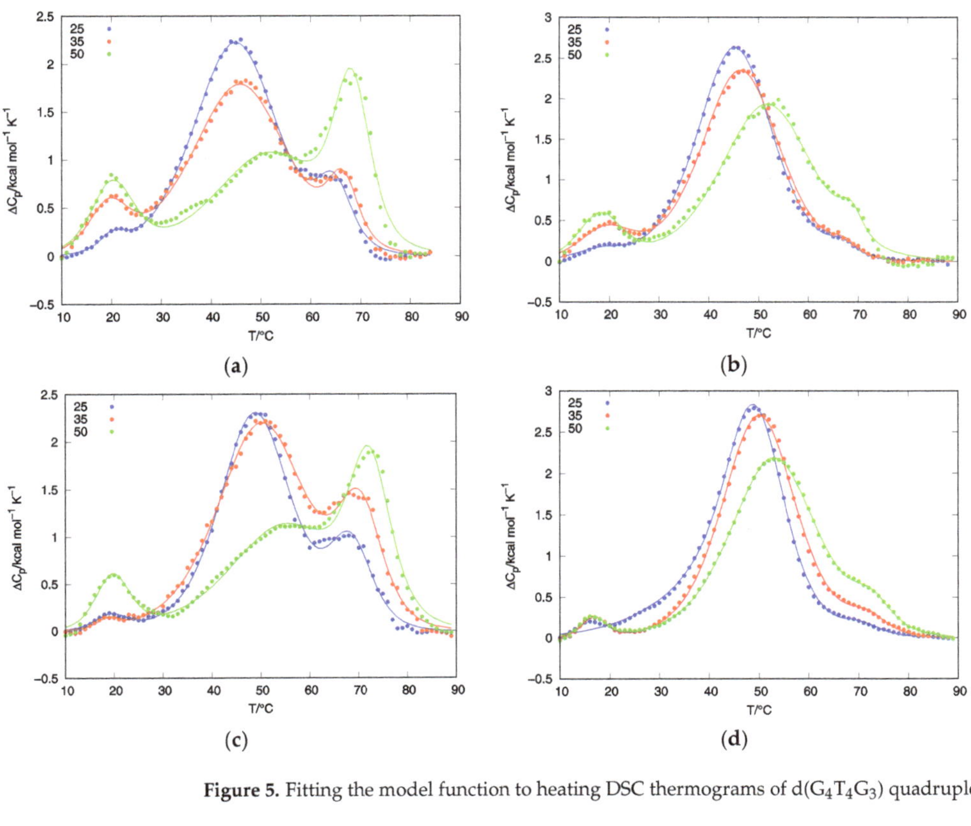

Figure 5. Fitting the model function to heating DSC thermograms of $d(G_4T_4G_3)$ quadruplexes at

different rates in the presence of 25 (blue color), 35 (red color) and 50 (green color) mM K^+ ions: (**a**) at the heating rate of 0.5 °C/min for samples prepared at the cooling rate of 0.05 °C/min; (**b**) at the heating rate of 0.5 °C/min for samples prepared at the cooling rate of 1.0 °C/min; (**c**) at the heating rate of 2.0 °C/min for samples prepared at the cooling rate of 0.05 °C/min; (**d**) at the heating rate of 2.0 °C/min for samples prepared at the cooling rate of 1.0 °C/min. In all panels full lines represent the model—based curves calculated from (1).

Table 1. Calculated adjustable parameters for the kinetic model in Figure 1 which were used to describe thermally induced folding/unfolding transitions. The unit of $e^{A'_{ij}}$ is s^{-1} for unfolding and s^{-1} M^{-1} for folding transitions of A_2, B_2 and C_2 and unit of E_{ij} is cal/mol. Experimentally determined enthalpies of transition, ΔH (cal/mol single strands) were as follows: $\Delta H_A = \Delta H_B = \Delta H_C = (3.1 \pm 0.2)\cdot 10^4$, $\Delta H_T = (3.6 \pm 0.2)\cdot 10^4$.

Parameter	C_{K+} = 25 mM	C_{K+} = 35 mM	C_{K+} = 50 mM
A'_{BS}	74 ± 5	79 ± 5	81 ± 5
A'_{SB}	19 ± 3	19 ± 3	19 ± 3
A'_{AS}	49 ± 2	49 ± 2	49 ± 2
A'_{TS}	100 ± 10	100 ± 10	100 ± 10
A'_{CS}	54 ± 2	53 ± 2	50 ± 2
A'_{BA}	15.5 ± 0.8	15.0 ± 0.8	14.9 ± 0.8
A'_{AB}	7.2 ± 0.4	7.2 ± 0.4	7.2 ± 0.4
A'_{BT}	66 ± 4	66 ± 4	66 ± 4
A'_{BC}	24 ± 2	25 ± 2	30 ± 2
E_{BS}	$(3.9 \pm 0.2) \times 10^4$	$(4.1 \pm 0.2) \times 10^4$	$(4.0 \pm 0.2) \times 10^4$
E_{AS}	$(3.2 \pm 0.1) \times 10^4$	$(3.2 \pm 0.1) \times 10^4$	$(3.2 \pm 0.1) \times 10^4$
E_{TS}	$(6.0 \pm 0.2) \times 10^4$	$(6.0 \pm 0.2) \times 10^4$	$(6.0 \pm 0.2) \times 10^4$
E_{CS}	$(3.7 \pm 0.1) \times 10^4$	$(3.6 \pm 0.1) \times 10^4$	$(3.7 \pm 0.1) \times 10^4$
E_{BA}	$(7.0 \pm 0.5) \times 10^3$	$(7.0 \pm 0.5) \times 10^3$	$(7.0 \pm 0.5) \times 10^3$
E_{BT}	$(3.2 \pm 0.1) \times 10^4$	$(3.2 \pm 0.1) \times 10^4$	$(3.2 \pm 0.1) \times 10^4$
E_{BC}	$(1.5 \pm 0.1) \times 10^4$	$(1.4 \pm 0.1) \times 10^4$	$(1.7 \pm 0.1) \times 10^4$

By comparing the parameters in Table 1 at different potassium concentrations, we can observe an increase of parameter A'_{BC}, which is connected to the frequency factor for the transition from structure B_2 to structure C_2 by equation $A_{BC} = e^{A'_{BC}}$. Also, we can observe a decrease of parameter A'_{CS}, which is connected to frequency factor for transition from structure B_2 to structure C_2 by equation $A_{BC} = e^{A'_{BC}}$. Frequency factors are proportional to the rate constant, k_{ij}, which means that increase in potassium concentration will increase the population of structure C_2. This behavior of model-based parameters correlates nicely with the experimental data which show higher populations of parallel quadruplex at C_{K+} = 50 mM after slow and fast rates of cooling. Calculated population distributions of bimolecular structures A_2 and C_2 at slower and at higher cooling rates at different potassium concentrations are shown in Figure 6. By comparing the data from Figures 4–6 we can see that increase of the third peak in DSC thermograms correlates to the increase of the ratio between the calculated fractions of bimolecular quadruplex species C_2 and A_2. This means that the position of the third peak in DSC heating thermograms, which increases in intensity when slow cooling or high concentrations of potassium ions are used, corresponds to the unfolding of structure C_2.

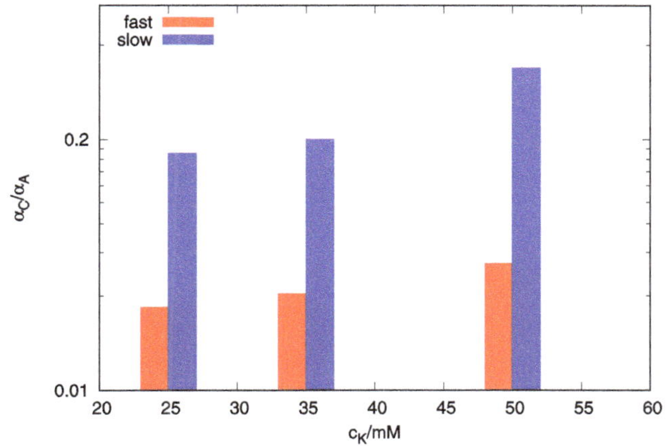

Figure 6. Ratio between calculated fractions of bimolecular quadruplex species C_2 and A_2 at different K^+ concentrations prepared at cooling rates of 0.05 and 1.0 °C/min on a logarithm scale.

The correlation between fitting parameters was calculated for all potassium concentrations (Tables S1–S3). Most of the parameters show weak correlation but stronger correlation is observed between parameters A'_{ij} and E_{ij}. Since the choice of baseline is the largest source of errors accompanying the DSC measurements, the errors were estimated by changing the baseline and plotting "top and bottom" DSC curves which served as confidence limits for our experiment (Figure S1). Applying the calculated errors simultaneously to highly correlated parameters resulted in poorer agreement between the model function and the experiment (data not shown). This suggests that within the estimated error margins the observed higher correlation between parameters A'_{ij} and E_{ij} has no significant effect on their reliability.

Fitting the model function to experimental DSC data yields temperature dependence of population of predicted quadruplex structures involved in the measured folding/unfolding process. In order to successfully interpret the observed DSC data, we had to include one additional bimolecular quadruplex structure B_2 in our model (Figure 1). The calculated population of quadruplex structure B_2 is negligibly small at all temperatures, suggesting that B_2 acts as an unstable intermediate. Even though the population of B_2 is negligibly small, its presence in our model is necessary to fit the entire cycle of DSC curves at selected potassium concentration (two different cooling rates and three different heating rates) with a single set of parameters. Efforts were made to fit the experimental data without quadruplex structure B_2, but this less complex model was only able to satisfactorily describe either the melting or cooling curves.

The calculated populations of other predicted structures can be used to construct a CD-melting curve and compare it to the experimental CD-curve. Figure 7 shows normalized CD melting curves of $d(G_4T_4G_3)$ quadruplexes obtained from experiments and the CD-melting curves calculated from the concentrations of predicted structures in the solution. Good agreement between the calculated and experimental CD-melting curves can be observed at 25 and 35 mM K^+ concentrations (Figure 7a,b). However, the agreement of model melting curve with the experimental melting curve at 50 mM K^+ concentration is not so good but still acceptable (Figure 7c). We attribute this discrepancy to the error in estimated heating rate during CD-measurements.

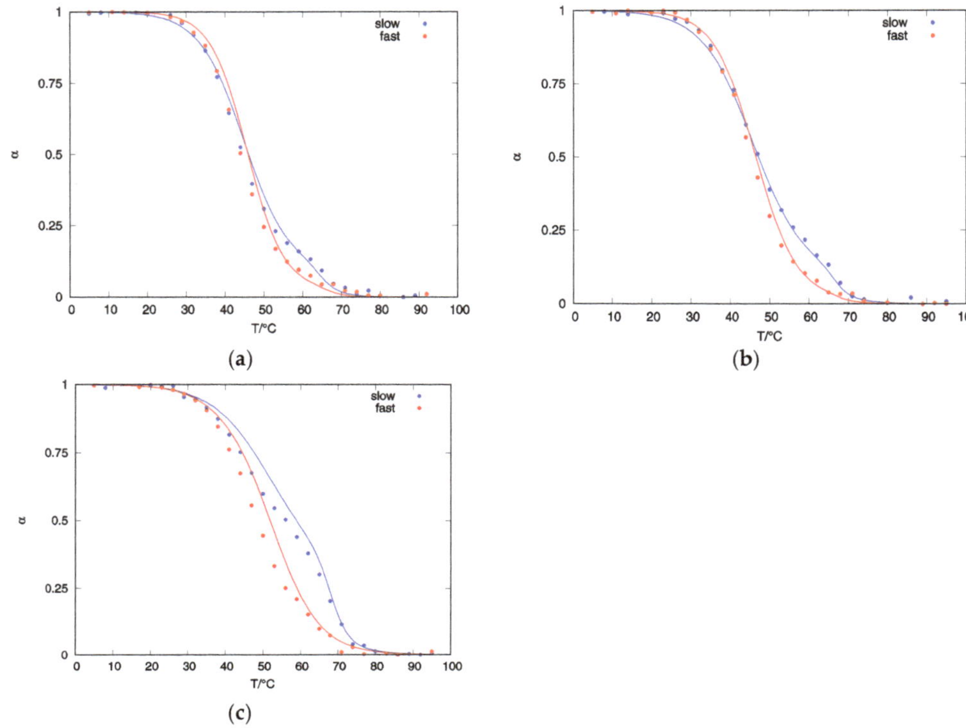

Figure 7. Normalized CD melting curves at λ = 290 nm of $d(G_4T_4G_3)$ quadruplexes obtained from experiments (dots) and calculated from fractions of quadruplex species present in the solution (α_A, α_B, α_C and α_T) (full line) after slow (blue color) and fast (red color) cooling: (**a**) in the presence of 25 mM K^+ ions; (**b**) in the presence of 35 mM K^+ ions; (**c**) in the presence of 50 mM K^+ ions.

Even though our proposed model is able to satisfactory describe experimental data, it is worth mentioning that the rate of quadruplex transitions can be slower than the DSC timescale [21,36]. Our DSC experiments were performed at different scanning rates, but other than that, we did not consider this issue in experimental setup and kinetic modeling. For example, if the transition between structures A_2 and C_2 is very slow, DSC would be unable to record it. Although the proposed kinetic model (Figure 1) allowed multiple pathways, only a minimum number of parameters was used to obtain a good agreement between model prediction and experimental results. The elimination of parameters might be interpreted as the nonexistence of corresponding transitions whereas in reality, the transitions were too slow to be recorded by DSC in dynamic mode or the change in enthalpy was close to zero. Additional experiments should be performed to quantify the real number of species in mixture and to define which (if any) are kinetically trapped intermediates. Fully matched quadruplex assemblies are the most stable structures but mismatched species can form at a faster rate and have long lifetimes [37–39]. For example, if we use data from Table 1 at $C_{K^+} = 25$ mM and T = 281 K, we can calculate the formation and dissociation rates of structures A_2 ($k_{BA} = 19.6\ s^{-1}$, $k_{AS} = 2.6{\cdot}\Delta 10^{-4}\ s^{-1}$) and C_2 ($k_{BC} = 0.05\ s^{-1}$, $k_{CS} = 5.0{\cdot}\Delta 10^{-6}\ s^{-1}$). Since $k_{BA} > k_{BC}$ and $k_{AS} < k_{CS}$, we may conclude that although folding of structure A_2 is faster, C_2 is thermodynamically more stable. When the concentration of potassium is increased ($C_{K^+} = 50$ mM), the folding rate of A_2 is decreased ($k_{BA} = 10.7\ s^{-1}$) while the folding rate of C_2 is increased ($k_{BC} = 0.66\ s^{-1}$). At the same time, the increased potassium concentration further stabilizes C_2 by decreasing the dissociation rate ($k_{CS} = 9.2 \times 10^{-8}\ s^{-1}$). One might jump to a conclusion that A_2 is kinetically trapped and C_2 is a thermodynamically stable structure, but further experiments on a longer timescale are needed to confirm this hypothesis.

4. Conclusions

The results of DSC, PAGE and CD experiments performed on $d(G_4T_4G_3)$ quadruplexes at different K^+ concentrations suggest a complex thermally induced folding/unfolding mechanism that includes the participation of at least three quadruplex structures. To describe the effects of cooling/heating rates and potassium concentration on thermally induced formation/disruption of $d(G_4T_4G_3)$ quadruplex structures, we developed a kinetic model that involves one parallel tetramolecular quadruplex structure, three bimolecular quadruplexes and the corresponding single strands. The model was able to describe an interesting behavior of the $d(G_4T_4G_3)$ quadruplex where the distribution of structures with different spectral and gel mobility properties could be manipulated by cooling rate and the concentration of potassium ions. We would like to emphasize that following thermally induced folding/unfolding transitions by DSC is a dynamic measurement. There could be additional transitions present in our system but if their rate is slow relative to the DSC timescale they remained undetected.

We believe that our research provides new insights into how conformational changes of highly ordered G-quadruplexes can be manipulated by the cooling rate of thermally induced folding and concentration of potassium ions. Also, we have shown how to use a kinetic model to describe the thermally induced folding/unfolding processes of a $d(G_4T_4G_3)_4$ quadruplex and obtain kinetic and thermodynamic parameters for individual elementary steps. It would be interesting to include the concentration of potassium ions as the fitting parameter in our model, although additional experiments would have to be performed at several other K^+ concentrations. Furthermore, additional experiments on a longer timescale should be performed to assess whether the system is at equilibrium and not kinetically trapped. However, this remains the focus of our future work.

Supplementary Materials: The following supporting information can be downloaded at: https://www.mdpi.com/article/10.3390/life12060825/s1, Figure S1. In order to estimate experimental error, two baselines were used to subtract DSC data. First baseline yielded largest overall area under thermogram (empty circles) and the second baseline yielded smallest overall area under thermogram (full circles). The Δc_p vs. T curves were measured at the heating rate of 1.0 °C/min for samples prepared at the cooling rate of 0.05 °C/min; (a) in 25 mM K^+ solution (b) in 35 mM K^+ solution; (c) in 50 mM K^+ solution, Figure S2. Non denaturing PAGE (20%) of $d(G_4T_4G_3)$ quadruplexes prepared at the cooling rate of either 1.0 °C/min (1st lane) or 0.05 °C/min (2nd lane) in 50 mM K^+ solution and followed by UV shadowing (λ = 254 nm) and Fluorescence (λ = 590 nm): (a) $d(G_4T_4G_3)$ quadruplexes measured at constant temperature of 5 °C; (b) $d(G_4T_4G_3)$ quadruplexes prepared at constant temperature of 20 °C. (c) $d(G_4T_4G_3)$ quadruplexes prepared at constant temperature of 35 °C. As control oligonucleotide double stranded d(5′-AGAAGAAAAGA-3′; 5′-TCTTTTCTTCT-3′) (3rd lane), single stranded d(5′-TCTTTTCTTCT-3′) (4th lane) and single stranded d(5′-AGAAGAAAAGA-3′) (5th lane) markers were used. Table S1: Correlation matrix corresponding to the model analysis of the DSC melting and cooling thermograms for concentration K^+ ions equal to 25 mM, Table S2: Correlation matrix corresponding to the model analysis of the DSC melting and cooling thermograms for concentration K^+ ions equal to 35 mM, Table S3: Correlation matrix corresponding to the model analysis of the DSC melting and cooling thermograms for concentration K^+ ions equal to 50 mM.

Author Contributions: Data curation, I.P.; Formal analysis, I.P. and T.U.; Funding acquisition, N.P.U.; Methodology, I.P. and T.U.; Software, I.P. and T.U.; Writing—original draft, I.P.; Writing—review & editing, T.U. and N.P.U. All authors have read and agreed to the published version of the manuscript.

Funding: This research was funded by the Slovenian Research Agency, grant numbers P4-0121 and P1-0201.

Institutional Review Board Statement: Not applicable.

Informed Consent Statement: Not applicable.

Data Availability Statement: The data presented in this study are available on request from the corresponding author.

Acknowledgments: Nataša Poklar Ulrih would like to thank Kenneth J. Breslauer for giving me the opportunity to learn from the best.

Conflicts of Interest: The authors declare that they have no conflict of interest.

References

1. Bruylants, G.; Wouters, J.; Michaux, C. Differential Scanning Calorimetry in Life Science: Thermodynamics, Stability, Molecular Recognition and Application in Drug Design. *Curr. Med. Chem.* **2005**, *12*, 2011–2020. [CrossRef] [PubMed]
2. Ouldridge, T.E. The Importance of Thermodynamics for Molecular Systems, and the Importance of Molecular Systems for Thermodynamics. *Nat. Comput.* **2018**, *17*, 3–29. [CrossRef] [PubMed]
3. Rhodes, D.; Lipps, H.J. Survey and Summary G-Quadruplexes and Their Regulatory Roles in Biology. *Nucleic Acids Res.* **2015**, *43*, 8627–8637. [CrossRef] [PubMed]
4. Abiri, A.; Lavigne, M.; Rezaei, M.; Nikzad, S.; Zare, P.; Mergny, J.L.; Rahimi, H.R. Unlocking G-Quadruplexes as Antiviral Targets. *Pharmacol. Rev.* **2021**, *73*, 897–923. [CrossRef]
5. Sasisekharan, V.; Zimmerman, S.; Davies, D.R. The Structure of Helical 5′-Guanosine Monophosphate. *J. Mol. Biol.* **1975**, *92*, 171–179. [CrossRef]
6. Gellert, M.; Lipsett, M.N.; Davies, D.R. Helix Formation BY Guanylic Acid. *Proc. Natl. Acad. Sci. USA* **1962**, *48*, 2013–2018. [CrossRef]
7. Largy, E.; Mergny, J.L.; Gabelica, V. Role of Alkali Metal Ions in G-Quadruplex Nucleic Acid Structure and Stability. In *Metal Ions in Life Sciences*; Springer: Berlin/Heidelberg, Germany, 2016; Volume 16, pp. 203–258.
8. Kettani, A.; Gorin, A.; Majumdar, A.; Hermann, T.; Skripkin, E.; Zhao, H.; Jones, R.; Patel, D.J. A Dimeric DNA Interface Stabilized by Stacked A·(G·G·G·G)·A Hexads and Coordinated Monovalent Cations. *J. Mol. Biol.* **2000**, *297*, 627–644. [CrossRef]
9. Strahan, G.D.; Keniry, M.A.; Shafer, R.H. NMR Structure Refinement and Dynamics of the K^+-$[d(G_3T_4G_3)]_2$ Quadruplex via Particle Mesh Ewald Molecular Dynamics Simulations. *Biophys. J.* **1998**, *75*, 968–981. [CrossRef]
10. Grün, J.T.; Schwalbe, H. Folding Dynamics of Polymorphic G-Quadruplex Structures. *Biopolymers* **2022**, *113*, e23477. [CrossRef]
11. Bončina, M.; Lah, J.; Prislan, I.; Vesnaver, G. Energetic Basis of Human Telomeric DNA Folding into G-Quadruplex Structures. *J. Am. Chem. Soc.* **2012**, *134*, 9657–9663. [CrossRef]
12. Čeru, S.; Šket, P.; Prislan, I.; Lah, J.; Plavec, J. A New Pathway of DNA G-Quadruplex Formation. *Angew. Chem.* **2014**, *126*, 4981–4984. [CrossRef]
13. Marquevielle, J.; Robert, C.; Lagrabette, O.; Wahid, M.; Bourdoncle, A.; Xodo, L.E.; Mergny, J.L.; Salgado, G.F. Structure of Two G-Quadruplexes in Equilibrium in the KRAS Promoter. *Nucleic Acids Res.* **2020**, *48*, 9336–9345. [CrossRef] [PubMed]
14. Greco, M.L.; Kotar, A.; Rigo, R.; Cristofari, C.; Plavec, J.; Sissi, C. Coexistence of Two Main Folded G-Quadruplexes within a Single G-Rich Domain in the EGFRpromoter. *Nucleic Acids Res.* **2017**, *45*, 10132–10142. [CrossRef] [PubMed]
15. Lim, K.W.; Lacroix, L.; Yue, D.J.E.; Lim, J.K.C.; Lim, J.M.W.; Phan, A.T. Coexistence of Two Distinct G-Quadruplex Conformations in the HTERT Promoter. *J. Am. Chem. Soc.* **2010**, *132*, 12331–12342. [CrossRef]
16. Bhattacharyya, D.; Arachchilage, G.M.; Basu, S. Metal Cations in G-Quadruplex Folding and Stability. *Front. Chem.* **2016**, *4*, 38. [CrossRef]
17. Gray, R.D.; Trent, J.O.; Chaires, J.B. Folding and Unfolding Pathways of the Human Telomeric G-Quadruplex. *J. Mol. Biol.* **2014**, *426*, 1629–1650. [CrossRef]
18. Marchand, A.; Gabelica, V. Folding and Misfolding Pathways of G-Quadruplex DNA. *Nucleic Acids Res.* **2016**, *44*, 10999–11012. [CrossRef]
19. Stadlbauer, P.; Kührová, P.; Banáš, P.; Koča, J.; Bussi, G.; Trantírek, L.; Otyepka, M.; Šponer, J. Hairpins Participating in Folding of Human Telomeric Sequence Quadruplexes Studied by Standard and T-REMD Simulations. *Nucleic Acids Res.* **2015**, *43*, 9626–9644. [CrossRef]
20. Miller, M.C.; Le, H.T.; Dean, W.L.; Holt, P.A.; Chaires, J.B.; Trent, J.O. Polymorphism and Resolution of Oncogene Promoter Quadruplex-Forming Sequences. *Org. Biomol. Chem.* **2011**, *9*, 7633–7637. [CrossRef]
21. Grün, J.T.; Blümler, A.; Burkhart, I.; Wirmer-Bartoschek, J.; Heckel, A.; Schwalbe, H. Unraveling the Kinetics of Spare-Tire DNA G-Quadruplex Folding. *J. Am. Chem. Soc.* **2021**, *143*, 6185–6193. [CrossRef]
22. Zhang, A.Y.Q.; Balasubramanian, S. The Kinetics and Folding Pathways of Intramolecular G-Quadruplex Nucleic Acids. *J. Am. Chem. Soc.* **2012**, *134*, 19297–19308. [CrossRef] [PubMed]
23. Nguyen, T.Q.N.; Lim, K.W.; Phan, A.T. Folding Kinetics of G-Quadruplexes: Duplex Stem Loops Drive and Accelerate G-Quadruplex Folding. *J. Phys. Chem. B* **2020**, *124*, 5122–5130. [CrossRef] [PubMed]
24. Prislan, I.; Lah, J.; Milanic, M.; Vesnaver, G. Kinetically Governed Polymorphism of d(G4T4G 3) Quadruplexes in K+ Solutions. *Nucleic Acids Res.* **2011**, *39*, 1933–1942. [CrossRef] [PubMed]
25. Prislan, I.; Lah, J.; Vesnaver, G. Diverse Polymorphism of G-Quadruplexes as a Kinetic Phenomenon. *J. Am. Chem. Soc.* **2008**, *130*, 14161–14169. [CrossRef]
26. Dalgleish, D. Biophysical Chemistry: Part III 'The Behaviour of Biological Macromolecules. *Biochem. Educ.* **1981**, *9*, 157. [CrossRef]

27. Seiler, M.C.; Seiler, F.A. Numerical Recipes in C: The Art of Scientific Computing. *Risk Anal.* **1989**, *9*, 415–416. [CrossRef]
28. Rujan, I.N.; Meleney, J.C.; Bolton, P.H. Vertebrate Telomere Repeat DNAs Favor External Loop Propeller Quadruplex Structures in the Presence of High Concentrations of Potassium. *Nucleic Acids Res.* **2005**, *33*, 2022–2031. [CrossRef]
29. Kypr, J.; Kejnovská, I.; Renčiuk, D.; Vorlíčková, M. Circular Dichroism and Conformational Polymorphism of DNA. *Nucleic Acids Res.* **2009**, *37*, 1713–1725. [CrossRef]
30. Chang, C.C.; Chien, C.W.; Lin, Y.H.; Kang, C.C.; Chang, T.C. Investigation of Spectral Conversion of d(TTAGGG)4 and d(TTAGGG)13 upon Potassium Titration by a G-Quadruplex Recognizer BMVC Molecule. *Nucleic Acids Res.* **2007**, *35*, 2846–2860. [CrossRef]
31. Luu, K.N.; Phan, A.T.; Kuryavyi, V.; Lacroix, L.; Patel, D.J. Structure of the Human Telomere in K+ Solution: An Intramolecular (3 + 1) G-Quadruplex Scaffold. *J. Am. Chem. Soc.* **2006**, *128*, 9963–9970. [CrossRef]
32. Sun, X.; Cao, E.; He, Y.; Qin, J. Fluorescence Studies on the Interaction of Ethidium Bromide with Duplex, Triplex and Quadruplex DNA Structures. *Sci. China Ser. B Chem.* **1999**, *42*, 62–69. [CrossRef]
33. Petraccone, L.; Pagano, B.; Esposito, V.; Randazzo, A.; Piccialli, G.; Barone, G.; Mattia, C.A.; Giancola, C. Thermodynamics and Kinetics of PNA-DNA Quadruplex-Forming Chimeras. *J. Am. Chem. Soc.* **2005**, *127*, 16215–16223. [CrossRef] [PubMed]
34. Ren, J.; Qu, X.; Trent, J.O.; Chaires, J.B. Tiny Telomere DNA. *Nucleic Acids Res.* **2002**, *30*, 2307–2315. [CrossRef] [PubMed]
35. Jin, R.; Gaffney, B.L.; Wang, C.; Jones, R.A.; Breslauer, K.J. Thermodynamics and Structure of a DNA Tetraplex: A Spectroscopic and Calorimetric Study of the Tetramolecular Complexes of d(TG3T) and d(TG3T2G3T). *Proc. Natl. Acad. Sci. USA* **1992**, *89*, 8832–8836. [CrossRef]
36. Carrino, S.; Hennecker, C.D.; Murrieta, A.C.; Mittermaier, A. Frustrated Folding of Guanine Quadruplexes in Telomeric DNA. *Nucleic Acids Res.* **2021**, *49*, 3063–3076. [CrossRef]
37. Long, X.; Stone, M.D. Kinetic Partitioning Modulates Human Telomere DNA G-Quadruplex Structural Polymorphism. *PLoS ONE* **2013**, *8*, e83420. [CrossRef]
38. Müller, D.; Bessi, I.; Richter, C.; Schwalbe, H. The Folding Landscapes of Human Telomeric RNA and DNA G-Quadruplexes Are Markedly Different. *Angew. Chem. Int. Ed.* **2021**, *60*, 10895–10901. [CrossRef]
39. Bardin, C.; Leroy, J.L. The Formation Pathway of Tetramolecular G-Quadruplexes. *Nucleic Acids Res.* **2008**, *36*, 477–488. [CrossRef]

life

MDPI

Article

The Pressure Dependence of the Stability of the G-quadruplex Formed by d(TGGGGT)

Nabeel Tariq, Takuma Kume, Ujala N. Feroze and Robert B. Macgregor, Jr. *

Graduate Department of Pharmaceutical Sciences, Leslie Dan Faculty of Pharmacy, University of Toronto, Toronto, ON M5G1M2, Canada; nabeel.tariq@mail.utoronto.ca (N.T.); tkume@wisc.edu (T.K.); ujala.feroze@mail.utoronto.ca (U.N.F.)
* Correspondence: rob.macgregor@utoronto.ca; Tel.: +1-416-978-7332

Abstract: The G-quadruplex (GQ), a tetrahelix formed by guanine-rich nucleic acid sequences, is a potential drug target for several diseases. Monomolecular GQs are stabilized by guanine tetrads and non-guanine regions that form loops. Hydrostatic pressure destabilizes the folded, monomolecular GQ structures. In this communication, we present data on the effect of pressure on the conformational stability of the tetramolecular GQ, d[5′-TGGGGT-3′]$_4$. This molecule does not have loops linking the tetrads; thus, its physical properties presumably reflect those of the tetrads alone. Understanding the properties of the tetrads will aid in understanding the contribution of the other structural components to the stability of GQ DNA. By measuring UV light absorption, we have studied the effect of hydrostatic pressure on the thermal stability of the tetramolecular d[5′-TGGGGT-3′]$_4$ in the presence of sodium ions. Our data show that, unlike monomolecular GQ, the temperature at which d[5′-TGGGGT-3′]$_4$ dissociates to form the constituent monomers is nearly independent of pressure up to 200 MPa. This implies that there is no net molar volume difference (ΔV) between the GQ and the unfolded random-coil states. This finding further suggests that the large negative ΔV values for the unfolding of monomolecular GQ are due to the presence of the loop regions in those structures.

Keywords: G-quadruplex; thermodynamics; hydrostatic pressure; volume; CD spectroscopy

Citation: Tariq, N.; Kume, T.; Feroze, U.N.; Macgregor, R.B., Jr. The Pressure Dependence of the Stability of the G-quadruplex Formed by d(TGGGGT). *Life* **2022**, *12*, 765. https://doi.org/10.3390/life12050765

Academic Editor: Paolo Mariani

Received: 1 April 2022
Accepted: 18 May 2022
Published: 21 May 2022

Publisher's Note: MDPI stays neutral with regard to jurisdictional claims in published maps and institutional affiliations.

1. Introduction

The guanine-quadruplex (GQ) is a secondary structure of DNA and RNA formed by molecules with sequences rich in guanines in the presence of some cations, such as sodium or potassium. The cation coordinates a tetrad of mutually hydrogen-bonded guanine residues that arrange in a plane, and these tetrads stack on one another to form the quadruplex. Intramolecular GQs form by the folding of a single DNA or RNA molecule, whereas intermolecular GQs contain two or four guanine-rich molecules [1–3]. Guanine-rich DNA sequences that form GQs in vitro are implicated in a number of human diseases and are situated in conserved regions in the genome across species [4–11]. The ability of GQs to self-assemble has led to their use in nanotechnology applications based on DNA [12–14].

Stability studies relying on changes in temperature, pH, and solvent parameters for GQ-containing solutions provide insight on the energetics of the folding process [15]. By studying the effect of pressure on GQ-containing systems, we can understand the role of water in the folding event [15–19]. The data to date show that the formation of monomolecular GQ structures is accompanied by a net release of water molecules into the bulk from the dehydration of metal cations and the formation of void volume by tetrad stacking [18,20,21]. The water-excluded voids, in turn, lead to a structure that is sensitive to pressure changes [15,22]. By analyzing the effect of pressure on the equilibrium between the folded and unfolded forms of the molecule, we can obtain the volumetric parameters that characterize the process [15,23–25]. Volumetric studies of GQs are important for understanding the hydration of these structures and the role the intracellular milieu may play in stabilizing or destabilizing these structures in vivo [26–29].

The effect of hydrostatic pressure on the conformational stability of monomolecular GQs has received some attention. In all cases, increasing the pressure leads to a destabilization of the folded structure to an oligonucleotide with no distinct secondary structure [15]. In addition, the role of the loops in the observed destabilization caused by pressure has also been investigated [21,30].

In this manuscript, we present data on the effect of hydrostatic pressure on the tetramolecular GQ formed by the hexanucleotide d[5′-TGGGGT-3′] (TG4T). The four-stranded quadruplex, d[5′-TGGGGT-3′]$_4$ (TG4T-GQ), does not have any loop structures and only consists of four G-tetrads with single thymine residues at the 5′- and 3′- ends. Thus, the volumetric properties of this quadruplex should be representative of those of the G-tetrads that stabilize the structure. Our data show, rather surprisingly, that in contrast to the behaviour observed for monomolecular GQs, the thermal stability of the d[5′-TGGGGT-3′]$_4$ structure is independent of pressure up to at least 200 MPa.

2. Results

2.1. Formation of TG4T-GQ

The formation of TG4T-GQ from its constituent oligonucleotides is slow [31,32]. We monitored the formation of TG4T-GQ in a solution containing 100 μM oligo concentration and 100 mM NaCl at room temperature and observed continued formation up until the last day the sample was measured at 22 days (data not shown). To facilitate the formation of TG4T-GQ for the purposes of studying the pressure dependence of its thermal stability, we increased the NaCl concentration to 1000 mM. Figure 1a presents 92 circular dichroism (CD) spectra taken at 15-min intervals over 1380 min (23 h) to demonstrate the formation of the quadruplex. The maximum in the CD spectrum, at 263 nm, is characteristic of the formation of the four-stranded structure, and a plot of the intensity of the CD signal at 263 nm as a function of time is shown in Figure 1b. The signal at 263 nm leveled off by the end of this incubation period of 23 h. It is important to note the results in Figure 1b are qualitative in nature, as we fit them to a simple exponential without consideration of a fourth-order reaction mechanism. Rather than elucidate the kinetics parameters, we were interested in determining a starting point for our experiments.

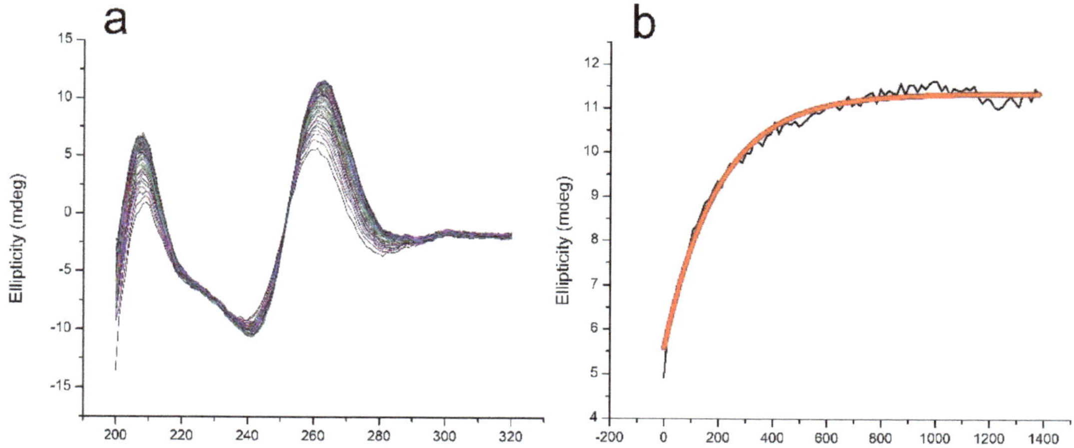

Figure 1. Kinetics of the formation of TG4T-GQ in aqueous solution containing 100 μM oligo and 1000 mM NaCl, 10 mM tris, pH 7.5. (**a**) The evolution of the CD spectrum of TG4T-GQ at 25 °C over

time; a spectrum was acquired every 15 min for 23 h. (**b**) Plot of the ellipticity at 263 nm from (**a**) as a function of time, showing the approach to equilibrium in the formation of TG4T-GQ. The half-time of the formation process is approximately 140 min. It must be noted that these kinetic results are merely a means of estimating the time at which most of the sample has folded into a GQ to establish a starting point for our pressure experiments. The data points were fit to a simple exponential without consideration of the dead time or fourth-order reaction mechanism. Refer to the paper by Mergny and colleagues on tetramolecular kinetics for a more comprehensive kinetic analysis of four-stranded GQ [31].

2.2. Measurements at Elevated Pressures

Next, we investigated the effect of pressure on the melting temperature ($T_{1/2}$) of TG4T-GQ. Using the Clausius–Clapeyron equation, we calculated the volume change associated with the unfolding of the structure. According to the results shown in Figure 2 and Table 1, there was only a very modest change in the $T_{1/2}$ of TG4T-GQ at pressures as high as 200 MPa. Based on these data, we calculated a molar volume change, ΔV, equal to -1.5 ± 2.3 cm^3 mol^{-1} for the dissociation of TG4T-GQ into its four constituent oligonucleotides, d[5′-TGGGGT-3′]. Analysis of the enthalpy change upon dissociation assuming a single-step reaction is also reported in Table 1.

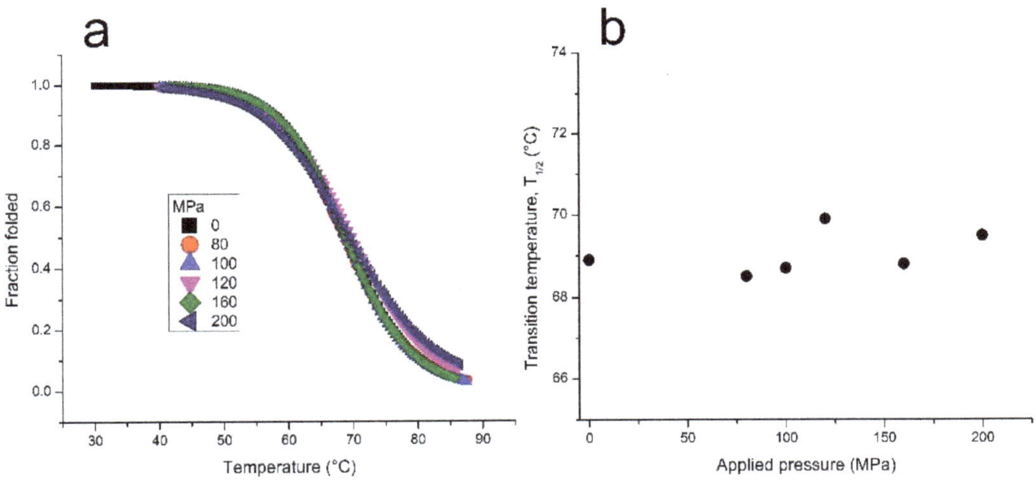

Figure 2. (**a**) Thermal denaturation of TG4T-GQ at different pressures. Fresh samples were used for each measurement. The samples contained 100 μM TG4T, 1000 mM NaCl, 10 mM tris, pH 7.5. (**b**) The pressure dependence of the melting temperature for the TG4T-GQ. The error in the $T_{1/2}$ measurements is ±0.5 °C.

The four oligonucleotides in TG4T-GQ are oriented parallel to one another. We examined another parallel-stranded GQ formed from the oligo Pu22-T12T13, a variant of a 22-mer sequence proximal to the VEGF transcription initiation site [33,34]. We studied the dependence of the stability of the Pu22-T12T13-GQ on hydrostatic pressure. In comparison to the results for TG4T-GQ, the $T_{1/2}$ of the parallel-stranded, monomolecular Pu22-T12T13-GQ showed significant destabilization with increasing pressure (Table 1). From our data, we calculated a molar volume change (ΔV) for the unfolding of Pu22-T12T13-GQ equal to -37.9 ± 10 cm^3 mol^{-1}.

Table 1. The effect of hydrostatic pressure on the thermal denaturation of TG4T-GQ [a] and the G-quadruplex formed by Pu22-T12T13 [b].

GQ	Applied Pressure	$T_{1/2}$ (°C) *	ΔH (kJ)	ΔV (cm^3 mol^{-1})
TG4T	0	68.9	240 ± 14	−1.5 ± 2.3
	80	68.5	270 ± 16	
	100	68.7	300 ± 19	
	120	69.9	250 ± 19	
	160	68.8	280 ± 25	
	200	69.5	270 ± 18	
Pu22-T12T13	0	71.8	190 ± 23	−37.9 ± 10 **
	80	69.9	120 ± 19	
	180	60.7	130 ± 26	

[a] TG4T samples: 100 μM TG4T, 1000 mM NaCl, 10 mM tris, pH 7.5. [b] Pu22-T12T13 samples: 100 μM Pu22-T12T13 in 2 mM KCl, 10 mM phosphoric acid, 0.1 mM EDTA, pH 7.0. * TG4T samples were heated at 0.9 °C/min, while Pu22-T12T13 samples were heated at 0.5 °C/min. Heating rates of 0.1, 0.5, and 0.9 °C/min did not influence the results for TG4T-GQ (unpublished). We present the data obtained with a 0.9 °C/min heating rate simply because it was the most complete set of data for TG4T. ** This error is an estimate based on prior publications with our pressure instrument [21].

3. Discussion

The number of studies that explore the effect of hydrostatic pressure on DNA G-quadruplex (GQ) structures is limited. In contrast to double-stranded conformations of nucleic acids, the G-quadruplexes have structural domains that may exhibit differential responses as a function of pressure. G-quadruplexes are stabilized by G-tetrads, which consist of four mutually hydrogen-bonded guanine residues that stack upon each other in the folded GQ structure. In monomolecular G-quadruplexes, successive G-tetrads are connected by nucleobases that are not guanine or cytosine residues. These bases that connect the stacked G-tetrads form the loops, which generally consist of two or three bases although loops with other numbers of bases are also possible. The bases in the loops do not form base-pairing interactions with other bases, and their existence leads to void volumes in folded G-quadruplex structures. The final component of the G-quadruplex structure are the cations that are coordinated by the O6 oxygen atoms of the guanine tetrads. When complexed with the G-tetrads, these ions are dehydrated, but upon denaturation of the G-quadruplex structure, they become rehydrated. The existence of structural voids and the rehydration of the cations released upon denaturation of the G-quadruplex structure are considered to be two of the primary causes of the destabilization of G-quadruplexes by pressure.

We can write the volume change arising from the denaturation of a monomolecular G-quadruplex as follows:

$$\Delta V_{(\text{mmGQ})} = \Delta V_{(\text{tetrad})} + \Delta V_{(\text{loops})} + \Delta V_{(\text{ion})}$$

where $\Delta V_{(\text{mmGQ})}$ is the observed volume change for a monomolecular G-quadruplex such as HTel (human telomeric sequence) or VEGF (vascular endothelial growth factor); $\Delta V_{(\text{tetrad})}$ is the volume change that arises from the denaturation of the G-tetrads; $\Delta V_{(\text{loops})}$ is the contribution of the loops to the observed value of $\Delta V_{(\text{mmGQ})}$; and $\Delta V_{(\text{ion})}$ is the volume change that arises from the rehydration of the ions coordinated by the O6 oxygen atoms of the G-tetrads from the native structure.

The role of the loops in the sensitivity of G-quadruplexes to pressure has been explored [21,30]. These studies show that the loops play an important role in the magnitude of the observed behaviour. Simple modifications to the loop regions result in significant changes in the unfolding volume of the monomolecular GQs formed by HTel and TBA (thrombin binding aptamer) GQs [15,21,30,35]. However, in all cases, regardless of the sequences of the loop, the volume change arising from the unfolding of the G-quadruplex is negative, i.e., the structure is destabilized by pressure. Upon the unfolding of the G-quadruplex structure, the bases that comprise the loops undergo a large change in the solvent-accessible surface area. In addition, the imperfect packing that arises from the structure of the loops in the folded structure leads to void volumes. These void volumes are lost upon the unfolding of the G-quadruplex structure.

To assess the contribution of the G-tetrads and ion release to the pressure dependence of the stability of G-quadruplexes, we have investigated the four-stranded complex formed by TG4T, which associates to form TG4T-GQ. The complex does not have loops; therefore, the observed volume change may be written as:

$$\Delta V_{\text{(TG4T-GQ)}} = \Delta V_{\text{(tetrad)}} + \Delta V_{\text{(ion)}}$$

where $\Delta V_{\text{(TG4T-GQ)}}$ is the observed volume change in the measurements carried out as a function of hydrostatic pressure. With knowledge of $\Delta V_{\text{(TG4T-GQ)}}$ and $\Delta V_{\text{(ion)}}$, we may then calculate the contribution of the G-tetrads to the measured volume change, $\Delta V_{\text{(TG4T-GQ)}}$.

TG4T-GQ has slow folding kinetics. Figure 1a,b show that in 1000 mM NaCl, a solution containing 100 μM TG4T formed over the course of 24 h with a half-time of formation of approximately 140 min. After incubating for 23 h, the CD spectrum was consistent with that of a G-quadruplex structure, and the spectra no longer changed with time. Because of the slow rate of formation of TG4T-GQ, each sample was subjected to only a single thermal denaturation.

When the TG4T-GQ samples were studied at different hydrostatic pressures, we did not observe a change in the stability of TG4T-GQ under the conditions of our measurements (see Figure 1b and Table 1). That is, we did not observe a statistically significant change in the $T_{1/2}$ as a function of pressure (p) (i.e., $\Delta T_{1/2}/\Delta p$ ~0 °C/MPa). This result suggests that either (1) the volume change arising from rehydration of the ions has nearly the same magnitude but opposite sign from the volume change arising from the disruption of the G-tetrads, or (2) the volume change of both processes is negligible.

The denaturation of the G-tetrads entails the loss of the hydrogen bonds between the guanine residues, the unstacking of the G-tetrads, and the release of bound cations. The guanine–guanine hydrogen bonds lost upon denaturation will be replaced by hydrogen bonds with water. However, the unstacking of the G-tetrads will cause an increase in the solvent-accessible surface area (SASA). This change in SASA will involve exposure of the aromatic bases to water and a negative volume change. The release of the cations bound to the O6 oxygen atoms of the guanine residues will lead to the formation of interactions between the ions and water. This is also a process that proceeds with a negative volume change.

Thus, $\Delta V_{\text{(tetrad)}}$ and $\Delta V_{\text{(ion)}}$ are both anticipated to be negative. The observation that the $T_{1/2}$ of TG4T-GQ is independent of hydrostatic pressure implies that $\Delta V_{\text{(tetrad)}} \approx -\Delta V_{\text{(ion)}}$. In the absence of loops, which give rise to void volumes, most of the effect of hydrostatic pressure on the stability of TG4T-GQ will arise from the changes in the interactions of water with the associated and dissociated states of this system. The properties of water, including its partial molar volume, are sensitive to temperature. For example, the partial molar volume passes from negative values at temperatures below ~50 °C to small positive values above this temperature [36]. From this, we infer that the interactions between the released ions and water are weaker in the higher temperature regime. Similar considerations apply to the interactions of the aromatic bases with water. Taken together, our findings suggest that possibility (2), the volume change of both processes is negligible, is relevant in this system, that is, the volume change arising from the hydration of the released ions and the exposed bases tend toward zero at the $T_{1/2}$ of TG4T-GQ.

The tetramolecular G-quadruplex structure formed from TG4T has strands in an all-parallel conformation, whereas the monomolecular quadruplex formed from HTel is non-parallel. For this reason, we also present data on the effect of pressure on the conformational stability of the G-quadruplex formed by Pu22-T12T13 (Table 1). Pu22-T12T13 forms a monomolecular structure with parallel strand orientation. The data show that the molar volume change of unfolding this G-quadruplex is large and negative, similar to the other monomolecular structures that have been studied [15]. The results from Pu22-T12T13 and other parallel-stranded GQs from the literature demonstrate that the parallel strand orientation is not a cause for the near-zero ΔV of TG4T-GQ [37]. Thus, we conclude that the behaviour exhibited by TG4T-GQ is not related to the orientation of the strands.

Extending the present findings to the behaviour of single-stranded oligonucleotides that can form tetrahelical GQ conformations suggests that the pressure dependence of these monomolecular structures arises mostly from the differential hydration of the loops in these cases. It would be useful to explore the behaviour of TG4T-GQ stabilized by other ions or other tetramolecular G-quadruplexes, such as d([5′-TGGGT-3′])$_4$ or d([5′-TGGGGGT-3′])$_4$.

In conclusion, we present data that show that the thermal denaturation of the four-stranded G-quadruplex, TG4T-GQ, does not depend on pressure in solutions containing 1000 mM sodium chloride. We attribute the finding that, within experimental error, $\Delta T_{1/2}/\Delta p$ = 0 cm^3 mol^{-1} to volume changes of both the disruption of the G-tetrads and the rehydration of the released cation tending toward zero at the temperature of the denaturation (~68 °C). This result suggests that the large, negative value of ΔV for the thermal denaturation of monomolecular G-quadruplex arises predominately from the presence of the loops in these structures and the void volumes to which they give rise.

4. Materials and Methods

We purchased the oligodeoxyribonucleotides TG4T (d[5′-TGGGGT-3′]) and Pu22-T12T13 (d[5′-CGGGGCGGGCCTTGGGCGGGGT-3′]) from ACGT (Toronto, ON, Canada). DNA was suspended in and dialyzed against Milli-Q® ultra-purified water (Millipore Milli-Q Biocel Water Purification System, Sigma-Aldrich, Oakville, ON, Canada), vacuum-dried, and stored at −20 °C. DNA concentration was determined by measuring the absorbance at 260 nm using a molar extinction coefficient of 57,800 M^{-1} cm^{-1} for TG4T and 200,400 M^{-1} cm^{-1} for Pu22-T12T13 (Cary model 300 Bio spectrophotometer, Varian Canada, Inc., Mississauga, ON, Canada) [31,38–40]. Stock 100 mM tris(hydroxymethyl)aminomethane (tris, min. 99.5%, Bioshop, Burlington, ON, Canada) buffer adjusted to pH 7.5 with HCl was prepared, filter-sterilized (0.22 μM pore size), and diluted to 10 mM as the solvent for the TG4T and NaCl (1000 mM sodium chloride; Bioshop, Burlington, ON, Canada) solutions. TG4T-GQ was formed by heating the DNA samples dissolved in buffer to 95 °C for five minutes and then allowing the sample to cool to room temperature overnight in a 1-L Dewar flask fitted with a loose lid. Pu22-T12T13 was also heated and cooled overnight. Pu22-T12T13 was prepared in a solution of 2 mM KCl (potassium chloride, min 99.5%), 10 mM phosphoric acid, and 0.1 mM EDTA (ethylenediaminetetraacetic acid), titrated to pH 7.0 with TBAOH (tetrabutylammonium hydroxide), all purchased from Sigma-Aldrich, Oakville, ON, Canada. Note that 2 mM KCl was chosen because Pu220T12T13 formation is complete at this concentration [34].

Circular dichroism (CD) spectra were collected on a JASCO model J-1100 CD spectropolarimeter (Jasco, Easton, MD, USA) at 25 °C. The average of at least two scans are reported for each spectrum. The kinetics results in Figure 1 were prepared by first adding salt to the GQ at room temperature in a 1 mm quartz cuvette, placing the sample in the CD instrument, heating to 95 °C and holding for 5 min, then cooling to 25 °C at a rate of 1 °C/minute. The spectra were then collected at 15-min intervals for 23 h.

For the pressure experiments, a 350-mL sample cuvette was positioned in an optical high-pressure cell. Silicon oil was used as the pressure-transmitting liquid [41]. The light absorption of the sample in the pressure cell was monitored on a Uvikon model 860 spectrophotometer (Kontron, Inc., Everett, MA, USA). The temperature of the pressure cell was maintained by a brass thermal jacket with a programmable circulating water bath. The temperature was increased at 0.5 °C/min or 0.9 °C/minute, and the sample temperature was measured with a thermocouple inserted into the pressure cell. Instrument control and data acquisition were achieved using a Windows PC running a Python program. The stability of the G-quadruplex structure was observed at several static pressures between atmospheric pressure and 200 MPa. As described in previous work, we calculated the melting temperature (T_m) as the mid-point of the normalized temperature-induced transition, and the enthalpy from the slope of the normalized transition using the van't Hoff equation; this assumes that the unfolding is a single-step mechanism [42,43]. In the case of tetramolecular GQs, it is reported as the apparent melting temperature, $T_{1/2}$ [31,44].

The molar volume change (ΔV) of the unfolding of GQs is calculated using the Clausius–Clapeyron equation,

$$\Delta V = \frac{\Delta H}{T_m} \frac{\Delta T_m}{\Delta p}$$

where ΔH is the change in enthalpy at the transition, and p is the hydrostatic pressure [15,37,45].

Throughout this manuscript, we refer to the single-stranded oligonucleotide, d[5′-TGGGGT-3′], as TG4T. The quadruplex formed from the association of four strands of TG4T, (d[5′-TGGGGT-3′])$_4$, is referred to as TG4T-GQ.

Author Contributions: Conceptualization, N.T. and R.B.M.J.; formal analysis, N.T. and T.K.; investigation, N.T., T.K. and U.N.F.; methodology, N.T. and R.B.M.J.; validation, N.T. and T.K.; visualization, N.T.; supervision, R.B.M.J.; writing, N.T. and R.B.M.J. All authors have read and agreed to the published version of the manuscript.

Funding: This work was supported by a grant from the Natural Science and Engineering Council of Canada (to R.B.M.J.).

Institutional Review Board Statement: Not applicable.

Informed Consent Statement: Not applicable.

Acknowledgments: We thank Tigran Chalikian for the use of the circular dichroism instrument.

Conflicts of Interest: The authors declare no conflict of interest.

References

1. Víglaský, V.; Bauer, L.; Tlučková, K. Structural Features of Intra- and Intermolecular G-Quadruplexes Derived from Telomeric Repeats. *Biochemistry* **2010**, *49*, 2110–2120. [CrossRef] [PubMed]
2. Spiegel, J.; Adhikari, S.; Balasubramanian, S. The Structure and Function of DNA G-Quadruplexes. *Trends Chem.* **2020**, *2*, 123–136. [CrossRef] [PubMed]
3. Burge, S.; Parkinson, G.N.; Hazel, P.; Todd, A.K.; Neidle, S. Quadruplex DNA: Sequence, Topology and Structure. *Nucleic Acids Res.* **2006**, *34*, 5402–5415. [CrossRef] [PubMed]
4. Wu, Y.; Brosh, R.M. G-Quadruplex Nucleic Acids and Human Disease: G-Quadruplex Nucleic Acids and Human Disease. *FEBS J.* **2010**, *277*, 3470–3488. [CrossRef] [PubMed]
5. Maizels, N. G4-associated Human Diseases. *EMBO Rep.* **2015**, *16*, 910–922. [CrossRef] [PubMed]
6. Kharel, P.; Balaratnam, S.; Beals, N.; Basu, S. The Role of RNA G-quadruplexes in Human Diseases and Therapeutic Strategies. *Wiley Interdiscip. Rev. RNA* **2020**, *11*, e1568. [CrossRef]
7. Capra, J.A.; Paeschke, K.; Singh, M.; Zakian, V.A. G-Quadruplex DNA Sequences Are Evolutionarily Conserved and Associated with Distinct Genomic Features in Saccharomyces Cerevisiae. *PLoS Comput. Biol.* **2010**, *6*, e1000861. [CrossRef]
8. Huppert, J.L.; Balasubramanian, S. G-Quadruplexes in Promoters throughout the Human Genome. *Nucleic Acids Res.* **2007**, *35*, 406–413. [CrossRef]
9. Nakken, S.; Rognes, T.; Hovig, E. The Disruptive Positions in Human G-Quadruplex Motifs Are Less Polymorphic and More Conserved than Their Neutral Counterparts. *Nucleic Acids Res.* **2009**, *37*, 5749–5756. [CrossRef]
10. Eddy, J.; Maizels, N. Gene Function Correlates with Potential for G4 DNA Formation in the Human Genome. *Nucleic Acids Res.* **2006**, *34*, 3887–3896. [CrossRef]
11. Verma, A.; Halder, K.; Halder, R.; Yadav, V.K.; Rawal, P.; Thakur, R.K.; Mohd, F.; Sharma, A.; Chowdhury, S. Genome-Wide Computational and Expression Analyses Reveal G-Quadruplex DNA Motifs as Conserved *Cis*- Regulatory Elements in Human and Related Species. *J. Med. Chem.* **2008**, *51*, 5641–5649. [CrossRef] [PubMed]
12. Mergny, J.-L.; Sen, D. DNA Quadruple Helices in Nanotechnology. *Chem. Rev.* **2019**, *119*, 6290–6325. [CrossRef] [PubMed]
13. Davis, J.T. G-Quartets 40 Years Later: From 5′-GMP to Molecular Biology and Supramolecular Chemistry. *Angew. Chem. Int. Ed.* **2004**, *43*, 668–698. [CrossRef] [PubMed]
14. Rodrigues Pontinha, A.D.; Chiorcea-Paquim, A.-M.; Eritja, R.; Oliveira-Brett, A.M. Quadruplex Nanostructures of d(TGGGGT): Influence of Sodium and Potassium Ions. *Anal. Chem.* **2014**, *86*, 5851–5857. [CrossRef]
15. Chalikian, T.V.; Macgregor, R.B. Volumetric Properties of Four-Stranded DNA Structures. *Biology* **2021**, *10*, 813. [CrossRef]

16. Miller, M.C.; Buscaglia, R.; Chaires, J.B.; Lane, A.N.; Trent, J.O. Hydration Is a Major Determinant of the G-Quadruplex Stability and Conformation of the Human Telomere 3′ Sequence of d($AG_3(TTAG_3)_3$). *J. Am. Chem. Soc.* **2010**, *132*, 17105–17107. [CrossRef]
17. Aslanyan, L.; Ko, J.; Kim, B.G.; Vardanyan, I.; Dalyan, Y.B.; Chalikian, T.V. Effect of Urea on G-Quadruplex Stability. *J. Phys. Chem. B* **2017**, *121*, 6511–6519. [CrossRef]
18. Miyoshi, D.; Karimata, H.; Sugimoto, N. Hydration Regulates Thermodynamics of G-Quadruplex Formation under Molecular Crowding Conditions. *J. Am. Chem. Soc.* **2006**, *128*, 7957–7963. [CrossRef]
19. Li, K.; Yatsunyk, L.; Neidle, S. Water Spines and Networks in G-Quadruplex Structures. *Nucleic Acids Res.* **2021**, *49*, 519–528. [CrossRef]
20. Kankia, B.I.; Marky, L.A. Folding of the Thrombin Aptamer into a G-Quadruplex with Sr^{2+}: Stability, Heat, and Hydration. *J. Am. Chem. Soc.* **2001**, *123*, 10799–10804. [CrossRef]
21. Li, Y.Y.; Dubins, D.N.; Le, D.M.N.T.; Leung, K.; Macgregor, R.B. The Role of Loops and Cation on the Volume of Unfolding of G-Quadruplexes Related to HTel. *Biophys. Chem.* **2017**, *231*, 55–63. [CrossRef] [PubMed]
22. Fan, H.Y.; Shek, Y.L.; Amiri, A.; Dubins, D.N.; Heerklotz, H.; Macgregor, R.B.; Chalikian, T.V. Volumetric Characterization of Sodium-Induced G-Quadruplex Formation. *J. Am. Chem. Soc.* **2011**, *133*, 4518–4526. [CrossRef] [PubMed]
23. Akasaka, K. Probing Conformational Fluctuation of Proteins by Pressure Perturbation. *Chem. Rev.* **2006**, *106*, 1814–1835. [CrossRef] [PubMed]
24. Smeller, L. Pressure–Temperature Phase Diagrams of Biomolecules. *Biochim. Biophys. Acta BBA Protein Struct. Mol. Enzymol.* **2002**, *1595*, 11–29. [CrossRef]
25. Meersman, F.; McMillan, P.F. High Hydrostatic Pressure: A Probing Tool and a Necessary Parameter in Biophysical Chemistry. *Chem. Commun.* **2014**, *50*, 766–775. [CrossRef] [PubMed]
26. Matsumoto, S.; Tateishi-Karimata, H.; Takahashi, S.; Ohyama, T.; Sugimoto, N. Effect of Molecular Crowding on the Stability of RNA G-Quadruplexes with Various Numbers of Quartets and Lengths of Loops. *Biochemistry* **2020**, *59*, 2640–2649. [CrossRef] [PubMed]
27. Petraccone, L.; Pagano, B.; Giancola, C. Studying the Effect of Crowding and Dehydration on DNA G-Quadruplexes. *Methods* **2012**, *57*, 76–83. [CrossRef]
28. Strychalski, W.; Guy, R.D. Intracellular Pressure Dynamics in Blebbing Cells. *Biophys. J.* **2016**, *110*, 1168–1179. [CrossRef]
29. Jiang, H.; Sun, S.X. Cellular Pressure and Volume Regulation and Implications for Cell Mechanics. *Biophys. J.* **2013**, *105*, 609–619. [CrossRef]
30. Takahashi, S.; Sugimoto, N. Volumetric Contributions of Loop Regions of G-Quadruplex DNA to the Formation of the Tertiary Structure. *Biophys. Chem.* **2017**, *231*, 146–154. [CrossRef]
31. Mergny, J.-L. Kinetics of Tetramolecular Quadruplexes. *Nucleic Acids Res.* **2005**, *33*, 81–94. [CrossRef] [PubMed]
32. Bardin, C.; Leroy, J.L. The Formation Pathway of Tetramolecular G-Quadruplexes. *Nucleic Acids Res.* **2008**, *36*, 477–488. [CrossRef] [PubMed]
33. Agrawal, P.; Hatzakis, E.; Guo, K.; Carver, M.; Yang, D. Solution Structure of the Major G-Quadruplex Formed in the Human VEGF Promoter in K+: Insights into Loop Interactions of the Parallel G-Quadruplexes. *Nucleic Acids Res.* **2013**, *41*, 10584–10592. [CrossRef] [PubMed]
34. Kim, B.G.; Long, J.; Dubins, D.N.; Chalikian, T.V. Ionic Effects on VEGF G-Quadruplex Stability. *J. Phys. Chem. B* **2016**, *120*, 4963–4971. [CrossRef] [PubMed]
35. Olsen, C.M.; Lee, H.-T.; Marky, L.A. Unfolding Thermodynamics of Intramolecular G-Quadruplexes: Base Sequence Contributions of the Loops. *J. Phys. Chem. B* **2009**, *113*, 2587–2595. [CrossRef] [PubMed]
36. Marcus, Y. The Standard Partial Molar Volumes of Ions in Solution. Part 4. Ionic Volumes in Water at 0–100 °C. *J. Phys. Chem. B* **2009**, *113*, 10285–10291. [CrossRef]
37. Molnár, O.R.; Somkuti, J.; Smeller, L. Negative Volume Changes of Human G-Quadruplexes at Unfolding. *Heliyon* **2020**, *6*, e05702. [CrossRef]
38. Warshaw, M.M.; Tinoco, I. Optical Properties of Sixteen Dinucleoside Phosphates. *J. Mol. Biol.* **1966**, *20*, 29–38. [CrossRef]
39. Cantor, C.R.; Warshaw, M.M.; Shapiro, H. Oligonucleotide Interactions. III. Circular Dichroism Studies of the Conformation of Deoxyoligonucleolides. *Biopolymers* **1970**, *9*, 1059–1077. [CrossRef]
40. Cavaluzzi, M.J.; Borer, P.N. Revised UV Extinction Coefficients for Nucleoside-5′-Monophosphates and Unpaired DNA and RNA. *Nucleic Acids Res.* **2004**, *32*, e13. [CrossRef]
41. Wu, J.Q.; Macgregor, R.B. A Temperature-Regulated Iso-Hyperbaric Spectrophotometer: Construction and Performance Characteristics. *Anal. Biochem.* **1993**, *211*, 66–71. [CrossRef] [PubMed]
42. Mergny, J.; Lacroix, L. UV Melting of G-Quadruplexes. *Curr. Protoc. Nucleic Acid Chem.* **2009**, *37*. [CrossRef] [PubMed]
43. Marky, L.A.; Breslauer, K.J. Calculating Thermodynamic Data for Transitions of Any Molecularity from Equilibrium Melting Curves. *Biopolymers* **1987**, *26*, 1601–1620. [CrossRef] [PubMed]
44. Wyatt, J.R.; Davis, P.W.; Freier, S.M. Kinetics of G-Quartet-Mediated Tetramer Formation. *Biochemistry* **1996**, *35*, 8002–8008. [CrossRef] [PubMed]
45. Takahashi, S.; Sugimoto, N. Effect of Pressure on Thermal Stability of G-Quadruplex DNA and Double-Stranded DNA Structures. *Molecules* **2013**, *18*, 13297–13319. [CrossRef]

life

MDPI

Article

Revisiting DNA Sequence-Dependent Deformability in High-Resolution Structures: Effects of Flanking Base Pairs on Dinucleotide Morphology and Global Chain Configuration

Robert T. Young [1], Luke Czapla [1,2], Zoe O. Wefers [1], Benjamin M. Cohen [1] and Wilma K. Olson [1,*]

[1] Department of Chemistry & Chemical Biology, Center for Quantitative Biology, Rutgers, The State University of New Jersey, Piscataway, NJ 08854, USA; ryoung.2011@rutgers.edu (R.T.Y.); czaplaluke@gmail.com (L.C.); zoewefers@comcast.net (Z.O.W.); b.cohen479@gmail.com (B.M.C.)

[2] Memorial Sloan Kettering Cancer Center, New York, NY 10065, USA

* Correspondence: wilma.olson@rutgers.edu

Abstract: DNA carries more than the list of biochemical ingredients that drive the basic functions of living systems. The sequence of base pairs includes a multitude of structural and energetic signals, which determine the degree to which the long, threadlike molecule moves and how it responds to proteins and other molecules that control its processing and govern its packaging. The chemical composition of base pairs directs the spatial disposition and fluctuations of successive residues. The observed arrangements of these moieties in high-resolution protein–DNA crystal structures provide one of the best available estimates of the natural, sequence-dependent structure and deformability of the double-helical molecule. Here, we update the set of knowledge-based elastic potentials designed to describe the observed equilibrium structures and configurational fluctuations of the ten unique base-pair steps. The large number of currently available structures makes it possible to characterize the configurational preferences of the DNA base-pair steps within the context of their immediate neighbors, i.e., tetrameric context. Use of these knowledge-based potentials shows promise in accounting for known effects of sequence in long chain molecules, e.g., the degree of curvature reported in classic gel mobility studies and the recently reported sequence-dependent responses of supercoiled minicircles to nuclease cleavage.

Keywords: DNA sequence-dependent structure; DNA deformability; DNA sequence context; DNA curvature; DNA minicircles

Citation: Young, R.T.; Czapla, L.; Wefers, Z.O.; Cohen, B.M.; Olson, W.K. Revisiting DNA Sequence-Dependent Deformability in High-Resolution Structures: Effects of Flanking Base Pairs on Dinucleotide Morphology and Global Chain Configuration. *Life* **2022**, *12*, 759. https://doi.org/10.3390/life12050759

Academic Editors: Tigran Chalikian and Jens Völker

Received: 1 April 2022
Accepted: 15 May 2022
Published: 20 May 2022

1. Introduction

Encoded in the strings of DNA bases that make up the genomes of living species are codes that underlie an assortment of biological processes. The underpinnings of these codes lie in the base sequence-dependent energetic and structural features of DNA, which dictate the degree to which the long, threadlike molecule fluctuates and how it responds to the proteins and other molecules involved in its activity and packaging. The preferred arrangements of base pairs determine the natural folding of individual sequences, as well as the ease with which these folds deform from their equilibrium rest states.

DNA structure depends upon both the underlying base sequence and the local chemical environment. Variations in the environment introduce large-scale rearrangements of the canonical right-handed double helix with 10 base pairs per turn [1] to under- and overwound structures of the same helical sense with, respectively, more or fewer base pairs per turn [2,3]. These changes in helical state are coupled to changes in the orientation and displacement of successive base pairs and to rearrangements of the intervening sugar-phosphate backbone [4]. Whereas the planes of base pairs pass through and stack roughly perpendicular to the global helical axis of canonical B DNA, they deviate from this alignment in under- and overwound forms. The changes of helical state introduce slight bends between successive base pairs and lateral movements that open channels through the center of the double helix and expose or hide different nucleotide atoms [5]. For example, the underwound A form of DNA is compacted relative to B DNA with atoms on the minor-groove edges of base pairs, i.e., the edges of the bases containing the pyrimidine O2 and the purine N3 atoms (Figure 1), much more exposed to the surrounding chemical environment than in B DNA. By contrast, the major-groove edge, on the opposite side of the base pairs, becomes more accessible in the overwound C form of DNA.

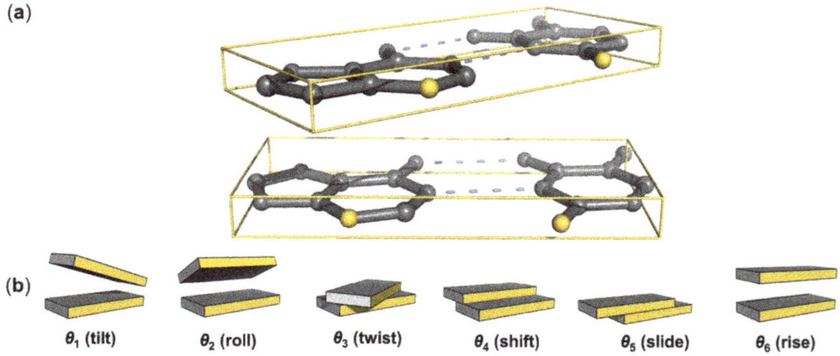

Figure 1. (**a**) All-atom model highlighting, in gold, the purine N3 and pyrimidine O2 atoms on the minor-groove edge of a DNA base-pair step. The orientation and displacement of successive base pairs, enclosed in rectangular slabs, are described in terms of six rigid-body parameters. (**b**) Schematics illustrating positive values of each parameter with the sequence-bearing strand at the left and the minor groove edge (gold) facing the reader.

The sequence introduces an even finer level of detail in DNA organization. The base pairs found in high-resolution structures deviate from the ideal, planar hydrogen-bonded arrangements anticipated by Watson and Crick [6] and do not align one above another in perfectly ordered arrays. The chemical composition of the bases directs the spatial disposition of successive residues [7,8], playing a critical role in the overall pathway of the chain and in processes involving DNA recognition. The helix tends to unwind at certain base-pair steps with accompanying changes in bending and displacement of the sort found in A DNA [9]. Other steps exhibit a propensity to overwind and adopt conformational states more characteristic of C DNA [10]. The shifting of base pairs between the different helical forms introduces local kinks in the DNA that have pronounced effects on the overall molecular pathway [11,12].

Analysis of high-resolution DNA structures has revealed subtle sequence-dependent irregularities in the apparent rest state and fluctuations of successive base pairs [13]. The observed correlations in base-pair orientation and displacement in crystalline complexes with proteins have led, in turn, to the determination of a set of knowledge-based elastic energy functions widely used for understanding the nucleic acid machinery. The average values and correlations in observed structural parameters have also provided useful benchmarks for checking state-of-the-art, atomic-level DNA calculations and have stimulated the extraction of similar elastic functions from the features of base pairs in large ensembles of computer-simulated molecules [14]. Compared with the experimental data, the computed datasets include vastly larger numbers of configurational 'snapshots' and describe DNA behavior in specific local chemical environments. The analysis of DNA variability within ensembles of protein–DNA structures assumes that different proteins impose different sorts of forces on DNA, that these forces effectively cancel one another (major vs. minor groove bending, etc.), and that the natural conformational response of DNA surfaces after averaging over the dataset [13]. On the other hand, the features of simulated DNA molecules depend upon the number of computed configurations and the reliability of the force fields used to describe atomic-level interactions [15].

The small number of high-resolution structures limited early analyses of DNA sequence-dependent structure and deformability to the dimer level, i.e., the spatial orientation and displacement of successive base pairs. Various properties of DNA—such as the preferred positioning of nucleosomes on DNA [16,17], the cutting patterns of the DNA backbone by hydroxyl radicals [18,19], the mobilities of synthetic oligonucleotides on gels [20], the profiles of DNase I digestion on restriction fragments and phage promoters [21], etc.—depend upon sequence content, i.e., the identities of the base pairs flanking a dimer. Moreover, analyses of simulated DNA structures point to variability in the rest states and deformations of a given dimer within different tetranucleotide environments [22–28], i.e., the identities of the immediately preceding and following base pairs. The substantial number of high-resolution nucleic acid structures now available makes it possible to examine the corresponding effects in protein–DNA crystal complexes. The data include a much wider variety of DNA-bound proteins than originally examined, with 40-fold or more structural examples of each unique dimer (see results below).

This article starts with an update of the apparent equilibrium rest states and intrinsic deformations of successive base pairs in protein–DNA complexes, highlighting trends in the data accumulated since determination of the first set of knowledge-based potentials. Next follows an overview of the effects of tetrameric context and resolution on the structural features of the base-pair steps in the collected data. The discussion focuses on the twist between successive base pairs and the consequent sequence-dependent under- and overwinding of DNA, as well as on the relative deformability of the base-pair steps. The narrative then turns to an examination of the extent to which the current dimer and tetrameric models of sequence-dependent DNA structure and deformability take account of known effects of sequence on apparent DNA curvature in classic gel mobility studies [20,29–31]. The test of the data involves comparison of the reported degree of curvature in selected series of concatenated oligonucleotide sequences with the ring closure propensities of modeled DNA minicircles of identical composition and chain length, i.e., number of base pairs (bp). The paper concludes with predictions of the effect of sequence on the preferred configurations of a well-characterized 336-bp minicircle and the extent to which the updated models take account of known sites of enzyme cutting on the DNA [32].

2. Materials and Methods

2.1. Dataset

The features of DNA base-pair steps reported herein are based on an ensemble of configurational states found in a collection of 3971 protein–DNA crystal structures extracted from the Protein Data Bank (pdb) [33] in February 2022. The dataset (see Table S1 in Supporting Material) excludes redundant structures, such as those solved independently under slightly different crystallographic conditions, with modifications of a few base pairs, with a mutant protein in place of the wild-type protein, etc. These structures are identified with a new automated procedure, which uses ECOD (evolutionary classifications of domains) identifiers [34,35] reported within each pdb file in combination with DNA sequence matching. The sequences in a pair of structures are taken as matched if the longest stretch of identical base pairs in the two structures is more than 70% of the total DNA length, and pairs of structures are deemed redundant if sequences so matched associate with a protein with the same ECOD identifiers, specifically the same number and types of domains (F-group/H-group names). The structure of better resolution is added to the dataset and the remaining structure is discarded. Duplicate helices in symmetric structures, e.g., two of the three helices comprising a three-armed junction, are also excluded. The resulting, nearly random sample of protein-DNA structures removes bias associated with the repetition of nearly identical structures, thereby allowing for a more uniform exploration of DNA configuration space.

2.2. Configurational States

The configurational states of base-pair steps within the selected structures are expressed in terms of the six rigid-body parameters—tilt, roll, twist, shift, slide, rise (Figure 1)—commonly used to describe the relative orientation and displacement of coordinate frames on successive base pairs [36]. The base-pair frames are located on the mid-frame between complementary bases, following the fitting procedure and rotational scheme used in 3DNA [5]. The angular parameters are extracted from the Euler angles used in the rotational scheme and the translational parameters from the vector that connects base-pair origins, when expressed in the mid-frame (see [37] for mathematical details). Numerical values are collected for all base-pair steps in a given structure and placed into 10 unique dimer groups, taking account of the sign differences of tilt and shift in non-unique dimers compared to those on the complementary unique base-pair steps [36]. The step parameters are further classified in terms of tetrameric context, with the 256 possible combinations of four successive base pairs reduced to 136 unique values, i.e., the 16 possible combinations of base pairs flanking each of the six unique non-self-complementary base-pair steps and the 10 possible combinations flanking each of the four self-complementary steps ($136 = 16 \times 6 + 10 \times 4$; see Figure S1). Base pairs at the ends of chains or adjacent to 'melted' (unpaired) steps are placed into separate groupings, which are not considered here. Each set of tetramers is then subjected to a culling procedure that excludes outlying states of extreme deformation in a stepwise fashion until there are no base-pair step parameters more than three standard deviations from their average values. In practice, quasi-Gaussian distributions of rigid-body parameters are obtained after 3–15 rounds of such culling. Even though the culling is restricted to the six rigid-body parameters, the procedure eliminates almost all non-canonical base pairs, e.g., wobble or Hoogsteen pairs with distinctly different interbase arrangements. The base pairs in the collected set of structures exhibit minor fluctuations about the ideal, Watson–Crick configuration [38], with occasional occurrences (<2% depending upon tetrameric context) of partially melted states with missing hydrogen bonds or slight in- and/or out-of-plane deformations. The six base-pair parameters—so-called buckle, propeller, opening, shear, stretch, stagger [36]—adopt values similar to those previously reported for double-helical structures [39,40]. Small uncertainties in the positions of individual atoms have limited effect on the computed values of both the base-pair and the base-pair-step parameters [41].

2.3. Knowledge-Based Potentials

Average step parameters are determined for the 10 unique dimers in different sequence, temporal, and resolution contexts. Dimer averages are determined in two ways—averages of step parameters over all structural examples and weighted averages over all possible tetrameric contexts, i.e., averages of the mean step parameters of the 16 combinations of flanking base pairs. Tetramer averages, i.e., the mean configurational parameters of dimers in a specific tetramer context, are determined over the available structures. Knowledge-based elastic energy functions are generated, as described previously [13], from the mean values and dispersion of the step parameters of dimers and tetramers in the collected data. The deformability of base-pair steps is reported in terms of the average volume of configuration space $\langle V_{\text{step}} \rangle$ accessible to the different steps. Values are determined from the product of the square roots of the eigenvalues of the covariance matrix, i.e., the 6×6 matrix with elements corresponding to the differences between the mean products and products of mean values $\langle \Delta\theta_i \Delta\theta_j \rangle = \langle \Delta\theta_i \theta_j \rangle - \langle \theta_i \rangle \langle \theta_j \rangle$ of all combinations of step parameters, where θ_i (i = 1–6) refers to one of the rigid-body parameters. Temporal averages are based on datasets collected up to and including a given year. Resolution averages are based on structures at or better than a specified limit.

Properties of a generic MN base-pair step are described in terms of both sequence-dependent averages, obtained by equally weighting the average step parameters of the 16 possible base-pair steps, and structure-based averages, evaluated over sets of available structures. In order to remove bias associated with the unequal number of examples of different base-pair steps in the collected data, the latter averages are based on the same number of examples of each type of dimer step, here 70% of the number of examples of the least represented dimer. The data are collected by randomly sampling a subset of the configurations associated with each of the 10 unique base-pair steps. The configurations of non-unique complementary steps are described by the same structural examples with requisite changes in the signs of tilt and shift (θ_1, θ_4) [36]. Self-complementary steps include the step parameters of both strands so that the averages values of tilt and shift are null. The MN parameters reflect the combined subsets of configurations, i.e., the set of step parameters and the configurational volumes collected for all 16 base-pair steps. The process is repeated several hundred times in the context of the year in which the data were available and the resolution of the most recently collected data.

2.4. Energy Optimization

The sequence-dependent configurations of 150-bp DNA minicircles are obtained using emDNA, new software that optimizes the energy of a collection of base pairs, in which the first and last pairs are held fixed [42,43]. The DNA is described at the level of base-pair steps in terms of the six rigid-body parameters and guided by the knowledge-based potentials described above. The configuration of the DNA as a whole is monitored by a second set of variables that keep track of the vectorial displacements of successive base pairs in a global reference frame. The introduction of the latter quantities makes it possible to take direct account of the spatial constraints imposed on the DNA and to use unconstrained numerical optimization methods. The linking number Lk is controlled by the twist assigned to uniformly spaced base pairs in the initial circular starting structure, here approximated by the rigid-body parameter of the same name (Figure 1) and assigned values based on the expected total twist of a relaxed chain. Supercoiled chains are assigned differences in linking number ΔLk relative to this reference. The total twist, or the total number of turns of helix in the starting structure, is obtained by dividing the sum of the assumed equilibrium twist angles, in degrees, by 360°. The total twist of the optimized structure is measured in terms of the twist of supercoiling, a quantity that takes account of both the rotational and the translational contributions to the wrapping of DNA strands about one another [44,45] as opposed to the step parameter used herein to characterize and build three-dimensional DNA models. The two twists are nearly identical in the optimized structures, where lateral displacements of successive base pairs are minimal. A Debye–Hückel term is used to prevent the self-contact of DNA residues separated by 11 bp or more. The charge on each phosphate group is placed on the base-pair center and assigned a value –0.24 esu in accordance with the predictions of counterion condensation theory [46]. The dielectric medium is taken to be that of a 100 mM aqueous monovalent salt solution.

2.5. Ring-Closure Propensities

The simulated ring-closure propensities, or *J*-factors, of short minicircles are compared with reported values of DNA curvature. The ease of cyclization is estimated from the statistical weights of the energy-optimized configurations. This treatment ignores other features of the system that might contribute to the free energy, e.g., base-pair melting, long-range attractive forces, room-temperature fluctuations, etc. DNA curvature is estimated by interpolation of the ratios of the apparent chain lengths of multimer sequences, determined from comparison with size markers on polyacrylamide gels, in published figures [20,29–31]. The predicted sequence-dependent uptake of twist in different topoisomers of optimized DNA minicircles is compared with observed hotspots of enzymatic cleavage [32].

3. Results

3.1. Base-Pair Steps within High-Resolution Structures

The original knowledge-based description of DNA sequence-dependent structure and deformability derived from a hand-curated dataset of 92 non-redundant protein-DNA crystal structures with $\sim$100 examples of each unique base-pair step [13]. There are now in excess of 5000 examples of each step, including more than 7000 examples of each of the four self-complementary steps, in the current collection of computationally curated structures (Figure 2a). The latter counts include steps from both strands of the sampled DNA structures, given that the signs of these parameters differ when expressed in terms of the leading or complementary chain [36]. Although the curation of structures excludes 48 of the 3971 selected protein–DNA complexes, the build-up over time of the number of examples of the 10 unique base-pair steps has been exponential, save for a pandemic-related leveling off of new entries in 2020–2022. The most and least represented dimers in the current collection are CG/CG and AG/CT steps with 7670 and 5078 examples, respectively (see Table S2 for a complete enumeration of structural counts).

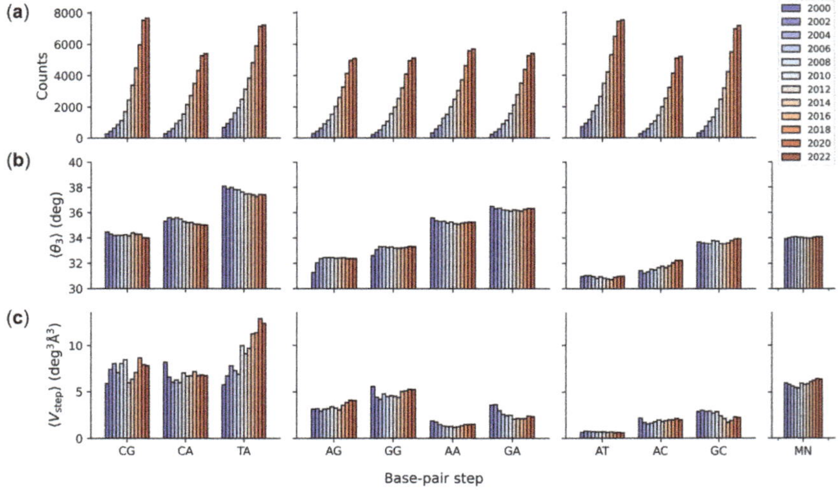

Figure 2. Color-coded histograms illustrating sequence-dependent features of DNA found within high-resolution protein–DNA structures collected over the last two decades: (**a**) number of base-pair steps; (**b**) intrinsic dimer structure measured in terms of the average twist angle $\langle V_{\text{step}} \rangle$ between successive base pairs; (**c**) dimer deformability measured in terms the average volume of configuration space $\langle V_{\text{step}} \rangle$ accessed by individual steps. MN parameters for a generic MpN step based on equal weighting of the average parameters of the 16 common dimers. Base-pair steps grouped by chemical class (pyrimidine–purine, purine–purine, and purine–pyrimidine). See Table S2 for numerical values and Methods for details.

Many of the sequence-dependent features of DNA base-pair steps found in the original set of protein–DNA crystal structures have not significantly changed over time. The average twist of individual dimers has levelled off to characteristic values, within ~2° of the original values (see Table S2 for details). Moreover, the same trends in the relative magnitude of twist deduced from early biophysical studies of DNA in gels and in solution [47] and found in the first few high-resolution structures [8,13] also persist (Figure 2b). That is, the twist of pyrimidine–purine, purine–purine, and purine–pyrimidine steps continues to increase in the same order—CG < CA < TA, AG < GG < AA < GA, and AT < AC < GC, respectively. Moreover, the 34.1° twist of a generic MN dimer, obtained by equally weighting the average twist values of the 16 possible base-pair steps, shows remarkable agreement with the 10.6 base-pair helical repeat of mixed sequence DNA found in pioneering micrococcal nuclease cutting [48] and electrophoresis gel band-shift [49,50] measurements, i.e., 360°/turn÷34.1°/bp = 10.6 bp/turn.

The deformability of base-pair steps, as measured in terms of the average volume of configuration space $\langle V_{\text{step}} \rangle$ enclosing the principal axes of dimeric distortion, has also remained much the same over time, save for appreciable growth in the apparent mobility of TA steps (Figure 2c). The volume occupied by current TA structural examples is roughly double that reported originally and appreciably greater than that of any other base-pair step. Moreover, the range of TA movement does not yet appear to have reached an asymptotic limit with the addition of most recent structural examples. The AT step, by contrast, remains the stiffest dimer in terms of $\langle V_{\text{step}} \rangle$, with less than a 10th of the volume accessible to the TA step and very limited change in magnitude as new structures have accumulated (see Table S2 for numerical details). The plotted deformability of a generic MN step is an average over the configurational volumes of the 16 possible base-pair steps. The resulting values of $\langle V_{\text{step}} \rangle$ have remained relatively constant over time, with magnitudes comparable to those of CA base-pair steps. The structure-based values of $\langle V_{\text{step}} \rangle$, obtained from random subsets of structures available in the database in a specified year (see Methods), are ~85% of the sequence-based volumes (see Table S2 for comparative values).

Structures of 3.0 Å or better resolution make up roughly two-thirds of the most recently accumulated dataset with mean values of twist not substantially different from those of dimers in the complete set of structures (Figure 3a,b). The average twist of base-pair steps in the ~15% best resolved structures, with 2 Å or better resolution, show somewhat larger differences in value (as much as 1°) from those of the complete dataset. The variation in the relative magnitudes of twist in the smaller dataset also differs slightly from that noted above, with the twist of CA steps slightly lower than those of CG steps. The deformability of base-pair steps is fairly sensitive to resolution, with individual steps in the 3.0 Å subset of structures occupying ~60% of the volume accessible to the corresponding dimers in the full dataset (Figure 3c). The accessible volume is even smaller in the set of best-resolved structures (<2 Å), with the volume occupied by the best-resolved GC steps less than 10% of that in the full dataset (see Table S3 for details). The configurational volumes of generic MN steps show similar decreases in value with improved resolution. The structure-derived values of $\langle V_{\text{step}} \rangle$ again fall short of the sequence-averaged values (see Table S3 for numerical comparisons).

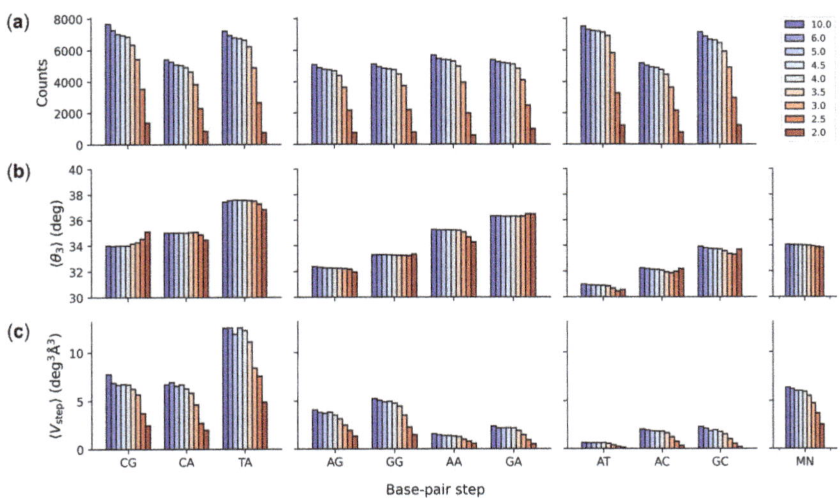

Figure 3. Color-coded histograms illustrating sequence-dependent features of DNA found in high-resolution protein–DNA structures of specified resolution accumulated as of February 2022: (**a**) number of base-pair steps; (**b**) intrinsic dimer structure measured in terms of the average twist angle $\langle \theta_3 \rangle$ between successive base pairs; (**c**) dimer deformability measured in terms the average volume of configuration space $\langle V_{\text{step}} \rangle$ accessed by individual steps. See Table S3 for numerical values.

3.2. Effects of Sequence Context on Base-Pair Structure and Deformability

The complete set of collected structures includes 160 or more examples of each DNA base-pair step in all possible tetrameric contexts, with a maximum of 869 examples of the configuration of an AT dimer flanked on both sides by a G·C base pair, i.e., the GATG tetramer (see Figure 4a). As anticipated from atomic-level simulations [22–28], the average structure of successive base pairs is sensitive to the surrounding nucleotide environment. The mean twist angles of a specific base-pair step vary over a range of 3–4° depending upon the identities of the surrounding base pairs. For example, tetramers with a central CG or GC step are equally likely to be slightly under- or overtwisted relative to the 10.6 helical repeat of mixed-sequence DNA, with a nearly equal mix of respective blue and red entries in Figure 4b. The central dimers within other tetramers tend to be either under- or overtwisted relative to the 10.6 reference regardless of the surrounding base pairs, e.g., primarily blue untwisted AG, AC, AT, GG steps vs. largely red overtwisted AA, GA, CA, TA steps. The TAAG step stands out in being highly undertwisted compared to other steps sharing a central AA dimer, with a value of twist characteristic of an 11-fold helix as opposed to the 10-fold structures adopted by the majority of AA dimers (see Table S4 for numerical values). The extremes of twisting occur in AT and TA dimers in the context of CATA and ATAG sequences with average values of 29.3° and 39.5°, respectively. The dimeric averages in the figure inset, which are based on equal weighting of the average twist of each base-pair step in all 16 possible tetrameric context, differ slightly ($\leq$0.1°) from the numbers reported in Figure 2 for the same February 2022 dataset. The latter values are averages evaluated over all structural examples of a given base-pair step, regardless of sequence context.

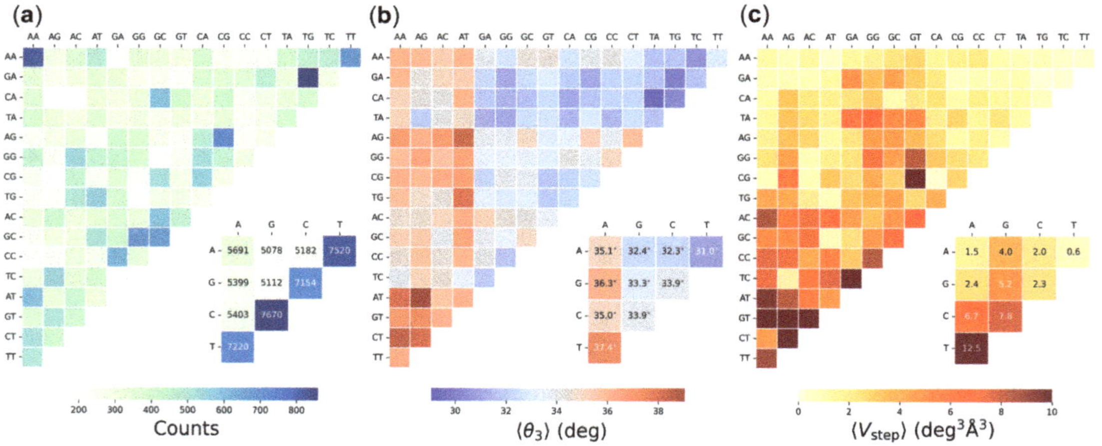

Figure 4. Color-coded heat maps of (**a**) the number of base-pair steps, (**b**) the average twist angle $\langle\theta_3\rangle$ between successive base pairs, and (**c**) the average volume of configuration space $\langle V_{step}\rangle$ accessible to DNA dimers in all unique tetrameric contexts. Data organized such that purine–purine steps lie at the top left of each grid, purine–pyrimidine steps at the lower left, and pyrimidine–purine steps at the top right. Values collected from 3923 high-resolution protein–DNA structures available in February 2022. Dimer values in numbered boxes are averages of the mean values found for each base-pair step in all 16 tetrameric contexts. See Table S4 for numerical values and Methods for details.

The collected data also point to examples of dimer deformability influenced by sequence context. Whereas pyrimidine-purine steps are generally much more deformable than purine–purine and purine–pyrimidine steps in terms of spanning a broader range of configuration space, the CA steps within GCAT and TCAG tetramers are surprisingly stiff, with very small values of $\langle V_{step}\rangle$ compared to other sequence contexts (CA entries, respectively color-coded beige and red/maroon in Figure 4c). A few purine–pyrimidine steps are extremely stiff, with configurational volumes substantially smaller than those of other steps sharing the same central base pairs. For example, the replacement of thymine by cytosine in TGCA compared to CGCA introduces a 30-fold decrease in $\langle V_{step} \rangle$ (neighboring tan vs. pale beige GC entries in Figure 4c corresponding to the greatest and least deformable GC steps). Among purine–purine steps, AG and GG dimers show greater sensitivity to sequence context than AA and GA dimers, which are nearly as stiff as purine–pyrimidine dimers. The context-averaged dimeric values of $\langle V_{step} \rangle$ reported in the figure inset show similar trends in magnitude but appreciable numerical differences from the structure-averaged configurational volumes reported in Table S2, particularly in the highly deformable TA and CG steps (see Table S4 for numerical values). The numerical differences are unsurprising given both the various averages contributing to $\langle V_{step} \rangle$ (see Methods) and the different number of examples of each tetrameric context. The structure-based dimer averages in Table S2 do not consider these differences.

Sequence context has much the same effect on base-pair step configuration and deformability in the subset of recently collected structures with 3.0 Å or better resolution. The latter data include 91 or more examples of each base-pair step in all tetrameric contexts, with 789 examples of the AT dimers within a GATG tetramer (the same sequence context found in greatest number in the complete set of recent structures regardless of resolution). The least populated tetrameric sequence, CGAG, is also common to both sets of structures. Although numerical values of average twist differ in individual cases, the trends in relative magnitude persist. For example, AT and TA dimers with respective twists of 28.4° in CATA tetramers and 39.9° in ATAG tetramers remain at the extremes of under- and overtwisting relative to the same 10.6 helical repeat of mixed-sequence DNA. The differences in average twist between the smaller and larger datasets range between −1.4° and +1.9° with an average magnitude of 0.3°. The trends in relative dimer deformability also persist despite the average 60% drop in $\langle V_{step} \rangle$ over all tetrameric contexts. The base-pair steps within some tetramers show little change in deformability between the two datasets (e.g., CTAA, TCGA, ACGT, TAGT) whereas others show very substantial drops in configurational volume (e.g., more than an order of magnitude decrease in CA deformability in the very stiff GCAT sequence noted above). See Table S5 for numerical details.

3.3. Sequence-Dependent DNA Curvature

The sequence-dependent structure and deformability of successive base pairs underlie larger-scale features of double-helical DNA, such as the intrinsic curvature associated with repeated tracts of A·T pairs separated by segments of G+C-rich DNA [51]. Estimates of DNA ring closure guided by the current set of knowledge-based potentials show remarkable agreement with the apparent curvature of assorted sequences determined in classic gel mobility studies (Figure 5a). The predicted cyclization propensities of a collection of 150-bp sequences increase in the same order as the reported ratios of apparent molecular size, based on observed electrophoresis markers of chain length, to the actual chain length [20,29–31]. The more easily closed sequences with larger *J*-factors match the more strongly curved sequences with larger apparent molecular sizes. Moreover, the potentials take account of the influence of the A-tract repeating length (A_jN_{10-j}, where N is G or C), the relative effects of specific bases within or at the ends of A tracts, and the contribution of A-tract polarity to the observed degree of curvature (see Figure 5a and Table S6 for details of sequence acronyms and numerical values). The estimated *J*-factors of energy-optimized circles of 147- and 168-bp chains with an $A_6N_4A_6N_5$ repeat, however, exceed experimentally measured values (~1×10^{-4} M) [20], with the treatment incorporating tetrameric context yielding higher ring-closure propensities (~1×10^{-1} M) than those based on dimer structure and deformability alone (~6×10^{-2} M). The optimization procedure used here does not consider fluctuations in DNA structure, which may contribute to the free energy of cyclization.

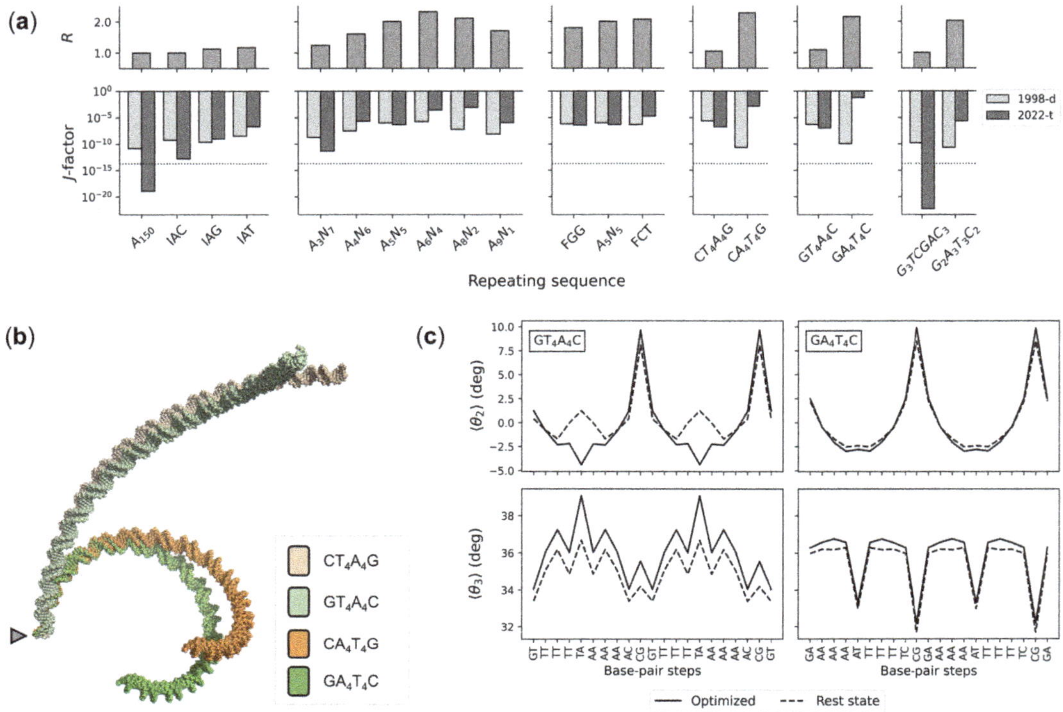

Figure 5. Sequence-dependent estimates of DNA ring closure compared with the degree of curvature determined in classic gel electrophoresis studies [20,29–31]. (**a**) Variation in DNA curvature, measured in terms of the ratio R of the apparent chain length to the true chain length, and in the computed J-factors of covalently closed 150-bp polymers bearing various 10-bp repeating sequences. J-factors extracted from the energies of minicircles optimized with the specified knowledge-based potentials: 1998-d (dimeric model reported in [13]); 2022-t (updated model that takes account of tetrameric context); ········· (ideal DNA). (**b**) Superimposed molecular images of the intrinsically straight and curved pathways of chains with NT_4A_4N and NA_4T_4N repeating sequences, where N is G or C, found with the tetrameric model. Pathways described in the frame of the first base pair (triangle). (**c**) Concerted changes in local DNA structure that underlie the predicted J-factors of the GT_4A_4C and GA_4T_4C sequences. Average values of roll $\langle\theta_2\rangle$ and twist $\langle\theta_3\rangle$ of dimer steps along a 21-bp stretch of the intrinsic linear structures are compared with those on the 150-bp minicircles optimized with the tetrameric model. Trends are similar for CT_4A_4G and CA_4T_4G sequences. See Table S6 for clarification of sequence acronyms and numerical data.

The updated set of knowledge-based potentials takes correct account of gel mobility data missed with the original set of potentials (labeled 1998-d in Figure 5a). For example, optimized minicircles with $N_{5-j}A_jT_jN_{5-j}$ repeating sequences now appear to be more curved than those made up of $N_{5-j}T_jA_jN_{5-j}$ sequences with higher computed J-factors. As noted above, the high-resolution structures of protein-bound DNA accumulated as of February 2022 show distinct differences in the preferred arrangements and deformations of AT vs. TA base-pair steps, with the former steps stiff and overtwisted and the latter steps highly deformable and undertwisted. The changes in DNA twisting occur in concert with well-known changes in the bending and lateral displacement of successive base pairs via roll and slide, respectively [52]. The stiffness of the AT steps restricts the slide to negative values ($-0.7_{\pm 0.1}$ Å) regardless of sequence context but allows for subtle, context-dependent variation in the sign of average roll. The TA steps, by contrast, vary widely, with the roll values spanning roughly twice the range of states sampled by the AT steps and the slide more likely to adopt mean positive values in most tetrameric contexts. The linear equilibrium structures of $N_{5-j}A_jT_jN_{5-j}$ sequences are both more curved than those of $N_{5-j}T_jA_jN_{5-j}$ sequences and more easily closed into circular configurations with the updated potentials (see the molecular images and changes in step-parameters that effect ring closure in Figure 5b,c). Whereas the intrinsic step parameters of the rest states in the former sequences closely match those in the circular structures, the base-pair steps along the latter sequence must overtwist and the five steps within the TTTAAA stretch must adopt more negative values of roll and/or slide in order to bring the chain ends into perfect register. This mechanical description of DNA curvature differs from conventional interpretations, which focus on static structural features that might underlie experimental observations—e.g., hypothesized wedges between successive A·T base pairs [53,54], differences in overall helical structure between A tracts and intervening G+C-rich linker segments [20,31], compensatory directions of bending at AT vs. TA steps [55–57]. Detailed maps of the collective patterns of base-pair structure and deformability will be presented elsewhere.

3.4. Sequence-Dependent Twist Uptake in DNA Minicircles

The updated potentials also provide a rationalization behind the sequence-dependent response of designed 336-bp DNA minicircles to nucleases known to cleave segments of 'melted' DNA [32,58]. The torsional stress associated with ring closure builds up non-uniformly in structures optimized on the basis of the context-dependent potentials, even in the most relaxed state where the linking number is 32, the total twist is 31.8 helical turns, and ΔLk is taken as zero. Although the twisting at individual base-pair steps deviates very slightly on average (–0.04°) from the intrinsic values along this pathway, the DNA over- and undertwists substantially at selected base-pair steps (see the computed variation in local twist in Figure 6a). Indeed, the twist changes by as much as $\pm 2°$ and the elastic energy jumps sharply at two TA steps, found in the context of CTAT and TTAC tetramers located, respectively, at residues 144–147 and 193–196 along the published sequence. Moreover, these sites absorb a disproportionately large degree of the twist introduced in the minicircle upon supercoiling. The highly overtwisted TA step within the CTAT tetramer on the relaxed topoisomer becomes substantially undertwisted in the ΔLk = –1 topoisomer while the highly undertwisted step within the TTAC tetramer on the relaxed topoisomer becomes substantially overtwisted in the ΔLk = +1 topoisomer (note the differences in the signs of $\Delta\theta_3$ at the sites marked, respectively, by a triangle and a star in the figure). The former step lies within the hotspot for Bal-31 endonuclease cleavage found in negatively supercoiled minicircles, while the latter step abuts the 5′-end of the strong S1 nuclease cleavage site in the same topoisomers [32,58]. Both steps are substantially higher in elastic energy than other steps along the modeled structures with deformation scores 7–10 times larger than the average values. The high energies point to sites of likely helical deformation and the different nature of the torsionally stressed steps to different modes of DNA distortion. Although the two enzymes have well-proven utility for probing disruptions in double-helical DNA [59,60], the precise details of protein–DNA recognition remain unknown. The predicted sites of localized overtwisting likely convert to different distorted forms from the predicted sites of undertwisting. The current potential functions do not take account of distortions within individual base pairs and associated movements of the sugar-phosphate backbone that 'melt' double-helical DNA.

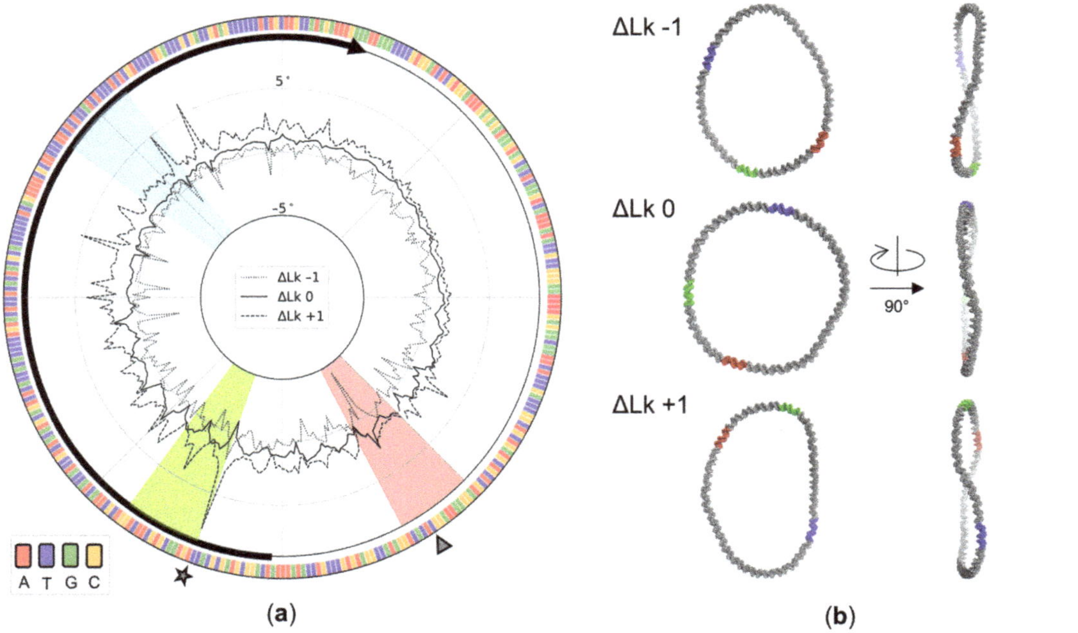

Figure 6. Effects of DNA supercoiling on the optimized configurations of a designed, 336-bp DNA minicircle. (**a**) Build-up of twist, $\Delta\theta_3$, in degrees, at individual base-pair steps along pathways optimized with the tetrameric model. The color-coded interior sectors correspond to reported hotspots of enzymatic cleavage [32] and the exterior triangle and star to sites of extreme twist uptake in the optimized structures. The heavy black curve highlights the 180-bp *att*R region that is a remnant of the λ integrase-mediated recombination process used to generate the minicircle [61], with the arrow denoting the 5′–3′ direction of the chain. The DNA base-pair sequence is color-coded along the outer edge of the figure. (**b**) Atomic-level representations of each optimized minicircle illustrating the changes in overall global shape and the relative locations of the hotspots in the frame of the two longest principal axes (*left*) and the frame of the longest and shortest axes (*right*).

The small changes in ΔLk have limited effect on the overall fold of the optimized minicircles. The optimized supercoiled structures adopt more elongated pathways with greater out-of-plane bending than the relaxed structure (Figure 6b). Although the configurations appear to be open from most perspectives, distant segments of the minicircles cross above or below one another in some viewpoints. Moreover, the relative locations of the enzymatic hotspots change with respect to the global features of the structure. The illustrated examples, depicted with two of the three principal axes of each configuration running along the horizontal and vertical directions of the page (axes 1, 2 in the left images and 1, 3 in the right), reveal the out-of-plane character of the supercoiled configurations compared to the relaxed ΔLk = 0 state, as well as the different directions of chain crossing in the positively vs. negatively supercoiled states (Figure 6b, *right*). The enzymatic hotspots rotate in a clockwise direction in the ΔLk + 1 topoisomer and a counterclockwise direction in the ΔLk − 1 topoisomer relative to the positions found for the relaxed state (Figure 6b, *left*).

4. Discussion

DNA base sequence carries a multitude of structural and energetic signals important to its biological activity and organization. Primary sequences of nucleic acid bases describe real three-dimensional structures with individual residues adopting characteristic spatial forms and macromolecular features that reflect the natural rest states and deformations of those structures. The first estimates of DNA intrinsic structure and deformability, extracted nearly 25 years ago from the fluctuations and correlations of the arrangements of successive base pairs within a small set of protein–DNA crystal structures [13], have provided useful insights into the role of sequence in DNA recognition and folding. As illustrated herein, the thousands of structures of protein–DNA complexes accumulated since then exhibit many of the sequence-dependent features of DNA base-pair steps found in the original 92 structures, including the same trends in the values and relative magnitudes of the twist angle between successive base pairs and the relative deformability of base-pair steps. The five remaining base-pair-step parameters either vary over broad ranges tightly coupled to the changes in twist, e.g., roll and slide [52], or exhibit very limited deformations within the set of structures, e.g., tilt, shift, and rise [62].

Although the current study does not consider whether the observed sequence-dependent propensities entail rearrangement of the intervening sugar-phosphate backbones, there are some striking similarities between the under- and overtwisted dimer steps reported here and the backbones found to connect the corresponding bases in a recent survey of high-resolution DNA structures [63]. For example, the CpC and GpG halves of the undertwisted GG dimer show higher than expected tendencies to adopt A-like conformational pathways while the backbones linking the overtwisted CA and TA steps show strong propensities to adopt the BII form characteristic of C DNA. How the sequence-dependent deformability of base-pair steps might be tied to the backbone linkages remains an open question.

The large dataset of currently available structures makes it possible to characterize the conformational preferences of the DNA base-pair steps within the context of their immediate neighbors, i.e., in the context of tetramers, for which there are now hundreds of structural examples of each of the 136 unique tetrameric settings. These data provide critical benchmarks for atomic-level simulations of double-helical DNA, as well as information potentially useful in interpreting the properties of specific DNA sequences. For example, the reported effects of sequence context on relative twist angles extracted from atomic-level simulations of short B DNA fragments [25,27] differ from those found here in high-resolution protein–DNA structures. The general trends in dimer deformability, as measured by the volume of accessible configuration space, show notable similarity, e.g., AT steps are the stiffest and YR steps the most flexible [25]. The configuration space sampled in the simulations, however, exceeds that found in the current set of crystal structures. The predicted effects of sequence on local DNA structure and deformability also depend on the simulation method, e.g., choice of force field [15].

The set of knowledge-based potentials extracted from the mean values and dispersion of base-pair-step parameters in the updated set of protein–DNA structures shows promise in accounting for known effects of sequence in long chain molecules. Estimates of DNA ring closure guided by the current set of potentials closely mirror the degree of curvature reported in classic gel mobility studies [20,29–31], including the effects of sequence polarity on curvature that the early set of structure-based potentials failed to match. The updated potentials also provide a rationalization behind the observed sequence-dependent response of designed DNA minicircles to nuclease cleavage [32,58]. The extreme build-up of twist in specific tetrameric contexts along supercoiled models occurs in the vicinity of observed hotspots of enzymatic cleavage. These torsionally stressed structures provide a useful starting point for studies of the 'melting' of the double helix necessary for generating the denatured forms of DNA—e.g., extrahelical bases, short single-stranded bubbles—thought to be recognized by the enzymes [58]. The data also provide a useful bridge between the atomic details of the molecular dynamics simulations [58] and the coarse-grained treatment of DNA [64] previously used to model the disruptions of base pairs and localized denaturation in the designed minicircle, revealing structural standards against which the former studies can be checked and identifying features in the observed base associations for improvement of the coarse-grained force field.

Supplementary Materials: The following supporting information can be downloaded at: https://www.mdpi.com/article/10.3390/life12050759/s1. This includes a .pdf document containing descriptions of the Supplemental Materials and Supplemental Figure S1 and a .xls file containing Supplemental Tables S1–S6 found on separate worksheets.

Author Contributions: Conceptualization, W.K.O. and R.T.Y.; methodology, R.T.Y. and W.K.O.; software, L.C., Z.O.W., and R.T.Y.; validation, R.T.Y.; formal analysis, R.T.Y. and B.M.C.; investigation, R.T.Y.; resources, L.C. and W.K.O.; data curation, L.C. and R.T.Y.; writing—original draft preparation, W.K.O.; writing—review and editing, R.T.Y. and W.K.O.; visualization, R.T.Y.; supervision, W.K.O.; funding acquisition, W.K.O. All authors have read and agreed to the published version of the manuscript.

Funding: This work was generously supported by the U.S. Public Health Service under research grant GM34809 and the Rutgers Center for Discrete Mathematics and Theoretical Computer Science (REU award to ZOW from National Science Foundation grant CCF-1852215).

Data Availability Statement: Data available upon reasonable request.

Conflicts of Interest: The authors declare no conflicts of interest.

References

1. Watson, J.D.; Crick, F.H.C. Molecular structure of nucleic acids. *Nature* **1953**, *171*, 737–738. [CrossRef] [PubMed]
2. Franklin, R.E.; Gosling, R.G. Molecular configuration in sodium thymonucleate. *Nature* **1953**, *171*, 740–741. [CrossRef] [PubMed]
3. Marvin, D.A.; Spencer, M.; Wilkins, M.H.F.; Hamilton, L.D. A new configuration of deoxyribonucleic acid. *Nature* **1958**, *182*, 387–388. [CrossRef] [PubMed]
4. Arnott, S.; Dover, S.D.; Wonacott, A.J. Least-squares refinement of the crystal and molecular structures of DNA and RNA from X-ray data and standard bond lengths and angles. *Acta Crystallogr.* **1969**, *B25*, 2192–2206. [CrossRef]
5. Lu, X.J.; Olson, W.K. 3DNA: A software package for the analysis, rebuilding, and visualization of three-dimensional nucleic acid structures. *Nucleic Acids Res.* **2003**, *31*, 5108–5121. [CrossRef]
6. Watson, J.D.; Crick, F.H.C. Genetical implications of the structure of deoxyribonucleic acid. *Nature* **1953**, *171*, 964–967. [CrossRef]
7. Calladine, C.R. Mechanics of sequence-dependent stacking of bases in B-DNA. *J. Mol. Biol.* **1982**, *161*, 343–352. [CrossRef]
8. Gorin, A.A.; Zhurkin, V.B.; Olson, W.K. B-DNA twisting correlates with base-pair morphology. *J. Mol. Biol.* **1995**, *247*, 34–48. [CrossRef]
9. Lu, X.J.; Shakked, Z.; Olson, W.K. A-form conformational motifs in ligand-bound DNA structures. *J. Mol. Biol.* **2000**, *300*, 819–840. [CrossRef]
10. Olson, W.K.; Zhurkin, V.B. Working the kinks out of nucleosomal DNA. *Curr. Opin. Struct. Biol.* **2011**, *21*, 348–357. [CrossRef]
11. Selsing, E.; Wells, R.D.; Alden, C.J.; Arnott, S. Bent DNA: Visualization of a base-paired and stacked A-B conformational junction. *J. Biol. Chem.* **1979**, *254*, 5417–5422. [CrossRef]
12. Marky, N.L.; Olson, W.K. Spatial translational motions of base pairs in DNA molecules: Application of the extended matrix generator method. *Biopolymers* **1994**, *34*, 121–142. [CrossRef] [PubMed]
13. Olson, W.K.; Gorin, A.A.; Lu, X.J.; Hock, L.M.; Zhurkin, V.B. DNA sequence-dependent deformability deduced from protein-DNA crystal complexes. *Proc. Natl. Acad. Sci. USA* **1998**, *95*, 11163–11168. [CrossRef]
14. Lankas, F.; Sponer, J.; Langowski, J.; Cheatham, T.E., 3rd. DNA basepair step deformability inferred from molecular dynamics simulations. *Biophys. J.* **2003**, *85*, 2872–2883. [CrossRef]
15. Ivani, I.; Dans, P.D.; Noy, A.; Pérez, A.; Faustino, I.; Hospital, A.; Walther, J.; Andrio, P.; Goñi, R.; Balaceanu, A.; et al. Parmbsc1: A refined force field for DNA simulations. *Nat. Methods* **2016**, *13*, 55–58. [CrossRef] [PubMed]
16. Satchwell, S.C.; Drew, H.R.; Travers, A.A. Sequence periodicities in chicken nucleosome core DNA. *J. Mol. Biol.* **1986**, *191*, 659–675. [CrossRef]
17. Chua, E.Y.D.; Vasudevan, D.; Davey, G.E.; Wu, B.; Davey, C.A. The mechanics behind DNA sequence-dependent properties of the nucleosome. *Nucleic Acids Res.* **2012**, *40*, 6338–6352. [CrossRef]
18. Burkhoff, A.M.; Tullius, T.D. The unusual conformation adopted by the adenine tracts in kinetoplast DNA. *Cell* **1987**, *48*, 935–943. [CrossRef]
19. Price, M.A.; Tullius, T.D. How the structure of an adenine tract depends on sequence context: A new model for the structure of T_nA_n DNA sequences. *Biochemistry* **1993**, *32*, 127–136. [CrossRef]
20. Koo, H.S.; Crothers, D.M. Calibration of DNA curvature and a unified description of sequence-directed bending. *Proc. Natl. Acad. Sci. USA* **1988**, *85*, 1763–1767. [CrossRef]
21. Brukner, I.; Sanchez, R.; Suck, D.; Pongor, S. Sequence-dependent bending propensity of DNA as revealed by DNase I: Parameters for trinucleotides. *EMBO J.* **1995**, *14*, 1812–1818. [CrossRef] [PubMed]
22. Packer, M.J.; Dauncey, M.P.; Hunter, C.A. Sequence-dependent DNA structure: Tetranucleotide conformational maps. *J. Mol. Biol.* **2000**, *295*, 85–103. [CrossRef] [PubMed]

23. Beveridge, D.L.; Barreiro, G.; Byun, K.S.; Case, D.A.; Cheatham, T.E., 3rd; Dixit, S.B.; Giudice, E.; Lankas, F.; Lavery, R.; Maddocks, J.H.; et al. Molecular dynamics simulations of the 136 unique tetranucleotide sequences of DNA oligonucleotides. I. Research design and results on d(CpG) steps. *Biophys. J.* **2004**, *87*, 3799–3813. [CrossRef] [PubMed]
24. Dixit, S.B.; Beveridge, D.L.; Case, D.A.; Cheatham, T.E., 3rd.; Giudice, E.; Lankas, F.; Lavery, R.; Maddocks, J.H.; Osman, R.; Sklenar, H.; et al. Molecular dynamics simulations of the 136 unique tetranucleotide sequences of DNA oligonucleotides. II: Sequence context effects on the dynamical structures of the 10 unique dinucleotide steps. *Biophys. J.* **2005**, *89*, 3721–3740. [CrossRef] [PubMed]
25. Fujii, S.; Kono, H.; Takenaka, S.; Go, N.; Sarai, A. Sequence-dependent DNA deformability studied using molecular dynamics simulations. *Nucleic Acids Res.* **2007**, *35*, 6063–6074. [CrossRef] [PubMed]
26. Lavery, R.; Zakrzewska, K.; Beveridge, D.; Bishop, T.C.; Case, D.A.; Cheatham, T.E., 3rd; Dixit, S.; Jayaram, B.; Lankas, F.; Laughton, C.; et al. A systematic molecular dynamics study of nearest-neighbor effects on base pair and base pair step conformations and fluctuations in B-DNA. *Nucleic Acids Res.* **2010**, *38*, 299–313. [CrossRef]
27. Pasi, M.; Maddocks, J.H.; Beveridge, D.; Bishop, T.C.; Case, D.A.; Cheatham, T., 3rd; Dans, P.D.; Jayaram, B.; Lankas, F.; Laughton, C.; et al. μABC: A systematic microsecond molecular dynamics study of tetranucleotide sequence effects in B-DNA. *Nucleic Acids Res.* **2014**, *42*, 12272–12283. [CrossRef]
28. Balaceanu, A.; Buitrago, D.; Walther, J.; Hospital, A.; Dans, P.D.; Orozco, M. Modulation of the helical properties of DNA: Next-to-nearest neighbour effects and beyond. *Nucleic Acids Res.* **2019**, *47*, 4418–4430. [CrossRef]
29. Hagerman, P.J. Sequence dependence of the curvature of DNA: A test of the phasing hypothesis. *Biochemistry* **1985**, *24*, 7033–7037. [CrossRef]
30. Hagerman, P.J. Sequence-directed curvature of DNA. *Nature* **1986**, *321*, 449–450. [CrossRef]
31. Koo, H.S.; Wu, H.M.; Crothers, D.M. DNA bending at adenine·thymine tracts. *Nature* **1986**, *320*, 501–506. [CrossRef] [PubMed]
32. Fogg, J.M.; Judge, A.K.; Stricker, E.; Chan, H.L.; Zechiedrich, L. Supercoiling and looping promote DNA base accessibility and coordination among distant sites. *Nat. Commun.* **2021**, *12*, 5683. [CrossRef] [PubMed]
33. Berman, H.M.; Westbrook, J.; Feng, Z.; Gilliland, G.; Bhat, T.N.; Weissig, H.; Shindyalov, I.N.; Bourne, P.E. The Protein Data Bank. *Nucleic Acids Res.* **2000**, *28*, 235–242. [CrossRef] [PubMed]
34. Cheng, H.; Schaeffer, D.; Liao, Y.; Kinch, L.N.; Pei, J.; Shi, S.; Kim, B.H.; Grishin, N.V. ECOD: An evolutionary classification of protein domains. *PLoS Comput. Biol.* **2014**, *10*, e1003926. [CrossRef]
35. Schaeffer, R.D.; Liao, Y.; Cheng, H.; Grishin, N.V. ECOD: New developments in the evolutionary classification of domains. *Nucleic Acids Res.* **2017**, *45*, D296–D302. [CrossRef]
36. Dickerson, R.E.; Bansal, M.; Calladine, C.R.; Diekmann, S.; Hunter, W.N.; Kennard, O.; von Kitzing, E.; Lavery, R.; Nelson, H.C.M.; Olson, W.K.; et al. Definitions and nomenclature of nucleic acid structure parameters. *Nucleic Acids Res.* **1989**, *17*, 1797–1803. [CrossRef]
37. Coleman, B.D.; Olson, W.K.; Swigon, D. Theory of sequence-dependent DNA elasticity. *J. Chem. Phys.* **2003**, *118*, 7127–7140. [CrossRef]
38. Olson, W.K.; Bansal, M.; Burley, S.K.; Dickerson, R.E.; Gerstein, M.; Harvey, S.C.; Heinemann, U.; Lu, X.J.; Neidle, S.; Shakked, Z.; et al. A standard reference frame for the description of nucleic acid base-pair geometry. *J. Mol. Biol.* **2001**, *313*, 229–237. [CrossRef]
39. Olson, W.K.; Colasanti, A.V.; Lu, X.J.; Zhurkin, V.B., Watson–Crick base pairs: Character and recognition. In *Wiley Encyclopedia of Chemical Biology*; John Wiley & Sons, Inc.: New York, NY, USA, 2008. [CrossRef]
40. Olson, W.K.; Esguerra, M.; Xin, Y.; Lu, X.J. New information content in RNA base pairing deduced from quantitative analysis of high-resolution structures. *Methods* **2009**, *47*, 177–186. [CrossRef]
41. Peckham, H.E.; Olson, W.K. Nucleic-acid structural deformability deduced from anisotropic displacement parameters. *Biopolymers* **2011**, *95*, 254–269. [CrossRef]
42. Clauvelin, N.; Olson, W.K. Synergy between protein positioning and DNA elasticity: Energy minimization of protein-decorated DNA minicircles. *J. Phys. Chem. B* **2021**, *125*, 2277–2287. [CrossRef] [PubMed]
43. Young, R.T.; Clauvelin, N.; Olson, W.K. emDNA— A tool for modeling protein-decorated DNA loops and minicircles at the base-pair step level. *J. Mol. Biol.* **2022**, 167558. . [CrossRef] [PubMed]
44. Britton, L.; Olson, W.K.; Tobias, I. Two perspectives on the twist of DNA. *J. Chem. Phys.* **2009**, *131*, 245101. [CrossRef] [PubMed]
45. Clauvelin, N.; Tobias, I.; Olson, W.K. Characterization of the geometry and topology of DNA pictured as a discrete collection of atoms. *J. Chem. Theor. Comp.* **2012**, *8*, 1092–1107. [CrossRef] [PubMed]
46. Manning, G.S. The molecular theory of polyelectrolyte solutions with applications to the electrostatic properties of polynucleotides. *Q. Rev. Biophys.* **1978**, *11*, 179–246. [CrossRef]
47. Kabsch, W.; Sander, C.; Trifonov, E.N. The ten helical twist angles of B-DNA. *Nucleic Acids Res.* **1982**, *10*, 1097–1104. [CrossRef]
48. Rhodes, D.; Klug, A. Helical periodicity of DNA determined by enzyme digestion. *Nature* **1980**, *286*, 573–578. [CrossRef]
49. Wang, J.C. Helical repeat of DNA in solution. *Proc. Natl. Acad. Sci. USA* **1979**, *76*, 200–203. [CrossRef]
50. Peck, L.J.; Wang, J.C. Sequence dependence of the helical repeat of DNA in solution. *Nature* **1981**, *292*, 375–378. [CrossRef]
51. Marini, J.C.; Levene, S.D.; Crothers, D.M.; Englund, P.T. Bent helical structure in kinetoplast DNA. *Proc. Natl. Acad. Sci. USA* **1983**, *80*, 7664–7678; Correction ibid **1983**, *80*, 7678. [CrossRef]
52. Tolstorukov, M.Y.; Colasanti, A.V.; McCandlish, D.; Olson, W.K.; Zhurkin, V.B. A novel roll-and-slide mechanism of DNA folding in chromatin: Implications for nucleosome positioning. *J. Mol. Biol.* **2007**, *371*, 725–738. [CrossRef] [PubMed]

53. Trifonov, E.N.; Sussman, J. The pitch of chromatin DNA is reflected in its nucleotide sequence. *Proc. Natl. Acad. Sci. USA* **1980**, *77*, 3816–3820. [CrossRef] [PubMed]
54. Ulanovsky, L.E.; Trifonov, E.N. Estimation of wedge components in curved DNA. *Nature* **1987**, *326*, 720–722. [CrossRef] [PubMed]
55. Hagerman, P.J. Evidence for the existence of stable curvature of DNA in solution. *Proc. Natl. Acad. Sci. USA* **1984**, *81*, 4632–4636. [CrossRef]
56. Zhurkin, V.B. Sequence-dependent bending of DNA and phasing of nucleosomes. *J. Biomol. Struct. Dyn.* **1985**, *2*, 785–804. [CrossRef]
57. Stefl, R.; Wu, H.; Ravindranathan, S.; Sklenár, V.; Feigon, J. DNA A-tract bending in three dimensions: solving the dA_4T_4 vs. dT_4A_4 conundrum. *Proc. Natl. Acad. Sci. USA* **2004**, *101*, 1177–1182. [CrossRef]
58. Irobalieva, R.N.; Fogg, J.M.; Catanese, D.J.; Sutthibutpong, T.; Chen, M.; Barker, A.K.; Ludtke, S.J.; Harris, S.A.; Schmid, M.F.; Chiu, W.; et al. Structural diversity of supercoiled DNA. *Nat. Commun.* **2015**, *12*, 8440. [CrossRef]
59. Gray, H.B., Jr.; Lu, T. The BAL 31 nucleases (EC 3.1.11). *Methods Mol. Biol.* **1993**, *16*, 231–251.
60. Chaudhry, M.A.; Weinfeld, M. Induction of double-strand breaks by S1 nuclease, mung bean nuclease and nuclease P1 in DNA containing abasic sites and nicks. *Nucleic Acids Res.* **1995**, *23*, 3805–3809. [CrossRef]
61. Grindley, N.D.F.; Whiteson, K.L.; Rice, P.A. Mechanisms of site-specific recombination. *Annu. Rev. Biochem.* **2006**, *75*, 567–605. [CrossRef]
62. Olson, W.K. Simulating DNA at low resolution. *Curr. Opin. Struct. Biol.* **1996**, *6*, 242–256. [CrossRef]
63. Černý, J.; Božíková, P.; Svoboda, J.; Schneider, B. A unified dinucleotide alphabet describing both RNA and DNA structures. *Nucleic Acids Res.* **2020**, *48*, 6367–6381. [CrossRef] [PubMed]
64. Wang, Q.; Irobalieva, R.N.; Chiu, W.; Schmid, M.F.; Fogg, J.M.; Zechiedrich, L.; Pettitt, B.M. Influence of DNA sequence on the structure of minicircles under torsional stress. *Nucleic Acids Res.* **2017**, *45*, 7633–7642. [CrossRef] [PubMed]

life

Article

The Bootstrap Model of Prebiotic Networks of Proteins and Nucleic Acids

Thomas Farquharson [1], Luca Agozzino [2] and Ken Dill [2,3,*]

1 Department of Chemistry, Stony Brook University, Stony Brook, NY 11794, USA; thomas.farquharson@stonybrook.edu
2 Laufer Center for Physical and Quantitative Biology, Stony Brook University, Stony Brook, NY 11794, USA; luca.agozzino@alumni.stonybrook.edu
3 Department of Physics and Astronomy, Stony Brook University, Stony Brook, NY 11794, USA
* Correspondence: dill@laufercenter.org

Abstract: It is not known how life arose from prebiotic physical chemistry. How did fruitful cell-like associations emerge from the two polymer types—informational (nucleic acids, xNAs = DNA or RNA) and functional (proteins)? Our model shows how functional networks could bootstrap from random sequence-independent initial states. For proteins, we adopt the foldamer hypothesis: through persistent nonequilibrium prebiotic syntheses, short random peptides fold and catalyze the elongation of others. The xNAs enter through random binding to the peptides, and all chains can mutate. Chains grow inside colloids that split when they're large, coupling faster growth speeds to bigger populations. Random and useless at first, these folding and binding events grow protein—xNA networks that resemble today's protein–protein networks.

Keywords: origins of life; DNA-protein networks; protocells

Citation: Farquharson, T.; Agozzino, L.; Dill, K. The Bootstrap Model of Prebiotic Networks of Proteins and Nucleic Acids. *Life* **2022**, *12*, 724. https://doi.org/10.3390/life12050724

Academic Editors: Tigran Chalikian and Jens Völker

Received: 31 March 2022
Accepted: 10 May 2022
Published: 12 May 2022

1. Introduction

How did life originate 3.5 billion years ago from the prebiotic world before it? This puzzle is made more challenging by entangling three mysterious complexities together: diverse functional molecules (mostly proteins), molecules that store information and memory (in xNAs, i.e., DNA and RNA) and encapsulation of biomolecules inside cells.

There have been speculations about *what came first?* like the chicken-and-egg problem. Did life start as an RNA world [1–3]? Or, did metabolic reactions precede the enzymes that could catalyze them [4,5]? Or, was encapsulation first in a "Lipid World" [6]?

On the one hand, a Something-Came-First World would certainly be a wonderful convenience for modelers, requiring the fewest assumptions and parameters, at least for that step of early origins. On the other hand, some form of cooperativity must have been crucial to the story of the origins of life. Furthermore, what is convenient for modelers is not necessarily what happened in reality. An alternative view is that biology originated through the co-origination of multiple molecule types together, such as RNA and proteins (along with small molecules) [7–10].

The attractiveness of the multi-molecule world lies in the fact that it does not require an explanatory mechanism for the evolution of another molecule type. Both informational and functional molecules exist and evolve concurrently. The importance and evidence of mutually fruitful interactions between RNA and proteins at life's origin has been recently elucidated [11,12]. Cationic proto-peptides, synthesized under plausible prebiotic conditions, were shown to react directly with RNA to produce mutually stabilizing partnerships: proto-peptides had longer lifespans and RNA duplexes had enhanced thermal stability.

The modeling challenge we take up here is not to seek a simpler problem that avoids the multi-molecule complexity, but rather to confront the more complex challenge of assimilating all three components—function, information and encapsulation—into a single model. We posit a model in which peptides and xNAs are produced and elongated inside protocells. The synthesis of new chains is coupled to the protocell growth rate and protocells grow and split when they become large. The xNAs and peptides can interact. When these biomolecules form complexes that accelerate chain elongation, protocell growth accelerates, leading to increased populations. This process is self-sustaining and grows functional biochemical networks that further increase protocell growth rates.

The present model builds upon the foldamer hypothesis [13]. In short, the hypothesis outlines a dynamical mechanism describing two features: (a) the physical basis for how short chains became longer chains with specific sequences; and (b) a structural and plausible kinetic basis for a prebiotic autocatalytic system. Furthermore, the present modeling is itself characterized as an autocatalytic system. Biomolecules in the system are sustained by what is in the environment and each reaction is catalyzed by a biomolecule type produced inside of the system. Autocatalytic systems are known to be important for life's origin and have been characterized and explored extensively [14]. More specifically, previous work has indicated that autocatalytic sets could spontaneously develop in an RNA–peptide world [15]. We remark here two principal properties which differentiate the present model from previously elucidated autocatalytic sets: (a) catalysis, via the HP foldcat mechanism [13], has a basis in molecular structure and physical chemistry; and (b) the resulting chemical networks that form have topological properties with dependencies like those in today's biology.

2. The Background and the Model

2.1. The Premises and Assumptions

We describe a speculative mechanism for how chain elongations of proteins and xNAs inside protocells otherwise acting stochastically could bootstrap prebiotic chemistry to grow and sustain increasingly complex interaction networks. The premises are found below:

- **A nonequilibrium driver.** Because life is now, and always must have been, out of equilibrium, we are at liberty to suppose some persistent nonequilibrium drive was present. There are many potential sources. Here, we assume the availability of amino acids and nucleic acids, and that both are persistently being polymerized. At first, these would produce only short-chain random sequences of xNAs or peptides. Such plausible syntheses have previously been demonstrated [16–19];
- **A propagation principle.** Today, life as a whole sustains, and never dies out, due to the survival-of-the-fittest propagation principle. Moreover, it is resourceful, creative and innovative, due to its ability to search and choose by mutation and selection. Without it, there is no biology. It results because changes in biomolecules lead to changes in cell growth rates, which lead to changes in cell populations. Here, we assume a simple physical precursor dynamic. We assume peptides and xNAs are encapsulated and polymerize inside colloids or vesicles, causing such protocells to grow and to divide by known surface-to-volume effects [20–23]. We assume, as others have done, that the amino acid and nucleic acid monomers from the surroundings can pass freely into the protocells, but that the chains inside are too long to pass back out [24–26];
- **Funneling in the molecule space**. We believe life originated more as a disorder-to-order process, and less as specific sequence actions or specific binding actions or specific recognition between polymers (such as the genetic code). Rather, we believe such specificity must have emerged from the propagation mechanism (see above) acting on random molecules.

Two of the premises, the propagation principle and funneling in the molecule space, bear a striking resemblance to the idea of reciprocally coupled XNOR gating which allows one to filter and link emergent life properties by interchanging antecedents and consequents in a "strange loop" [27].

A key property of both proteins and xNA molecules is simply the lengths and growth rates of their chains. We assume that nucleic acids polymerize to have chain length distributions that resemble most known polymerization processes [13,28–30], while peptides polymerize into differently shaped chain length distributions by virtue of their ability to collapse in water into compact—sometimes uniquely folded—structures and have sequence-dependent abilities for functionality. We accept that peptides are hydrophobic–polar (HP) polymers given by the previously elucidated foldamer hypothesis [13,31].

We note here that the difference whereby proteins are functional and xNAs are informational must have come early in the origins process; living systems need both functional and informational biomolecules. Proteins can fold so sequence determines structure. In contrast, xNAs are relatively stiff and rigid, so their properties are relatively independent of their sequences. The main distinction here is that functional biomolecule types would have needed to have a strong sequence–structure–function interdependence, whereas the informational biomolecule types required a strong *independence* of sequence to structure and function. Memory storage must be able to store any sequence without bias, so that all sequences, in principle, could be searched and sampled by mutation. To that point, functional biomolecules would therefore make for poor information storage units because different sequences have different physical properties, thus biasing which sequences would be searched and sampled by mutation.

Our present model recognizes the above distinction between molecule types. Foldability is a property of peptides and proteins. Furthermore, while some xNAs chains can also fold, they are stiffer chains, so they fold less frequently and without the sequence–structure relationship. However, we acknowledge that certain xNA sequences can fold, especially RNA sequences, and are capable of functional activity (i.e., ribozymes). However, at the current level of granularity of the present model, we do not expect the presence of ribozymes to dominate or change the observed conclusions. Ribozymes play a small role in contemporary biology; thus, we suppose they play a role similar in magnitude here.

2.2. The Growth and Split Mechanism

The main property that we require here is the length distributions of xNA chains and HP peptides. We consider the synthesis rate of a peptide of length L: If $k(L)$ is the rate at which a molecule of length L is produced it is therefore simply inversely proportional to the length itself. If k_X is the rate of elongation for polymers of type X, we have $k(L) = k_X/L$; this is simply obtained by assuming a constant time interval between each monomer addition. Now we consider two different elongation rates for the informational molecule, k_I, and for the functional molecules, k_F. We can now write the rate equations for the two types of molecules as:

$$\frac{dm_\mu}{dt} = \frac{k_I}{L_\mu}, \tag{1}$$

$$\frac{dn_j}{dt} = \frac{k_F}{L_j}, \tag{2}$$

where m_μ and n_j are the copy numbers for each different type of NA chain of type μ and HP chain of type j, respectively. Here, and in the rest of this paper, we will use Greek indices for I-type molecules and Latin indices for F-type ones. So $n_j(t)$ represents the number of functional molecules inside the droplet of type j, hence with length L_j, at time t, and similarly for $m_\mu(t)$. In this mechanism, the sole factor that determines the growth rates of the vesicle/droplet/protocell is simply how fast the nucleic acids and amino acid chains are elongating inside it. A visual schematic for protocell growth and splitting, due to chain elongation, is shown in Figure 1.

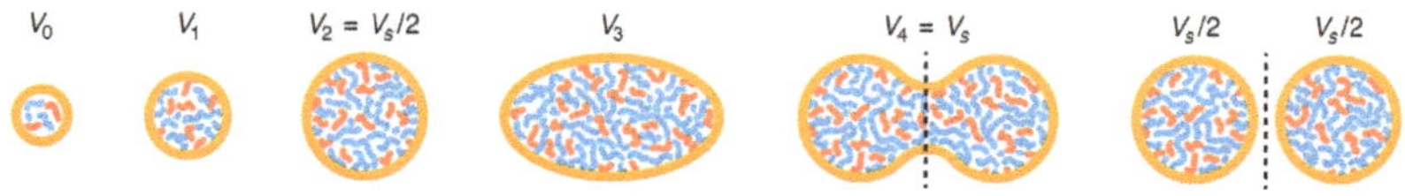

Figure 1. Protocells grow (and split) through chain elongation of its polymer inside. (Blue) Peptides as functional molecules. (Red) Nucleic acids as informational polymers. They grow in mass and length through synthesis. (Orange) The surface-to-volume ratio of the protocell decreases as new polymer chains are elongated inside. Subsequently, the protocell splits to produce two "daughter" protocells.

In water, nonpolar matter forms droplets due to the oil–water forces [32,33]. When droplets grow big, they split into two. We define V_s to be the average droplet volume at the splitting point. We define T_s to be the average time required to reach the splitting volume. Now, taking v_X to be the volume increase due to the addition of a monomer to the molecule type X, and N and M to be the total number of different types of peptides and xNAs, respectively, we get:

$$V_s = \sum_{j=1}^{N} n_j(T_s) v_F L_j + \sum_{\mu=1}^{M} m_\mu(T_s) v_I L_\mu. \tag{3}$$

The time scale for the elongation of a given molecule is faster than the protocell growth rate. Therefore, we assume that molecules which are incompletely synthesized (i.e., molecules shorter than the mature length of either L_j and L_μ) do not have an impact on the overall growth. Now we can solve Equations (1) and (2); and place them into Equation (3):

$$n_j(t) = n_j^0 + \frac{k_F}{L_j} t, \tag{4}$$

$$m_\mu(t) = m_\mu^0 + \frac{k_I}{L_\mu} t, \tag{5}$$

where $n_j^0 \equiv n_j(0)$ and $m_\mu^0 \equiv m_\mu(0)$. Now we take the intial time to be one of the splitting events, when the volume of the droplet is exactly $V_0 = V_s/2$, and calculate the time at which the total population is sufficient to double this volume. Therefore, we have:

$$2V_0 = V_0 + (v_F k_F N + v_I k_I M) T_s, \tag{6}$$

where $V_0 = \sum_{j=1}^{N} n_j^0 v_F L_j + \sum_{\mu=1}^{M} m_\mu^0 v_I L_\mu$ is the reference volume; it is the volume of the protocell right after a splitting event. From this expression we can calculate what effectively is the growth rate of a protocell with a given composition of initial populations n_j^0 and m_μ^0 as the inverse of the splitting time:

$$r_0 = \frac{1}{T_s} = \frac{v_F k_F N + v_I k_I M}{\sum_{j=1}^{N} n_j^0 v_F L_j + \sum_{\mu=1}^{M} m_\mu^0 v_I L_\mu}, \tag{7}$$

This is the reference growth rate for a system of protocells whose growing mechanism is controlled solely by chain elongation. Parameters used to calculate the reference growth rate are found in Table A1, Equations (A9)–(A11).

2.3. *Intermolecular Interactions Drive Network Formation*

In this model, proteins and xNA molecules in water interact through hydrophobic/polar interactions. xNAs are known to act not only by hydrogen bonding base pairing, but also by hydrophobic base stacking [34]. Here, when a hydrophobic chain monomer is exposed to water, it attracts other exposed hydrophobic monomers from either type of chain. We note that this binary interaction is just a simplification for the present modeling. Since many of the 20 amino acids were likely present at the origin of life, the full complexity of catalytic activities and binding interactions could have begun early. For our simple model here, our HP coding is a stand-in for how these early simple polymers could fold, recognize, bind, catalyze and react with one another.

Through this interaction mechanism, some random chains will associate with each other. This means that interaction networks can form. Here, we call this a protein–informational interaction (PII) network: every node represents one of the two molecule types. A link between two nodes corresponds to some form of interaction between the two corresponding molecules. A link can exist both between molecules of the same type (i.e., $i - j$ or $\mu - \nu$ for protein–protein and information–information molecules, respectively) or between molecules of different type (i.e., $j - \mu$ for protein–information interactions). Interactions can be of various types. While one outcome is aggregation (nonspecific interactions), another outcome is protein machines that have some primitive functional activity analogous to more contemporary enzymatic functions. The focus of this study is to explore a primitive form of functional interaction, and aggregation is only treated as an average.

Here we single out for special focus those proteins that are information copy machines (polymerase-like proteins) and those that are information-to-protein copy machines (ribosome-like proteins). We call them xNA copiers and protein copiers, respectively. Visual representations of the network's nodes and subgraphs can be seen in Figure 2. Protein copiers are 3-molecule subgraphs; a protein that reads an xNA and produces another protein. xNA copiers are 2-molecule subgraphs; a protein that approximately duplicates an I-molecule. When such machines are catalytically active inside the protocell, they can increase its growth rate. Copy machines are considered "catalytically active" when the relevant completed replication or translation subgraph has been formed in the PII network. The discovery of new interactions is a consequence of changes in the sequence structure of a molecule type during foldamer-catalyzed elongation. We refer to these changes as "mutations." Below we describe the term on a granular level and then show how it is represented in the present coarse-grain model.

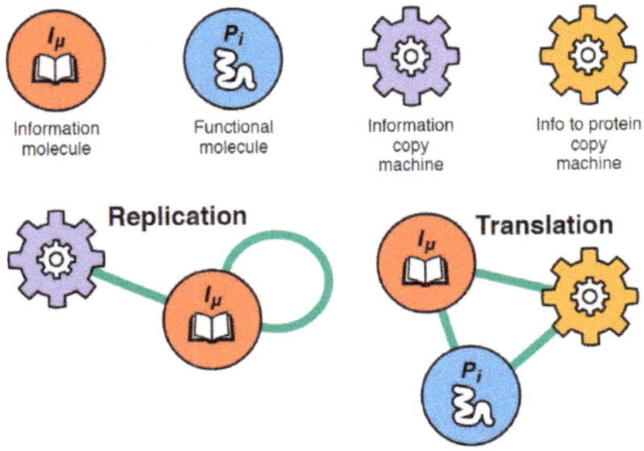

Figure 2. Defining the symbols representing the model's PII nodes and important subgraphs. (Top row, left to right) Four major types of molecules exist in the PII interaction network: xNA molecules, protein molecules, ribosome-like proteins, and polymerase-like proteins. μ and i are iterators used to further distinguish between the different types of xNAs and proteins in the network (i.e., molecules which differ in sequence structure). Nodes represent the entire molecular population of a given type of molecule. (Bottom row, left to right) The two important types of subgraphs which denote primitive replicative function in the model: xNA copier subgraph and protein copier subgraph (see text).

In this present model, mutations are a consequence of life's origin lacking specific binding actions or specific molecular recognition. We accept the foldamer hypothesis which posits that foldamers in autocatalytic sets can cross-catalyze foldamers of a different molecular sequence or can catalyze the elongation of a foldamer of the same sequence [13]. However, without molecular specificity, these elongation processes could have resulted in foldamer variants being synthesized. If we consider the following scenario: (a) foldamers of type A are responsible for catalyzing the elongation of a protein/xNA molecule of type B; (b) a mutation in the molecular space causes the synthesis of foldamers of type A to be polymerized as variants, A^V; (c) foldamers of type A^V now catalyze a new variant of protein/xNA molecule of type B, hence B^V; and (d) B-type molecules are no longer synthesized, the population now reflects molecules of type B^V; we can see how mutations provide new sequence variations to molecule types existing in the PII network.

The consequence of the mutation is what is simulated in the present model. When new molecular variants are synthesized, they will either gain or lose an interaction with a pre-existing molecule type accounted for in the model. The mutational effects are subtle; only a single interaction can be lost or gained at a time, per a single mutation. Mutations were random events. A randomly selected A_{ij} matrix element was selected and changed to its opposite value: $1 \rightarrow 0$ or $0 \rightarrow 1$. Additionally, we did not track which of the two molecule types was the new variant, simply we simulate whether the mutation led to the loss or discovery of a molecular interaction between the two molecule types.

Most of the copy machines within the molecular population will not be functional; we represent this by an effectiveness parameter (α and β below). Effectiveness represents the catalytic accuracy of a population of a given copy machine type. It is assigned via a randomized process; an integer between 0 and 1 is chosen at random and assigned to a machine type. Larger effectiveness values (i.e., close to 1) indicate that the copy machines are efficient in synthesizing new polymers with little error in the sequence. Lower effectiveness values (close to 0) indicate that copy machines were error prone; only a few machines produce polymers having the intended sequence structure. When machines are effective, they can boost the production of polymers. When an F-molecule is interacting with a protein copier, its rate of production is subject to the effectiveness parameter. So, now the chain elongation rate is:

$$\frac{dn_j}{dt} = \frac{k_F}{L_j} + \alpha \sum\nolimits_{k\mu} A_{jk} A_{k\mu} n_j n_k m_\mu \delta(L_j, L_\mu), \tag{8}$$

where α is a parameter that measures how effective the protein copier is, and A_{jk} and $A_{k\mu}$ are the elements of the adjacency matrix of the network. These elements are 1 if there is a link between the two index nodes or 0 otherwise, and the Kronecker delta function only enforces that the transfer of information is possible if the lengths are the same. Similarly, when an xNA copier interacts with an I-molecule the elongation rate is given by:

$$\frac{dm_\mu}{dt} = \frac{k_I}{L_\mu} + \beta \sum\nolimits_{k} A_{\mu\mu} A_{\mu k} n_k m_\mu^2, \tag{9}$$

where β is the effectiveness parameter of the xNA copier.

Protein Copiers as Peptides

We remark here that the protein copy machines in the present model are peptides, which goes against the known structure of contemporary ribosomes [35] and the supposed structures of primordial ribosomes (primarily RNA) [36,37]. We make three points here to justify our treatment of protein copy machines below:

- The RNA fraction in contemporary ribosomes ranges from 1/3rd–2/3rd in different organisms [38,39], indicating that the requirement of RNA as part of the machine is not a strong constraint;
- Contemporary organisms are known to have some nonribosomal peptide syntheses which are facilitated by other protein structures (i.e., nonribosomal peptide synthetase) [40]. As far as we know, there are no known equivalents for xNAs being duplicated by solely other xNAs;

- At the current level of coarseness of the present modeling, we simply approximate ribosomes as being catalytic elongators. Therefore, the network structure would not differ much from the observed results.

2.4. Computing the Growth Dynamics

Now, with these growth laws, we can determine the time the protocell would take to reach its splitting volume. These coupled differential equations can be solved numerically. However, it is possible to solve in the case of a single graph of the types in Figure 2 and then extrapolate the results in the case of many graphs of such type. In Equation (A5) we show that the overall protocell growth rate can be written as:

$$r \approx r_0 + \Delta r, \tag{10}$$

where Δr is a function of the topology of the interaction network. The process we now model is that of networks that change through mutations of the molecular sequences, leading to appearances or disappearances of links in the PII network. Mutations are random and can occur either in peptides or xNAs.

2.5. Mutations Drive the Network to Discover New Functional Relations, Affecting the Protocell's Growth Rate

When the consequences of a mutation are modeled through the PII network (i.e., the appearance or loss of an interaction), the system's discovery of a new interaction can lead to an increase in the growth rate of the protocell. Specifically, if the mutation leads to the formation of the appropriate subgraph necessary to represent xNA or protein copier function, the growth rate of the mutant-type protocell increases. Mutant-type protocells which discover these copy machine functions have enhanced growth rates. The new growth rate of the mutant is given by:

$$r \approx r_0 + \Delta r_R + \Delta r_C, \tag{11}$$

where Δr_R is the change in the growth rate due to the introduction of a protein copier, whereas Δr_C is that due to the introduction of a xNA copier. Details of their expressions are given in Appendix A Equations (A1)–(A6). The growth rate r represents the growth rate of the protocell system where foldamer, xNA copier and protein copier catalysis all contribute to polymer elongation.

2.6. Polymer Aggregation Decrease Proto-Cellular Growth Rate

Polymer aggregation is a consequence of promiscuous interactions. When polymers aggregate, we predict there is a decrease in the growth rate. Aggregation removes polymers which are participating in replication or translation subgraphs and those used as templates in foldamer catalyzed elongation reactions. Consequently, the growth rate contributions from foldamer catalysis, protein copiers and xNA copiers would be less than their idealized calculated values. To reflect this feature, we include an aggregation cost:

$$r = (r_0 + \Delta r_R + \Delta r_C) - \textstyle\sum_k g(k), \tag{12}$$

where $g(k)$ is the aggregation cost for each polymer type summed across all polymer types k in the model. The aggregation cost is given by:

$$g(k) = \begin{cases} 0, & d(k) < D \\ \delta\, d(k), & d(k) \geq D \end{cases}, \tag{13}$$

where the aggregation cost for a polymer type is zero if its number of links $d(k)$ is less than the threshold aggregation parameter, D; or its aggregation cost is given by $\delta\, d(k)$ if the number of links to the polymer type exceeds or equals the threshold parameter. δ is a static scaling parameter used to calibrate the aggregation cost to the magnitude of the growth rate. Parameters were set to $D = 5$ and $\delta = 0.005$ to reflect the network size used and the magnitude of the reference growth rate.

2.7. Mutations Can Be Advantageous or Noise

The present model predicts that protocell populations evolve through two mechanisms which resemble natural selection and genetic drift: some changes in the distribution of polymers in a protocell have a relevant effect on the overall duplication rate, resulting in protocells with a higher chance to become common in a population; other changes have minimal to no effect, increasing the diversity of polymers distribution.

When an individual protocell, existing in a system of protocells, undergoes a mutation, there is a change in the sequence of one of the polymer types in its interaction network. There are three possible outcomes. First, if the mutation results in the network discovering some activity—the completion of either a xNA copier or protein copier subgraph—the cell growth rate and fitness increase. As the protocell population grows and reproduces, lineages from the mutant protocell have greater reproductive success than wild-type protocells. Consequently, each generation of protocells progressively look more and more like the mutant than the wild-type. In this way, beneficial mutations ultimately become fixed within the population. Second, other mutations can be deleterious, decreasing the growth rates of those protocells. Or, third, a mutation can be neutral, having no effect.

2.8. Mutations of the Individual Cells Propagate through the Population

In order to determine how likely a new mutation is to be selected by evolution, hence fixed in the population, it is necessary to consider a model of natural selection. The probability of fixation in such cases can be assumed to be given by Motoo Kimura's expression [41], which simply expresses the probability that a given mutation with some selective advantage will ultimately be present in the entire population:

$$\mu = \frac{1 - e^{-2s}}{1 - e^{-4Ns}}, \tag{14}$$

$$s = \log r_{\mathrm{mut}} - \log r_{\mathrm{wt}}, \tag{15}$$

where s is the change in fitness due to a mutation, assuming that fitness is given by the log of the growth rate. N is the size of the protocell system and a parameter in the simulation. Simulated evolution trajectories used $N = 100{,}000$ protocells. Neutral mutations are fixed with a probability of [41]:

$$\mu = \frac{1}{2N}. \tag{16}$$

2.9. Computer Simulations of the Model

The initial wild-type PII was a randomly generated A_{ij} symmetric adjacency matrix with a size given by $N + M$. Network sparsity was determined by an adjustable probability whereby a given A_{ij} matrix element is assigned a zero or a one. The adjacency matrix is mapped into the corresponding adjacency graph, which is the first wild-type PII interaction network for a protocell in a system of identical wild-type protocells; see Figure 3. The nodes on that graph represent different types of functional polymers (blue), informational polymers (red), a xNA copier (purple) and a protein copier (yellow). Each type of polymer has an initial population size and an assigned length. A link (edge) in the graph indicates an intermolecular interaction between polymers of either the same type (self-loop) or different types.

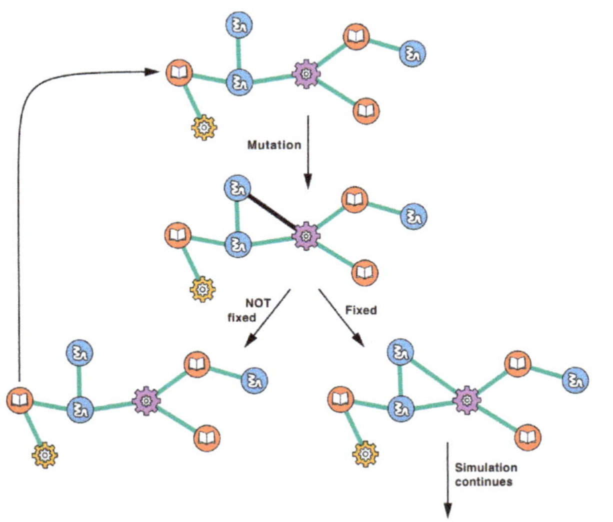

Figure 3. The available actions at each time step. (Top) A mutation can add a link (black line) or remove one. (Middle) That link either becomes fixed into the network (bottom right, green link) or not fixed (bottom left). The network now proceeds to the next time step.

The growth rate of a protocell with the given wild-type PII network was calculated using Equation (7). The first mutation is then introduced into the system and a link/interaction is either lost or discovered. Here the growth rate of a protocell with the mutant PII network is calculated using Equation (12). The log of the growth rate for both the wild-type network and mutant network are taken and further evaluated using Equation (15) to give the selective advantage, s. If $s = 0$, then the fixation of the mutation in the population is driven by genetic drift. If $s \neq 0$, fixation is driven by natural selection. If it is not fixed by either evolution force, then the wild-type network "wins out" over the mutant network (left path in Figure 3). The mutant protocell's mutation falls out of the population after multiple generations of growth and selection with the wild-type PII network being the only available alternative. The process then repeats again with the wild-type network. If the mutation is fixed by either evolution force, then the mutant network becomes the dominant network-type in the population (right path in Figure 3) after multiple generations of growth and selection. The previous wild-type network is lost, and the mutant becomes the wild-type network. The process then repeats until a preset number of mutations have been introduced. Seven evolution trajectories were simulated for a protocell system containing 100,000 individuals. For each simulation 1,000,000 mutations were introduced into the system and the simulation ended at the 1,000,000th mutation.

3. Results and Discussion

3.1. When a Network Discovers Complete Copier Subgraphs, Its Protocell Grows Faster

When the above processes are modeled, the model predicts survival-of-the-fittest behavior. Figure 4 shows a time graph simulating introduced mutations in a protocell population. When the network of a mutant protocell discovers a beneficial mutation (i.e., the discovery of either complete copy machine subgraph), it wins out against the alternative wild-type in the population. This feature is highlighted by the ever-increasing growth rate of the population. Protocell generations resemble parents that had discovered additional ways to elongate their polymer chains. We can relate this to a simple tournament bracket analogy. When we compare the growth rate of a wild-type and mutant-type protocell, the one with the higher growth rate will be nonrandomly selected for and its lineage continues onward to become the new wild-type. Another mutation arises in the population and the new mutant and wild-type are compared, with nonrandom selection once again favoring the protocell with the higher growth rate. This cycle repeats with the result being a maximization of proto-cellular fitness.

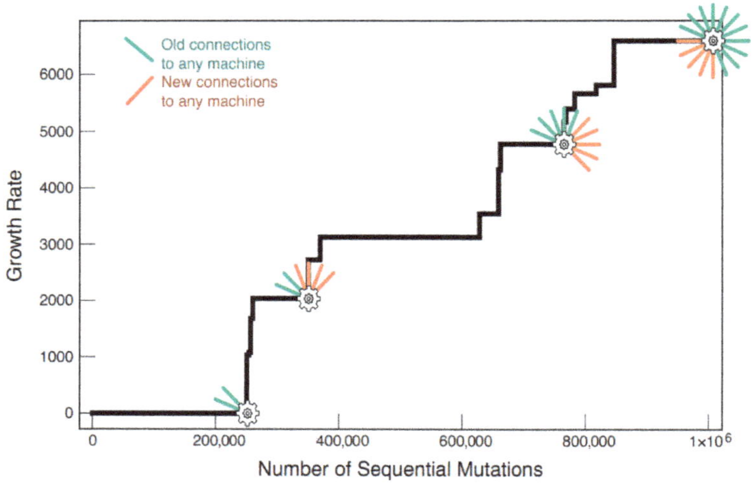

Figure 4. A protocell's growth rate ratchets up over time as random mutations in the network happen to discover and lock in the specific interactions that specify the subnet machines in Figure 2.

We note here that periods of no change in the growth rate do no imply stagnation in the evolution of the protocell population. Mutants with neutral mutations can still win out over the wild-type variant, but this due simply to chance. Given the network size chosen in our simulations, on average, <0.005% of all mutations out of 1,000,000 were beneficial. The remaining 95.995% of mutations were neutral. The frequency by which neutral mutations occurred does not diminish their importance. Neutral mutations are important for discovering the requisite interactions for completion of either copy machine subgraph.

As a general principle of the model, novel protein copier links are harder to discover and subsequently fix compared to links discovered for the xNA copier. The reason is two-fold: the subgraph depiction for protein copier function in this model requires one additional link than that of xNA copier function (Figure 2), and there is a length requirement for primitive translation between the functional molecule and the informational molecule. To maintain simplicity in the model we have assumed that the earliest form of the genetic code had a one-to-one correspondence between an amino acid and a nucleic acid. In essence, the functional molecule and the informational molecule in the translation subgraph must be the same length. The addition of the length requirement increases the time it takes for the first translation subgraphs to appear in the network. However, once a few of them have been established, subsequent interactions among participating functional and informational molecules in the established subgraphs become more facile.

Another key feature of the bootstrap model is that it simulates a form of cooperativity that is known to occur in today's cellular protein–protein interaction (PPI) networks [42]. In short, bigger subgraphs in PPI networks have higher probability of forming an added link than smaller subgraphs have. In the present model, this applies to the two types of subgraphs: the 2-link xNA copier (transcription) and the 3-link protein copier (translation). When a 2-link translation or 1-link transcription subgraph is present, the probability that the subgraphs will grow into their respective 3-link and 2-link subgraphs is enhanced. An interaction which completes a copier subgraph brings with it an increase in growth rate. Copier subgraphs which have some, but not all, of the requisite interactions for completion bootstrap the formation and fixation of the remaining interaction(s) which complete it.

This form of cooperativity is exemplified with the protein copier. When a few of the 3-link translation subgraphs have already been discovered, the subsequent discovery for more is enhanced. Consider the case shown in Figure 5.

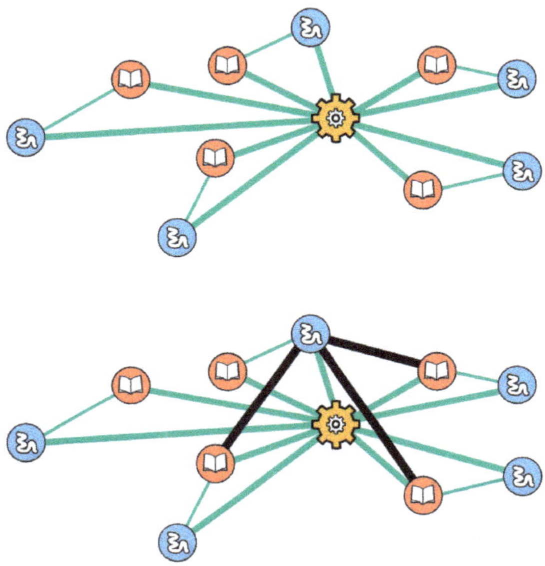

Figure 5. Cooperativity: when a network has at least two protein copier subgraphs, subsequent formation of more becomes more facile; see text.

Functional and informational molecules participating in complete subgraphs have pre-existing interactions with the protein copier. A new interaction which arises between a functional and informational polymer on two different 3-link subgraphs can lead to the immediate formation of another 3-link translation subgraph. Consequently, there is an increase in growth rate. This observed cooperativity suggests that interactions between functional and informational polymers in different protein copier subgraphs are more facile than similar interactions elsewhere in the network.

3.2. Networks Evolve to Become Bigger and More Complex

Figure 6 shows an example trajectory of an evolving PII network. It grows in nodes and edges. The network begins with only a few interactions. The average number of initial, randomly generated interactions in the starting network was 64 ± 6. In a network size of 50 different polymer types, split evenly between functional and information polymers, the average interaction per polymer was 2 ± 2. After 1,000,000 mutations, where some were fixed or lost, through processes resembling natural selection and genetic drift, the final network averaged 160 ± 16 total interactions. Here, each polymer type had on average 6 ± 3 interactions. Novel interactions between a polymer and either copy machine made up a fifth of all newly discovered interactions. Calculations regarding the number of interactions per polymer and the total initial and final network sizes were computed averages taken from the results of seven simulations.

The giant component of the network starts out sparse, with only a few interactions existing between polymer types. A component is defined as a group of nodes which are connected either indirectly or directly. Therefore, we define the giant component as the network component with the larger proportion of polymer types in it [43]. At time t = 0, when no mutations have been introduced into the system, a protocell's growth rate is dictated solely by the chain elongation processes that occur from foldamer catalysis. To reflect this, our initial networks did not contain any copy machine subgraphs. Networks also started with a slight degree of fragmentation. Some polymer types were not connected with the giant component either through direct or indirect interaction. These polymer types became connected later in evolution as new interactions were discovered.

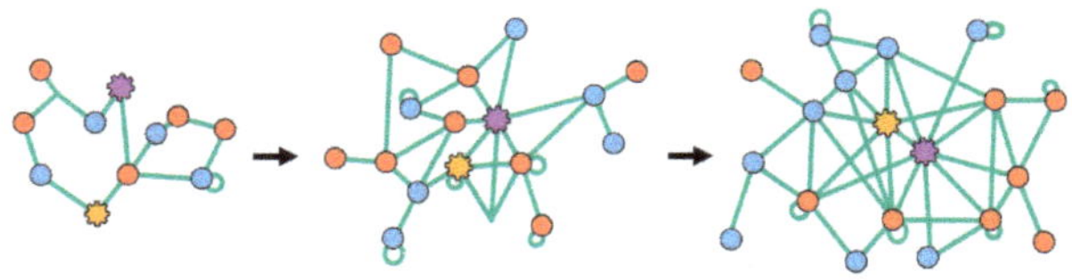

Figure 6. Example of a one-time evolution trajectory of a network. Initial network (left) is small and sparsely linked. Later networks have grown and changed through mutations discovered and lost, increasing the network's complexity and size.

Mutations and selection result in an increase of network complexity. Interactions that are growth-rate neutral are observed most frequently. Protein copier and xNA copiers also discover fruitful interactions. The completion of a translation or transcription sub-graphs provides substantial growth rate increases. Networks continue to grow in size as nonconnected polymers discover interactions with the giant component.

3.3. Bootstrap Model Network Topologies Resemble Today's PPI Networks

The bootstrap model predicts how simple initial networks grow into more complex structures later in evolution. The structure and complexities of networks can be characterized by their topological features. Three are considered here: degree centrality, betweenness centrality and closeness centrality. The mathematical definition for all three centralities can be found in Appendix B, either written in text or shown in Equations (A7) and (A8). Figure 7 shows that these features predicted from the bootstrap model resemble the corresponding features of protein–protein interaction networks in present-day cells [42]. This comparison is made by comparing the present model's topological features in a dynamic setting to those of static, fully evolved and simulated PPI network topologies. An important distinction we make here is that while the size of our PII network does not allow for a direct one-to-one comparison with known PPI networks, irrespective of that, the observed topologies of each share similar dependencies.

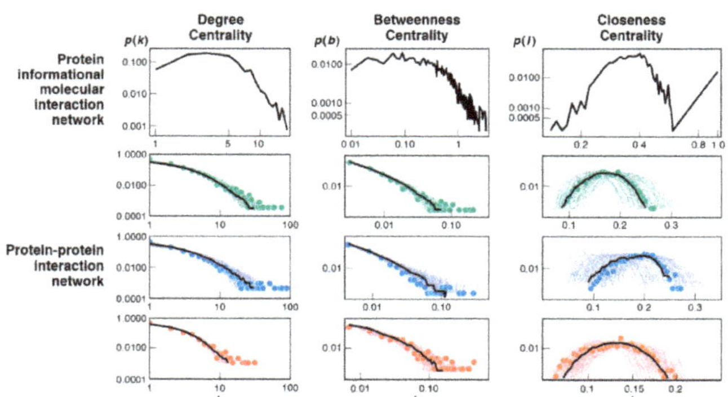

Figure 7. Topological node distributions of the bootstrap model networks resemble today's cell PPI networks. (Top row) Predicted from this bootstrap model for protein–informational interactions (PPI). (Bottom) Measured protein–protein interaction (PPI) networks of humans (green), yeast (blue) and fruit flies (red); adapted with permission from [42], © 2012, Peterson et al. Since the size of the present simulation is far from a real-world PPI, direct comparison cannot be drawn; however the general behavior of these topological features shows that the present bootstrap mechanism gives a plausible evolutionary route to current cellular networks; see text.

Here are the interpretations. First, if networks had many hubs—like many big cities in a traffic network—then the degree centrality plot would show large $p(k)$ values at large values of k on these figures. However, that is not the case either from the bootstrap model or the experimental PPI data. Most proteins are connected to relatively few other proteins. There are very few hubs; they are mostly copy machines. Second, the between centrality reflects the number of bridges in the network. These are situations in which one molecule is a go-between linking two other molecules. In both the model and the PPI data, few molecules are bridging any two other molecules. In the model, the bridging molecules are largely the copy machines. Third, the closeness centrality shows the number of molecules that are either highly centralized (close to many other proteins) or highly decentralized, far away from other proteins. It measures the extent to which a molecule can interact with all other molecules in the network. The peaks in these plots indicate that most molecules are neither particularly isolated from others, nor crowded together with others. Our copy machines have high closeness values because of their hub-like nature. Direct interactions lead to many indirect interactions. A given molecule can make interactions with all other molecules without specificity in this model. Molecules that are nodal neighbors to the copy machines drive copy machines to become more centralized when they discover interactions to periphery molecules in the network.

Lastly, we briefly comment on the nature of other types of interactions networks, mainly DNA–protein networks, and RNA–protein networks. For brevity, we classify these as xNA–protein networks. At present, sufficient data is not available to compare xNA–protein network topology with the PII network. While recent efforts have been made to elucidate xNA–protein interactions [44–47], the databases housing the data do not provide graphical construction of the full interactome. Over the last decade, there has been a standardization in the data format and quality of PPI network data, but such standards do not exist for xNA–protein networks. Nevertheless, we are aware of a topological analysis on one graphical dataset [48] of a noncoding RNA–protein interaction network in yeast, but it only includes a small subset of interactions and not the full topological comparisons we seek here.

4. Conclusions

We develop here the bootstrap model for how proteins and nucleic acids might have evolved fruitful relationships in the origins of life. It is based on premises that we regard as plausible physical chemistry and maximal initial randomness. It supposes that xNA and protein molecules occupy vesicles. Since life requires nonequilibrium, our NEQ premise is the availability of persistent random short-chain syntheses of both polymers. Since life cannot exist without some form of survival-of-the-fittest propagation dynamic, we assume protocell colloids grow from the growing chains inside, and split, converting cell growth rates to cell populations. We suppose that the peptides are hydrophobic–polar (HP) polymers, and accept the previously elucidated foldamer hypothesis, wherein short HP peptides collapse hydrophobically in water, expose hydrophobic binding sites, and could, in principle, accelerate chain elongations with primitive ribosome-like and polymerase-like functionality. Random mutations can lead to growth advantages, spontaneous propagation and biochemical networks that have growing complexity. The biochemical network topologies predicted by the bootstrap model resemble those of today's PPI networks in living cells.

Author Contributions: Conceptualization, L.A. and K.D.; formal analysis, T.F.; methodology, T.F.; software, T.F.; supervision, K.D.; writing—original draft, T.F.; writing—review and editing, L.A. and K.D. All authors have read and agreed to the published version of the manuscript.

Funding: This research received no external funding and was supported by the Laufer Center, Stony Brook University.

Institutional Review Board Statement: Not applicable.

Informed Consent Statement: Not applicable.

Data Availability Statement: The computational code used for this model is available from the corresponding author on reasonable request.

Acknowledgments: We dedicate this paper to Ken Breslauer in this Festschrift in honor of his 75th birthday. Through many years together, KD has long admired Ken Breslauer's contributions to calorimetry and biomolecular thermodynamics, his personal warmth and charm, and his extraordinary savvy and scientific leadership.

Conflicts of Interest: The authors declare no conflict of interest.

Appendix A

Here we give an approximate solution to the growth rate Equations (8) and (9). The full solution of these coupled differential equations is not readily obtainable, at least by us. However, it is possible to solve them for each single graph of the types in Figure 2 and we can then extrapolate the results in the case of many graphs of such type.

For a single graph of the functional type, the amount of protein p would be given by the following:

$$n_p(t) = n_p^0 \exp\left(\int_0^t F(t\prime)dt\prime\right) + \frac{k_F}{L_p}t, \tag{A1}$$

where $F(t) = \alpha\left(n_r^0 m_\mu^0 + \left(k_F m_\mu^0 + k_I n_r^0\right)t + k_F k_I t^2\right)$. Now we assume that the splitting time τ_s is small enough that a first order approximation is valid. The result can be shown to be:

$$\tau_s \approx \frac{T_s}{1 + \alpha\kappa_p n_r^0 m_\mu^0}, \tag{A2}$$

$$\kappa_p = \frac{v_F n_p^0 L_p}{v_F k_F N + v_I k_I M}. \tag{A3}$$

Since the second term in the numerator $\alpha\kappa_p n_r^0 m_\mu^0 > 0$ we see how the new splitting time is always smaller than the splitting time in absence of such machine T_s. This testifies that a single machine of this type is sufficient to boost the growth rate and hence the fitness of the protocell. We now extrapolate this result, assuming that this linear approximation holds on this case as well (true for sufficient small times) and we obtain the following expression for the growth rate of the entire system:

$$r \approx r_0 + \sum\nolimits_{p\mu r} A_{pr}A_{r\mu}\frac{\alpha r_0 L_p}{k_F N + k_I M v_I / v_F} n_p^0 m_\mu^0 n_r^0, \tag{A4}$$

where $\sum_{p\mu r} A_{pr}A_{r\mu}$ indicates the sum over all graphs of the functional type. Similarly for the xNA copiers, we can obtain an expression for the modified growth rate by solving the corresponding differential equation for a single graph and then extrapolating to an arbitrary number of graphs. The growth rate r summed over all possible graphs is:

$$\begin{aligned} r \approx r_0 + \sum_{p\mu r} A_{pr}A_{r\mu}\tfrac{\alpha r_0 L_p}{k_F N + k_I M v_I / v_F} n_p^0 m_\mu^0 n_r^0 \\ \rightarrow + \tfrac{v_I}{v_0}\sum\nolimits_{pv} A_{vv}A_{vp}\beta L_v \left(m_v^0\right)^2 n_p^0, \end{aligned} \tag{A5}$$

$$\equiv r_0 + \Delta r_R + \Delta r_C, \tag{A6}$$

where $\sum_{pv} A_{vv}A_{vp}$ indexes all graphs of the xNA type. A convenient definition of fitness is the logarithm of the growth rate. Hence the rate can be decomposed into a component due to protein copiers, Δr_R and a component due to xNA copiers, Δr_C.

Appendix B

The degree centrality is the measure of links, k coming off a given node. The betweenness centrality is calculated using:

$$b = \sum \frac{\sigma_{st}(v)}{\sigma_{st}}, \tag{A7}$$

where σ_{st} represents the totality of shortest path routes extending from arbitrary node s to node t. The term of $\sigma_{st}(v)$ represents the number of paths which pass through vertex v. The closeness centrality is calculated using:

$$l = \frac{N-1}{\sum_s d(t,s)}, \tag{A8}$$

where $s \neq t$, $d(t,s)$ is the length of the shortest path between nodes t and s in the network and N is the total number of nodes.

Reported graphs for degree, betweenness and closeness represent the dynamic topologies for the network as it evolves over time. Topological values are recorded for a network every 10,000 mutations. These values are compiled at the end of the simulation and a probability distribution for each topological feature is generated. The generated graphs are the median values of the probability distribution for each given value of k, b and l across all simulations.

Reported closeness values of $l = 0$ are functions of randomized network generation and fragmentation. Occassionally some nodes in the graph at $t = 0$ form a subgraph that is disconnected from the giant component. Due to the very small size of the disconnected subgraph, nodes will report high values for l because a centralized node might be connected directly to every node in the outlying subgraph.

Appendix C

The magnitude of the reference growth rate was dependent on both static and dynamic parameters. Static parameters were not changed between simulations. These parameters included: N, M, v_F, v_I, k_F, and k_I. The preset values for each of these parameters can be seen in Table A1.

Table A1. Static parameters for the reference growth rate.

	N	M	v_F	v_I	k_F	k_I
Static Presets	25	25	0.143	0.303	10	10

Total polymer types for proteins and xNAs were limited to 25. This preset was selected as it offered good predictive power in relation to computational simulation time. Average volumes were computed from published literature volumes [49,50] and converted from A^3 to nm^3 to ensure poper parametric scaling. We assume that the polymer elongation rate for proteins and xNAs are equivalent, regardless of the mechanism behind polymerization. The preset for elongation rates of proteins and xNAs had been kept at 10 to keep the reference growth rate sufficiently small.

Dynamic parameters included: n_j^0, L_j m_μ^0, and L_μ. These parameters were reset and reassigned at the beginning of each new protocell evolution trajectory. Lengths were assigned to each type of protein and xNA molecule using an exponential length distribution function adapted from the modeled data in the foldamer hypothesis [13]:

$$d(l) = 0.067e^{-0.106x}, \tag{A9}$$

where x were chain lengths ranging from 10 monomeric units to 50 monomeric units. Subsequently, a list of 1000 pseudorandom variates were generated from the distribution and randomly assigned to each type of protein and xNA molecule in the model. The initial population size for each protein type and xNA type in the model were calculated using:

$$n_j^0 = \frac{0.067e^{-0.106(L_j)}}{0.067e^{-0.106(length\ of\ shortest\ protein)}}(n_j^0\ of\ shortest\ protein), \tag{A10}$$

$$m_\mu^0 = \frac{0.067e^{-0.106(L_\mu)}}{0.067e^{-0.106(length\ of\ shortest\ xNA)}}(m_\mu^0\ of\ shortest\ xNA). \tag{A11}$$

The initial population sizes of the shortest length protein type and xNA type were set to 1000 for every simulation. Lengths of L_j and L_μ in Equations (A10) and (A11) corresponded to the length of the biomolecule for which the initial population size was being calculated for.

References

1. Gilbert, W. Origin of life: The RNA world. *Nature* **1986**, *319*, 618. [CrossRef]
2. Orgel, L.E. Prebiotic chemistry and the origin of the RNA world. *Crit. Rev. Biochem. Mol. Biol.* **2004**, *39*, 99–123. [CrossRef] [PubMed]
3. Crick, F.H. The origin of the genetic code. *J. Mol. Biol.* **1968**, *38*, 367–379. [CrossRef]
4. Wächtershäuser, G. Before enzymes and templates: Theory of surface metabolism. *Microbiol. Rev.* **1988**, *52*, 452–484. [CrossRef]
5. Dyson, F.J. *Origins of Life*, 2nd ed.; Cambridge University Press: Cambridge, UK, 1985.
6. Segré, D.; Ben-Eli, D.; Deamer, D.W.; Lancet, D. The lipid world. *Orig. Life Evol. Biosph.* **2001**, *31*, 119–145. [CrossRef]
7. Carter, C.W. What RNA World? Why a Peptide/RNA Partnership Merits Renewed Experimental Attention. *Life* **2015**, *5*, 294–320. [CrossRef]
8. Bowman, J.C.; Hud, N.V.; Williams, L.D. The ribosome challenge to the RNA world. *J. Mol. Evol.* **2015**, *80*, 143–161. [CrossRef]
9. Carter, C.W., Jr. An Alternative to the RNA World. *Nat. Hist.* **2016**, *125*, 28–33.
10. Cech, T.R. Crawling out of the RNA world. *Cell* **2009**, *136*, 599–602. [CrossRef]
11. Frenkel-Pinter, M.; Haynes, J.W.; Mohyeldin, A.M.; Martin, C.; Sargon, A.B.; Petrov, A.S.; Krishnamurthy, R.; Hud, N.V.; Williams, L.D.; Leman, L.J. Mutually Stabilizing Interactions Between Proto-Peptides and RNA. *Nat. Commun.* **2020**, *11*, 3137. [CrossRef]
12. Frenkel-Pinter, M.; Haynes, J.W.; Martin, C.; Petrov, A.S.; Burcar, B.T.; Krishnamurthy, R.; Hud, N.V.; Leman, L.J.; Williams, L.D. Selective Incorporation of Proteinaceous over Nonproteinaceous Cationic Amino Acids in Model Prebiotic Oligomerization Reactions. *Proc. Natl. Acad. Sci. USA* **2019**, *116*, 16338–16346. [CrossRef] [PubMed]
13. Guseva, E.; Zuckermann, R.N.; Dill, K.A. Foldamer Hypothesis for the Growth and Sequence Differentiation of Prebiotic Polymers. *Proc. Natl. Acad. Sci. USA* **2017**, *114*, 7460–7468. [CrossRef] [PubMed]
14. Hordijk, W. A History of Autocatalytic Sets. *Biol. Theory* **2019**, *14*, 224–246. [CrossRef]
15. Smith, J.I.; Steel, M.; Hordijk, W. Autocatalytic Sets in a Partitioned Biochemical Network. *J. Syst. Chem.* **2014**, *5*, 2. [CrossRef] [PubMed]
16. Wu, M.; Higgs, P.G. Origin of Self-Replicating Biopolymers: Autocatalytic Feedback Can Jump-Start the RNA World. *J. Mol. Evol.* **2009**, *69*, 541–554. [CrossRef]
17. Tkachenko, A.V.; Maslov, S. Onset of autocatalysis of information-coding polymers. *arXiv* **2014**, arXiv:1405.2888. [CrossRef]
18. Lee, D.; Granja, J.; Martinez, J.; Severin, K.; Ghadiri, M.R. A Self-Replicating Peptide. *Nature* **1996**, *382*, 525–528. [CrossRef]
19. Rubinov, B.; Wagner, N.; Rapaport, H.; Ashkenasy, G. Self-Replicating Amphiphilic β-Sheet Peptides. *Angew. Chem. Int. Ed.* **2009**, *48*, 6683–6686. [CrossRef]
20. Szostak, J.W.; Bartel, D.P.; Luisi, P.L. Synthesizing Life. *Nature* **2001**, *409*, 387–390. [CrossRef]
21. Monnard, P.A.; Deamer, D.W. Membrane Self-Assembly Processes: Steps towards the First Cellular Life. *Anat. Rec.* **2002**, *268*, 196–207. [CrossRef]
22. Hanczyc, M.M.; Fujikawa, S.M.; Szostak, J.W. Experimental Models of Primitive Cellular Compartments: Encapsulation, Growth, and Division. *Science* **2003**, *302*, 618–622. [CrossRef] [PubMed]
23. Chen, I.A.; Roberts, R.W.; Szostak, J.W. The Emergence of Competition Between Model Protocells. *Science* **2004**, *305*, 1474–1476. [CrossRef] [PubMed]
24. Joyce, G.F.; Szostak, J.W. Protocells and RNA Self-Replication. *Cold Spring Harb. Perspect. Biol.* **2018**, *10*, a034801. [CrossRef] [PubMed]
25. Walde, P.; Goto, A.; Monnard, P.A.; Wessicken, M.; Luisi, P.L. Oparins Reactions Revisted-Enzymatic-Synthesis of Poly (adenylic acid) in MICELLES and Self-Reproducing Vesicles. *J. Am. Chem. Soc.* **1994**, *116*, 7541–7547. [CrossRef]
26. Chakrabarti, A.C.; Deamer, D.W. Permeation of Membranes by the Neutral Form of Amino-acids and Peptides—Relevance to the Origin of Peptide Translocation. *J. Mol. Evol.* **1994**, *39*, 1–5. [CrossRef]
27. Carter, C.W., Jr.; Wills, P.R. Reciprocally-Coupled Gating: Strange Loops in Bioenergetics, Genetics, and Catalysis. *Biomolecules* **2021**, *11*, 265. [CrossRef]

28. Horowitz, E.D.; Engelhart, A.E.; Chen, M.C.; Quarles, K.A.; Smith, M.W.; Lynn, D.G.; Hud, N.V. Intercalation as a Means to Suppress Cyclization and Promote Polymerization of Base-Pairing Oligonucleotides in a Prebiotic World. *Proc. Natl. Acad. Sci. USA* **2010**, *107*, 5288–5293. [CrossRef]
29. Jain, S.S.; Anet, F.A.L.; Stahle, C.J.; Hud, N.V. Enzymatic Behavior by Intercalating Molecules in a Template-Directed Ligation Reaction. *Angew. Chem. Int. Ed.* **2004**, *43*, 2004–2008. [CrossRef]
30. Hud, N.V.; Cafferty, B.J.; Krishnamurthy, R.; Williams, L.D. The Origin of RNA and "My Grandfather's Axe". *Biol. Chem.* **2013**, *20*, 466–474. [CrossRef]
31. Dill, K.A.; Agozzino, L. Driving Forces in the Origins of Life. *Open Biol.* **2021**, *11*, 200324. [CrossRef]
32. Maibaum, L.; Dinner, A.R.; Chandler, D. Micelle Formation and the Hydrophobic Effect. *J. Phys. Chem.* **2004**, *108*, 6778–6781. [CrossRef]
33. Tanford, C. Theory of Micelle Formation in Aqueous Solutions. *J. Phys. Chem.* **1974**, *78*, 2469–2479. [CrossRef]
34. Feng, B.; Sosa, R.P.; Martensson, A.K.F.; Jiang, K.; Tong, A.; Dorfman, K.D.; Takahashi, M.; Lincoln, P.; Bustamante, C.J.; Westerlund, F.; et al. Hydrophobic Catalysis and a Potential Biological Role of DNA Unstacking Induced by Environment Effects. *Proc. Natl. Acad. Sci. USA* **2019**, *116*, 17169–17174. [CrossRef] [PubMed]
35. Marat, Y.M.; Gulnara, Y.Z.; Baucom, A.; Lieberman, K.; Earnest, T.N.; Cate, J.H.D.; Noller, H.F. Crystal Structure of the Ribosome at 5.5 Å Resolution. *Science* **2001**, *292*, 883–896. [CrossRef]
36. Moore, P.; Steitz, T. The Involvement of RNA in Ribosome Function. *Nature* **2002**, *418*, 229–235. [CrossRef]
37. Campbell, J.H. An RNA Replisome as the Ancestor of the Ribosome. *J. Mol. Evol.* **1991**, *32*, 3–5. [CrossRef]
38. Ferreira-Cerca, S.; Pöll, G.; Gleizes, P.E.; Tschochner, H.; Milkereit, P. Roles of Eukaryotic Ribosomal Proteins in Maturation and Transport of Pre-18S rRNA and Ribosome Function. *Mol. Cell* **2005**, *20*, 263–275. [CrossRef]
39. Venema, J.; Tollervey, D. Ribosome Synthesis in *Saccharomyces cerevisiae*. *Annu. Rev. Genet.* **1999**, *33*, 261–311. [CrossRef]
40. Reimer, J.M.; Haque, A.S.; Tarry, M.J.; Schmeing, M.T. Piecing Together Nonribosomal Peptide Synthesis. *Curr. Opin. Struct. Biol.* **2018**, *49*, 104–113. [CrossRef]
41. Kimura, M. On the Probability of Fixation of Mutant Genes in a Population. *Genetics* **1962**, *47*, 713–719. [CrossRef]
42. Peterson, G.J.; Pressé, S.; Peterson, K.S.; Dill, K.A. Simulated Evolution of Protein-Protein Interaction Networks with Realistic Topology. *PLoS ONE* **2012**, *7*, e39052. [CrossRef] [PubMed]
43. Erdős, P.; Rényi, A. On the Evolution of Random Graphs. *Publ. Math. Inst. Hung. Acad. Sci.* **1960**, *7*, 17–61.
44. Fujimori, S.; Hino, K.; Saito, A.; Miyano, S.; Miyamoto-Sato, E. PRD: A Protein-RNA Interaction Database. *J. Bioinform.* **2012**, *8*, 729–730. [CrossRef] [PubMed]
45. Yi, Y.; Zhao, Y.; Li, C.; Zhang, L.; Huang, H.; Li, Y.; Liu, L.; Hou, P.; Cui, T.; Tan, P.; et al. RAID v2.0: An Updated Resource of RNA-Associated Interactions Across Organisms. *Nucleic Acids Res.* **2016**, *45*, D115–D118. [CrossRef]
46. Yi, Y.; Zhao, Y.; Huang, Y.; Wang, D. A Brief Review of RNA-Protein Interaction Database Resources. *Noncoding RNA* **2017**, *3*, 6. [CrossRef]
47. Kirsanov, D.D.; Zanegina, O.N.; Aksianov, E.A.; Spirin, S.A.; Karyagina, A.S.; Alexeevski, A.V. NPIDB: Nucleic Acid-Protein Interaction Database. *Nucleic Acids Res.* **2013**, *41*, D517–D523. [CrossRef]
48. Panni, S.; Prakash, A.; Bateman, A.; Orchard, S. The Yeast Noncoding RNA Interaction Network. *RNA* **2017**, *23*, 1479–1492. [CrossRef]
49. Perkins, S.J. The Calculations of Partial Specific Volumes, Neutron Scattering Matchpoints and 280-nm Absorption Coefficients for Proteins and Glycoproteins from Amino Acid Sequences. *Eur. J. Chem.* **1986**, *157*, 169–180. [CrossRef]
50. Voss, N.R.; Gerstein, M. Calculation of Standard Atomic Volumes for RNA and Comparison with Proteins: RNA Is Packed More Tightly. *J. Mol. Biol.* **2005**, *346*, 477–492. [CrossRef]

life

Article

The Effects of Flexibility on dsDNA–dsDNA Interactions

Chuanying Chen and B. Montgomery Pettitt *

Department of Biochemistry and Molecular Biology, Sealy Center for Structural Biology and Molecular Biophysics, University of Texas Medical Branch, Galveston, TX 77555, USA; ch2chen@utmb.edu
* Correspondence: mpettitt@utmb.edu

Abstract: A detailed understanding of the physical mechanism of ion-mediated dsDNA interactions is important in biological functions such as DNA packaging and homologous pairing. We report the potential of mean force (PMF) or the effective solvent mediated interactions between two parallel identical dsDNAs as a function of interhelical separation in 0.15 M NaCl solution. Here, we study the influence of flexibility of dsDNAs on the effective interactions by comparing PMFs between rigid models and flexible ones. The role of flexibility of dsDNA pairs in their association is elucidated by studying the energetic properties of Na^+ ions as well as the fluctuations of ions around dsDNAs. The introduction of flexibility of dsDNAs softens the vdW contact wall and induces more counterion fluctuations around dsDNAs. In addition, flexibility facilitates the Na^+ ions dynamics affecting their distribution. The results quantify the extent of attraction influenced by dsDNA flexibility and further emphasize the importance of non-continuum solvation approaches.

Keywords: molecular dynamics simulation (MD); potential of mean force (PMF); dsDNA–dsDNA interactions; flexibility

Citation: Chen, C.; Pettitt, B.M. The Effects of Flexibility on dsDNA–dsDNA Interactions. *Life* **2022**, *12*, 699. https://doi.org/10.3390/life12050699

Academic Editors: Tigran Chalikian and Jens Völker

Received: 7 April 2022
Accepted: 6 May 2022
Published: 7 May 2022

Publisher's Note: MDPI stays neutral with regard to jurisdictional claims in published maps and institutional affiliations.

1. Introduction

It has been shown experimentally [1–3] that ion-mediated attraction happens between like-charged polyelectrolytes such as dsDNA. The attraction is not captured by the well-known Poisson–Boltzmann (PB) theory and hence other approaches have been proposed. An extension of condensation theory [4] on two infinite line charges in a highly dilute 1:1 salt solution proposed that the attraction comes from the increase of translational entropy of condensed counterions. As two dsDNAs approach in an intermediate region, the volume of the condensation region increases. Another possible explanation of attractive interactions is counterion correlation, which can be explained by two mechanisms. One [5] involves counterions that reposition themselves and form a strongly correlated liquid on the surface of dsDNAs (similar to a Wigner lattice) due to strong interactions with polyelectrolytes, and with each other. The other explanation [6,7] considers the attractions that originate from charge fluctuations along the rods. The movement of condensed counterions introduces the fluctuations of the local charges on monomers along the polymer, resulting in dipole–dipole attractions and even higher-order multipole interactions. In addition, a more structural argument posits that an ideal alignment of dsDNA pairs produces an "electrostatic zipper" [8], in which positive counterions in the grooves of one dsDNA can interact with the negatively charged phosphate groups of the neighboring dsDNA. This idea also demonstrates a helical-specific recognition of dsDNA–dsDNA interactions.

In previous work, we found that the localization of ions near charged groups can give rise to a local energy minimum at an optimal short separation due to the formation of a hydrogen bond (HB) network among Na^+ ions, water and phosphate atoms of dsDNAs [9]. The studies mentioned above are mostly based on rigid bodies. Although the rigid structures used were relaxed [10] to improve the theoretical model, possible physical mechanisms responsible for the attraction are complicated and collectively influenced by the spatial distributions of counterions, solvent and dsDNAs.

Interplay between DNA flexibility and the surrounding counterions has been studied experimentally and theoretically [11–15]. A long dsDNA is a semi-flexible polyelectrolyte with persistence length. Its overall flexibility can be roughly described by a worm-like chain model (WLC) [16,17] in which the chain is relatively rigid at a small length scale but turns flexible over a longer length. Its rigidity is balanced by electrostatic effects (like-charge repulsion) and various elastic effects (e.g., base pair stacking), and strongly depends on the counterion types, valence, shape and concentration [14,18]. The counterion density and fluctuations can reduce the chain persistence length and could lead to instability of rodlike chain conformations [19]. For a short dsDNA fragment it is arguable among the experimental, theoretical and simulation studies that the duplex is much softer than predicted by a WLC model [20–22]. Thus, there exists flexibility in a short dsDNA fragment, which in turn influences the distributions of counterions and could lead to differences in correlations between counterions and DNA and among counterions themselves.

In this paper, we present a comparison, using oriented flexible structures of identical dsDNA models, of the potential of mean force (PMF) of the pairing processes along the interhelical separation. We wish to consider the effects of local dsDNA flexibility on parallel strands and investigate the radial and azimuthal angular dependence of the forces. The PMF calculations were carried out utilizing the adaptive biasing force (ABF) method [23,24] sampled via molecular dynamics (MD) simulations. The simulations sampled forces between dsDNA pairs in three different helical alignments in 0.15 M NaCl salt solution. Although the ion-mediated interaction of infinitely dilute dsDNA pairs is thermodynamically unfavorable in this monovalent salt solution, an attractive component exists at short surface separation (within 8 Å of contact), which comes from the collective correlations of the counterions and water and the relative helical alignment of dsDNA pairs [9]. We investigated the effects of dsDNA flexibility on dsDNA–dsDNA interactions considering the surrounding counterions, particularly by studying energetic and dynamic behaviors of Na^+ ions around the dsDNAs. The calculations provide the extent of attraction influenced by dsDNA flexibility when compared to rigid models and further emphasize the importance of counterion correlations that are not captured in continuum PB theory.

2. Methods

2.1. MD Simulations

Three models of dsDNA pairs having different helical alignments were considered (Figure 1). The first one (Model-0) has two identical sequences of 30-base-pair DNAs (DNA1 and DNA2) parallel to each other sampled at various interhelical separations, *d*. The helical axis is along the z-axis. In the other two models, DNA2 was rotated about its helical axis by 72 degrees (Model-72) and 180 degrees (Model-180), respectively.

Each model was then solvated in a TIP3P [25] water box with dimensions of 99 × 99 × 103 Å. Na^+ and Cl^- ions were randomly added to both neutralize the system and set the salt concentration at 0.15 M. The resulting system contained about 101,000 atoms. Each DNA duplex is effectively infinite as each DNA duplex strand is covalently bonded to itself through the periodic boundary.

MD simulations were performed using NAMD 2.14 software [26] with the CHARMM36 force field parameter set [27]. NBFIX corrections [28] were applied to Lennard–Jones interaction potentials between Na^+ ions and Cl^- ions and phosphate oxygens. Such corrections are used to fit to osmotic pressure for concentrated aqueous solutions or confined systems [28,29]. Particle mesh Ewald [30] was used to calculate long-range electrostatic interactions, and van der Waals interactions were truncated at 12 Å. All bonds were constrained using the SETTLE algorithm [31] and equations of motion were integrated with a time step of 2 fs. Temperature was controlled with Langevin coupling with a damping coefficient of 1/ps. After 50K steps of energy minimization, the system was heated up from 0 K to 310 K with restraints on the DNAs with a force constant of 500 kcal/(mol·Å^2). The system was then switched to the NVT ensemble for equilibration for over 100 ns until the salt and solvent distributions were stable.

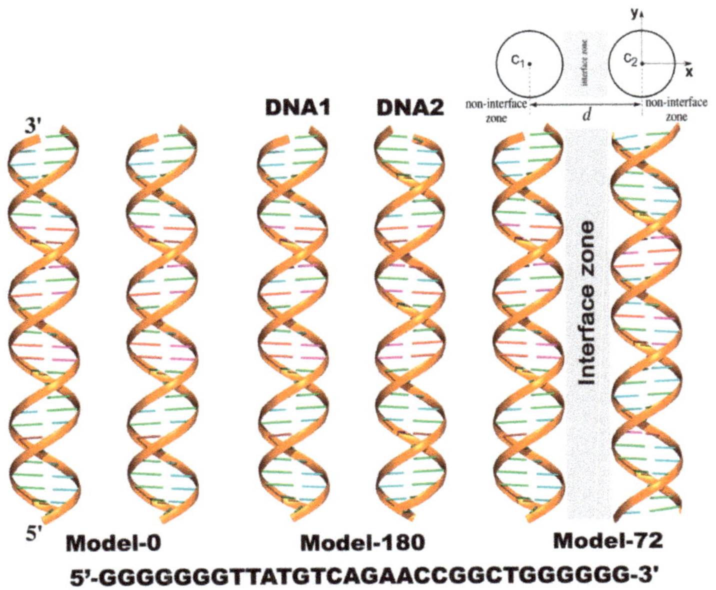

Figure 1. Three models of parallel 30-bp dsDNA pairs in the simulations. The helical axis is along the z-axis. From the top view of the systems, $c_1(c_{1x},c_{1y},c_{1z})$ and $c_2(c_{2x},c_{2y},c_{2z})$ represent the helical centers of DNA1 and DNA2, respectively. d is the interhelical distance. Any point (x,y,z) is considered inside dsDNA when its radial distance from the corresponding c_1 or c_2 is smaller than the radius of the dsDNA, $R_{DNA} = 10$ Å. Non-interface zone includes each cylindrical region of dsDNA extending to the bulk solution ($x > c_{1x}$ or $x < c_{2x}$).

2.2. ABF Calculations

We calculated the PMFs on the three geometric models using two DNA flexibility settings: rigid body and flexible structures, resulting in a total of six systems. The PMF calculations were carried out utilizing the ABF method implemented in the Colvars module [24] of NAMD. In the ABF calculation, the orientations of the dsDNAs were constrained. DNA1 was fixed in its global position, and DNA2 was permitted to diffuse along the reaction coordinate of interhelical distance d by a force that is adjusted via ABF in response to effective energetic barriers. The biasing force acting on DNA2 is equal in magnitude and opposite in sign to the mean force on DNA2. Instantaneous values of the force were accumulated in bins of 0.1 Å along d from 31 to 21 Å center to center. To enhance sampling of the distribution of configurations and increase the efficiency of the calculations, the PMF pathway was divided into three consecutive, non-overlapping windows.

The ABF calculations started at 31.0 Å for both flexible and rigid models sharing the same structures. Then different collective variable restraints were introduced by means of harmonic potentials with corresponding force constants, k: (i) positional restraints of DNA1 with k = 80 kcal/(mol·Å^2), (ii) rotational restraints of DNA1 and DNA2 with k = 2000 kcal/(mol·deg^2); (iii) positional restraint of DNA2 to restrict its movement in the yz-plane, k = 80 kcal/(mol·Å^2). Those restraints allow the dsDNA atoms to be locally flexible and yet maintain their separation and orientation features. For the rigid body systems, we introduced strong restraints on root-mean-square deviation (RMSD) of both dsDNAs with k = 2000 kcal/(mol·Å^2).

The PMF uncertainty was estimated using the method proposed by Henin and Chipot [23]. The PMF was considered converged once the distribution of the instantaneous force at each bin followed a Gaussian distribution and sampling was reasonably uniform along the reaction coordinate in each window (Figure S1). For each system the estimate of the sampling time required to complete the entire calculation is ~400 ns.

2.3. Differential Entropy from Density Fluctuations

To investigate possible ion correlations, we estimated the entropy change associated with the flexibility influence on the statistics of sodium ion density fluctuations. Theoretically, fluctuations of water/ions density in small microscopic volumes obey Gaussian statistics [32,33]. Following the coarse-grained density field method [34,35], we can compute the density field at a series of spatial grid points **r** at time *t*:

$$\overline{\rho(\mathbf{r},t)} = \sum_i \phi(|\mathbf{r} - r_i(t)|;\xi) \tag{1}$$

and

$$\phi(\mathrm{r}) = \left(2\pi\xi^2\right)^{-d/2} exp\left(-r^2/2\xi^2\right) \tag{2}$$

where r is the distance of the ith particle of interest (Na^+ ion here) from **r**, ξ is the Gaussian width and chosen to be 3.0 Å, and d is dimensionality. The sum is over all Na^+ ions in the whole space.

For Gaussian density functions, entropy has an analytic form proportional to the determinant of the covariance matrix. We estimate the correlation of multivariate Gaussians, which can be quantified by a differential entropy $S(x)$ [36,37],

$$S(x) = -\int_{-\infty}^{+\infty} p(x)\ln[p(x)]dx \tag{3}$$

where x is a random variable with an expected value of μ and a continuous density function $p(x)$. $p(x)$ is a multivariate Gaussian density function given by

$$p(x) = \frac{1}{(2\pi)^{N/2}|\Sigma|^{1/2}} exp\left[-\frac{1}{2}(x-\mu)^T\Sigma^{-1}(x-\mu)\right] \tag{4}$$

in which Σ is the covariance matrix. In this study, $\Sigma_{ij} = \delta\rho_i\delta\rho_j = (\rho-\overline{\rho})_i(\rho-\overline{\rho})_j$ and $\overline{\rho_i}$ is the time averaged density of Na^+ ions at point i. Then the final entropy can be deduced from

$$S = const + \frac{k_B}{2}\ln(|\Sigma|) \tag{5}$$

where $|\Sigma|$ is the determinant of covariance matrix, and k_B is the Boltzmann constant. The constant term in Equation (5) depends on grid spacing and the number of grid points studied. Under the same conditions, we estimate the difference of entropy, ΔS = *S(flexible)* − *S(rigid)*, from density fluctuations between the rigid and flexible systems.

2.4. Localization of Diffusion Coefficients of Na^+ Ions

The local diffusion constant was calculated by using a finite difference expression [38,39] with a grid spacing of 1.0 Å.

$$6D_{uvw} = \frac{1}{t_2 - t_1} < |(r(t_2) - r(t_0))^2 - (r(t_1) - r(t_0))^2| > \tag{6}$$

where $r(t_0)$ is the initial position at instant t_0; D_{uvw} is the local diffusion coefficient at a grid point uvw and was computed whenever $|r(t_0) - r_{uvw}| < 1.0$ Å; t_1 and t_2 were fixed at 1 ps and 2 ps, respectively, assuming the diffusion regime would be reached after 1 ps [40] and would not diffuse beyond 3 Å from $r(t_0)$.

3. Results and Discussion

3.1. Potential of Mean Force

In Figure 2, we observe that for all six systems their PMF profiles along d deviate from the continuum picture or PB theory. The PMF profiles can be divided into three regions: (i) in the region of $d > 26$ Å, the PMFs cannot be distinguished, which reflects the formally dominating long-ranged repulsive electrostatic interactions between two like-charged polyelectrolytes; (ii) in a range of ~24 Å < *d* < 26 Å, the PMFs establish a plateau or local minimum as has been seen before [9], an indication of the existence of attractive components of the interactions; (iii) in a region of *d* < ~24 Å, the PMF profiles display remarkable differences depending on the helical alignments of dsDNA pairs. The values of the PMFs in Model-72 and Model-180 are much lower than those in Model-0. This further demonstrates specific recognition of dsDNA–dsDNA in term of helical alignment [8,9,41].

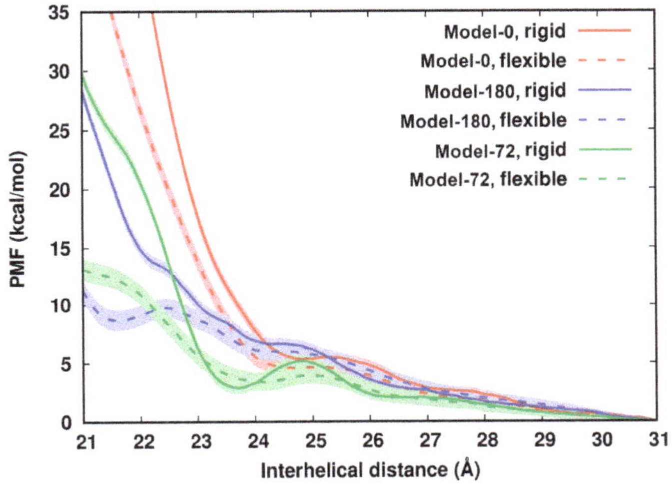

Figure 2. Potential of mean force of pairing two identical dsDNAs with difference helical features along the interhelical distance. The uncertainty (in shade) is up to 1.0 kcal/mol. The curves are arbitrarily set to zero at 31 Å.

Comparison of rigid and flexible systems shows that dsDNA flexibility softens the contact wall of the PMFs of dsDNA pairing in all three models. The influence from flexibility becomes apparent when two dsDNAs get close. In Model-72 as an example: from *d* = 23 Å to *d* = 22.0 Å, the difference of the PMFs between the flexible to the rigid system increases by 15 times from 0.64 kcal/mol to 9.7 kcal/mol, many times larger than thermal energy. In addition, we observed that in the distance range between 23.5 Å and 25.5 Å, the rigid Model-72 forms an energy barrier of about 2.3 kcal/mol. DNA flexibility lowers the energy barrier by as much as 1.3 kcal/mol and flattens the PMF curve.

We wish to investigate how flexibility enhances the pairing process. We might posit that the enhancement could be contributed by positive configurational entropy of dsDNA. Due to the constraint on dsDNAs to remain a helical arrangement in this study, during the whole simulation the RMSDs of all atoms from corresponding initial structures for all flexible systems are ~1.0 Å, much smaller than 3~9 Å which was inferred from an experiment [42] for dsDNAs in free motion. The RMS fluctuations (RMSF) of base groups are about 0.48~0.71 Å, 0.72~0.84 Å for sugar groups, and 0.85~1.2 Å for phosphate groups, respectively. The RMSFs are all slightly smaller than the corresponding groups of dsDNA in free motion [43]. Both RMSDs and RMSFs of the dsDNAs in our model systems are independent of distance spacing, so the configurational entropy contribution is not considered in this study.

To further investigate influences of flexibility on dsDNA–dsDNA pairing, we chose ten configurations at several distance separations and extended MD simulations for 18 ns more. The trajectories were saved every 0.1 ps for analysis. According to the PMF profiles, we chose 22.4 Å (a local maximum) and 24.4 Å (in the plateau) for Model-180; 22.4 Å, 23.7 Å (a local minimum for the rigid system and in the plateau for the flexible system), 24.9 Å (in a plateau) for Model-72.

3.2. vdW Interactions

When surface separation between two dsDNAs is as short as ~2.4 Å, smaller than the diameter of a water molecule, we investigated the role of the vdW interactions in the dsDNAs pairing process. The vdW interaction was modeled by the Lennard–Jones (LJ) potential [27]. The differences of LJ potential energies of water with dsDNAs between the rigid and flexible structures in all models were estimated to be less than 1.0 kcal/mol, so we mainly focus on the dsDNA–dsDNA interactions with the participation of counterions. In Figure 3, we observed that the LJ energies of two dsDNAs, $E_{LJ,DNA\text{-}DNA}$, show little difference between the rigid and flexible systems. In addition, the total LJ energies, $E_{LJ,total}$, in all cases are attractive, and are more favorable in the flexible structures than in the rigid ones. The total energy difference, $\Delta E_{LJ,total} = E_{LJ,total}(\text{flexible}) - E_{LJ,total}(\text{rigid})$, is mainly contributed by the interactions of the Na^+ ions with the dsDNAs. In the rigid models, in events where Na^+ ions are moving in close proximity of dsDNA atoms, vdW clash could happen, which would result in more repulsion between the Na^+ ions and dsDNAs. In contrast, flexibility provides sufficient relaxation of the structures and allows the vdW contact wall to shift and consequently reduce the clashes of counterions with the DNA atoms.

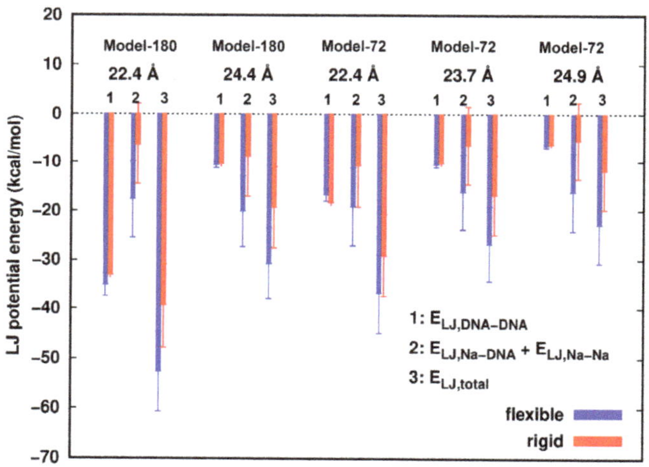

Figure 3. Lennard–Jones (LJ) potential energies of Na^+ ions participated dsDNA–dsDNA interactions. Error bars represent standard deviations.

Although the total LJ component is attractive for the approach of two dsDNAs with each other, the total interactions are energetically unfavorable, indicating that the electrostatic components are dominantly repulsive between isolated dsDNAs given the surrounding ions and solvent.

3.3. Ion Distributions and Electrostatic Energy of Ions around dsDNA Pairs

A previous study [9] revealed that close proximity of dsDNAs to each other has essentially no influence on the fraction of charge inside the grooves. The charge fraction around one dsDNA is larger than the prediction of Manning counterion condensation theory [44,45] for a single dsDNA. The large value of charge fraction here is due to the combination of Na^+ ions correlated with the dsDNA, and the Na^+ ion atmosphere shared with the neighboring dsDNA.

Considering the ion mobility near condensed dsDNA molecules and the heterogeneous environment of dsDNA, we discretized the space into a three-dimensional grid with a spacing of 1.0 Å. At a voxel we counted the number density of Na^+ ions and calculated the electrostatic energy of this Na^+ ion with the dsDNAs and the other Na^+ ions using the standard Ewald summation [46,47]. In this way, the electrostatic energy contains information on local variations of Na^+ ions and local correlations among the Na^+ ions and the dsDNA charges.

From the number density we calculated the radial distribution of the Na^+ ions from the helical center of the dsDNAs. For the non-interface zone (Figure 1), to avoid mutual interference in the interface from each dsDNA, we restricted the grid points to each dsDNA hemicylinder extending into the bulk solution. For the interface zone between two dsDNAs, a grid point was assigned to DNA1 when its x-coordinate is less than the x-coordinate of the midpoint of both dsDNA centers; otherwise it was assigned to DNA2. Radial number density distributions of Na^+ ions in the interface and non-interface zone were calculated by averaging densities within the cylindrical shells from the center of each dsDNA with a layer spacing of 0.5 Å.

Figure 4A,C displays the radial distributions of Na^+ ions number density and electrostatic energy, respectively, for Model-180 with d = 22.4 Å as an example. We observed that the density profile is correlated to the electrostatic energy profile as expected, particularly within 6 Å from the helical center of the dsDNAs, where a higher density is found in the rigid model than in the flexible one. The electrostatic energy reaches a minimum value at ~4.5 Å but with a large uncertainty. Possible explanations of such difference are that (i) due to flexibility, the electrostatic potentials at some local regions are averaged, resulting in the change of the electrostatic environment of the dsDNAs. In the rigid model, the sodium ions are likely to have stronger interactions with dsDNAs in some limited locations. In particular, they could have higher probabilities of direct contact with dsDNA (Figure S2) with a long lifetime. Direct contacts give rise to more favorable electrostatic energy at short range. Alternatively, (ii) dsDNA flexibility induces higher mobility of sodium ions. Even though direct contacts are possible, the densities are averaged in a fixed shell to yield a lower density distribution.

A radial distribution describes an average quantity in space, so we further compared the number density and electrostatic energy of sodium ions at cross sections between the rigid and flexible structures (Figure 4B,D). In both structures, the lowest electrostatic energy regions around dsDNA are found in the grooves and interface region. These lower electrostatic energies correspond to higher sodium densities in the corresponding regions. In addition, the distribution is more structured along the minor groove of the flexible systems, indicating flexibility of dsDNAs indeed influences the dynamic averaging of the Na^+ ions. When the mobility is relatively low, Na^+ ions are concentrated in a local region with a long lifetime. When the mobility is relatively high (e.g., in the minor groove) in the flexible model, the densities or electrostatic energies in a local region are smeared with a short lifetime. The density and electrostatic energy distribution profiles of Na^+ ions in Model-72 are similar and displayed in Figure S3 in the SI text.

The rigid systems have higher Na^+ ion densities and electrostatic potential energies inside dsDNAs within 6 Å from the helical axis and in the interface zone, which classically would indicate more screening of the dsDNA charges. We see the flexibility at short range dominating these screening effects. In addition, $\Delta E_{LJ,total}$ are similar at about −10 kcal/mol in all configurations studied. Thus, there must be other influences from flexibility when $d < 23$ Å. We found that the electrostatic energy distribution shows non-mean field interactions of Na^+ ions with dsDNAs, which has implications for ion fluctuations and different dynamic behaviors of ions.

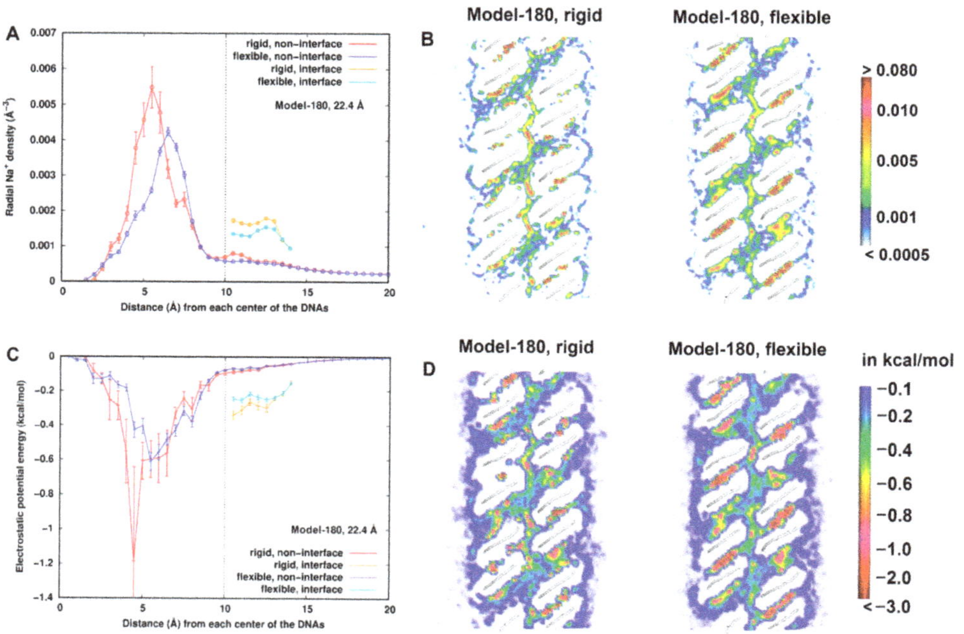

Figure 4. At d = 22.4 Å for Model 180, (**A**) radial number density distribution of Na^+ ions from the helical center of the dsDNAs. (**B**) Comparison of number densities of Na^+ ions between the flexible and rigid system. The number density is in the unit of $Å^{-3}$. (**C**) Radial distribution of electrostatic energy of Na^+ ions from the helical center of the dsDNAs. (**D**) Comparison of electrostatic potential energy of Na^+ ions with the dsDNA pairs between the flexible and rigid system. A cross section (x–z plane) is displayed.

3.4. Differential Entropy from Density Fluctuations

Counterions could exhibit correlations in fluctuations in distinct local volumes. Due to the different spatial limits around the two dsDNAs, we focus on three regions: the minor groove, major groove and interface. Applying a coarse-grained procedure to each of the three regions with the collection of $\delta\rho(\mathbf{r}) = \left(\rho(\mathbf{r}) - \overline{\rho(\mathbf{r})}\right)$ and producing the observed correlation of $\delta\rho_i\delta\rho_j$, we calculated the entropy change of density fluctuations of the ions near flexible structures relative to the rigid ones. If Na^+ ions are restricted in the rigid model, we expect low fluctuations and a smaller magnitude of entropy as well. For all configurations, we used the same grid spacing of 1.0 Å and for the configurations having the same relative helical angle we restricted the same number of grid points in the minor and major groove region, respectively. As a result, the constants involved in S (Equation (5)) are canceled in calculations of the entropy difference ΔS.

In Figure 5, we observe that in all cases the values of ΔS in the three regions are positive, indicating higher density fluctuations in the flexible systems. In a confined volume close to the dsDNAs, flexibility of the dsDNAs confers movements to the Na^+ ions, resulting in relatively large fluctuations in density. Positive entropy also suggests a favorable free energy contribution to the interactions of dsDNA–dsDNA.

Although we restrict the study of entropy to the limited spaces, we cannot exclude possible correlations among these regions and the dsDNAs. Ha and Liu [6] showed that an increase of charge fluctuation of DNAs not only helps screen the electrostatic repulsion but also helps contribute to the dipole, quadrupoles and even higher-order multipoles along the DNAs. These multipoles can interact attractively with other multipoles (or monopoles), either on the same rod or on neighboring rods. Assuming there were little ion correlations,

ΔS in the minor groove or major groove would be the same along d as well as in the interface zone sharing the same number of grids. However, we observed that for Model-180, ΔS in the minor groove/major groove at d = 22.4 Å is more positive than that at d = 24.4 Å. Similarly for Model-72, ΔS in the interface zone at d = 22.4 Å is more positive than that at d = 23.7 Å, indicating that ion fluctuations and correlations are strongly influenced by the close proximity of two dsDNAs.

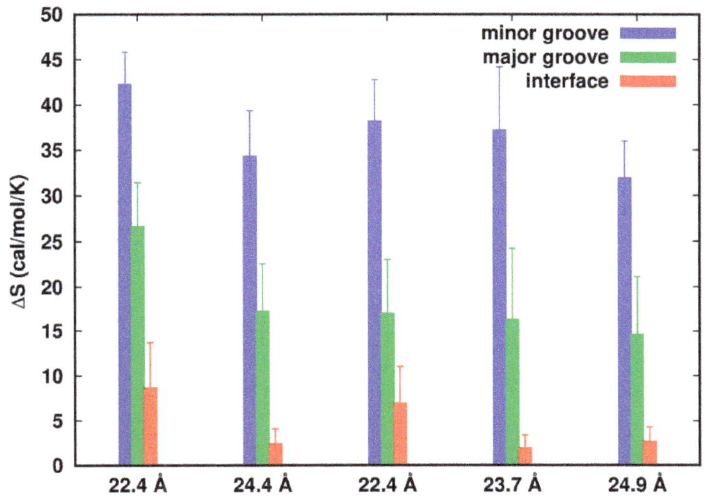

Figure 5. Difference of entropy from density fluctuations of the flexible systems relative to the rigid ones. The error bars represent the standard errors.

3.5. Localization of Diffusive Dynamics of Na^+ Ions

As suggested above, "smeared" density and electrostatic energy in the minor groove implies high mobility of the Na^+ ions. So, we studied the mobility of Na^+ ions by localizing diffusion coefficients of the Na^+ ions around the dsDNAs. The radial distribution of the diffusion coefficient of the Na^+ ions (Figure 6A,B) shows that the calculated diffusion coefficient in the bulk is 0.21 $\pm$ 0.01 Å^2/ps. This is slightly higher than another simulated result (Na^+ ions: 0.17 Å^2/ps [38]) and the experimental data (0.12 Å^2/ps [48]). It could be an artifact of the force field or using a damping coefficient of 1.0 ps^{-1} in temperature control in the simulation. The lowest spatially localized averaged diffusion constants are identified with the values of ~0.06 Å^2/ps at around ~5.5 Å, which are mainly located in the minor groove. The Na^+ ions in the interface are less diffusive compared to those in the other similar shell distances when mediating the interactions between the two dsDNAs.

Figure 6A,B also shows that the Na^+ ions are slightly more mobile around the DNAs in the flexible systems than in the rigid ones for all the systems, suggesting different dynamic behaviors of Na^+ ions between the rigid and flexible systems. Diffusion coefficients were categorized into three groups, the minor groove, major groove and interface of the dsDNAs. To remove the noise due to low populations at some locations, we divided the 18-ns trajectory into nine blocks, calculated the diffusion coefficient at each grid point for each block, extracted the points which are ~70% overlapped, and finally smoothed the points by weighted averaging of the six closest neighbors until the separation of the sites was larger than 2.8 Å. Such a method has been successfully applied to identify hydration sites and effective sodium sites around proteins and DNA [9,49].

Comparison between the rigid and flexible systems is displayed in Figure 6C,D for Model-180 and Model-72, respectively. The mobilities of the Na^+ ions increase in the order of minor groove < major groove $\leq$ interface for all systems, reflecting the different dynamic behaviors around the dsDNAs. It can be inferred from the figures that the datasets have a non-normal distribution and contain some extreme values. To compare the diffusion coefficients between the rigid and flexible systems, we performed Wilcoxon rank sum test [50,51] with the null hypotheses being equal distribution of the rigid and flexible systems at the 5% level of significance. Along with the p-value of the test, we estimated the effective size that describes the magnitude of the difference. We used Cliff's delta [52], which is the probability that a value from one group (e.g., rigid system) is greater than a value from the other (e.g., flexible system) group, minus the reverse probability.

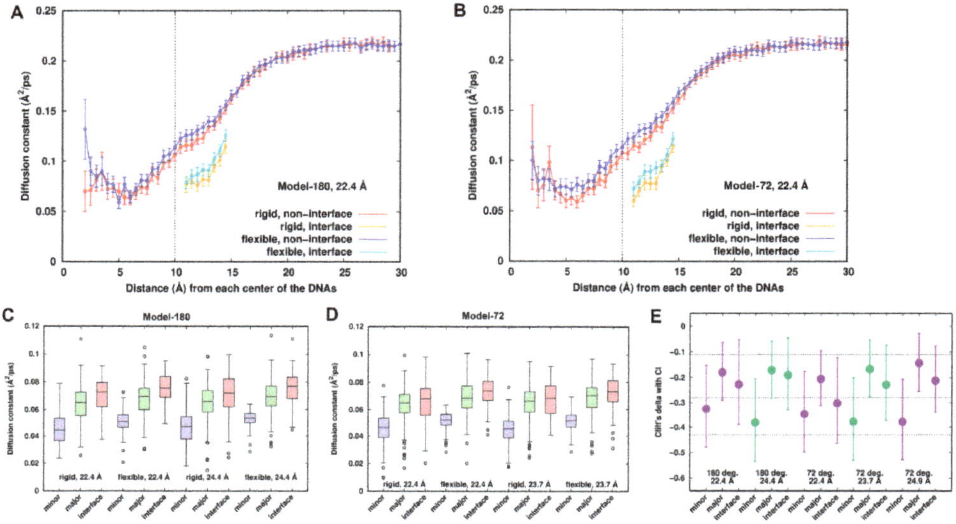

Figure 6. Comparison of local diffusion constants of sodium ions between the rigid and flexible systems. (**A**,**B**) Radial distribution of diffusion constants in the interface and non-interface zone for Model-180 and Model-72 at 22.4 Å, respectively. (**C**,**D**) Boxplots of local diffusion coefficients of Na^+ ions around the DNA pairs for Model-180 and Model-72. (**E**) Cliff's delta effective sizes with the error bars of 95% confidence interval. An effect size of +1.0 or −1.0 indicates the absence of overlap between the two groups, whereas 0.0 indicates that group distributions overlap completely. Generally, Cliff's delta effect sizes of 0.11, 0.28 and 0.43 correspond to small, medium and large effects, respectively, displayed in the dash line.

Figure 6E depicts a representation of Cliff's delta effect sizes and their 95% confidence interval. The p-values from Wilcoxon rank sum test are all less than the significance level of 0.05. Thus, we can conclude that diffusion coefficients in the rigid systems are significantly different from those in the flexible systems. For the minor groove, the effect size is medium, and there is a ~70% chance (Table S1) that a location randomly chosen from the flexible systems has a higher diffusion coefficient than a location randomly chosen from the rigid systems. Such a chance is lower for both the major groove and the interface, having a value of ~60% with a small effect size.

In addition to spatial heterogeneity of the Na^+ ions in terms of diffusion coefficients, we investigated the extent of the timescale during which the Na^+ ions can reside in a local position or site. Localized residence times around the dsDNAs are displayed in Figure S4. Local residence times decrease in the order of minor groove > major groove $\geq$ interface region. The decreasing order is consistent with the increasing order of local diffusion coefficients. Almost all the sites having residence time longer than 1 ns are in the minor groove of the rigid systems. At those sites, the Na^+ ions are partially dehydrated to directly contact nucleobase groups and/or O4′ of the sugar groups, and they are also in strong contacts indirectly via water with the DNAs. While in the similar locations in the flexible system, the residence times shrink to a few tens or hundreds of pico-seconds. In the interface zone, the majority of residence times become shorter, around 10~30 ps, in the flexible systems, while they are 20~50 ps in the rigid systems.

4. Conclusions

In this paper we focus on the influence of flexibility on dsDNA–dsDNA interactions by comparing rigid systems with flexible ones in the dsDNA pairing process in 0.15 M NaCl solution. The calculated PMF curves along the interhelical distance between two parallel dsDNAs indicate that flexibility enhances the dsDNA pairing process, particularly in a short distance region (d < 23 Å). Local flexibility of a molecule easily allows sub-Angstrom displacements in response to strong perturbing forces, such as the electrostatic field and vdW sphere contact of other surrounding molecular atoms. Thus, flexibility affects not only energetic properties of Na^+ ions but also dynamic behaviors associated with close motion around dsDNAs, which in turn will affect dsDNA–dsDNA interactions.

It has been proposed that local alignment and pairing of dsDNAs in a "protein-free" environment is an initial step in homologous recombination [53–57]. Sequence-dependent attractive interactions are governed by local attractive interactions. A mutual electrostatic complementarity model [58] was provided to interpret the mechanism of homologous pairing. However, the calculations were electrostatically mean-field in a continuum solvent using torsionally rigid DNA, and so may be insensitive to some local features of the dsDNA. Our simulations demonstrate strong correlated interactions involved in the dsDNA–dsDNA pairing process. Considering monoatomic counterions like Na^+, K^+ and Mg^{2+}, the sequence-dependent specificity found is of an electrostatic nature, and we expect that non-mean-field interactions play a role in the recognition preference in dsDNA–dsDNA interactions.

Supplementary Materials: The following supporting information can be downloaded at: https://www.mdpi.com/article/10.3390/life12050699/s1, Figure S1: Distribution of instantaneous force at chosen separation distances for both flexible and rigid systems; Figure S2: Minimum distance distribution of Na^+ ions from heavy atoms of dsDNAs in the rigid and flexible structures of Model-180 with the inter-helical distance at 22.4 Å; Figure S3: From left to right are radial distribution of Na^+ ions number density from the helical center of the dsDNAs, radial distribution of electrostatic potential energy of Na^+ ions with dsDNAs, a cross section of electrostatic potential energy in rigid structures and flexible structures, respectively; Figure S4: The effective sodium sites in the grooves and interface zone in the rigid and flexible Model180 and Model-72 with the inter-helical distance d = 22.4 Å, respectively; Table S1: Probabilities that a location randomly chosen from the flexible system has higher diffusion coefficient than a location randomly chosen from the rigid one.

Author Contributions: Both authors conceptualized the research and wrote the manuscript. C.C. did the computations. All authors have read and agreed to the published version of the manuscript.

Funding: We acknowledge partial support from the Robert A. Welch Foundation (H-0013). The research was carried out through the Texas Advanced Computing Center (TACC) at The University of Texas at Austin using the Extreme Science and Engineering Discovery Environment (XSEDE), which is supported by National Science Foundation grant number ACI-1548562 and the Computing Resource at the Sealy Center for Structural Biology and Molecular Biophysics at The University of Texas Medical Branch at Galveston.

Data Availability Statement: Data available upon reasonable request.

Acknowledgments: We thank Ka-yiu Wong and Gillian Lynch for helpful discussions. BMP acknowledges helpful conversations with M. Prentiss in the initial stages of the project.

Conflicts of Interest: The authors declare no conflict of interest.

References

1. Qiu, X.; Kwok, L.W.; Park, H.Y.; Lamb, J.S.; Andresen, K.; Pollack, L. Measuring inter-DNA potentials in solution. *Phys. Rev. Lett.* **2006**, *96*, 138101. [CrossRef] [PubMed]
2. Baldwin, G.S.; Brooks, N.J.; Robson, R.E.; Wynveen, A.; Goldar, A.; Leikin, S.; Seddon, J.M.; Kornyshev, A.A. DNA double helices recognize mutual sequence homology in a protein free environment. *J. Phys. Chem. B* **2008**, *112*, 1060–1064. [CrossRef] [PubMed]
3. Rau, D.C.; Parsegian, V.A. Direct measurement of the intermolecular forces between counterion-condensed DNA double helices. Evidence for long range attractive hydration forces. *Biophys. J.* **1992**, *61*, 246–259. [CrossRef]
4. Ray, J.; Manning, G.S. An attractive force between two rodlike polyions mediated by the sharing of condensed counterions. *Langmuir* **1994**, *10*, 2450–2461. [CrossRef]
5. Grønbech-Jensen, N.; Mashl, R.J.; Bruinsma, R.F.; Gelbart, W.M. Counterion-induced attraction between rigid polyelectrolytes. *Phys. Rev. Lett.* **1997**, *78*, 2477–2480. [CrossRef]
6. Ha, B.Y.; Liu, A.J. Counterion-mediated attraction between two like-charged rods. *Phys. Rev. Lett.* **1997**, *79*, 1289–1292. [CrossRef]
7. Podgornik, R.; Parsegian, V.A. Charge-fluctuation forces between rodlike polyelectrolytes: Pairwise summability reexamined. *Phys. Rev. Lett.* **1998**, *80*, 1560–1563. [CrossRef]
8. Kornyshev, A.A.; Leikin, S. Electrostatic zipper motif for DNA aggregation. *Phys. Rev. Lett.* **1999**, *82*, 4138–4141. [CrossRef]
9. Lai, C.L.; Chen, C.; Ou, S.C.; Prentiss, M.; Pettitt, B.M. Interactions between identical DNA double helices. *Phys. Rev. E* **2020**, *101*, 032414. [CrossRef]
10. Lee, D.J.; Wynveen, A.; Kornyshev, A.A.; Leikin, S. Undulations enhance the effect of helical structure on DNA interactions. *J. Phys. Chem. B* **2010**, *114*, 11668–11680. [CrossRef]
11. Kam, Z.; Borochov, N.; Eisenberg, H. Dependence of laser light scattering of DNA on NaCl concentration. *Biopolymers* **1981**, *20*, 2671–2690. [CrossRef] [PubMed]
12. Sobel, E.S.; Harpst, J.A. Effects of Na^+ on the persistence length and excluded volume of T7 bacteriophage DNA. *Biopolymers* **1991**, *31*, 1559–1564. [CrossRef] [PubMed]
13. Baumann, C.G.; Smith, S.B.; Bloomfield, V.A.; Bustamante, C. Ionic effects on the elasticity of single DNA molecules. *Proc. Natl. Acad. Sci. USA* **1997**, *94*, 6185–6190. [CrossRef] [PubMed]
14. Savelyev, A. Do monovalent mobile ions affect DNA's flexibility at high salt content? *Phys. Chem. Chem. Phys.* **2012**, *14*, 2250–2254. [CrossRef] [PubMed]
15. Guilbaud, S.; Salome, L.; Destainville, N.; Manghi, M.; Tardin, C. Dependence of DNA persistence length on ionic strength and ion type. *Phys. Rev. Lett.* **2019**, *122*, 28102. [CrossRef]
16. Bustamante, C.; Marko, J.F.; Siggia, E.D.; Smith, S. Entropic elasticity of lambda-phage DNA. *Science* **1994**, *265*, 1599–1600. [CrossRef]
17. Bouchiat, C.; Wang, M.D.; Allemand, J.; Strick, T.; Block, S.M.; Croquette, V. Estimating the persistence length of a worm-like chain molecule from force-extension measurements. *Biophys. J.* **1999**, *76*, 409–413. [CrossRef]
18. Savelyev, A.; Materese, C.K.; Papoian, G.A. Is DNA's rigidity dominated by electrostatic or nonelectrostatic interactions? *J. Am. Chem. Soc.* **2011**, *133*, 19290–19293. [CrossRef]
19. Golestanian, R.; Liverpool, T.B. Conformational instability of rodlike polyelectrolytes due to counterion fluctuations. *Phys. Rev. E* **2002**, *66*, 51802. [CrossRef]
20. Seol, Y.; Li, J.; Nelson, P.C.; Perkins, T.T.; Betterton, M.D. Elasticity of short DNA molecules: Theory and experiment for contour lengths of 0.6–7 mm. *Biophys. J.* **2007**, *93*, 4360–4373. [CrossRef]
21. Mazur, A.K. Wormlike chain theory and bending of short DNA. *Phys. Rev. Lett.* **2007**, *98*, 218102. [CrossRef] [PubMed]
22. Mazur, A.K.; Maaloum, M. DNA flexibility on short length scales probed by atomic force microscopy. *Phys. Rev. Lett.* **2014**, *112*, 068104. [CrossRef] [PubMed]
23. Henin, J.; Chipot, C. Overcoming free energy barriers using unconstrained molecular dynamics simulations. *J. Chem. Phys.* **2004**, *121*, 2904–2914. [CrossRef] [PubMed]
24. Fiorin, G.; Klein, M.L.; Hénin, J. Using collective variables to drive molecular dynamics simulations. *Mol. Phys.* **2013**, *111*, 3345–3362. [CrossRef]
25. Neria, E.; Fischer, S.; Karplus, M.J. Simulation of activation free energies in molecular systems. *J. Chem. Phys.* **1996**, *105*, 1902–1921. [CrossRef]
26. Phillips, J.C.; Wang, W.; Gumbart, J.; Tajkhorshid, E.; Villa, E.; Chipot, C.; Skeel, R.D.; Kalé, L.; Schulten, K. Scalable molecular dynamics with NAMD. *J. Comput. Chem.* **2005**, *26*, 1781–1802. [CrossRef] [PubMed]
27. Hart, K.; Foloppe, N.; Baker, C.M.; Denning, E.J.; Nilsson, L.; MacKerell, A.D., Jr. Optimization of the charmm additive force field for DNA: Improved treatment of the BI/BII conformational equilibrium. *J. Chem. Theory Comput.* **2012**, *8*, 348–362. [CrossRef]
28. Yoo, J.; Aksimentiev, A. Improved parametrization of Li^+, Na^+, K^+, and Mg^{2+} ions for all-atom molecular dynamics simulations of nucleic acid systems. *J. Phys. Chem. Lett.* **2012**, *3*, 45–50. [CrossRef]
29. Luo, Y.; Roux, B. Simulation of osmotic pressure in concentrated aqueous salt solutions. *J. Phys. Chem. Lett.* **2010**, *1*, 183–189. [CrossRef]
30. Darden, T.; York, D.; Pedersen, L. Particle mesh ewald: An N·log(N) method for Ewald sums in large systems. *J. Chem. Phys.* **1993**, *98*, 10089–10092. [CrossRef]

31. Miyamoto, S.; Kollman, P.A. SETTLE: An analytical version of the shake and rattle algorithm for rigid water models. *J. Comput. Chem.* **1992**, *13*, 952–962. [CrossRef]
32. Chandler, D. Gaussian field model of fluids with an application to polymeric fluids. *Phys. Rev. E* **1993**, *48*, 2898–2905. [CrossRef] [PubMed]
33. Hummer, G.; Garde, S.; Garcia, A.E.; Pohorille, A.; Pratt, L.R. An information theory model of hydrophobic interactions. *Proc. Natl. Acad. Sci. USA* **1996**, *93*, 8951–8955. [CrossRef] [PubMed]
34. Willard, A.P.; Chandler, D. Instantaneous liquid interfaces. *J. Phys. Chem. B* **2010**, *114*, 1954–1958. [CrossRef]
35. Patel, A.J.; Varilly, P.; Chandler, D. Fluctuations of water near extended hydrophobic and hydrophilic surfaces. *J. Phys. Chem. B* **2010**, *114*, 1632–1737. [CrossRef]
36. Ahmed, N.A.; Gokhale, D.V. Entropy expressions and their estimators for multivariated distributions. *IEEE Trans. Inf. Theory* **1989**, *35*, 688–692. [CrossRef]
37. Garbaczewski, P. Differential entropy and time. *Entropy* **2005**, *253*, 253–299. [CrossRef]
38. Makarov, V.A.; Feig, M.; Andrews, B.K.; Pettitt, B.M. Diffusion of solvent around biomolecular solutes: A molecular dynamics simulation study. *Biophys. J.* **1998**, *75*, 150–158. [CrossRef]
39. Lounnas, V.; Pettitt, B.M.; Phillips, G.N., Jr. A global model of the protein-solvent interface. *Biophys. J.* **1994**, *66*, 601–614. [CrossRef]
40. Lounnas, V.; Pettitt, B.M. Distribution function implied dynamics versus residence times and correlations: Solvation shells of myoglobin. *Proteins Struct. Funct. Bioinform.* **1994**, *18*, 148–160. [CrossRef]
41. Luan, B.; Aksimentiev, A. DNA attraction in monovalent and divalent electrolytes. *J. Am. Chem. Soc.* **2008**, *130*, 15754–15755. [CrossRef] [PubMed]
42. Podgornik, R.; Rau, D.C.; Parsegian, V.A. The action of interhelical forces on the organization of DNA double helices: Fluctuation-enhanced decay of electrostatic double-layer and hydration forces. *Macromolecules* **1989**, *22*, 1780–1786. [CrossRef]
43. Pan, Y.; MacKerell, A.D., Jr. Altered structural fluctuations in duplex RNA versus DNA: A conformational switch involving base pair opening. *Nucleic Acids Res.* **2003**, *31*, 7131–7140. [CrossRef] [PubMed]
44. Manning, G.S. Limiting laws and counterion condensation in polyelectrolyte solutions I. Colligative properties. *J. Chem. Phys.* **1969**, *51*, 924–933. [CrossRef]
45. Manning, G.S. Counterion condensation revisited. *J. Biomol. Struct. Dyn.* **1998**, *16*, 461–467. [CrossRef]
46. Ewald, P. The calculation of optical and electrostatic grid potential. *Ann. Phys.* **1921**, *64*, 253–287. [CrossRef]
47. de Leeuw, S.; Perram, J.; Smith, E. Simulation of electrostatic systems in periodic boundary conditions. I. lattice sums and dielectric constants. *Proc. R. Soc. Lond.* **1980**, *A373*, 27–56.
48. Tyrrell, H.J.V.; Harris, K.R. Diffusion in electrolytes. In *Diffusion in Liquids*; Butterworths: Oxford, UK, 1984; pp. 387–437.
49. Chen, C.; Beck, B.W.; Krause, K.; Pettitt, B.M. Solvent participation in Serratia marcescens endonuclease complexes. *Proteins* **2006**, *62*, 982–995. [CrossRef]
50. Wilcoxon, F. Individual comparisons by ranking methods. *Biom. Bull.* **1945**, *1*, 80–83. [CrossRef]
51. Mann, H.B.; Whitney, D.R. On a test of whether one of two random variables is stochastically larger than the other. *Ann. Math. Statist.* **1947**, *18*, 50–60. [CrossRef]
52. Hess, M.R.; Kromrey, J.D. Robust confidence intervals for effect sizes: A comparative study of Cohen's d and Cliff's delta under non-normality and heterogeneous variances. In Proceedings of the Annual Meeting of the American Educational Research Association, San Diego, CA, USA, 12–16 April 2004.
53. Inoue, S.; Sugiyama, S.; Travers, A.A.; Ohyama, T. Self-assembly of double-stranded DNA molecules at nanomolar concentrations. *Biochemistry* **2007**, *46*, 164–171. [CrossRef] [PubMed]
54. Danilowicz, C.; Lee, C.H.; Kim, K.; Hatch, K.; Coljee, V.W.; Kleckner, N.; Prentiss, M. Single molecule detection of direct, homologous, DNA/DNA pairing. *Proc. Natl. Acad. Sci. USA* **2009**, *106*, 19824–19829. [CrossRef] [PubMed]
55. Lee, D.J.; Danilowicz, C.; Rochester, C.; Kornyshev, A.; Prentiss, M. Evidence of protein-free homology recognition in magnetic bead force–extension experiments. *Proc. Math. Phys. Eng. Sci.* **2016**, *472*, 20160186. [CrossRef]
56. Gladyshev, E.; Kleckner, N. Direct recognition of homology between double helices of DNA in Neurospora crassa. *Nat. Commun.* **2014**, *5*, 3509–3519. [CrossRef]
57. Gladyshev, E.; Kleckner, N. Recombination-independent recognition of DNA homology for repeat-induced point mutation. *Curr. Genet.* **2017**, *63*, 389–400. [CrossRef]
58. Kornyshev, A.A.; Leikin, S. Sequence recognition in the pairing of DNA duplexes. *Phys. Rev. Lett.* **2001**, *86*, 3666–3669. [CrossRef]

life

Article

Thermodynamic Factors That Drive Sequence-Specific DNA Binding of Designed, Synthetic Minor Groove Binding Agents

Ananya Paul [1], Abdelbasset A. Farahat [1,2], David W. Boykin [1] and W. David Wilson [1,*]

[1] Department of Chemistry and Center for Diagnostics and Therapeutics, Georgia State University, Atlanta, GA 30303, USA; apaul@gsu.edu (A.P.); a4f2002@mans.edu.eg (A.A.F.); dboykin@gsu.edu (D.W.B.)
[2] Department of Pharmaceutical Organic Chemistry, Faculty of Pharmacy, Mansoura University, Mansoura 35516, Egypt
* Correspondence: wdw@gsu.edu; Tel.: +1-404-413-5503; Fax: +1-404-413-5505

Abstract: Ken Breslauer began studies on the thermodynamics of small cationic molecules binding in the DNA minor groove over 30 years ago, and the studies reported here are an extension of those ground-breaking reports. The goals of this report are to develop a detailed understanding of the binding thermodynamics of pyridine-based sequence-specific minor groove binders that have different terminal cationic groups. We apply biosensor-surface plasmon resonance and ITC methods to extend the understanding of minor groove binders in two directions: (i) by using designed, heterocyclic dicationic minor groove binders that can incorporate a G•C base pair (bp), with flanking AT base pairs, into their DNA recognition site, and bind to DNA sequences specifically; and (ii) by using a range of flanking AT sequences to better define molecular recognition of the minor groove. A G•C bp in the DNA recognition site causes a generally more negative binding enthalpy than with most previously used pure AT binding sites. The binding is enthalpy-driven at 25 °C and above. The flanking AT sequences also have a large effect on the binding energetics with the -AAAGTTT- site having the strongest affinity. As a result of these studies, we now have a much better understanding of the effects of the DNA sequence and compound structure on the molecular recognition and thermodynamics of minor groove complexes.

Keywords: DNA minor groove binder; mixed base-pair DNA sequences; sequence selectivity; ligand–DNA complex thermodynamic; molecular curvature; heterocyclic diamidine; biosensor; calorimetry

Citation: Paul, A.; Farahat, A.A.; Boykin, D.W.; Wilson, W.D. Thermodynamic Factors That Drive Sequence-Specific DNA Binding of Designed, Synthetic Minor Groove Binding Agents. *Life* **2022**, *12*, 681. https://doi.org/10.3390/life12050681

Academic Editors: Tigran Chalikian and Jens Völker

Received: 30 March 2022
Accepted: 29 April 2022
Published: 4 May 2022

1. Introduction

In the 1980s, Ken Breslauer and some excellent coworkers initiated a series of fundamental studies on the thermodynamics of small-molecule, minor groove agents, and intercalators, binding to different DNA sequences [1–7]. Among other techniques, they used batch calorimetry in ground-breaking investigations of the DNA complexes of these compounds. They established methods for these types of studies that have continued to influence thermodynamic analysis of small-molecule-DNA binding to this day, including the work reported here. They introduced concepts and methods such as entropy-enthalpy compensation that masked driving forces when looking at free energy alone [5]. They used structural studies to bring an understanding of the structural and solution properties that influence the thermodynamics of DNA complex formation [4]. In these studies, they began to develop a microscopic understanding of the experimental macroscopic thermodynamic results. They started the research with state-of-the-art batch calorimeters, which, from personal experience with one of my early colleagues, Harry Hopkins, require a thermodynamics artist's touch to bring forth beautiful thermodynamics pictures. They later moved to much-improved titration calorimeters from MicroCal but continued with detailed thermodynamics studies of DNA interactions that now are going on around the world. It is a pleasure to write this paper in honor of Ken's 75 birthday.

The Breslauer and other laboratories began studies with available, classical minor groove binders from polyamides, such as netropsin, to other types of cationic heterocycles, such as berenil and Hoechst dyes, which were all specific for binding to AT DNA sequences (Figure 1A) [1–8]. These are all uniformly concave-shaped compounds that fit snugly into the minor groove in A-tract sequences and have groups that can H-bond with the N3 of dA or O2 of dT at the floor of the groove.

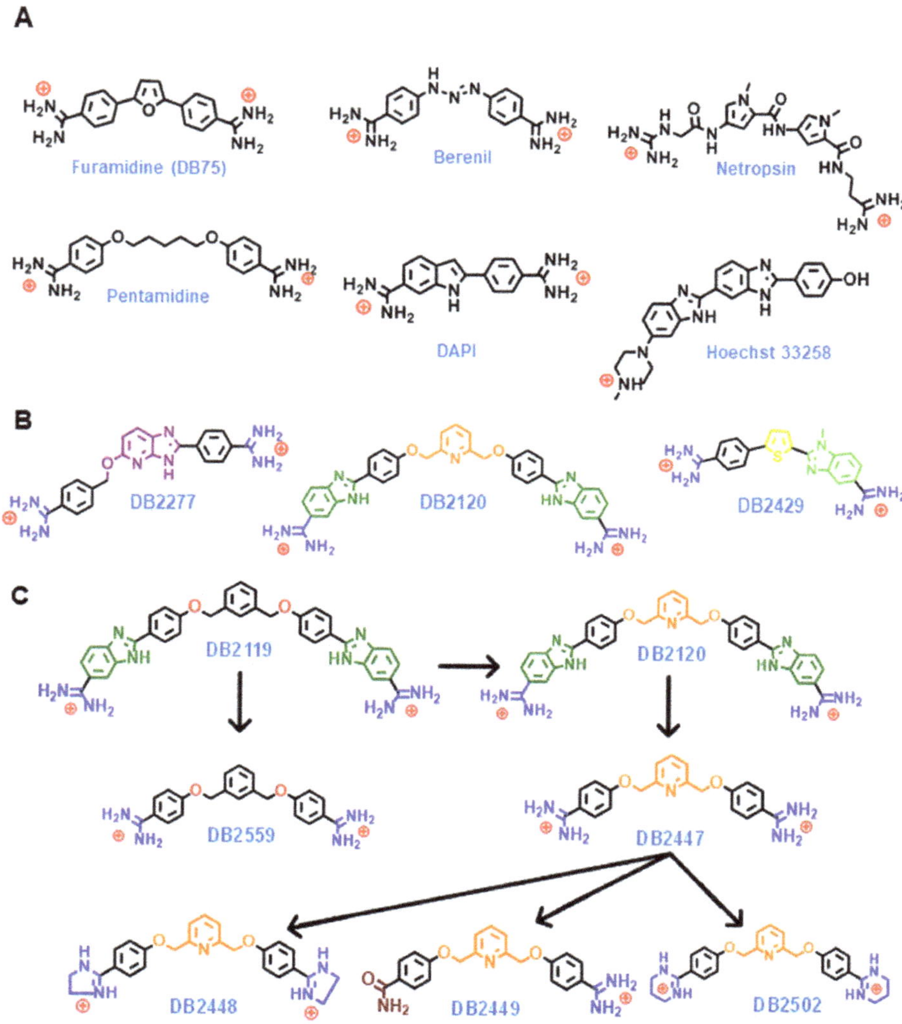

Figure 1. (**A**) Structures of classical AT-specific DNA minor groove binders; (**B**) structures of three very different designed mixed-sequence DNA minor groove binders; (**C**) schematic representation of the development of AT-specific DNA binding compounds from DB2119 to DB2559 and from mixed DNA sequence-specific compounds DB2120 to DB2447 and the analogs for this study. Color schemes used in this figure denote different functional groups.

In addition to forming an H-bond with the cytosine C=O in the minor groove, the 2-amino group of G projects an -N-H into the groove and presents a steric block to classical AT-specific minor groove binders [9]. In spite of the many successes with these early minor groove binders, the lack of a broad selection of sequence-specific compounds was a limitation for expanded applications. Crystal structures of DNA and complexes suggested that compounds with H-bond acceptor groups could bind to the free G-NH in the minor groove to give GC binding specificity [9–14]. This concept was successful with polyamides but has not been broadly applied to the other minor groove binders such as those listed above (Figure 1).

As heterocyclic amidine minor groove binders such as 4′,6-diamidino-2-phenylindole (DAPI) are widely used as cellular nuclear stains and others such as pentamidine, furamidine, and berenil (Figure 1A) have been used or tested in humans and/or animals for therapeutic applications [15–19], we have used the heterocyclic amidine template as the platform for incorporation of single hydrogen bond-accepting groups in generating the first set of compounds with increased specificity. The goal with these first compounds was to recognize a GC base pair in an AT sequence. In this way, it would be possible to create modules that could be combined to give broad DNA sequence recognition. Successful compounds with azabenzimidazole, N-alkylbenzimidazole, and pyridine H-bond acceptor groups for G recognition were created (Figure 1B) [20–25]. Extensive elaborations of the N-alkylbenzimidazole module have significantly improved the affinity and selectivity of that module. With a thiophene adjacent to the imidazole of N-alkylbenzimidazole, a preorganized structure for minor groove recognition was created [25–28].

The original pyridyl-linked amidine-benzimidazole-phenyl compound, DB2120, binds strongly with the single G•C bp-containing -A4GT4- target sequence with a $K_D < 0.1$ nM. DB2120, however, has poor solution properties and aggregation difficulties even at low concentrations under standard experimental conditions for DNA complexes. Because of its size and AT binding benzimidazole units, it also has less selectivity than optimum for use in most applications. To increase the possible uses of the pyridine series of compounds, it is, thus, essential to develop smaller molecules that will have better solution and sequence recognition characteristics. Hence, in the studies reported here, in-depth experiments were conducted with 12 primary DNA hairpin duplex sequences (Figure 2) four of which contain pure AT sequences in their recognition sites (AAAAAA, AAATTT, AATAAT, and ATATAT), four have mixed AT and single GC bp DNA sequences (AAAGAAA, AAAGTTT, AATGAAT, and ATAGTAT), and four have AT sequences with two G•C bps (AAAGCAAA, AAAGCTTT, AATGCAAT, and ATAGCTAT). Three pyridine compounds that are smaller than DB2120, as well as a phenyl compound, were synthesized with the smaller compound design ideas (Figure 1C). The smaller compounds are easier to synthesize, have improved solution properties, and have strong and specific binding to DNA sequences with a single G•C bp. The results with these new pyridine and phenyl compounds are reported here.

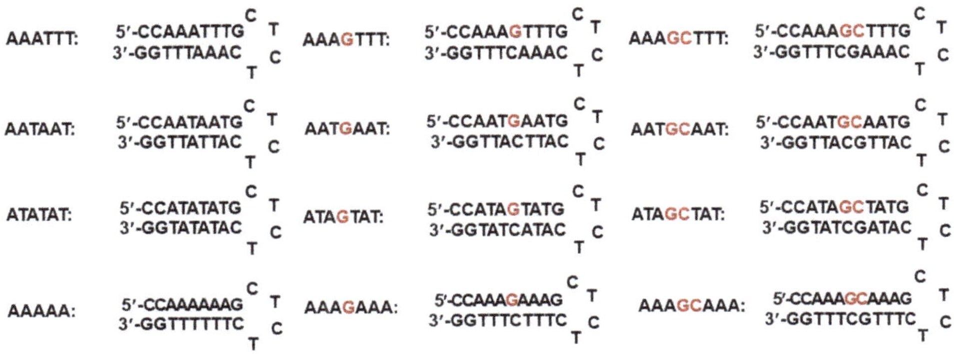

Figure 2. The DNA sequences used in this study; DNA sequences used for Surface Plasmon Resonance (SPR) studies were labeled with 5′-biotin.

2. Compound Design

The goal with these compounds was to make a series of DB2120 analogs of reduced size but better solution properties and with strong and selective binding to target single G•C bp-containing sequences [23,24]. The compounds would retain the DB2120 core structure but with different terminal cationic groups. Dications with amidine, imidazoline, and tetrahydropyrimidine were successfully prepared along with a monocation with terminal amidine and amide groups (Figure 1C). As a control, an amidine analog with a central phenyl ring that cannot form the H-bond required for GC specificity was also prepared. By using our relative curvature determination method [28] (Results), the compounds were found to have an optimum curvature for the minor groove. Results with these five compounds and the target and control DNA sequences described above (Figure 2) are reported here. For complete studies, target sequences have a single G•C bp in an AT sequence context, while the control sequences have no G•C bp or a two G•C bp insert.

3. Results

3.1. Compound Synthesis

Scheme 1 describes the synthesis of the final amidines **4**. Cyanophenol derivatives **2** were allowed to react with bis(chloromethyl)pyridine **1** in anhydrous dimethylformamide in the presence of potassium carbonate as a base. The formed bisnitriles were converted to the final amidines by applying Pinner reaction conditions [29,30], where the bisnitriles were converted to the intermediate imidate ester hydrochloride by stirring in dry ethanolic HCl. The formed imidate ester was converted to the corresponding amidine by stirring in dry ethanolic ammonia. Finally, the amidines were purified by conversion to the free base using sodium hydroxide and then formation of the hydrochloride salt by stirring in ethanolic HCl. The characterization of the final compounds have described in Experimental Methods.

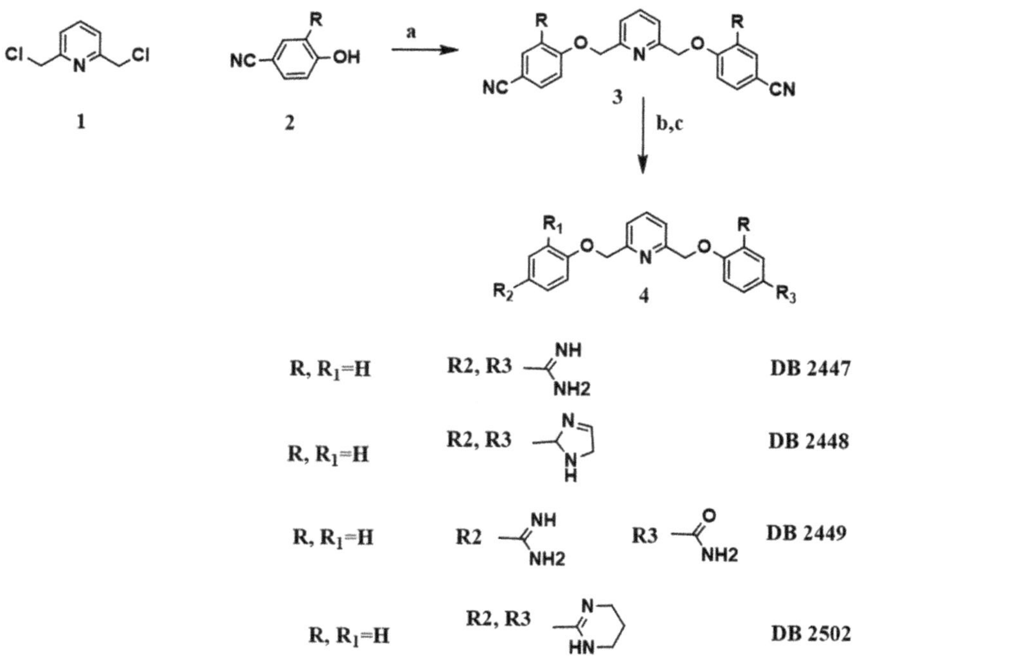

Scheme 1. Reagents and conditions: (**a**) K_2CO_3, DMF, 45 °C, 4 h; (**b**) ethanol-HCl, 0 °C—r.t.; (**c**) ethanol, ammonia or amine, r.t or reflux.

3.2. DNA Thermal Melting (ΔT_m): Screening for Relative Affinity and Sequence Selectivity for Target and Control DNA Binding

Changes in DNA thermal melting temperature (ΔT_m) provide an initial ranking of compounds for binding affinity and relative sequence specificity with the target and control DNA sequences [31,32]. The binding affinities of the pyridyl-centered heterocyclic cations were tested with pure AT sequences, which are the primary sites of most known minor groove binders from netropsin to furamidine (Figure 1A) [33–35]. The target DNAs of primary interest for the compounds in this research have a single G•C bp with flanking AT sequences. The T_m experiments were carried out in the presence of nonalternating and alternating AT, and mixed flanking sequences, AAA-TTT, ATA-TAT, AAT-AAT, and AAA-AAA, Figure 2, where the dashes indicate zero, one, or two G•C bps. These are useful test sequences as each flanking AT sequence has different properties, including variations in the minor groove positions of the H-bond acceptor groups on the A•T bp. The number of G•C bp also varies in these selected mixed DNA sequences, which also gives significant differences in minor groove width (Figure 3).

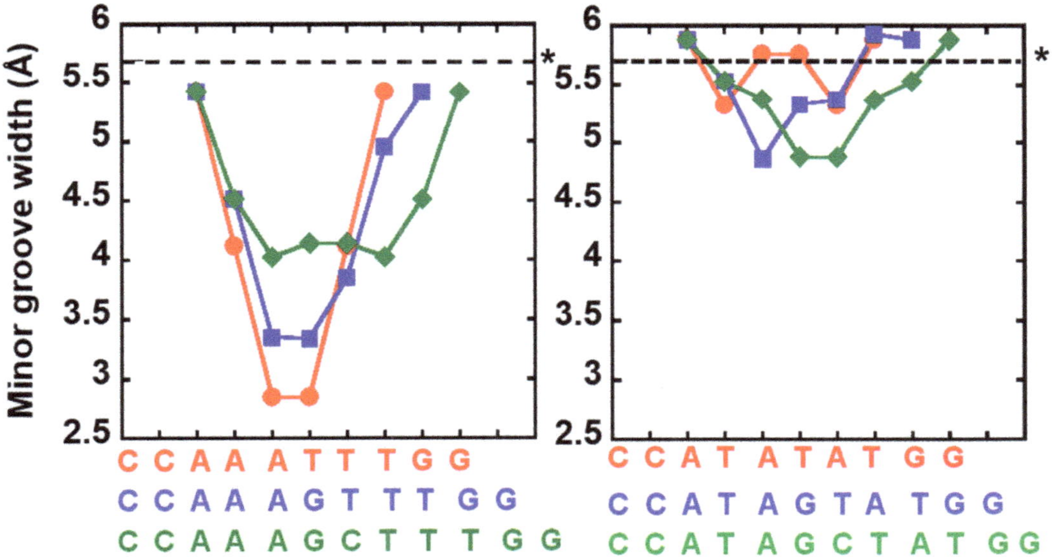

Figure 3. Minor groove width vs. target DNA sequences calculated from the online algorithm of Rohs and coworkers (*Nucleic Acids Res.* **2013**, *4*, W56-62). * indicates the minor groove width of standard *B*-form DNA. Groove width gives the perpendicular separation of helix strands drawn through phosphate groups, diminished by 5.8 Å to account for van der Waals radii of phosphate groups.

With AAA-TTT, a type of nonalternating AT sequence, DB2447, the direct truncated analog of the original pyridine, DB2120, resulted in an encouraging increase in the thermal stability of the single G•C bp-containing target site sequences (Figure 4 and Table S1). DB2447 also showed sequence selectivity, as expected from previous results and the compound design approach. The compound showed significantly lower thermal stability increases for the two G•C bps and all AT-containing sequences (Figure 4 and Table S1).

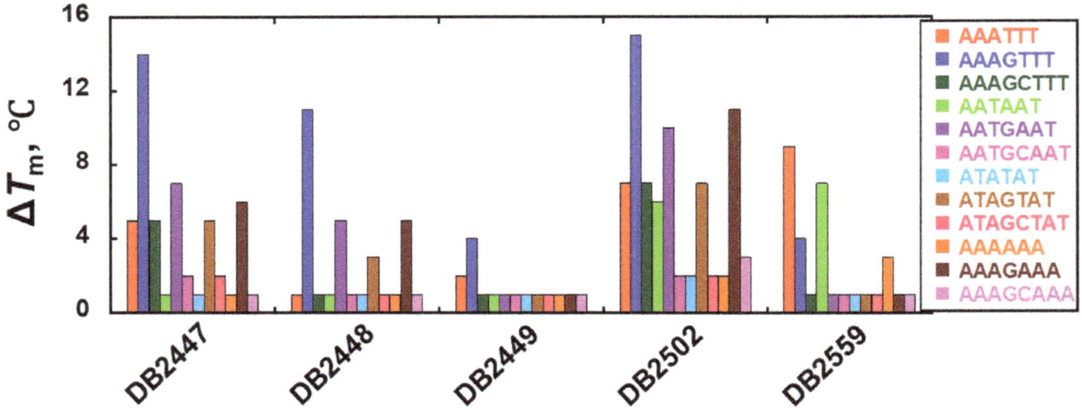

Figure 4. Comparison of relative binding affinities, ΔT_m, °C ($\Delta T_m = T_{m\,(\text{the complex})} - T_{m\,(\text{the free DNA})}$) by thermal melting experiments of the designed heterocyclic amidine compounds with pure AT base-pair and mixed single G•C base-pair-containing DNA sequences. The listed values are for the 2:1 [ligand]/[DNA] ratio and an average of two independent experiments with a reproducibility of ±0.5 °C. Full DNA sequences are as described in Figure 2.

Two analogs of DB2447 with terminal 4,5-dihydro-1-H-imidazole (*Im*) (DB2448) and 1,4,5,6-tetrahydro pyrimidine (*THP*) (DB2502) were synthesized and tested to determine how the molecular size and chemistry of terminal dications affect the sequence binding affinity and selectivity. The *Im* compound binds to a single G•C bp DNA somewhat more weakly than DB2447 but with significantly improved selectivity (Figure 4 and Table S1). The *THP* compound DB2502 binds to the single G•C bp target slightly more strongly than DB2447 and with similar selectivity. These results indicate that the terminal *Im* and *THP* analogs are useful additions to expand the chemical space of the parent, DB2447. The three compounds can be especially useful in biological test applications where their uptake into different cell types can be very different. In cellular applications where strong specificity is needed, the Im analog may be preferred. The compound DB2449, with one amidine replaced with an amide, has quite low ΔT_m values with all of the tested DNAs (Figure 4 and Table S1).

In addition to the compound structure and properties, the DNA sequence has a major effect on minor groove binding. The alternating DNAs ATA-ATA have a significant drop in ΔT_m relative to the target, AAA-TTT sequences with all compounds. This is expected based on the wider Minor Groove (MG) of the alternating sequence (Figures 3 and 4 and Table S1) [36,37]. The sequence with a pure A-tract AAA-AAA also has a surprisingly lower ΔT_m relative to the target, AAA-TTT sequences. The AAT-AAT sequences have an intermediate ΔT_m relative to the target, AAA-TTT sequences. In summary, the -AAAGTTT- sequence has the best ΔT_m of all of the flanking AT sequence variations. This can easily be seen in Figure 4, where the three pyridyl-diamidines with each DNA sequence group have the highest peaks in the histogram set. With DB2559, the -AAATTT- and -AATAAT- sequences have the highest ΔT_{ms} as expected for a central phenyl. Analysis of the minor groove widths of the different single G•C bp sequences showed that of all sequences, -AAAGTTT- has the most narrow minor groove and is most appropriate for binding compounds such as DB2447 with a connected aromatic system of approximately 3.4 A in width (Figure 3). The weakest binding is seen with the -ATAGTAT- sequence and it has the widest DNA minor groove, basically the same as a standard B-form minor groove. The groove widths for the -AAAGTTT- and -ATAGTAT are shown in Figure 3 for reference.

3.3. Biosensor-Surface Plasmon Resonance (SPR): High-Resolution Evaluation of Binding Affinity, Kinetics, Stoichiometry, and Cooperativity

Biosensor-SPR methods provide a high-resolution, label-free way to quantitatively evaluate the binding affinity, selectivity, and stoichiometry of a set of compounds with a spectrum of immobilized DNAs [38–40]. The parent pyridyl-linked-phenyl-amidine, DB2447, binds strongly with AAAGTTT, and global kinetics fitting defined a single binding site with $K_A = 5.5 \times 10^8$ M^{-1} ($K_D = 1.8$ nM) at 0.1 M NaCl (Figures 5 and 6, Table S2). The strong binding of DB2447 is the result of the rapid association that is at the instrumental limitation ($k_a = {\sim}4.4 \times 10^7$ $M^{-1}s^{-1}$) and a comparatively slow dissociation rate constant ($k_d = 7.8 \times 10^{-2}$ s^{-1}). DB2447 binds to the pure AT, -AAATTT-, sequence as a monomer complex with rapid dissociation and a 200-fold lower affinity compared to AAAGTTT. This result indicates that DB2447 maintains surprisingly high sequence selectivity for the single G•C bp sequence. The sensorgram of AAATTT shows an off-rate that is much faster, and complete dissociation from the complex occurs within the first few seconds of the dissociation phase (Figure S1). With the -AAAGCTTT- binding site, DB2447 shows a 30-fold weaker binding affinity than with -AAAGTTT- under the same experimental conditions (Figures 5 and S1, Table S2). It appears that DB2447 can induce a somewhat favorable minor groove site in -AAAGCTTT- for binding, but the extra GC is a mismatch that reduces the binding affinity. The phenyl derivative, DB2559, with a simple *-N-* to *-CH* conversion, has a stronger binding for pure AT sequences, over ten times stronger than with the single G•C sequence, as with most classical minor grooves binders (Figures 5 and S2, Table S2). This result is as expected for compounds without an H-bond acceptor group in a position to bind to the G-NH that faces into the minor groove.

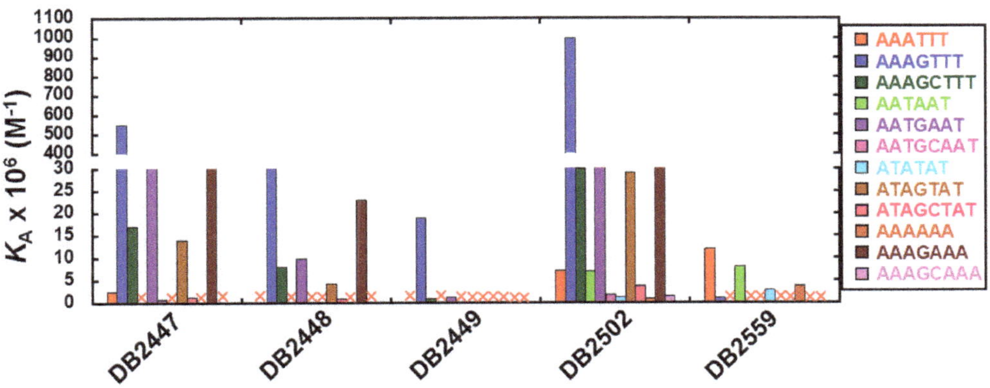

Figure 5. Comparison of equilibrium binding constants (K_A, M^{-1}) of phenyl and pyridine analogs with pure AT and mixed single G•C base-pair-containing DNA sequences. "X" represents no measurable K_A under our experimental conditions. The listed binding affinities are an average of two independent experiments carried out with two different sensor chips, and the values are reproducible within a 10% experimental error. Full DNA sequences are as described in Figure 2.

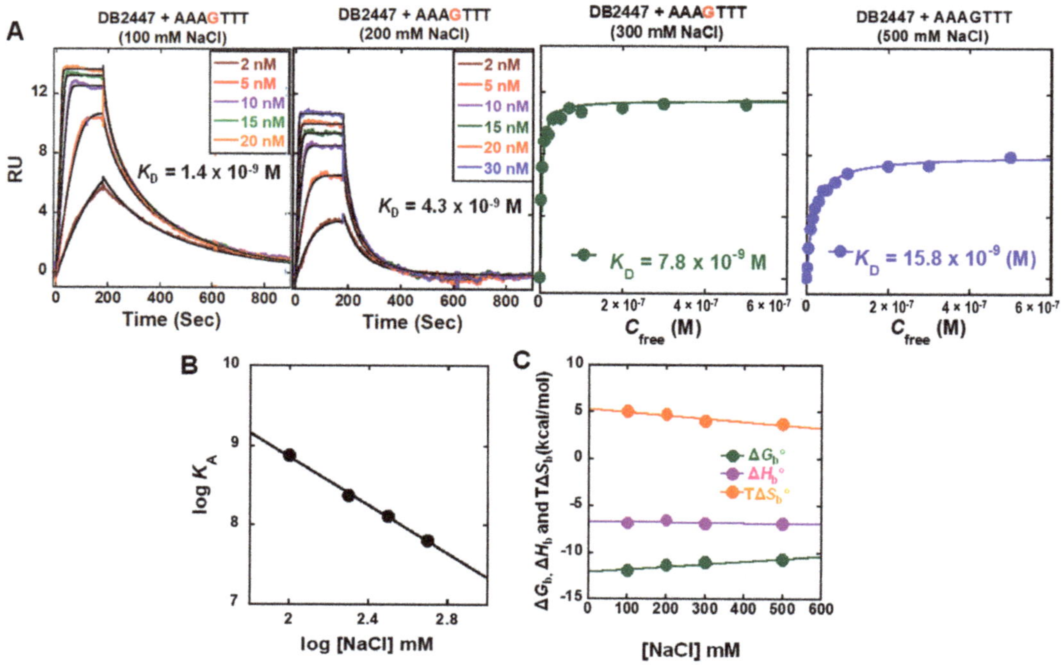

Figure 6. (**A**) SPR sensorgrams with kinetics fits and steady-state binding plots for DB2447 with the AAAGTTT DNA sequence at different salt concentrations; (**B**) salt dependence of K_A for DB2447 binding as determined by SPR. The K_A values were obtained by global kinetics (at two lower salt concentrations) and steady-state fits (at two higher salt concentrations); (**C**) plot of $\Delta G_b°$, $\Delta H_b°$, and $T\Delta S_b°$ versus salt concentrations for DB2447 with AAAGTTT sequence at 25 °C. The listed binding affinities are standard thermodynamic values and are an average of two independent experiments carried out with two different sensor chips. The values are reproducible within a 10% experimental error; RU = Response Unit based on amounts of bound compounds on the immobilized DNA.

The derivative, DB2448, with an imidazoline terminal group has a K_D value about half as strong as DB2447 for the single G•C bp sequence, but it has negligible binding to AAATTT under our experimental conditions, an impressive improvement in sequence selectivity (Figures 5 and S3, Table S2). DB2502, on the other hand, with six-atom terminal cations has a K_D value that is about twice as strong as that of DB2447. Unfortunately, its K_D value for the pure AT sequence is about three times stronger than for DB2447. With the AAAGCTTT sequence, DB2448 has a K_D value over twice that of the DB2447 constant but the DB2502 value is below half of the DB2447 K_D value. The *Im* substitution thus has a promising increase in overall selectivity while the results with DB2502 are a disappointment in selectivity for this series (Figures 5 and S4, Table S2). As with the T_m experiments, DB2449, the monocation, has relatively weak binding to all tested DNA sequences except -AAAGTTT with a K_D of 52 nM (Figures 5 and S5, Table S2). This compound has very different solution properties relative to all other compounds and may be an advantage in cell studies.

With the other three sets of flanking sequences, the single G•C bp sequences always have the strongest binding (Figures 5 and S1–S5, Table S2), as with -AAAGTTT-, but the binding is weaker than with -AAAGTTT. With -AAAGAAA- and AATGAAT, the binding with DB2447 is about a factor of ten weaker than with -AAAGTTT-. With the fully alternating sequence -ATAGTAT-, the binding is reduced by close to a factor of 40. Similar reductions in affinity are seen with the other compounds and these DNA sequences

(Figure 5, Table S2). With the Figure 5 histograms, the single G•C set of sequences have the highest plot in each set of DNAs. It should also be noted, however, that the GC sequences have a stronger-than-expected binding.

3.4. Molecular Curvature Determination

Molecular curvature plays a crucial role in sequence-selective DNA minor groove recognition in conjunction with DNA minor groove binders' molecular functionality and stacking surface. Correct curvature is important for strong H-bonding interaction, charge interactions, and van der Waals stacking in the groove of the DNA. Our previous report [28] stated a graphical approach method to determine relative molecular curvature values for minor groove binding compounds. In this approach, the diamidines compounds are energy-minimized in the SPARTAN software package using the DFT/B3LYP theory with the 6-31+G* basis set. The compounds are then matched up in a PowerPoint graphics package. A reference circle (black circle) is drawn through both amidine carbons, the center of the compound where the circle's periphery passes through the center point of the individual molecular unit of the entire molecule, illustrated in DB2447, DB2448, and DB2502 (Figure 7). Two straight lines (orange) are drawn from the circle point at the center of the molecule to the amidine carbons. The midpoint of these two lines defines the comparative curvature value for the diamidines compound. The curvature values are 147° for DB2447, 147° for DB2448, and 143° for DB2502. The curvature analysis of a library of strong DNA minor groove binding compounds by this method offers a standardization value of around 140–145° curvature angle. DB2447 and its analogs also show a similar ° of curvature angle, which supports the strong binding affinity of these diamidines compounds.

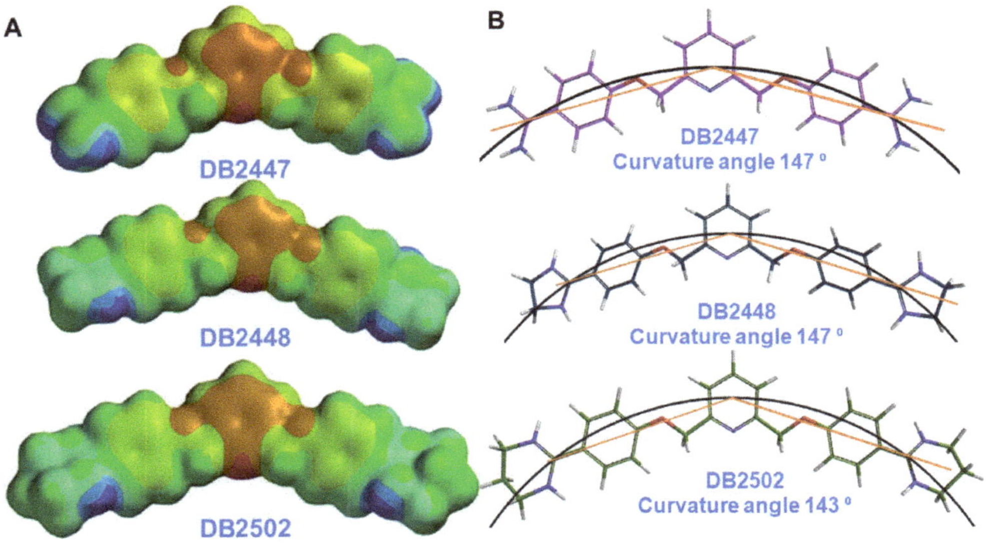

Figure 7. (**A**) Energy-minimized structures of DB2447, DB2448, and DB2502 at B3LYP/6-31G* (p,d) level of theory. In the electrostatic potential maps, red indicates high electronegativity and blue indicates electron-deficient/positively charged regions. (**B**) Molecular curvatures for DB2447, DB2448, and DB2502. Black circles represent reference circles and orange lines represent two straight lines drawn from the circle point at the center of the molecule to the amidine carbons.

3.5. Effect of Salt Concentration and Temperature on DB2447 Binding to Single G•C bp DNA Sequences

The T_m, structure evaluation, and SPR binding results indicate that DB2447 has an optimized length, curvature, and flexibility for effective and selective recognition of a single G•C bp in an AT minor groove sequence. The SPR results also indicate that DB2447 has improved solution properties relative to DB2120, which allows us to explore the thermodynamic behavior of this compound with single G•C bp and other sequences in more detail. These results help to provide a fundamental understanding of the molecular basis for specific recognition of the DNA minor groove.

To evaluate the effect of ionic strength on DB2447 binding affinity with the -AAAGTTT-sequence, SPR experiments were carried out from 100 to 500 mM NaCl concentrations at 25 °C (Figure 6, Table 1). The equilibrium binding constants (K_A) obtained either by global kinetics fits at low salt concentration or by steady-state fits at higher salt concentration are collected in Table 1 and Figure 6. Both theory and experiment suggest that the logarithm of the equilibrium binding constant K_A is a linear function of the logarithm of NaCl concentration for many organic cations binding to DNA [41,42]. For a typical DNA−cation complex, the equilibrium binding constant values decrease as the salt concentration increases with a slope that depends on the compound charge [42–46]. As seen in Figure 6, the $\log(K_A)$ versus $\log[Na^+]$ plot for DB2447 is linear with a slope of 1.6. The number of phosphate contacts (Z) between DB2447 and the AAAGTTT DNA sequence is predicted to be two and can be obtained in experiments from the slope/0.88, where 0.88 is the fraction of phosphate charge shielded by the total associated counterions. For the 11-base-pair synthetic oligomer AAAGTTT, the obtained Z is 1.8. These results indicate that the dicationic DB2447 releases 2 Na^+ ions when binding to the DNA minor groove. The enthalpy change, $\Delta H_b°$, for binding is essentially constant with salt concentration, while for both $\Delta G_b°$ and $T\Delta S_b°$, values decrease by one kilocalorie with the change in salt concentration from 100 to 500 mM NaCl.

Table 1. Kinetics [a] and steady-state [b] analysis of DB2447 with the AAAGTTT sequence at different salt concentrations in 50 mM Tris-HCl, 1 mM EDTA buffer, pH 7.4 at 25 °C.

NaCl (mM)	K_D (nM)	$\Delta G_b°$ (kcal/mol)	$\Delta H_b°$ (kcal/mol)	$T\Delta S_b°$ (kcal/mol)
100	1.4	−12.0	−6.8	5.2
200	4.3	−11.4	−6.5	4.9
300	7.8	−11.1	−6.9	4.2
500	15.8	−10.7	−6.9	3.8

[a] Kinetics analysis was performed by global fitting with a 1:1 binding model and [b] steady-state fits were carried out by using 1:1 binding equation model. $\Delta H_b°$ was determined in ITC experiments; $T\Delta S_b = -\Delta G_b° + \Delta H_b°$.

To obtain an additional understanding of the thermodynamic basis for DB2447 interactions with DNA, SPR experiments were conducted from 25 °C to 40 °C at 100 mM salt concentration. The SPR results reveal that temperature significantly affects the DB2447-DNA binding thermodynamics and kinetics, as shown in Figure 8 and Table 2. The ligand-DNA binding affinity decreases (K_A) with experimental temperature. However, the temperature changes have a smaller effect on the $\Delta G_b°$ as previously observed for other minor groove binders [42]. As can be seen in Figure 7, the enthalpy and entropy for binding have compensating decreases with increasing temperature. The $T\Delta S_b°$ and $\Delta H_b°$ values are similar at 20 °C but $T\Delta S_b°$ approaches zero at 45 °C, and $\Delta H_b°$ completely accounts for $\Delta G_b°$.

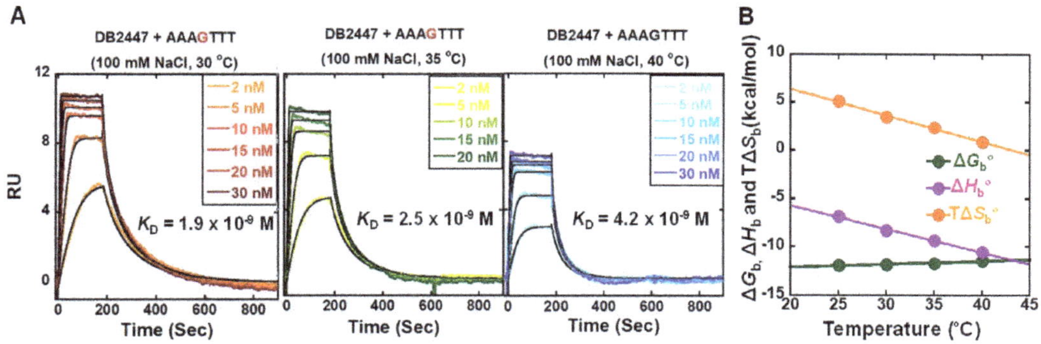

Figure 8. (**A**) SPR sensorgrams with kinetics fit for DB2447 with the AAAGTTT DNA sequence at different temperatures at 100 mM NaCl; (**B**) thermodynamic results, from Table 2 for binding of DB2447 to the AAAGTTT site at different temperatures. The listed binding affinities are standard thermodynamic values and binding affinities are an average of two independent experiments carried out with two different sensor chips, and the values are reproducible within a 10% experimental error; RU = Response Unit based on amounts of bound compounds on the immobilized DNA.

Table 2. Kinetics and thermodynamic results for DB2447 with the AAAGTTT sequence at different experimental temperatures at 50 mM Tris-HCl, 100 mM NaCl, 1 mM EDTA buffer at pH 7.4.

Temperature (°C)	K_D (nM)	$\Delta G_b{}^\circ$ (kcal/mol)	$\Delta H_b{}^\circ$ (kcal/mol)	$T\Delta S_b{}^\circ$ (kcal/mol)
25	1.4	−12.0	−6.8	5.2
30	2.0	−11.8	−8.3	3.5
35	2.5	−11.7	−9.3	2.4
40	4.2	−11.4	−10.5	0.9

3.6. Isothermal Titration Calorimetry (ITC) of Complex Formation: Effects of Salt Concentration and Temperature

Characterization of the full thermodynamics of cationic compound–DNA interactions is an essential component of any detailed analysis of DNA molecular recognition and rational drug design. Isothermal titration calorimetry (ITC) is the method of choice to analyze the energetic basis for the strong and selective binding of compounds with the DNA minor groove. It has been used with minor groove binders since the early studies of Ken Breslauer and colleagues and it is applied here to DB2447–DNA complexes. ITC provides a key component of the thermodynamic profile of ligand–DNA interactions by direct determination of the enthalpy ($\Delta H_b{}^\circ$) of binding. With the enthalpy and $\Delta G_b{}^\circ$ from biosensor experiments, it is possible to calculate the entropy of binding (ΔS_b) from the known thermodynamic relationships ($\Delta G_b{}^\circ = -\mathrm{RTln}K_A$) and ($\Delta G_b{}^\circ = \Delta H_b{}^\circ - T\Delta S_b$). Thermodynamic profiles are valuable in drug design because they provide quantitative data on drug–DNA interactions that cannot be obtained directly by structural or computational methods [47–51]. They also provide valuable ideas about the compound-DNA-water components of complex formation [52].

In the experiments reported here, ITC was used to monitor the heat released upon the binding of DB2447 to the target binding sites. For strong binding compounds such as DB2447, ITC experiments require considerably higher concentrations than the compound K_D for DNA interactions. In cases such as this, the enthalpy of binding can be determined quite accurately in the presence of excess DNA in the calorimetry cell such that the compound is fully bound to the DNA. In this model-free approach, the $\Delta H_b{}^\circ$ is simply determined from the average of $\Delta H_b{}^\circ$ versus the binding ratio below saturation binding. The binding constant must then be determined at lower concentrations by an alternative method such as SPR or fluorescence methods.

The ITC curves were fitted using Origin software to obtain the enthalpy at each titration point. The data (Figure 9) indicate an exothermic interaction after adding DB2447 to the solution containing -AAAGTTT-DNA at each NaCl concentration. The subtraction of the integrated peak areas for ligand/buffer titration from the ligand/DNA titration gives a direct determination of $\Delta H_b{}^\circ$ at each temperature. Figure 9 shows the titration of DB2447 into AAAGTTT with the blank buffer correction, and the $\Delta H_b{}^\circ$ value is -6.8 ± 0.2 kcal/mol at 25 °C at a 100 mM salt concentration. The favorable negative enthalpy change suggests strong H-bonding, electrostatic, and van der Waals interactions between DB2447 and AAAGTTT DNA. Interestingly, the complex formation of DB2447-AAAGTTT has an enthalpy at 25 °C that is larger than most of the A-tract minor groove binders reported in the literature [44,47–49]. This is expected from the known sequence-dependent differences in minor groove structure and hydration between the single G•C bp and all AT DNA sequences (Table 1).

To evaluate the relationship between thermodynamic measurements of the DB2447–AAAGTTT complex and experimental salt concentrations, the ITC experiments were carried out at 100–500 mM NaCl concentrations. The SPR results show that the binding constant decreases by almost ten times (Table 1) with increasing salt concentration. However, the ITC experiments show that salt concentrations have a much smaller effect on $\Delta H_b{}^\circ$ than on $\Delta G_b{}^\circ$ (Figure 9 and Table 1). This phenomenon reveals the enthalpy of complex formation for an energetic component, which is the sum of interactions such as hydrogen bond formation and van der Waals interactions and is essentially independent of salt concentration.

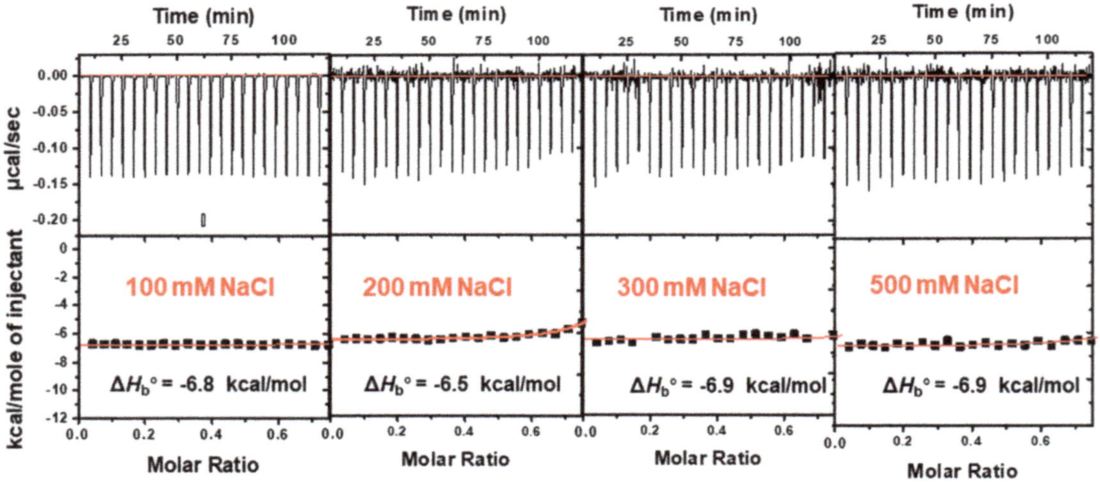

Figure 9. ITC data for the titration of DB2447 with AAAGTTT DNA at different salt concentrations at 25 °C. The listed binding enthalpies are an average of two independent experiments, and the values are reproducible within a 10% experimental error.

3.7. Determination of the Heat Capacity of the DB2447−AAAGTTT Complex

The ITC experiments of DB2447 with AAAGTTT were also carried out at different temperatures (15−40 °C) with a constant 100 mM NaCl concentration (Figure 10). The titration profiles indicate that the enthalpy of DB2447–DNA complex formation strongly depends on the experimental temperature and becomes more negative with increases in temperature (Table 2). The temperature-dependent differences in the binding enthalpy of DB2447–DNA complexes were used to calculate the heat capacity (ΔC_p) for binding from the slope of a linear least-squares fit of the plot of $\Delta H_b{}^\circ$ versus temperature, $\Delta C_p = -285$ cal/mol K (Figure 10). The temperature-dependent ITC results also show the effect of temperature on the entropy term. As the $\Delta G_b{}^\circ$ of binding is essentially constant with temperature (Figure 8,

Table 2), the subtraction of $\Delta H_b°$ from $\Delta G_b°$ yields results that show that the entropy of binding decreases as the temperature is increased (Table 2) and, as previously noted, $T\Delta S_b°$ approaches zero at 40–45 °C.

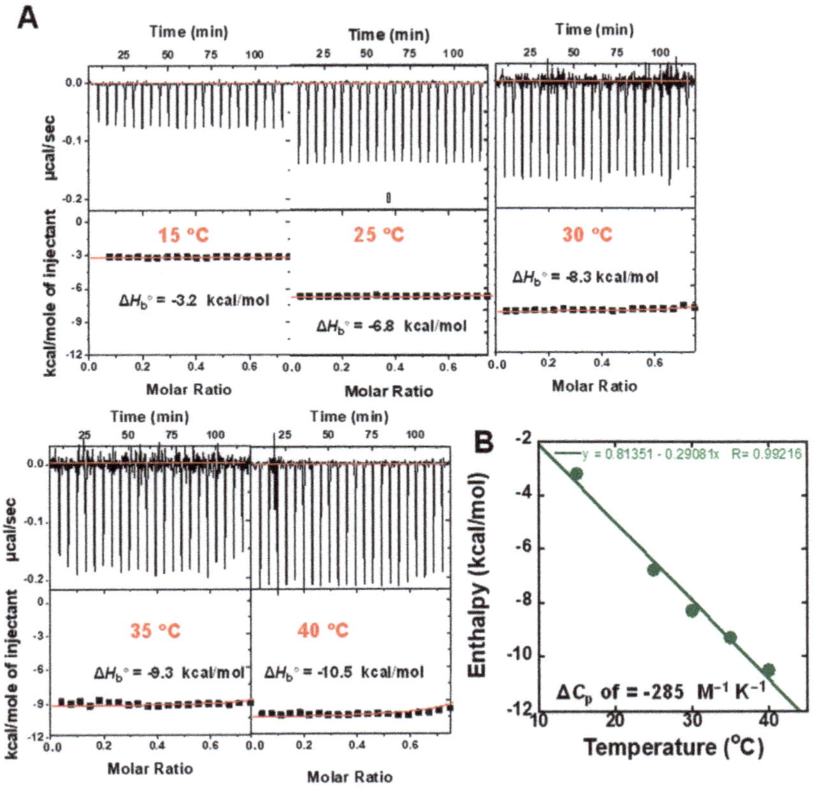

Figure 10. (**A**) ITC data for the titration of DB2447 with AAAGTTT DNA at different experimental temperatures by using 100 mM NaCl. (**B**) Plot of $\Delta H_b°$ versus temperature for DB2447 with AAAGTTT DNA, and the linear fit yields a ΔC_p of -285 M^{-1} K^{-1}. The listed binding enthalpies are an average of two independent experiments, and the values are reproducible within a 10% experimental error.

3.8. The Effects of AT Flanking Sequence Variations on DB2447 Binding Thermodynamics

In Figure 11, the effects of the four sequences with different AT sequences flanking the single G•C bp binding site are shown. Determination of the ITC $\Delta H_b°$ and the SPR $\Delta G_b°$ (Figure 4) allows calculation of the ΔS_b for each sequence (Figure 11). The -AAAGTTT- sequence has the most favorable $\Delta G_b°$ and with a substantial $\Delta S_b°$, which is why it is the best binding sequence. The AATGAAT sequence has a similar $\Delta H_b°$ but a smaller $\Delta S_b°$ with a decreased $\Delta G_b°$. With AAAGTTT and ATAGTAT, the binding $\Delta H_b°$ is decreased but the $\Delta S_b°$ is higher than with AATGAAT. With AAAGTTT, the binding becomes entropy-driven, and binding to AAATTT is shown as a strongly entropy-driven reference (Figure 11).

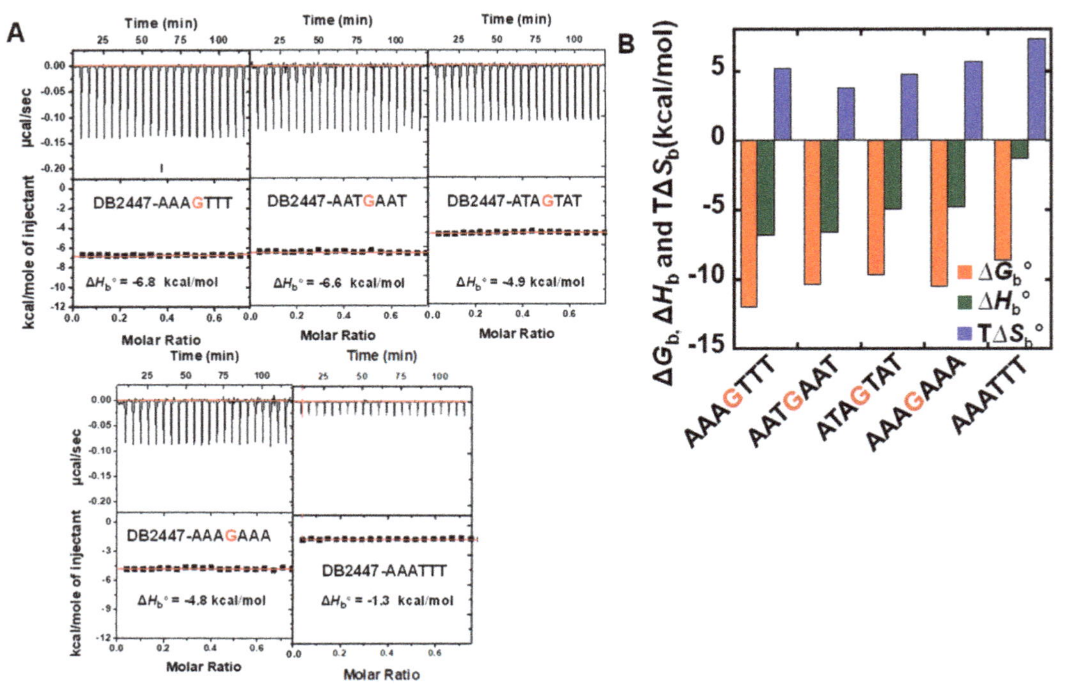

Figure 11. (**A**) ITC data for the titration of DB2447 with different single G•C base-pair-containing mixed DNA sequences in 50 mM Tris-HCl, 100 mM NaCl, 1 mM EDTA buffer, pH 7.4 at 25 °C. (**B**) Comparison of thermodynamic parameters, $\Delta G_b°$, $\Delta H_b°$, and $T\Delta S_b°$, of DB2447 with different single G•C base-pair-containing mixed DNA sequences. The plotted thermodynamic parameters are an average of two independent experiments, and the values are reproducible within a 10% experimental error. Full DNA sequences as described in Figure 2.

3.9. Thermodynamic Effects in the Binding of the Pyridyl Diamidine Compounds

The three compounds have similar thermodynamics with DB2502 having a slightly larger $\Delta G_b°$, while the value is smallest with DB2448 (Figure 12). DB2447 has the most favorable $\Delta H_b°$, suggesting that the unsubstituted amidines form the most favorable H-bonds. DB2502 has the most favorable $\Delta S_b°$, indicating that the large tetrahydropyrimidine displaces the most water from the minor groove on binding.

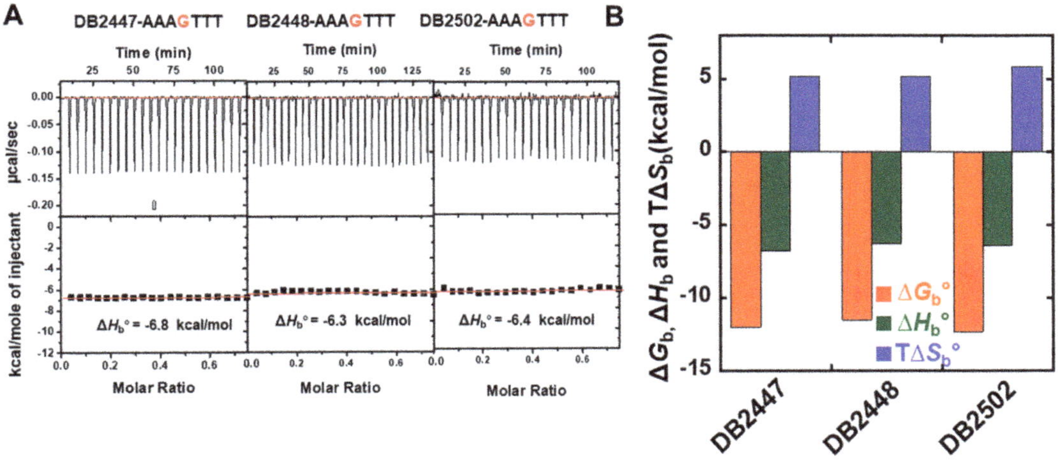

Figure 12. (**A**) ITC data for the titration of DB2447, DB2448, and DB2502 with AAAGTTT DNA sequence in 50 mM Tris-HCl, 100 mM NaCl, 1 mM EDTA buffer, pH 7.4 at 25 °C. (**B**) Comparison of thermodynamic parameters, $\Delta G_b°$, $\Delta H_b°$, and $T\Delta S_b°$, of DB2447, DB2448, and DB2502 with AAAGTTT. The plotted thermodynamic parameters are an average of two independent experiments, and the values are reproducible within a 10% experimental error.

3.10. Competition Electrospray Ionization Mass Spectrometry (ESI-MS) of DB2447

Competition MS allows high-throughput screening for the comparison of binding of compounds to a panel of DNA sequences for the evaluation of relative affinity and selectivity [53,54]. The dashes represent zero, one, or two G•C bps with the AAA-TTT sequences used. In Figure 13A, the free DNA peaks are shown for AAATTT (m/z 6684), AAAGTTT (m/z 7302), and AAAGCTTT (m/z 7921). After the addition of DB2447, the intensity of the peak for AAAGTTT (m/z 7302) is reduced with the appearance of a new peak at m/z 7672, which is the characteristic of a 1:1 AAAGTTT–DB2447 complex (Figure 13B). This is in agreement with the ΔT_m and SPR results that show stronger binding to the single G•C bp sequence. At the 1:1 ratio of compound-to-DNA in this experiment, only binding to -AAAGTTT- is seen. There is no appearance of other DNA–ligand complex peaks at this compound-to-DNA ratio. As the ratio is increased beyond that in Figure 13, binding to other, less favored sequences begins to be observed. The observed ESI-MS spectra strongly indicate the high sequence specificity and affinity of DB2447 for the single G•C bp sequence.

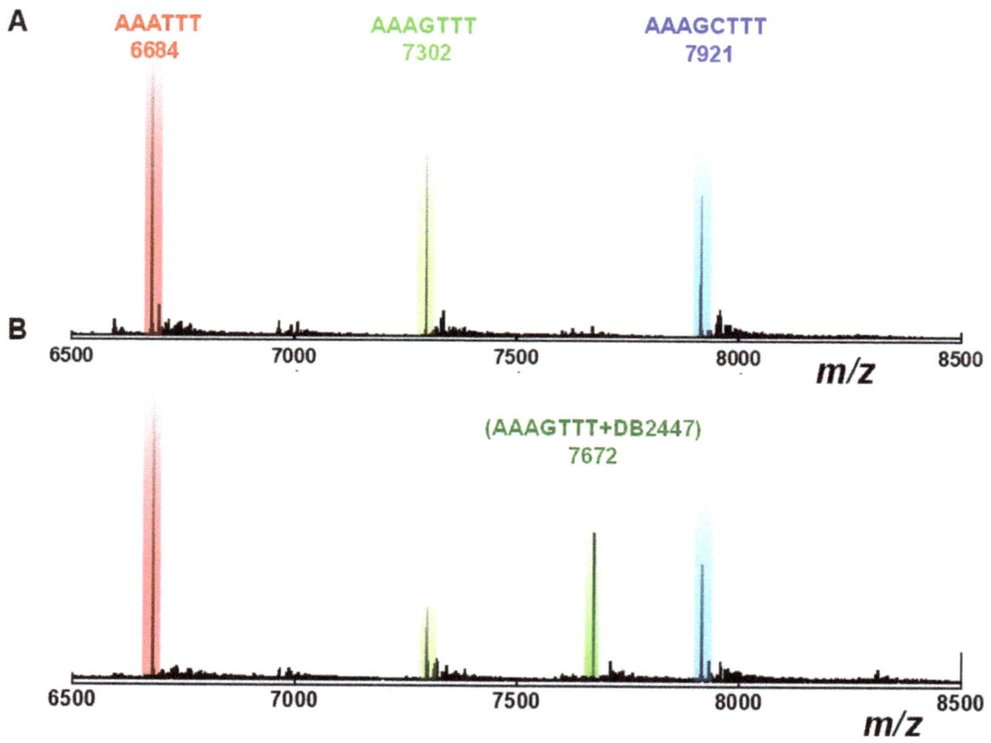

Figure 13. ESI-MS negative mode spectra of the competition binding of sequences AAATTT, AAAGTTT, and AAAGCTTT (10 μM each), with 30 μM DB2447 in buffer (100 mM ammonium acetate with 10% methanol (v/v), pH 6.8). (**A**) The ESI-MS spectra of free DNA mixtures; (**B**) the ESI-MS spectra of DNA mixtures with DB2447. The ESI-MS results are deconvoluted spectra, and molecular weights are shown with each peak.

4. Discussion

Over the last 50 years, there have been extensive studies on the interaction of a broad range of minor groove binders with DNA. Most of these compounds, especially those in the initial studies such as netropsin, DAPI, and Hoechst 33258 (Figure 1A), have been specific for binding to pure AT bp sequences, especially with A-tract type sequences. In the project described in this paper, the studies have been broadened to include new compounds from our laboratories that were designed to include a pyridine group with the goal to add a G•C bp to the traditional AT recognition sequence (Scheme 1, Figure 1C). The DNA sequences were also designed to include variations in the AT bp sequences that flank the G•C bp (Figure 3) to probe the effects of flanking sequences on affinity and specificity in binding. The pyridine group has been incorporated in the minor groove binders to complex with the G-NH that protrudes into the minor groove. One control compound has a phenyl in place of the pyridine to quantitatively evaluate the pyridine effects. The terminal cationic groups on the compounds (Figure 1) were varied to determine their role in DNA recognition. The compounds have either amidines, imidazolines, tetrahydropyridines, or, in one case, a combination of one amidine and one amide for a total of five compounds. The DNA samples included in the studies have zero, one, or two central GC base pairs with various flanking AT sequences (Figure 2). The combination gives five compounds and 12 DNA sequences for broad comparison with thermal melting, SPR, ITC, and combination MS methods under a variety of salt concentration and temperature conditions.

The ΔT_m results present a complete but low-resolution picture of all five compounds with the 12 DNA samples. For all of the pyridine derivatives, the single GC sequences have the highest ΔT_m values, and the highest of all is obtained with the -AAAGTTT- sequence (Figure 4, Table S1). It clearly has the optimum combination of groove width, curvature, and placement of H-bond acceptor groups in the minor groove to interact with the pyridine-substituted compounds (Figure 1C). With DB2447, for example, the ΔT_m with -AAAGTTT- is 14 °C, while the ΔT_m values for all other single G DNAs are between 5 and 7, about one-half the -AAAGTTT- ΔT_m. In all cases, the alternating AT sequence -ATAGTAT- has the lowest ΔT_m values with all DNA sequences. This sequence has a significantly wider minor groove width than -AATGTTT- and is a less favorable binding site for these types of minor groove binders (Figure 3). DB2448, with imidazoline cationic groups, binds a little weaker than the amidine, but it has excellent selectivity and only has a significant ΔT_m with single G•C sequences. DB2502 with tetrahydropyrimidine terminal cations binds more strongly than the other two compounds but with lower specificity. Ab initio calculations on the three pyridine dications indicate that DB2448 is the most planar structure while DB2502 is the least planar. DB2559 with the pyridine replaced with a phenyl binds best to pure AT sequences as expected. It has the best binding with -AAATTT-, while all other pure AT sequences only have about one-half of the ΔT_m of -AAATTT-.

Biosensor-SPR methods provide more quantitative binding results than the ΔT_m values but they are in qualitative agreement (Figures 4 and 5). The K_D for DB2447 with -AAAGTTT- is 1.8 nM, and results with -AAAGAAA- and AATGAAT are ten-fold higher, in agreement with their lower ΔT_m values. As with ΔT_m, the weakest binding is seen with -ATAGTAT- with a K_D of 68 nM. The same trend is observed with the other pyridine compounds with the ΔT_m results: DB2448 binds slightly weaker but with greater selectivity, while DB2502 binds slightly more strongly but with lower selectivity. DB2559 binds best to pure AT sequences and, as expected, it has the strongest binding to -AAATTT-.

Given that -AAAGTTT- is the best binding sequence, more detailed thermodynamic studies were conducted with it and DB2447 to determine what components are most important for complex formation. As is typical of many biological complexes that are formed from numerous relatively weak interactions, the pyridine DB2447 complex in this set has a large negative heat capacity for complex formation, −248 cal/mol deg. With DB2447, the $\Delta H_b°$ and $-T\Delta S_b$ values are −6.8 and −5.1 kcal/mol at 25 °C, respectively. As expected from the negative heat capacity, the values at 40 °C are −11.4 and −0.90 kcal/mol and the compensating changes maintain an almost constant $\Delta G_b°$. This agrees with the thermodynamics for other minor groove binders that interact with a G•C bp and indicates an enthalpy-driven complex stabilized by an array of H-bonding, van der Waals, and electrostatic interactions [22,55]. With compounds such as DB75 that recognize only AT bp sequences, the binding entropy is the dominant component of the complex formation due to the release of water from the minor groove AT sites [49]. With DB2559, for example, the $\Delta H_b°$ and $-T\Delta S_b°$ values are approximately −4 (Figure S6) and −6 kcal/mol at 25 °C, respectively, and the difference is even larger with DB75, approximately −2 and −7 kcal/mol, respectively [49]. The clear conclusion from the available results then is that adding a G•C bp to a minor groove binder recognition sequence significantly increases the binding enthalpy and reduces the binding entropy. A similar thermodynamic shift was seen with DB293 and DB2277 [22,55]. With an -AATT- sequence, however, DB293 binds as an entropy-driven complex [56]. In all of these systems, the binding energetics shift more to energetic emphasis on H-bond formation through the G-NH to compound acceptor group and more to water release and entropy in a pure AT minor groove sequence.

Strong support for the highly selective binding of DB2447 to -AAAGTTT- is seen in competition mass spectroscopy experiments (Figure 13). This is in agreement with the ΔT_m and SPR results that show stronger binding to the single G•C sequence. To help better understand the structural basis of molecular recognition of DNA sequences with a GC bp in an AT flanking sequence context, a molecular dynamics (MD) simulation for a complex of the pyridine compound, DB2447, with the DNA sequence *ds*[5′-CCAAAGTTTGG-

3′)(5′CCAAACTTTGG-3′)] was conducted (Figure 14). Force constants for DB2447 were determined as described previously and added to the force field for the simulation [57,58]. The MD simulation was performed by using Amber 16 in the presence of 0.15 M NaCl as previously described. The DB2447 complex can dynamically orient to provide favorable curvature to the DNA complex and interactions between the compound and DNA. The pyridine N and amidine -NH groups are positioned for strong H-bonds with the -G-NH (3.1 A) and -two T=O (2.9–3.0 A) groups at the floor of the minor groove (Figure 14). The H-bonding ability, stacking with the minor groove walls, and dynamics of the bound system help provide the high binding affinity of DB2447 to the -AAAGTTT- binding site. The strong G-NH to pyridine N H-bond provides high binding selectivity of DB2447 toward the AAAGTTT sequence, in agreement with the ESI-MS results. Additional selectivity in binding is provided by −CH groups of the two phenyls that point into the minor groove (Figure 14B). The −CH groups form a dynamic weak interaction with −dT=O that are adjacent to the central G·C bp. The combination of weak to strong interactions in the complex gives the large negative heat capacity and strong binding of DB2447.

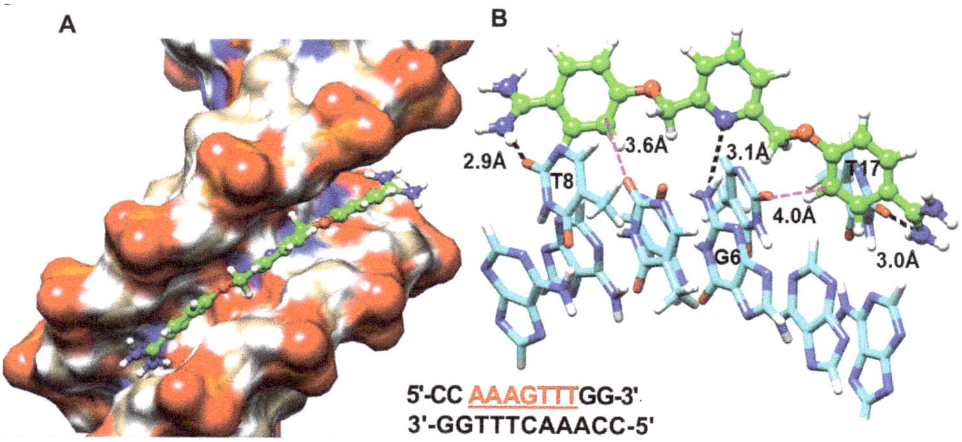

Figure 14. (**A**) A Snapshot of minor groove view of the DB2447-d[(5′-CCAAAGTTTGG-3′)(5′-CCAAACTTTGG-3′)] complex from MD simulations. The sphere model in a green-white-blue-red-yellow (C-H-N-O-S) color scheme represents DB2447. The DNA bases are represented in a ball and stick and ribbon model with a cyan-white-red-blue-orange (C-H-O-N-P) color scheme. (**B**) The important interactions between different sections of the DB2447–DNA complex are illustrated. DB2447 forms three direct H-bonds (black dashed lines) with DNA bases. Red lines indicate compound aromatic -CH interactions with -T=O groups.

For drug design, it is essential to understand the effects of solution conditions on minor groove binder–DNA complexes. The equilibrium constant, K_A, decreases as the salt concentration is increased as with other minor groove binders. As expected, the slope of a $\log K_A$ versus $\log[Na^+]$ plot is linear with a slope of 1.6 (Figure 6). The enthalpy change, $\Delta H_b{}^\circ$, is affected by a very small amount with changes in salt concentration. The effects of salt concentration on $\Delta G_b{}^\circ$ and $T\Delta S_b{}^\circ$ are complementary and amount to about a one-kilocalorie decrease as the salt concentration increases from 100 to 500 mM NaCl. As the temperature increases, however, $\Delta H_b{}^\circ$ becomes much more negative and the heat capacity for binding is a large negative value for binding a small molecule.

ITC experiments were also used to evaluate the effects of the DNA sequence on the binding of DB2447. The results show (Figure 11) that the -AAAGTTT- sequence has substantial $\Delta H_b{}^\circ$ and ΔS_b values and the most favorable $\Delta G_b{}^\circ$, which is why it is the best binding sequence. The -ATAGATA- sequence has a relatively small $\Delta H_b{}^\circ$ and ΔS_b that sum to give it the lowest $\Delta G_b{}^\circ$ value of all the single G sequences. Evaluation of the three pyrimidine diamidine compounds shows that they all have similar binding thermodynamics. DB2502 has the most favorable ΔS_b, indicating that the cyclic tetrahydropyrimidine group displaces the most water from the minor groove on binding. DB2448 has the lowest ΔS_b that sums with its $\Delta H_b{}^{\circ}$ to give it the lowest $\Delta G_b{}^{\circ}$ of the three compounds.

In summary, the results presented here show that minor groove binding thermodynamics depends on both compound structure and DNA sequence.

Supplementary Materials: The following supporting information can be downloaded at: https://www.mdpi.com/article/10.3390/life12050681/s1. References [21,22,24,28] are cited in the supplementary materials. Experimental details of synthesis, DNA melting experimental data table, SPR sensorgrams, and affinity binding curves for DB2447, DB2448, DB2449, DB2502, and DB2559; SPR equilibrium dissociation constants (K_D, nM) data table for DB2447 and analogs with pure A·T and mixed DNA sequences; ITC of DB2559 and DB2448 with AAATTT sequence; ^{1}H NMR spectra of final products. Figure S1: Comparison of equilibrium binding constants (K_D, M) of DB2447 with pure AT and mixed single/two G•C base pair(s) containing DNA sequences; Figure S2: Comparison of equilibrium binding constants (K_D, M) of DB2559 with pure AT and mixed single G•C base-pair containing DNA sequences; Figure S3: Comparison of equilibrium binding constants (K_D, M) of DB2448 with mixed single/two G•C base pair(s) containing DNA sequences; Figure S4: Comparison of equilibrium binding constants (K_D, M) of DB2502 with pure AT and single/two G•C base pair(s) containing DNA sequences; Figure S5: Comparison of equilibrium binding constants (K_D, M) of DB2449 with mixed single/two G•C base pair(s) containing DNA sequences; Table S1: Thermal Melting Studies (ΔT_m, [a] °C) of the designed heterocyclic amidine compounds with pure A·T and mixed DNA sequences; Table S2: Biosensor-SPR equilibrium dissociation constants (K_D, nM) of DB2447 and analogues with pure A·T and mixed DNA sequences; Figure S6: ITC data for the titration of DB559 and DB2448 with AAATTT DNA at 100 mM NaCl at 25 °C.

Author Contributions: W.D.W., D.W.B. and A.P. designed the study and experiments. A.A.F. and D.W.B. designed the novel compound syntheses. W.D.W., A.P. and A.A.F. wrote the manuscript. All authors have read and agreed to the published version of the manuscript.

Funding: National Institutes of Health Grant GM111749 (W.D.W. and D.W.B.).

Institutional Review Board Statement: Not applicable.

Informed Consent Statement: Not applicable.

Data Availability Statement: Not applicable.

Acknowledgments: We thank the National Institutes of Health Grant GM111749 (W.D.W. and D.W.B.) for financial support and Arvind Kumar for providing some compounds.

Conflicts of Interest: The authors declare no conflict of interest.

References

1. Marky, L.A.; Blumenfeld, K.S.; Breslauer, K.J. Calorimetric and spectroscopic investigation of drug-DNA interactions. I. The binding of netropsin to poly d(AT). *Nucleic Acids Res.* **1983**, *11*, 2857–2870. [CrossRef] [PubMed]
2. Marky, L.A.; Snyder, J.G.; Breslauer, K.J. Calorimetric and spectroscopic investigation of drug-DNA interactions: II. Dipyrandium binding to poly d(AT). *Nucleic Acids Res.* **1983**, *11*, 5701–5715. [CrossRef] [PubMed]
3. Marky, L.A.; Snyder, J.G.; Remeta, D.P.; Breslauer, K.J. Thermodynamics of drug-DNA interactions. *J. Biomol. Struct. Dyn.* **1983**, *1*, 487–507. [CrossRef] [PubMed]
4. Marky, L.A.; Breslauer, K.J. Origins of netropsin binding affinity and specificity: Correlations of thermodynamic and structural data. *Proc. Natl. Acad. Sci. USA* **1987**, *84*, 4359–4363. [CrossRef]
5. Breslauer, K.J.; Remeta, D.P.; Chou, W.Y.; Ferrante, R.; Curry, J.; Zaunczkowski, D.; Snyder, J.G.; Marky, L.A. Enthalpy-entropy compensations in drug-DNA binding studies. *Proc. Natl. Acad. Sci. USA* **1987**, *84*, 8922–8926. [CrossRef]
6. Jin, R.; Breslauer, K.J. Characterization of the minor groove environment in a drug-DNA complex: Bisbenzimide bound to the poly[d(AT)].poly[d(AT)]duplex. *Proc. Natl. Acad. Sci. USA* **1988**, *85*, 8939–8942. [CrossRef]
7. Marky, L.A.; Patel, D.; Breslauer, K.J. Effect of tetramethylammonium ion on the helix-to-coil transition of poly(deoxyadenylylthymidine): A nuclear magnetic resonance and calorimetric investigation. *Biochemistry* **1981**, *20*, 1427–1431. [CrossRef]
8. Pilch, D.S.; Kirolos, M.A.; Liu, X.; Plum, G.E.; Breslauer, K.J. Berenil [1,3-bis(4′-amidinophenyl)triazene] binding to DNA duplexes and to a RNA duplex: Evidence for both intercalative and minor groove binding properties. *Biochemistry* **1995**, *34*, 9962–9976. [CrossRef]

9. Kopka, M.L.; Goodsell, D.S.; Han, G.W.; Chiu, T.K.; Lown, J.W.; Dickerson, R.E. Defining GC-specificity in the minor groove: Side-by-side binding of the di-imidazole lexitropsin to C-A-T-G-G-C-C-A-T-G. *Structure* **1997**, *5*, 1033–1046. [CrossRef]
10. Reddy, B.S.; Sharma, S.K.; Lown, J.W. Recent developments in sequence selective minor groove DNA effectors. *Curr. Med. Chem.* **2001**, *8*, 475–508. [CrossRef]
11. O'Hare, C.C.; Mack, D.; Tandon, M.; Sharma, S.K.; Lown, J.W.; Kopka, M.L.; Dickerson, R.E.; Hartley, J.A. DNA sequence recognition in the minor groove by crosslinked polyamides: The effect of N-terminal head group and linker length on binding affinity and specificity. *Proc. Natl. Acad. Sci. USA* **2002**, *99*, 72–77. [CrossRef] [PubMed]
12. Zhang, Q.; Dwyer, T.J.; Tsui, V.; Case, D.A.; Cho, J.; Dervan, P.B.; Wemmer, D.E. NMR Structure of a cyclic polyamide-DNA complex. *J. Am. Chem. Soc.* **2004**, *126*, 7958–7966. [CrossRef] [PubMed]
13. Goodsell, D.S.; Ng, H.L.; Kopka, M.L.; Lown, J.W.; Dickerson, R.E. Structure of a dicationic monoimidazole lexitropsin bound to DNA. *Biochemistry* **1995**, *34*, 16654–16661. [CrossRef] [PubMed]
14. Quintana, J.R.; Lipanov, A.A.; Dickerson, R.E. Low-temperature crystallographic analyses of the binding of Hoechst 33258 to the double-helical DNA dodecamer C-G-C-G-A-A-T-T-C-G-C-G. *Biochemistry* **1991**, *30*, 10294–10306. [CrossRef]
15. Ming, X.; Ju, W.; Wu, H.; Tidwell, R.R.; Hall, J.E.; Thakker, D.R. Transport of dicationic drugs pentamidine and furamidine by human organic cation transporters. *Drug. Metab. Dispos.* **2009**, *37*, 424–430. [CrossRef]
16. Paine, M.F.; Wang, M.; Generaux, C.N.; Boykin, D.W.; Wilson, W.D.; Koning, H.P.; Olson, C.A.; Pohlig, G.; Burri, C.; Brun, R.; et al. Diamidines for human African trypanosomiasis. *Curr. Opin. Investig. Drugs* **2010**, *11*, 876–883.
17. Peregrine, A.S.; Mamman, M. Pharmacology of diminazene: A review. *Acta. Trop.* **1993**, *54*, 185–203. [CrossRef]
18. Elamin, E.A.; Homeida, A.M.; Adam, S.E.; Mahmoud, M.M. The efficacy of berenil (diminazene aceturate) against Trypanosoma evansi infection in mice. *J. Vet. Pharmacol. Ther.* **1982**, *5*, 259–265. [CrossRef]
19. Soeiro, M.N.; Werbovetz, K.; Boykin, D.W.; Wilson, W.D.; Wang, M.Z.; Hemphill, A. Novel amidines and analogues as promising agents against intracellular parasites: A systematic review. *Parasitology* **2013**, *40*, 929–951. [CrossRef]
20. Paul, A.; Guo, P.; Boykin, D.W.; Wilson, W.D. A new generation of minor-groove-binding-heterocyclic diamidines that recognize G·C base pairs in an AT sequence context. *Molecules* **2019**, *24*, 946. [CrossRef]
21. Chai, Y.; Paul, A.; Rettig, M.; Wilson, W.D.; Boykin, D.W. Design and synthesis of heterocyclic cations for specific DNA recognition: From AT-rich to mixed-base-pair DNA sequences. *J. Org. Chem.* **2014**, *79*, 852–866. [CrossRef] [PubMed]
22. Paul, A.; Chai, Y.; Boykin, D.W.; Wilson, W.D. Understanding mixed sequence DNA recognition by novel designed compounds: The kinetic and thermodynamic behavior of azabenzimidazole diamidines. *Biochemistry* **2015**, *54*, 577–587. [CrossRef] [PubMed]
23. Paul, A.; Nanjunda, R.; Kumar, A.; Laughlin, S.; Nhili, R.; Depauw, S.; Deuser, S.S.; Chai, Y.; Chaudhary, A.S.; David-Cordonnier, M.H.; et al. Mixed up minor groove binders: Convincing A·T specific compounds to recognize a G·C base pair. *Bioorg. Med. Chem. Lett.* **2015**, *25*, 4927–4932. [CrossRef]
24. Paul, A.; Kumar, A.; Nanjunda, R.; Farahat, A.A.; Boykin, D.W.; Wilson, W.D. Systematic synthetic and biophysical development of mixed sequence DNA binding agents. *Org. Biomol. Chem.* **2017**, *15*, 827–835. [CrossRef] [PubMed]
25. Guo, P.; Paul, A.; Kumar, A.; Farahat, A.A.; Kumar, D.; Wang, S.; Boykin, D.W.; Wilson, W.D. The Thiophene "Sigma-Hole" as a concept for preorganized, specific recognition of G·C base pairs in the DNA minor groove. *Chemistry* **2016**, *22*, 15404–15412. [CrossRef] [PubMed]
26. Guo, P.; Farahat, A.A.; Paul, A.; Harika, N.K.; Boykin, D.W.; Wilson, W.D. Compound shape Effects in minor groove binding affinity and specificity for mixed sequence DNA. *J. Am. Chem. Soc.* **2018**, *140*, 14761–14769. [CrossRef]
27. Guo, P.; Paul, A.; Kumar, A.; Harika, N.K.; Wang, S.; Farahat, A.A.; Boykin, D.W.; Wilson, W.D. A modular design for minor groove binding and recognition of mixed base pair sequences of DNA. *Chem. Commun.* **2017**, *53*, 10406–10409. [CrossRef]
28. Guo, P.; Farahat, A.A.; Paul, A.; Boykin, D.W.; Wilson, W.D. Engineered modular heterocyclic-diamidines for sequence-specific recognition of mixed AT/GC base pairs at the DNA minor groove. *Chem. Sci.* **2021**, *12*, 15849–15861. [CrossRef]
29. Boykin, D.W.; Kumar, A.; Xiao, G.; Wilson, W.D.; Bender, B.C.; McCurdy, D.R.; Hall, J.E.; Tidwell, R.R. 2,5-bis[4-(N-alkylamidino)phenyl]furans as anti-Pneumocystis carinii agents. *J. Med. Chem.* **1998**, *41*, 124–129. [CrossRef]
30. Das, B.P.; Boykin, D.W. Synthesis and antiprotozoal activity of 2,5-bis(4-guanylphenyl) furans. *J. Med. Chem.* **1977**, *20*, 531–536. [CrossRef]
31. Wilson, W.D.; Tanious, F.A.; Fernandez-Saiz, M.; Rigl, C.T. Evaluation of drug-nucleic acid interactions by thermal melting curves. *Methods Mol. Biol.* **1997**, *90*, 219–240. [PubMed]
32. Shi, X.C.; Chaires, J.B. Sequence- and structural-selective nucleic acid binding revealed by the melting of mixtures. *Nucleic Acids Res.* **2006**, *34*, e14. [CrossRef] [PubMed]
33. Nguyen, B.; Neidle, S.; Wilson, W.D. A role for water molecules in DNA-ligand minor groove recognition. *Acc. Chem. Res.* **2009**, *42*, 11–21. [CrossRef] [PubMed]
34. Wilson, W.D.; Nguyen, B.; Tanious, F.A.; Mathis, A.; Hall, J.E.; Stephens, C.E.; Boykin, D.W. Dications that target the DNA minor groove: Compound design and preparation, DNA interactions, cellular distribution and biological activity. *Curr. Med. Chem. Anticancer Agents* **2005**, *5*, 389–408. [CrossRef]
35. Munde, M.; Wang, S.; Kumar, A.; Stephens, C.E.; Farahat, A.A.; Boykin, D.W.; Wilson, W.D.; Poon, G.M. Structure-dependent inhibition of the ETS-family transcription factor PU.1 by novel heterocyclic diamidines. *Nucleic Acids Res.* **2014**, *42*, 1379–1390. [CrossRef]

36. Rohs, R.; West, S.M.; Sosinsky, A.; Liu, P.; Mann, R.S.; Honig, B. The role of DNA shape in protein-DNA recognition. *Nature* **2009**, *461*, 1248–1253. [CrossRef]
37. Zhou, T.; Yang, L.; Lu, Y.; Dror, I.; Dantas-Machado, A.C.; Ghane, T.; Di Felice, R.; Rohs, R. DNAshape: A method for the high-throughput prediction of DNA structural features on a genomic scale. *Nucleic Acids Res.* **2013**, *4*, W56–W62. [CrossRef]
38. Liu, Y.; Chai, Y.; Kumar, A.; Tidwell, R.R.; Boykin, D.W.; Wilson, W.D. Designed compounds for recognition of 10 base pairs of DNA with two AT binding sites. *J. Am. Chem. Soc.* **2012**, *134*, 5290–5299. [CrossRef]
39. Nguyen, B.; Tanious, F.A.; Wilson, W.D. Biosensor-surface plasmon resonance: Quantitative analysis of small molecule−nucleic acid interactions. *Methods* **2007**, *42*, 150–161. [CrossRef]
40. Nanjunda, R.; Munde, M.; Liu, Y.; Wilson, W.D. *Methods for Studying DNA/Drug Interactions*; Chapter 4; Wanunu, M., Tor, Y., Eds.; CRC Press-Taylor & Francis Group: Boca Raton, FL, USA, 2011.
41. DeHaseth, P.L.; Lohman, T.M.; Record, M.T., Jr. Nonspecific interaction of iac repressor with DNA: An association reaction driven by counterion release. *Biochemistry* **1977**, *16*, 4783–4790. [CrossRef]
42. Wang, S.; Kumar, A.; Aston, K.; Nguyen, B.; Bashkin, J.K.; Boykin, D.W.; Wilson, W.D. Different thermodynamic signatures for DNA minor groove binding with changes in salt concentration and temperature. *Chem. Commun.* **2013**, *49*, 8543–8545. [CrossRef] [PubMed]
43. Manning, G.S. The molecular theory of polyelectrolyte solutions with applications to the electrostatic properties of polynucleotides. *Q. Rev. Biophys.* **1978**, *11*, 179–246. [CrossRef] [PubMed]
44. Chaires, J.B. A thermodynamic signature for drug-DNA binding mode. *Arch. Biochem. Biophys.* **2006**, *453*, 26–31. [CrossRef] [PubMed]
45. Privalov, P.L.; Dragan, A.I.; Crane-Robinson, C. Interpreting protein/DNA interactions: Distinguishing specific from non-specific and electrostatic from non-electrostatic components. *Nucleic Acids Res.* **2011**, *39*, 2483–2491. [CrossRef]
46. Record, M.T., Jr.; Anderson, C.F.; Lohman, T.M. Thermodynamic analysis of ion effects on the binding and conformational equilibria of proteins and nucleic acids: The roles of ion association or release, screening, and ion effects on water activity. *Q. Rev. Biophys.* **1978**, *11*, 103–178. [CrossRef]
47. Munde, M.; Lee, M.; Neidle, S.; Arafa, R.; Boykin, D.W.; Liu, Y.; Bailly, C.; Wilson, W.D. Induced fit conformational changes of a "reversed amidine" heterocycle: Optimized interactions in a DNA minor groove complex. *J. Am. Chem. Soc.* **2007**, *129*, 5688–5698. [CrossRef]
48. Wang, S.; Aston, K.; Koeller, K.J.; Harris, G.D., Jr.; Rath, N.P.; Bashkin, J.K.; Wilson, W.D. Modulation of DNA-polyamide interaction by β-alanine substitutions: A study of positional effects on binding affinity, kinetics and thermodynamics. *Org. Biomol. Chem.* **2014**, *12*, 7523–7536. [CrossRef]
49. Liu, Y.; Collar, C.J.; Kumar, A.; Stephens, C.E.; Boykin, D.W.; Wilson, W.D. Heterocyclic diamidine interactions at AT base pairs in the DNA minor groove: Effects of heterocycle differences, DNA AT sequence and length. *J. Phys. Chem. B.* **2008**, *112*, 11809–11818. [CrossRef]
50. Buurma, N.J.; Haq, I. Calorimetric and spectroscopic studies of Hoechst 33258: Self-association and binding to non-cognate DNA. *J. Mol. Biol.* **2008**, *381*, 607–621. [CrossRef]
51. Ferreira, J.M.; Sheardy, R.D. Linking temperature, cation concentration and water activity for the B to Z conformational transition in DNA. *Molecules* **2018**, *23*, 1806. [CrossRef]
52. Kumar, S.; Xue, L.; Arya, D.P. Neomycin-neomycin dimer: An all-carbohydrate scaffold with high affinity for AT-rich DNA duplexes. *J. Am. Chem. Soc.* **2011**, *133*, 7361–7375. [CrossRef] [PubMed]
53. Laughlin, S.; Wilson, W.D. May the best molecule win: Competition ESI mass spectrometry. *Int. J. Mol. Sci.* **2015**, *16*, 24506–24531. [CrossRef] [PubMed]
54. Laughlin, S.; Wang, S.; Kumar, A.; Farahat, A.A.; Boykin, D.W.; Wilson, W.D. Resolution of mixed site DNA complexes with dimer-forming minor-groove binders by using electrospray ionization mass spectrometry: Compound structure and DNA sequence effects. *Chemistry* **2015**, *21*, 5528–5539. [CrossRef] [PubMed]
55. Tanious, F.; Wilson, W.D.; Wang, L.; Kumar, A.; Boykin, D.W.; Marty, C.; Baldeyrou, B.; Bailly, C. Cooperative dimerization of a heterocyclic diamidine determines sequence-specific DNA recognition. *Biochemistry* **2003**, *42*, 13576–13586. [CrossRef] [PubMed]
56. Wang, L.; Kumar, A.; Boykin, D.W.; Bailly, C.; Wilson, W.D. Comparative thermodynamics for monomer and dimer sequence-dependent binding of a heterocyclic dication in the DNA minor groove. *J. Mol. Biol.* **2002**, *317*, 361–374. [CrossRef] [PubMed]
57. Harika, N.K.; Wilson, W.D. Bound compound, interfacial water, and phenyl ring rotation dynamics of a compound in the DNA minor groove. *Biochemistry* **2018**, *57*, 5050–5057. [CrossRef]
58. Athri, P.; Wilson, W.D. Molecular dynamics of water-mediated interactions of a linear benzimidazole-biphenyl diamidine with the DNA minor groove. *J. Am. Chem. Soc.* **2009**, *131*, 7618–7625. [CrossRef]

life

Article

Structural Responses of Nucleic Acids to Mars-Relevant Salts at Deep Subsurface Conditions

Jim-Marcel Knop [1], Sanjib K. Mukherjee [1], Stewart Gault [2], Charles S. Cockell [2] and Roland Winter [1,*,†]

[1] Physical Chemistry I—Biophysical Chemistry, Faculty of Chemistry and Chemical Biology, TU Dortmund University, Otto-Hahn Street 4a, 44227 Dortmund, Germany; jim-marcel.knop@tu-dortmund.de (J.-M.K.); sanjib.mukherjee@tu-dortmund.de (S.K.M.)

[2] UK Centre for Astrobiology, SUPA School of Physics and Astronomy, University of Edinburgh, James Clerk Maxwell Building, Peter Guthrie Tait Road, Edinburgh EH9 3FD, UK; gaults@tcd.ie (S.G.); ccockell@exseed.ed.ac.uk (C.S.C.)

* Correspondence: roland.winter@tu-dortmund.de

† Dedicated to Professor Kenneth J. Breslauer on the occasion of his 75th birthday.

Abstract: High pressure deep subsurface environments of Mars may harbor high concentrations of dissolved salts, such as perchlorates, yet we know little about how these salts influence the conditions for life, particularly in combination with high hydrostatic pressure. We investigated the effects of high magnesium perchlorate concentrations compared to sodium and magnesium chloride salts and high pressure on the conformational dynamics and stability of double-stranded B-DNA and, as a representative of a non-canonical DNA structure, a DNA-hairpin (HP), whose structure is known to be rather pressure-sensitive. To this end, fluorescence spectroscopies including single-molecule FRET methodology were applied. Our results show that the stability both of the B-DNA as well as the DNA-HP is largely preserved at high pressures and high salt concentrations, including the presence of chaotropic perchlorates. The perchlorate anion has a small destabilizing effect compared to chloride, however. These results show that high pressures at the kbar level and perchlorate anions can modify the stability of nucleic acids, but that they do not represent a barrier to the gross stability of such molecules in conditions associated with the deep subsurface of Mars.

Keywords: B-DNA; DNA-hairpin; perchlorates; brines; pressure; Mars; habitability

Citation: Knop, J.-M.; Mukherjee, S.K.; Gault, S.; Cockell, C.S.; Winter, R. Structural Responses of Nucleic Acids to Mars-Relevant Salts at Deep Subsurface Conditions. *Life* **2022**, *12*, 677. https://doi.org/10.3390/life12050677

Academic Editors: Tigran Chalikian and Jens Völker

Received: 14 April 2022
Accepted: 29 April 2022
Published: 2 May 2022

1. Introduction

Investigating the chemical and physical limits of life and its associated biomacromolecules allows us to assess the habitability of extraterrestrial environments with respect to known life on Earth [1]. A large number of organisms thrive under extreme conditions on Earth, such as in the deep ocean, the subseafloor, in marine hydrothermal vents and volcanic environments, which are also suggested to be candidate locations for the origin of life. Extremophilic organisms on Earth, such as halophiles (salt-loving microorganisms), psychrophiles (cold-loving organisms) or piezophiles (pressure-loving microorganisms), as encountered in the deep sea where pressures up to about 1000 bar are encountered, are found in all three domains of life [1–5].

One of the requirements for life as we know it is the presence of liquid water. There is now abundant evidence of periods of surface water on ancient Mars, and more controversial suggestions of liquid water on present-day Mars, including in the deep subsurface [1,6–8]. In addition to any geochemical stressors, deep subsurface water on Mars would also be subjected to pressurization, which would be ~1 kbar at the base of the cryosphere if it reached a depth of 10 km [6]. Such deep subsurface environments have been proposed to harbor high concentrations of dissolved salts, yet we know little about how such brines shape the conditions for life. One particular anion found ubiquitously on Mars is perchlorate, ClO_4^- [9–11]. Perchlorate salts, such as $Mg(ClO_4)_2$, exhibit deep eutectic temperatures allowing for the presence of liquid water at temperatures as low as −70 °C. Perchlorates are known to act as chaotropes, perturbing the structure of water and its hydrogen bonding capacity [12], and might, hence, affect biomolecular hydration, structure, dynamics and function. To advance our understanding of the ability of subsurface environments of Mars to support life, we need to examine the combination of strong ionic effects imposed by these salts and high hydrostatic pressures (HHP).

Previous studies have demonstrated deleterious effects of perchlorates on the activity of enzymes such as α-chymotrypsin (α-CT) at ambient conditions [10,11,13]. Conversely, stabilizing salts, such as magnesium sulfate, $MgSO_4$, have been shown to increase the activity and structural stability of α-CT [11]. We demonstrated that high pressures increase the enzymatic activity of α-CT, even in the presence of high perchlorate concentrations, and the results suggested that HHP may increase the habitability of environments under perchlorate stress. In another study, we investigated the binding of the small ligand 8-anilinonaphthalene-1-sulfonic acid (ANS) to the protein bovine serum albumin (BSA) at ambient and low temperatures, and at high pressure conditions in the presence of ions associated with the surface and subsurface of Mars [14]. We found that salts such as $MgCl_2$ and $MgSO_4$ only slightly affect protein–ligand complex formation. In contrast, $Mg(ClO_4)_2$ strongly affects the interaction between ANS and BSA, leading to a change in stoichiometry and strength of ligand binding. Remarkably, both a decrease in temperature and an increase in pressure favor the ligand binding process. Ligand binding studies of the complex formation between the ligand thioflavin T (ThT) and tRNA in the presence of Martian salts showed a strong reduction in the binding constant as well [15]. This effect is largely due to the interaction of ThT with the salt anions, which leads to a strong decrease in the activity of the ligand. Remarkably, the pressure favored ligand binding regardless of the type of salt [15]. Studies of the effect of Martian-like salts on biomolecular protein condensates based on liquid-liquid phase separation (LLPS) phenomena, which might also have played a significant role during protocell formation under prebiotic conditions, showed that the driving force for phase separation of dense protein solutions is not only sensitively dictated by the amino acid sequence of the polypeptide, but also strongly influenced by the type of salt and its concentration [16]. We showed that at high salinity, as encountered in Martian soil, short-range interactions, ion correlation (e.g., ion pairing) effects and hydrophobic interactions can sustain LLPS for suitable polypeptide sequences. We have also seen that spatial confinement, such as narrow pores in sediments or water pools, can dramatically stabilize the droplet phase [16]. In a subsequent study of these Martian salts on lipid-based compartments, viz. model biomembranes, we could show that the fluidity, lateral organization and morphology of lipid membranes are largely affected under extreme salt and pressure conditions relevant to Mars-like environments [17].

The effect of high concentrations of Mars-like salts on the stability of nucleic acids is still largely unknown. Pressure is known to have a small stabilizing, destabilizing or no effect on the structure of the B-DNA, depending on the temperature and salt concentration [18–23]. Recently, it was found that non-canonical nucleic acid structure, such as DNA-hairpins (DNA-HP) and G-quadruplexes, are much more pressure-sensitive, however [24–32]. Single-molecule Förster resonance energy transfer (sm-FRET) measurements [33] were used to directly measure the population distribution of DNA-HPs at HHP conditions. The pressure sensitivity of such structures is due to a conformational transition from a closed state to an open state, which is accompanied by a volume decrease, ΔV, in the order of −10 to −30 mL mol^{-1} [28,32].

Here, we present first data on the effect of high concentrations of $Mg(ClO_4)_2$ compared to NaCl and $MgCl_2$ and high hydrostatic pressures on the conformational dynamics and stability of B-DNA and, as a representative of a non-canonical DNA structure, a DNA-HP. To this end, high-pressure UV-absorption spectroscopic and sm-FRET experiments were carried out.

2. Materials and Methods

The oligomeric nucleic acids used in this study were purchased from biomers.net (Ulm, Germany). The sequences are given below:

HP1: 5′-TGG CGA CGG CAG CGA GGC TTA GCG GCA $(A)_{30}$ AGC CGC X-3′ (X is T-Atto 550); HP2: 5′-GCC TCG CYG CCG TCG CCA-3′ (Y is T-Atto 647N); A: 5′-GGA CTA GTC TAG GCG AAC GTT TAA GGC GAT CTC TGT TTA CAA CTC CGA-3′; B: 5′-TCG GAG TTG TAA ACA GAG ATC GCC TTA AAC GTT CGC CTA GAC TAG TCC-3′.

TrisHCl, TrisBase, NaCl, $NaClO_4$, $MgCl_2$ and $Mg(ClO_4)_2$ were purchased from Sigma-Aldrich (Darmstadt, Germany). The double-stranded DNA samples were prepared by mixing 20 μL of 100 μM solution of strand A and the same amount of the complementary strand B with 360 mM buffer containing 20 mM TrisHCl at pH 7.5 and 15 mM NaCl. The solution was heated to 95 °C for 5 min and subsequently cooled down to 25 °C at a rate of 1 °C min^{-1} to ensure proper annealing. For the UV-absorption measurements, 100 μL of this solution was added to 300 μL of buffer containing the required amount of salt to achieve a final concentration of 250 mM (or 15 mM) salt and 1.25 μM double-strand DNA.

Annealing of the DNA-HP for the single-molecule FRET microscopy experiments was carried out in a similar way. One microliter of each strand (HP1 and HP2) at a concentration of 100 μM was diluted in 100 μL of 20 mM TrisHCl pH 7.5 buffer with 15 mM NaCl. The annealed strands were kept at −80 °C for storage. Before the measurements, the 1 μM annealed stock solution was diluted to yield the desired low concentration for the single-molecule measurements of about 50–100 pM in 20 mM TrisHCl pH 7.5 buffer including the required concentration of salts.

The fluorescence microscopy setup and pressure apparatus used for the sm-FRET measurements was described in detail elsewhere [32]. The UV-spectroscopic measurements were performed using a Perkin Elmer Lambda 25 spectrometer. For the temperature dependent measurements at different pressure points, a home-made cylindrical stainless-steel pressure cell with two quartz glass windows (1 cm thickness) and water-thermostat for temperature control (Julabo Hl F32) was used. The path-length of the sample cell was 1 mm. Pressure was applied via a piston-pump using water as pressurizing medium. To separate the sample from the pressurizing medium, a drop of oil was applied to the connection of the pressure pump. Before the sample measurements, the spectrometer was blanked with the corresponding buffer.

3. Results and Discussion

We show the combined effects of pressure and high concentrations of the salts $MgCl_2$ and $Mg(ClO_4)_2$ on the stability, as expressed in increases in the melting temperature, T_m, of two different classes of nucleic acid structures, a 48 bp B-DNA and an adenine DNA-HP. Figure 1A,B show, as representative examples, melting curves of the 48 bp B-DNA in buffer consisting of 15 mM NaCl, 20 mM Tris pH 7.5, at 1 bar and at 1500 bar, respectively. Figure 1C shows the pressure dependence of T_m-values obtained including experiments carried out at high salt concentrations, viz. 250 mM NaCl, 250 mM $MgCl_2$ and 250 mM $Mg(ClO_4)_2$. The addition of 250 mM NaCl increases T_m from 68.7 ± 0.1 °C to 80.0 ± 0.3 °C at ambient pressure. The salt $MgCl_2$, at 250 mM, shows a similar strong stabilizing effect (T_m = 78 ± 2 °C). The addition of 250 mM $Mg(ClO_4)_2$ increases T_m from 68.7 ± 0.1 °C to 73 ± 4 °C only (Figure 1C), indicating a decreased stability compared to the presence of 250 mM NaCl.

The salts NaCl and $Mg(ClO_4)_2$ show a similar pressure dependence of the T_m-values of the B-DNA (consisting of almost equal GC (23/48) and AT (25/48) bp) at 250 mM salt concentration. In both cases, T_m increases at a rate of about 10 °C/kbar. Conversely, in buffer solution we found the melting temperature to increase less dramatically with pressure (~3 °C/kbar) (Figure 1C). This finding is in good agreement with studies on poly(A-T) and poly(C-G) DNA, showing an increase in T_m of 3–4 °C/kbar [34].

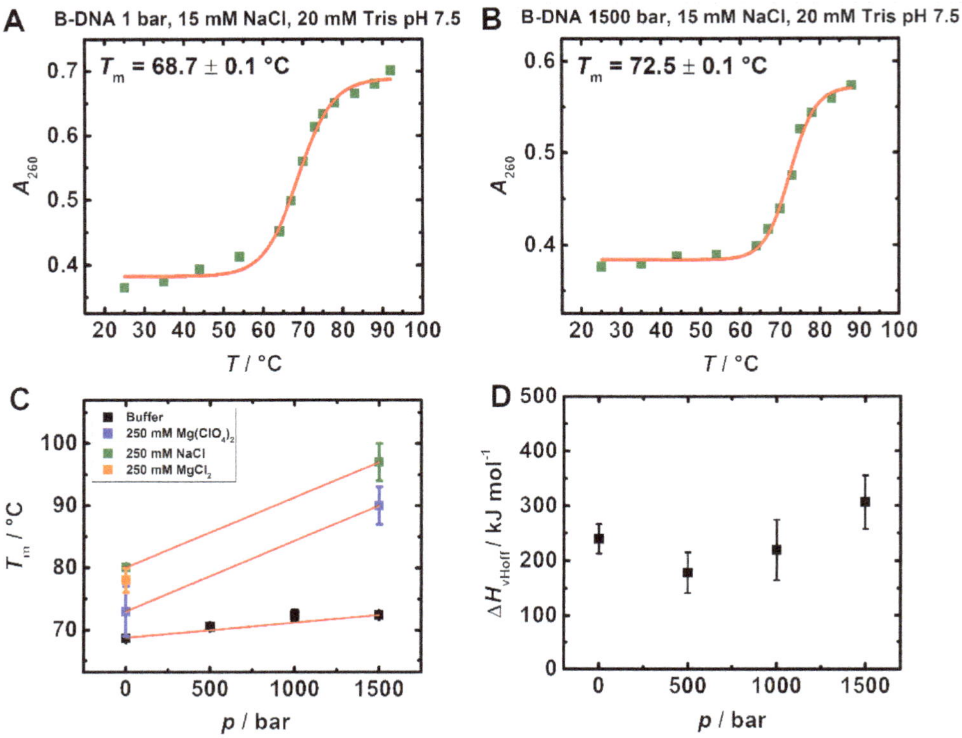

Figure 1. Pressure dependent melting curves of a 48 bp B-DNA. (**A**) shows the melting curve in buffer (15 mM NaCl, 20 mM Tris, pH 7.5) at 1 bar, (**B**) shows the melting curve at 1500 bar. (**C**) shows the pressure dependence of all T_m-values obtained from BOLTZMANN fits including experiments carried out at high salt contents (250 mM NaCl, 250 mM $MgCl_2$ and 250 mM $Mg(ClO_4)_2$). The maximum temperature measured was limited to 92 °C; therefore, T_m data at 1500 bar obtained from the Boltzmann fits can only be considered approximations, since a distinct second plateau was not reached. (**D**) shows the corresponding VAN'T HOFF melting enthalpies obtained for the measurement in buffer solution.

The VAN'T HOFF melting enthalpy, ΔH_{vHoff}, was obtained by using a Boltzmann fit of the melting curves for a two-state helix-to-coil transition. Figure 1D depicts the data for the measurement in buffer solution as a function of pressure, showing ΔH_{vHoff} values around 250 kJ mol^{-1} within the accuracy of the experiment (± 100 kJ mol^{-1}). Owing to the absence of nice S-shaped melting curves at high salt concentration due to the high melting temperatures reached at high pressures, ΔH_{vHoff} values could not be determined for these samples. Using the Clapeyron equation

$$\frac{dT_m}{dp} = \frac{\Delta V}{\Delta H / T_m} \tag{1}$$

The volume change, ΔV, upon melting of the B-DNA construct in pure buffer solution can be calculated. We obtained $\Delta V \approx +18$ mL mol^{-1}, which is about the molar volume of one water molecule, only. The positive sign of ΔV is in line with the observed stabilization of the folded duplex state upon pressurization. Such stabilization is quite expected as, in contrast to proteins, canonical DNA structures (having T_m-values > 50 °C) are generally stabilized by hydrostatic pressure [18,19] owing to their dense packing and stabilization via H-bonding and π-π-stacking interactions, which are marginally affected by pressure, only.

The volumetric properties of nucleic acids seem to be essentially determined by the state of hydration of the counterions that are accumulated in their vicinity, the specific contribution of the counterions depending on their identity (charge density, polarizability, size) and the structure of the nucleic acid sequence [18,19,23]. Significant stabilization was observed in the presence of high concentrations of the sodium and magnesium salts, which is even more pronounced at high pressure of 1500 bar, where the melting temperature reaches about ~97 °C in the presence of 250 mM NaCl. A similar observation was made in the presence of 250 mM $MgCl_2$ for which no melting of the B-DNA was observed below 100 °C at 1500 bar. These data indicate a similar strong stabilization effect of the B-DNA by Na^+ and Mg^{2+} with Cl^- as counterion at ambient pressure, a further increase in stability at high pressures for the Mg^{2+} cation, however. The comparison with the 250 mM $Mg(ClO_4)_2$ data indicate that the perchlorate anion, $ClO_4{}^-$, imposes a slight destabilizing effect compared to Cl^-. The similarity of Clapeyron slopes, $\mathrm{d}T_\mathrm{m}/\mathrm{d}p$, suggests similar volume and enthalpy changes upon melting, however. The increase in Clapeyron slope in the presence of salts can be explained by an increase in ΔV and/or decrease in ΔH, which could be due to differential hydration effects in these salt solutions.

To unveil the influence of typical Martian salts like perchlorates on the conformational dynamics of non-canonical nucleic acids at elevated pressure conditions, we studied their effect on the stability of a DNA-HP. DNA-hairpins are common secondary structure motifs that play important roles in gene expression, DNA recombination, and transposition [35]. To be able to differentiate between different conformational states of the system, we employed the single-molecule FÖRSTER resonance energy transfer (sm-FRET) technique, which also allows pressure dependent measurements to be carried out using special pressure-resistant quartz capillaries. For experimental details, please refer to [28–34]. Peaks in the recorded FRET efficiency histograms are related to conformations with different spatial separations, R, of the two attached fluorescent dyes, yielding different FRET efficiencies, E, according to $E = R_0^6 \cdot (R_0^6 + R^6)^{-1}$. The FÖRSTER radius, R_0, is the distance at which 50% of the excited donor molecules will be deactivated; here, R_0 = 6.5 nm for the fluorophores used, Atto 550 and Atto 647N. According to the FRET efficiency analysis, individual hairpins remain in a low ($E \approx 0.3$) or a high ($E \approx 0.90$) FRET state, which correspond to the open and the closed (native) conformation of the DNA-HP, respectively. We found that over the whole pressure range covered (1–1500 bar), the hairpin stays always in an equilibrium between the open and closed conformation in neat buffer solution. As seen in Figure 2A, upon pressurization, the structure of the DNA-HP gets destabilized, leading to a shift in the conformational equilibrium towards the open, unfolded state. At high hydrostatic pressure (1500 bar), the population of the open conformation increases up to ~60%, which corresponds to a volume change, ΔV, for the helix-to-coil transition of about -18 mL mol^{-1}, in agreement with the literature data [32]. Significant changes were observed in the presence of $MgCl_2$ and $Mg(ClO_4)_2$ salt. When adding even low salt concentrations such as 6 mM $MgCl_2$, only the native, closed conformation is detectable in the FRET histogram at 1 bar, and the conformation remains stable even up to pressures of 1500 bar (Figure 2B). Conversely, in the case of 6 mM $Mg(ClO_4)_2$, both open and closed conformers are detectable at ambient pressure, the population of the closed and open states being about 80% and 20%, respectively (Figure 2C). Furthermore, unlike the scenario in neat buffer condition, the fraction of closed conformation of the DNA-HP remains essentially constant up to the maximum pressure reached, indicating that stabilization of the DNA-HP structure is not only driven by the divalent cation, but also affected by the corresponding counterion. Furthermore, the perchlorate leads to a small destabilizing effect of the native conformation of the DNA-HP compared to $MgCl_2$, resulting in a small population of unfolded states. The population distribution does not change with pressure, suggesting that the magnitude of the volume change decreased, rendering ΔV negligible.

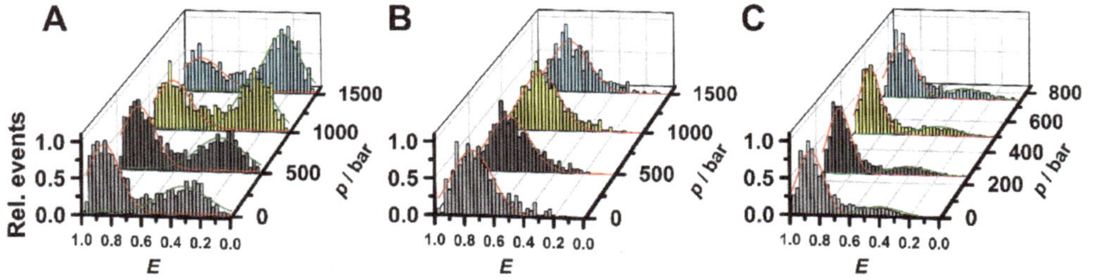

Figure 2. Pressure dependent FRET distribution histograms of the DNA-hairpin (**A**) in buffer, (**B**) in 6 mM $MgCl_2$ and (**C**) in 6 mM $Mg(ClO_4)_2$. The buffer was 20 mM TrisHCl, pH 7.5. The measurements were carried out at 25 °C. Please note, that sm-FRET measurements in the high-pressure capillary are a bit noisier and broadened compared to the ambient pressure measurements on a coverslip (Figure 3).

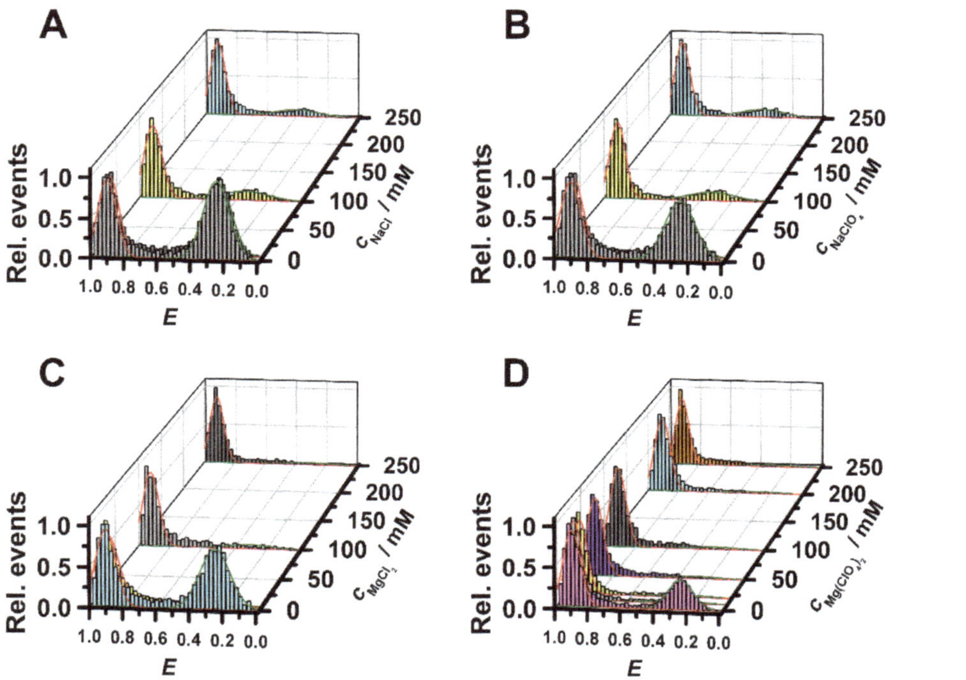

Figure 3. FRET histograms of the DNA-hairpin in the presence of different salt concentrations: (**A**) NaCl, (**B**) $NaClO_4$, (**C**) $MgCl_2$ and (**D**) 15 mM NaCl + $Mg(ClO_4)_2$ at ambient pressure and temperature (T = 25 °C).

Altogether, we found that the salt effects observed were dependent on the chemistry of the salt and the internal architecture of the nucleic acid structure. In the case of the canonical nucleic acid conformation, the double-stranded B-DNA, perchlorate anion lowers the stability of the double-strand compared to the chloride, probably due to weak (stacking) interactions of this highly polarizable, almost hydrophobic anion with nucleic acid bases. Furthermore, the stability in the presence of NaCl is higher compared to both $MgCl_2$ and $Mg(ClO_4)_2$ at the same salt concentration. Hence, the stability is not only dependent on the cation but also on its counterion. On the other hand, the folded state of the DNA-HP is more efficiently stabilized in the presence of $MgCl_2$ and $Mg(ClO_4)_2$ compared to the corresponding sodium salts, which suggests that the extra conformational stability is gained by the divalent cation, which is also in accordance with earlier studies [36] reporting that DNA and RNA helices are stabilized more effectively by Mg^{2+} than by Na^{+}. The salt-dependent sm-FRET data presented for the DNA-HP in Figure 3 clearly show that the addition of NaCl and $NaClO_4$ has a similar effect on the conformational stability of the DNA-HP, leaving a small population of unfolded states (~20%) still present at 250 mM salt concentration. Conversely, the $MgCl_2$ and $Mg(ClO_4)_2$ salts lead to a more effective stabilization of the folded structure of the DNA-HP.

4. Concluding Remarks

Taken together, we found that significant stabilization of the B-DNA and DNA-HP is gained in the presence of all salts, even those including the chaotropic perchlorate anion, the degree of stabilization depending both on the cation and its counterion. The perchlorate imposes a destabilizing effect compared to the chloride salts, which is still small at high (250 mM) salt concentrations. Pressure has a minor influence on the conformational dynamics of the DNA-HP in the presence of high salt concentration. The stability of the B-DNA in the presence of high salt increases still significantly at high pressures, in particular in the presence of $MgCl_2$. This suggests that the salt induces a more compact packing by reducing the repulsive interaction between the backbone's phosphate groups through effective charge screening and different hydration characteristics of the DNA. Hence, these results show that Mars-relevant salts confer small-scale changes in the stability of nucleic acids at pressures up to the kbar level, which may imply some biochemical adaptations to adjust, but that nucleic acids remain fundamentally stable under potential subsurface Martian conditions. This suggests that biochemical adaptation to high pressure, perchlorate rich environments should predominantly be seen in lipid membranes and proteins as they exhibit higher sensitivity to such conditions than nucleic acids.

Author Contributions: Conceptualization, R.W. and C.S.C.; experiments and data analysis, J.-M.K. and S.K.M.; discussion and writing, J.-M.K., S.K.M., S.G., C.S.C. and R.W. All authors have read and agreed to the published version of the manuscript.

Funding: R.W. received funding from the European Union's Horizon 2020 research and innovation programme under the Marie Skłodowska-Curie grant agreement No 801459—FP-RESOMUS and was funded by the Deutsche Forschungsgemeinschaft (DFG) under Germany's Excellence Strategy —EXC 2033—390677874—RESOLV. C.S.C. acknowledges support from the Science and Technology Facilities Council (grant ST/V000586/1). S.G. was supported by an EPSRC studentship.

Institutional Review Board Statement: Not applicable.

Informed Consent Statement: Not applicable.

Data Availability Statement: The data presented in this study are available on request from the corresponding author.

Conflicts of Interest: The authors declare no conflict of interest.

References

1. Cockell, C.S. *Astrobiology: Understanding Life in the Universe*; Wiley-Blackwell: Oxford, UK, 2020.
2. Harrison, J.P.; Gheeraert, N.; Tsigelnitskiy, D.; Cockell, C.S. The limits for life under multiple extremes. *Trends Microbiol.* **2013**, *21*, 204–212. [CrossRef] [PubMed]
3. Daniel, I.; Oger, P.; Winter, R. Origins of life and biochemistry under high-pressure conditions. *Chem. Soc. Rev.* **2006**, *35*, 858–875. [CrossRef] [PubMed]
4. Oger, M.; Jebbar, M. The many ways of coping with pressure. *Res. Microbiol.* **2010**, *161*, 799–809. [CrossRef] [PubMed]
5. Meersmann, F.; Daniel, I.; Bartlett, D.H.; Winter, R.; Hazael, R.; McMillan, P.F. High-Pressure Biochemistry and Biophysics. *Rev. Mineral. Geochem.* **2013**, *75*, 607–648. [CrossRef]
6. Clifford, S.M.; Lasue, J.; Heggy, E.; Boisson, J.; McGovern, P.; Max, M.D. Depth of the Martian cryosphere: Revised estimates and implications for the existence and detection of subpermafrost groundwater. *J. Geophys. Res.* **2010**, *115*, E07001. [CrossRef]

7. Orosei, R.; Lauro, S.E.; Pettinelli, E.; Cicchetti, A.; Coradini, M.; Cosciotti, B.; Di Paolo, F.; Flamini, E.; Mattei, E.; Pajola, M.; et al. Radar evidence of subglacial liquid water on Mars. *Science* **2018**, *361*, 490–493. [CrossRef]
8. Orosei, R.; Ding, C.; Fa, W.; Giannopoulos, A.; Hérique, A.; Kofman, W.; Lauro, S.E.; Li, C.; Pettinelli, E.; Su, Y.; et al. The global search for liquid water on Mars from orbit: Current and future perspective. *Life* **2020**, *10*, 120. [CrossRef]
9. Hecht, M.H.; Quinn, R.C.; West, S.J.; Young, S.M.M.; Ming, D.W.; Catling, D.C.; Clark, B.C.; Boynton, W.V.; Hoffman, J.; DeFlores, L.P.; et al. Detection of perchlorate and the soluble chemistry of martian soil at the Phoenix lander site. *Science* **2009**, *325*, 64–67. [CrossRef]
10. Laye, V.J.; DasSarma, S. An antarctic extreme halophile and its polyextremophilic enzyme: Effects of perchlorate salts. *Astrobiology* **2018**, *18*, 412–418. [CrossRef]
11. Gault, S.; Jaworek, M.W.; Winter, R.; Cockell, C.S. High pressures increase α-chymotrypsin enzyme activity under perchlorate stress. *Commun. Biol.* **2020**, *3*, 550. [CrossRef]
12. Lenton, S.; Rhys, N.H.; Towey, J.J.; Soper, A.K.; Dougan, L. Highly compressed water structure observed in a perchlorate aqueous solution. *Nat. Commun.* **2017**, *8*, 919. [CrossRef] [PubMed]
13. Gault, S.; Cockell, C.S. Perchlorate Salts Exert a Dominant, Deleterious Effect on the Structure, Stability, and Activity of α-Chymotrypsin. *Astrobiology* **2021**, *21*, 405–412. [CrossRef] [PubMed]
14. Jahmidi-Azizi, N.; Oliva, R.; Gault, S.; Cockell, C.S.; Winter, R. The Effects of Temperature and Pressure on Protein-Ligand Binding in the Presence of Mars-relevant Salts. *Biology* **2021**, *10*, 687. [CrossRef] [PubMed]
15. Jahmidi-Azizi, N.; Gault, S.; Cockell, C.S.; Oliva, R.; Winter, R. Ions in the Deep Subsurface of Earth, Mars and Icy Moons: Their Effects in Combination with Temperature and Pressure on tRNA-Ligand Binding, Int. *J. Mol. Sci.* **2021**, *22*, 10861. [CrossRef]
16. Fetahaj, Z.; Ostermeier, L.; Cinar, H.; Oliva, R.; Winter, R. Biomolecular Condensates under Extreme Martian Salt Conditions. *J. Am. Chem. Soc.* **2021**, *143*, 5247–5259. [CrossRef]
17. Kriegler, S.; Herzog, M.; Oliva, R.; Gault, S.; Cockell, C.S.; Winter, R. Structural Responses of Model Biomembranes to Mars-relevant Salts. *Phys. Chem. Chem. Phys.* **2021**, *23*, 14212–14223. [CrossRef]
18. Dubins, D.N.; Lee, A.; Macgregor, R.B.; Chalikian, T.V. On the stability of double stranded nucleic acids. *J. Am. Chem. Soc.* **2001**, *123*, 9254–9259. [CrossRef]
19. Son, I.; Shek, Y.L.; Dubins, D.N.; Chalikian, T.V. Hydration changes accompanying helix-to-coil DNA transitions. *J. Am. Chem. Soc.* **2014**, *136*, 4040–4047. [CrossRef]
20. Wilton, D.J.; Ghosh, M.; Chary, K.V.A.; Akasaka, K.; Williamson, M.P. Structural change in a B-DNA helix with hydrostatic pressure. *Nucleic Acids Res.* **2008**, *36*, 4032–4037. [CrossRef]
21. Girard, E.; Prangé, T.; Dhaussy, A.C.; Migianu-Griffoni, E.; Lecouvey, M.; Chervin, J.C.; Mezouar, M.; Kahn, R.; Fourme, R. Adaptation of the base-paired double-helix molecular architecture to extreme pressure. *Nucleic Acids Res.* **2007**, *35*, 4800–4808. [CrossRef]
22. Winter, R. Interrogating the Structural Dynamics and Energetics of Biomolecular Systems with Pressure Modulation. *Ann. Rev. Biophys.* **2019**, *48*, 441–461. [CrossRef] [PubMed]
23. Akasaka, K.; Matsuki, H. (Eds.) *High Pressure Bioscience*; Springer: New York, NY, USA, 2015.
24. Fan, H.Y.; Shek, Y.L.; Amiri, A.; Dubins, D.N.; Heerklotz, H.; MacGregor, R.B.; Chalikian, T.V. Volumetric characterization of sodium-induced G-quadruplex formation. *J. Am. Chem. Soc.* **2011**, *133*, 4518–4526. [CrossRef] [PubMed]
25. Takahashi, S.; Sugimoto, N. Effect of pressure on thermal stability of G-Quadruplex DNA and double-stranded DNA structures. *Molecules* **2013**, *18*, 13297–13319. [CrossRef] [PubMed]
26. Takahashi, S.; Sugimoto, N. Effect of pressure on the stability of G-quadruplex DNA: Thermodynamics under crowding conditions. *Angew. Chem. Int. Ed. Engl.* **2013**, *52*, 13774–13778. [CrossRef] [PubMed]
27. Takahashi, S.; Sugimoto, N. Pressure-dependent formation of i-motif and G-quadruplex DNA structures. *Phys. Chem. Chem. Phys.* **2015**, *17*, 31004–31010. [CrossRef] [PubMed]
28. Patra, S.; Anders, C.; Erwin, N.; Winter, R. Osmolyte Effects on the Conformational Dynamics of a DNA Hairpin at Ambient and Extreme Environmental Conditions. *Angew. Chem. Int. Ed.* **2017**, *56*, 5045–5049. [CrossRef]
29. Knop, J.-M.; Patra, S.; Harish, B.; Royer, C.; Winter, R. The Deep Sea Osmolyte TMAO and Macromolecular Crowders Rescue the Antiparallel Conformation of the Human Telomeric G-Quadruplex from Urea and Pressure Stress. *Chem. Eur. J.* **2018**, *24*, 14346–14351. [CrossRef]
30. Sung, H.L.; Nesbitt, D.J. Single-Molecule Kinetic Studies of DNA Hybridization under Extreme Pressures. *Phys. Chem. Chem. Phys.* **2020**, *22*, 23491–23501. [CrossRef]
31. Garcia, A.E.; Paschek, D. Simulation of the pressure and temperature folding/unfolding equilibrium of a small RNA hairpin. *J. Am. Chem. Soc.* **2008**, *130*, 815–817. [CrossRef]
32. Patra, S.; Anders, C.; Schummel, P.H.; Winter, R. Antagonistic effects of natural osmolyte mixtures and hydrostatic pressure on the conformational dynamics of a DNA hairpin probed at the single-molecule level. *Phys. Chem. Chem. Phys.* **2018**, *20*, 13159–13170. [CrossRef]
33. Wozniak, A.K.; Schröder, G.F.; Grubmüller, H.; Seidel, C.A.M.; Oesterhelt, F. Single-molecule FRET measures bends and kinks in DNA. *Proc. Natl. Acad. Sci. USA* **2008**, *105*, 18337–18342. [CrossRef] [PubMed]

34. Chen, Y.Z.; Prohofsky, E.W. Theory of pressure-dependent melting of the DNA double helix: Role of strained hydrogen bonds. *Phys. Rev. E* **1993**, *47*, 2100–2108. [CrossRef] [PubMed]
35. Sugimoto, N. *Chemistry and Biology of Non-Canonical Nucleic Acids;* Wiley: Weinheim, Germany, 2021.
36. Tan, Z.-J.; Chen, S.-J. Nucleic acid helix stability: Effects of salt concentration, cation valence and size, and chain length. *Biophys. J.* **2006**, *90*, 1175–1190. [CrossRef] [PubMed]

life

MDPI

Article

Self-Consistent Parameterization of DNA Residues for the Non-Polarizable AMBER Force Fields

Amelia L. Schneider [1], Amanda V. Albrecht [1], Kenneth Huang [1], Markus W. Germann [1,2,*] and Gregory M. K. Poon [1,3,*]

[1] Department of Chemistry, Georgia State University, Atlanta, GA 30303, USA; aschneider@gsu.edu (A.L.S.); aalbrecht1@gsu.edu (A.V.A.); khuang8@gsu.edu (K.H.)
[2] Department of Biology, Georgia State University, Atlanta, GA 30303, USA
[3] Center for Diagnostics and Therapeutics, Georgia State University, Atlanta, GA 30303, USA
* Correspondence: mwg@gsu.edu (M.W.G.); gpoon@gsu.edu (G.M.K.P.)

Abstract: Fixed-charge (non-polarizable) forcefields are accurate and computationally efficient tools for modeling the molecular dynamics of nucleic acid polymers, particularly DNA, well into the μs timescale. The continued utility of these forcefields depends in part on expanding the residue set in step with advancing nucleic acid chemistry and biology. A key step in parameterizing new residues is charge derivation which is self-consistent with the existing residues. As atomic charges are derived by fitting against molecular electrostatic potentials, appropriate structural models are critical. Benchmarking against the existing charge set used in current AMBER nucleic acid forcefields, we report that quantum mechanical models of deoxynucleosides, even at a high level of theory, are not optimal structures for charge derivation. Instead, structures from molecular mechanics minimization yield charges with up to 6-fold lower RMS deviation from the published values, due to the choice of such an approach in the derivation of the original charge set. We present a contemporary protocol for rendering self-consistent charges as well as optimized charges for a panel of nine non-canonical residues that will permit comparison with literature as well as studying the dynamics of novel DNA polymers.

Keywords: nucleic acids; DNA; charge; electrostatic potential; ab initio methods; NMR spectroscopy; molecular dynamics; forcefield; AMBER

Citation: Schneider, A.L.; Albrecht, A.V.; Huang, K.; Germann, M.W.; Poon, G.M.K. Self-Consistent Parameterization of DNA Residues for the Non-Polarizable AMBER Force Fields. *Life* **2022**, *12*, 666. https://doi.org/10.3390/life12050666

Academic Editors: Tigran Chalikian and Jens Völker

Received: 1 April 2022
Accepted: 27 April 2022
Published: 30 April 2022

1. Introduction

The commercial success of consumer-grade graphics processing units (GPUs) and their adoption by major molecular dynamics (MD) packages has rendered practical many atomistic explicit-solvent simulations on affordable commodity computers. In the case of DNA, the AMBER forcefield continues to enjoy widespread use more than twenty years since the release of the second generation by Cornell et al. [1]. This popularity is attributable to subsequent reparameterization of the parameter set (parm94) that captures dynamic behavior to the μs timescale [2–5]. These changes (the latest known as OL15 [6] and parmbsc1 [7] for DNA) have involved the complete parametrization of the backbone dihedral potentials while retaining the atomic charges in parm94. This evolution contrasts with the extensive reparameterization of the charge set for proteins post-ff94 [8]. There is consensus that, taken together, these refinements in OL15 and parmbsc1 represent the accuracy limits for classical DNA forcefields based on fixed-charge two-body interactions [2]. Efforts are underway to overcome the limitations of classical forcefields, such as by incorporating nuclear quantum effects in so-called ab initio MD or AIMD [9–11]. Currently, the computational demands of AIMD mostly limit its application to the detailed solvation chemistry of low-MW systems over the fs-ps timescale [12]. Interrogation of biomolecular polymers exhibiting ns-μs timescales dynamics, which for many purposes do not require quantum mechanical treatment, remain very amenable to classical forcefields. One may therefore expect continued utility of AMBER forcefields in molecular mechanics work of DNA for the foreseeable future.

The derivation of atomic charges is critical to correctly capture noncovalent interactions in a classical forcefield. For the AMBER series of biomolecular forcefields, atomic charges are fundamentally derived from fitting against a quantum mechanical (QM) model of the electrostatic potential (ESPs) at the molecular surface [13]. For parm94, ESP-fitted charges are computed using the 6-31G* basis set in the gas phase [1]. The 6-31G* basis set, which is known to overestimate bond polarity, is chosen deliberately for condensed-phase systems to balance water models (such as TIP3P and TIP4P) which are themselves hyperpolarized over the gas-phase value for water [14]. Error compensation in water-solute and water-water interactions is an inherent feature of fixed-charge forcefields that lack accounting polarizability and nuclear quantum effects [9,10]. To mitigate spurious sensitivity of ESP fitting to molecular conformation, a second model, known as restrained electrostatic potential (RESP) [15], was devised to "restrain" polarization of buried atoms, which are poorly determined by surface ESPs, towards a zero value during the fit. RESP fitting is a key finishing step in the parameterization of novel solutes [16] as well as building blocks for nucleic acids, proteins, and carbohydrates [17].

For polymeric solutes that exemplify biomolecular macromolecules such as nucleic acids, it is critical that all the residues be parameterized on an equivalent, self-consistent basis. Self-consistency is a specific concern when new residues are introduced and incorporated with existing residues in a mixed polymer. The original set of nucleic acid residues in AMBER contains only the canonical set of A, C, G, and T/U. Advances in solid-state phosphoramidite chemistry have greatly broadened the scope of nucleic acid residues, many of which have been parameterized for the AMBER forcefield. For DNA, they include non-canonical bases that occur naturally, such as hypoxanthine, epigenetically modified cytosines (e.g., 5-methylcytosine), diaminopurine (DAP or 2-aminoadenine, 2AA), and DNA damage products (e.g., 8-oxoguanine). In addition, many non-natural nucleobases, such as 2-aminopurine, are used as spectroscopic and chemical probes in molecular biology. Given that mixed sequences of new and canonical residues are typical, it is clearly of interest to parameterize novel residues to preserve self-consistency with the original canonical bases.

From a self-consistency perspective, RESP fitting is a critical step because it allows for globally fitting multiple species, with flexibility in fixing, sharing, and restricting charge assignments during the fit [18]. In the parm94 nucleic acids charge set, which were RESP fitted from ESPs computed at the HF/6-31G* level, values for the backbone (deoxyribose and phosphate) atoms are shared across the four canonical bases, with the exception of the C1′ and H1′ atoms. The latter two atoms float with the variable nucleobase atoms during the fit. In the literature, parameterization of new residues has generally adhered to this scheme. More critical and unfortunately less uniform, however, are the structures used to derive the ESP and RESP-fitted charges. Because atomic charges are fitted against the surface molecular potential, ESP fitting is highly sensitive to molecular conformation [15]. Although RESP is more robust to the statistical ill effects of buried atoms than ESP fitting [19], conformational effects on the derived atomic charges are general and reflect the molecular microenvironment. The need to control for conformational effects on charge derivation has spurred several innovations, such as the R.E.D. tools by Dupradeau, Cieplak, and coworkers [20], aimed at standardizing and automating the charge derivation workflow.

In the original charge derivation of the canonical nucleic acids in parm94 [18], structures were derived by molecular mechanics (MM) minimization using the previous-generation ff86 forcefield by Weiner et al. [21,22]. This choice was presumably due to the computational demand of the time for ab initio optimization of whole nucleosides. In contrast, contemporary charge parameterization typically begins with geometry optimization of de novo models at the HF/6-31G* level which is affordable nowadays [23]. In principle, QM optimization should yield physically more accurate structures, but the consistency of this contemporary practice with the parm94 charge set is not obvious and has never been clarified to our knowledge. If the QM-optimized structures do not sufficiently capture the peculiarity of the MM models (however flawed the latter may be relative to the former), the self-consistency of the forcefield with respect to a mixed polymer could be compromised.

How could this be tested? A major stated design principle of the AMBER forcefield is transferability. Adhering to this principle, factors that impact the parameterization of a new residue should similarly impact the canonical residues, whose RESP-fitted charges are known, i.e., parm94. An unambiguous approach to testing the self-consistency of a parameterization protocol with the forcefield is therefore to apply the protocol to extant residues in parm94. If the protocol is self-consistent with the derivation of ff94, it should naturally reproduce the atomic charges of the original bases in the forcefield. The purpose of this work is two-fold. (1) Determine a parameterization protocol for ff94 that best preserves self-consistency with the canonical bases in the forcefield. (2) Provide self-consistent parameters for a panel of non-canonical nucleobases, including several that are not yet reported. Here, we concentrate on DNA, but we expect the resultant principles and recommendations to apply to the parameterization of RNA residues as well.

2. Materials and Methods

Chemical structure optimizations. Initial atomic models were obtained from the ff86 forcefield or generated with GaussView (Version 5.0.9; Gaussian, Wallingford, CT, USA). Coordinates were parameterized to reflect point group symmetry, planarity, or specific conformations as described in the text. Geometry optimization and subsequent quantum mechanical calculations were performed in internal coordinates with Gaussian 16 (Revision A.03; Gaussian). The stationarity of the optimized structures was confirmed with a frequency calculation. MM energy minimization was performed in either AMBER5 or AMBER16 using the *sander* module.

Atomic charge fitting. Fitting to a QM electrostatic potential was performed on QM-optimized or otherwise specified structures with Gaussian 16 using the Merz–Singh–Kollman scheme (pop = MK). The ESP was computed at 4 layers (1.4, 1.6, 1.8, 2.0 $\times$ the van der Waals radius) and a nominal density of 1 point/$Å^2$ [13]. RESP fitting was performed per the reported two-step multi-molecular procedure [18] as described in Supplemental Methods.

NMR spectroscopy. Hairpin-forming oligodeoxynucleotides were synthesized by Integrated DNA Technologies (Coralville, IA, USA) by standard phosphoramidite synthesis. The DNA was adjusted to 0.5 M NaCl to dissociate ionic contaminants and purified by size-exclusion chromatography on a 5 $\times$ 5 mL HiTrap Desalting column on an ÄKTA instrument (Cytiva, Marlborough, MA, USA.) The desalted DNA was lyophilized and dissolved in 20 mM NaH_2PO_4/Na_2HPO_4 containing 50 mM NaCl and 0.5 mM EDTA. D_2O was added to 10% and the pH was adjusted to 6.40. NMR experiments were performed on a Bruker Avance I 500 spectrometer equipped with a TBI $^1H\{^{13}C, X\}$ z gradient probe. For monitoring imino proton resonances, a 1-1 jump and return sequence was used to record spectra from 288 K to 308 K. Phase-sensitive 1-1 jump and return NOESYs were collected at 288 K with 2048 $\times$ 800 data points in the two dimensions and 72 scans per t_1 increment using a 150 ms mixing time and a 1.0 s relaxation delay. Two-dimensional (2D) spectra were strip transformed and processed using a 4K $\times$ 2K matrix. Both dimensions were apodized with shifted $\sin(\pi/2)$ bell functions. Proton chemical shifts were referenced to internal 2,2-dimethyl-2-silapentane-5-sulfonate (DSS).

Molecular dynamics simulations. The conformations of an 18-nucleotide (nt) DNA hairpin designed to probe the effect of an internally positioned residue were sampled using the GROMACS 2022 package. The parmbsc1 update [7] of the ff94 forcefield was used. Following topology generation, each system was set up in dodecahedral boxes 1.0 nm wider than the longest dimension of the solute, solvated with TIP3P water, and neutralized with Na^+ and Cl^- to 0.05 M. Electrostatic interactions were handled by the P3M method [24] with a 1 nm distance cutoff. A timestep of 2 fs was used and bonds including hydrogens were constrained using LINCS. After the structures were energy-minimized, the *NVT* ensemble was equilibrated at 298 K (modified Berendsen thermostat) [25] for 1 ns to thermalize the system, followed by another 1 ns of equilibration of the *NPT* ensemble at 1 bar (stochastic cell rescaling) [26] and 298 K. The NPT ensemble was simulated at 298 K without restraints for 200 ns.

Computational analysis. Furanose ring puckering was computed according to the Cremer–Pople scheme [27] from ordered atomic coordinates (O4′→ C1′→ C2′→ C3′→ C4′) as described by Chan et al. [28]. For NMR chemical shift calculations, averaged structures were first optimized by energy minimization against parmbsc1 in TIP3P water with a nominal complement of 0.05 M Na^+ and Cl^- ions. Single-point calculations were performed using the GIAO method at the B3LYP/6-31G* level in implicit CPCM water. A calculation on the DSS anion, optimized at B3LYP/6-31 + G*, was performed to reference the computed isotropic shielding tensors.

3. Results and Discussion

MM-minimized structures are superior to QM-optimized models for ff94 charge parameterization. To determine whether QM-optimized models represented good structures for ff94 parameterization, we optimized the four canonical deoxynucleosides (DAN, DGN, DCN, and DTN) at HF/6-31G*. Optimizations were constrained only to enforce planarity of the purine or pyrimidine rings and a *trans* conformation for the 5′- and 3′-OH relative to the connected heavy atoms, the latter as indicated in the derivation of ff94 [18]. The resultant structures differed from the reported MM-optimized geometry [18], which was provided in summary form in terms of the sugar pucker, the backbone dihedral γ (O5′-C5′-C4′-C3′), and the *N*-glycosidic dihedral χ (O4′-C1′-N9-C4 for purines, O4′-C1′-N1-C2 for pyrimidines), by ~10% (Table 1).

For comparison, we optimized a parallel set of structures at the same level of theory and basis set with additional constraints imposed to match the reported geometry. We note that these geometric parameters do not specify an explicit structure due to the specification of sugar puckers as amplitude q and phase W according to the convention of Cremer and Pople [27]. Since the Cremer–Pople convention takes directly as input ordered atomic coordinates (O4′→ C1′→ C2′→ C3′→ C4′), the pucker parameters cannot be constrained directly in terms of functions of internal coordinates during optimization. By scanning the endocyclic dihedrals, we determined representative geometries matching the reported phases and amplitudes to within $W < 0.01$ Å (<3% deviation) in phase and $q < 1°$ (<0.7%) in amplitude. These ring dihedrals were then additionally constrained during optimization to yield a set of structures in much better agreement (~1% deviation) with the reference geometries.

To assess the impact of these discrepancies on the final atomic charges, we followed the originally described two-step multi-molecular RESP procedure (see Supplemental Methods) [18]. As our immediate objective is to determine how well the structures reproduce the parm94 charges, we did not fix any of the charges, only equivalencing rotatable hydrogen atoms and imposing targeted constraints in combining nucleosides and the phosphate analog (dimethylphosphate) as specified by Cieplak et al. [18]. The imposition of puckering and *N*-glycosidic geometries reported in Cieplak et al., as could be practically executed, yielded final RESP charge parameters in better agreement with parm94 values, albeit less than expected from the improved agreement in geometry (Table S1, Supplementary Materials). To test if agreement might be further improved with more polished structures, we fitted nucleosides optimized at the MP2/6-31G* level (Table S1). The MP2 models did not improve on the agreement with the parm94 charges over the HF/6-31G* models. For structure preparation in parm94 charge derivation, Hartree–Fock appeared to be near or at the limit of usefulness achievable by ab initio approaches.

Table 1. QM- and MM-optimized geometries of canonical DNA nucleosides. Reference values are quoted exactly as reported by Cieplak et al. [18]. Parametric values from QM (HF/6-31G*) and MM optimizations (against the ff86 forcefield) are given to one additional significant figure, with % deviation from the reference values in parenthesis. In all cases, H5′ and H3′ are fixed in *trans* with the bonded heavy atoms [18]. The constrained QM optimizations imposed additional dihedral constraints to satisfy the reference values in the initial structure. The corresponding geometry of 1-NH_2-deoxyribose is provided to assess the impact of the base in the nucleosides.

	q, Å	W, °	γ, °	χ, °
DAN				
Reference	0.38	151.9	58.5	210.0
QM	0.337 (−11.3%)	163.62 (7.7%)	51.87 (−11.3%)	227.63 (8.4%)
QM constrained	0.379 (−0.3%)	152.56 (0.4%)	58.50	210.00
MM	0.380	151.84 (−0.04%)	60.40 (3.2%)	206.69 (−1.6%)
DGN				
Reference	0.38	151.4	58.5	209.9
QM	0.338 (−11.0%)	164.37 (8.6%)	50.91 (−13.0%)	231.60 (10.3%)
QM constrained	0.376 (−1.1%)	152.01 (0.4%)	58.50	209.90
MM	0.378 (−0.5%)	151.23 (−0.1%)	59.56 (1.8%)	208.74 (−0.5%)
DCN				
Reference	0.38	149.2	58.9	209.7
QM	0.337 (−11.3%)	159.50 (6.9%)	54.40 (−7.6%)	205.82 (−1.9%)
QM constrained	0.379 (−0.3%)	149.74 (0.4%)	58.90	209.70
MM	0.383 (0.8%)	149.46 (0.2%)	60.85 (4.0%)	210.27 (0.2%)
DTN				
Reference	0.38	149.1	58.4	215.7
QM	0.345 (−9.2%)	159.73 (7.1%)	52.23 (−10.6%)	226.54 (5.0%)
QM constrained	0.383 (0.8%)	149.59 (0.3%)	58.40	215.70
MM	0.382 (0.5%)	149.62 (0.3%)	60.49 (3.6%)	215.43 (−0.1%)
1-NH_2-deoxyribose (QM)	0.332	163.03	51.54	

The results with HF/6-31G*-optimized structures as inputs for charge fitting prompted us to ask whether QM-based structures were necessarily better models for charge fitting over the MM-minimized structures (generated by the previous-generation ff86 forcefield) [18] used in the derivation of ff94. As the legacy AMBER4 program used to energy-minimize the structures for ff94 is no longer maintained in official AMBER repositories, we used AMBER5 (courtesy of Dr. Hector Baldoni, Universidad Nacional de San Luis) as the closest substitute. As inputs, we constructed deoxynucleosides based on the parameter set of ff86 (parm86). The resultant models were quite different in conformation and charge distribution from the specifications in Cieplak et al. [18]. Nevertheless, following energy minimization in the ff86 forcefield, the output structures were significantly closer in conformation than the HF/6-31G*-optimized models to the target geometries, even without the use of strong restraints to enforce agreement (Table S1). Subsequent charge fitting by ESP followed by multi-molecular RESP yielded atomic charges that agree better with parm94 by a factor of 3 (deoxythymidine) to over 6 (deoxycytidine) in RMSD over charges derived from HF/6-31G*-optimized structures (Figure 1A). The goodness-of-fit metrics were indistinguishable in all cases (Table S1), indicating that the differences in the charges did not originate with RESP fitting but were pre-existing in the structures.

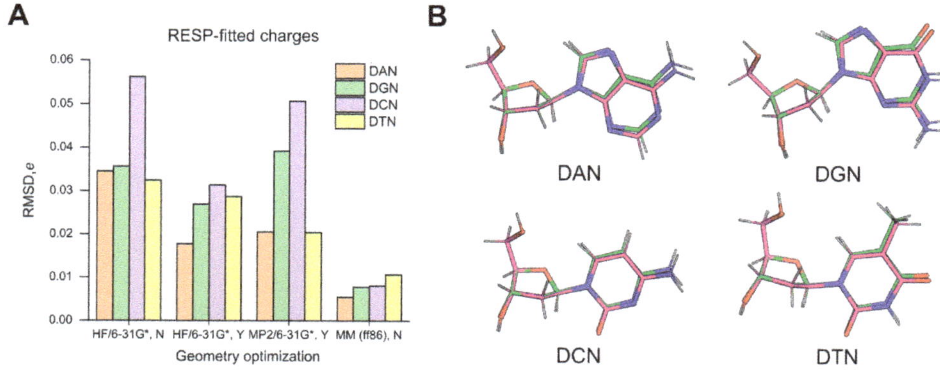

Figure 1. Comparison of geometry optimization schemes for ff94 charge parameterization. (**A**) Agreement with the parm94 charge set by RESP-fitted charges derived from structures optimized by ab initio (HF/6-31G* and MP2/6-31G*) and MM methods as described in the text. Y/N in the abscissa refers to whether the optimization was constrained (QM) or restrained (MM). Parametric values are provided in Table S1, Supplementary Materials. (**B**) Each pair of QM-optimized (magenta carbon) and MM-minimized (green carbon) structures were aligned by the five deoxyribose heavy ring atoms (C1′ to C4′ and O4′). The QM models were optimized with constraints targeted at the geometry specified for ff94 charge derivation. The MM minimization against the ff86 forcefield closely approached the geometric targets even without any strong restraints. See Table 1 for numerical values.

To better understand the structural basis of the differences, we examined the optimized structures by HF/6-31G* and energy minimization in ff86. For each nucleoside, we aligned the pair of structures by the five atoms of the deoxyribose ring (C1′ to C4′ as well as O4′) (Figure 1B). In all cases, the RMSD values for these endocyclic atoms were below 0.01 Å. Since ff86-minimization closely achieved the Cremer–Pople pucker values specified by Cieplak et al. (Table 1) [18], the slight deviations in the pucker of the constrained QM structures were not the main contributor to the discrepancy in RESP-fitted charges. Thus, the overall geometries used in ff94 derivation appeared to fundamentally deviate from those predicted quanta mechanically. Since ESP fitting is known to be highly sensitive to conformation [15], one expects structural differences to be amplified and passed on to RESP, which uses the ESP results directly. As a result, the selection of MM-minimized models as structures for QM-based charge parameterization in the original derivation of ff94 would be a consequential choice, with implications for the parameterization of new residues for use with this forcefield. The adage: "All models are wrong, but some are useful," famously attributed to the statistician George E.P. Box, appears to apply to this situation.

A contemporary protocol for consistently parameterized DNA residues for ff94. The better suitability of the AMBER-minimized models as input structures for QM-based charge fitting raises two potential challenges if they were to serve an accessible workflow for the community. The first is the general lack of availability of legacy versions of the AMBER software from the time that ff94 was developed. To overcome this, we verified that AMBER16 reproduced AMBER5 outputs to 10^{-3} or better when minimization is carried out in a flat dielectric corresponding to the gas phase (unit dielectric) (Table S2, Supplementary Materials).

The second challenge with generating appropriate MM-minimized models for ff94 concerns the input structures. Nucleoside structures are constructed in ff86 by a fragment-based approach that conjoins separately prepared bases and deoxyribose. The atomic charges were computed by ESP fitting from experimental structures of *N*9-methylpurine/*N*1-methylpyrimidine base and 1′-aminodeoxyribose analogs [13]. Sourcing appropriate experimental structures of such analogs would likely be generally problematic for novel entities. To overcome this, we asked whether the structures of the analogs could be adequately furnished by QM optimization of ab initio models. We therefore generated structures of various compounds in silico and compared their ESP-fitted atomic charges with reported values [13] computed from experimental structures. Charges in simple test compounds (water, formaldehyde, dimethyl ether, and methanol) were all well matched up to HF/6-31G** when the in silico structures were optimized at mp2/aug-cc-pVTZ (Table S3, Supplementary Materials). For nucleobases analogs (*N*9-methyladenine, *N*9-methylguanine, *N*1-methylcytosine, *N*1-methyladenine, and *N*1-methyluracil) and 1′-aminodeoxyribose, structures optimized at HF/6-311G++(3df,2p) level gave ESP-fitted charges at the HF/STO-3G level (which was used in deriving ff86) with comparable RMSDs (Table S4, Supplementary Materials). More computationally costly structures at a higher level of theory or larger basis set did not furnish significantly better matched charges. We therefore conclude that base fragments polished at the HF/6-311G++(3df,2p) level represented sufficiently accurate models for constructing nucleosides for the ff86 forcefield [21,22].

To summarize, our protocol for parameterizing new residues that retain maximum self-consistency with the AMBER charge set from parm94 up to the most current (parmbsc1 and OL15) is as follows (Figure 2A). Perform HF optimization of the methyl base analog at as large a basis set as practical, e.g., HF/6-31G++(3df,p), followed by ESP fitting at HF/STO-3G. Based on these ESP charges, generate an ff86-compatible structure using charges from parm86 for the deoxyribose. A detailed description of this procedure is provided in Supplemental Methods. Energy-minimize this structure in a flat gas-phase dielectric while applying restraints on H5′ and H3′ to remain in *trans* with bonded heavy atoms. Continue with ESP and RESP as usual [18]. We note that the steps prior to ESP are straightforward and computationally inexpensive relative to ab initio optimization of whole nucleosides at comparable levels of theory and size of basis set.

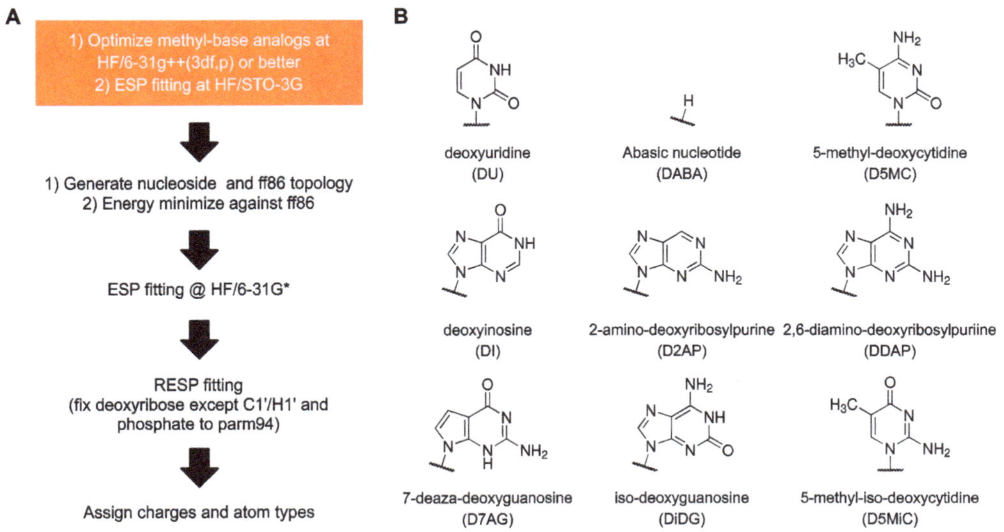

Figure 2. Self-consistent parameterization of DNA residues for ff94 and derived (parmbsc1, and OL15) AMBER forcefields. (**A**) Workflow of the procedure. The steps in the orange box are described in the main text. A detailed summary of generating the ff86 topology is provided in Supplemental Methods. The remaining steps are exactly as practiced elsewhere [18]. (**B**) Residues derived using this protocol. Final RESP-fitted charges are listed in Table S5 together with a comparison with previously reported values in the literature.

The limitations of our recipe are the same as those for the ff94-based forcefields. The parm94 charge set used only a single conformation of the deoxyribose (C2′-endo) and backbone torsions. For self-consistency with the canonical residues, we did not incorporate multiple conformations in the RESP fitting [20,23]. One could of course parameterize the canonical and new residues completely anew with multi-conformational fitting, but the resultant charges and simulations would not be consistent with those using the authentic parm94-derived parameter set. Finally, residues with radically novel structures may not be adequately supported by the existing set of bond and dihedral types in the ff86 forcefield and require additional parameterization regardless of the method of charge derivation.

Having established a protocol for setting the atomic charges of residues to maximize consistency with the existing residues in ff94, we set out to parameterize a panel of useful non-canonical DNA residues (Figure 2B). Since *N*1-methyluracil had been parameterized for uridine for ff86 [13], this recipe immediately yielded deoxyuridine. The recipe also yielded an abasic residue (hydrogen-capped C1′), another experimentally useful construct, by using CH_4 as a methyl analog of a hydrogen atom. Other methyl analogs were drawn from residues that have previously been parameterized as well as commercially available residues: inosine, 2-aminopurine (2AP), 2,6-diaminopurine (DAP), 5-methylcytosine (5-MC), iso-guanine (iso-dG), and 5-methyl-iso-cytosine (5-methyl-iso-dC). All *N*-methyl-base analogs were optimized at the HF/6-311++g(3df,2p) level and conjoined with 1-aminodeoxyribose exactly as described for ff86 [22] (see Supplemental Methods). Roughly, the excess partial charges incurred from removing the methyl and amino substituents from the base and deoxyribose were absorbed into N1/N9 and C1′ atoms. All nucleosides generated were associated with the same set of charges for the sugar atoms for energy minimization. Following ESP fitting at the HF/STO-3G level, the new residues were RESP fitted globally with the four canonical DNA residues with the charges of the phosphate and deoxyribose (except for C1′and H1′) fixed to parm94 values. The chemical modifications in the presented non-canonical bases were assigned by analog using existing bond and dihedral types. The full set of RESP-fitted charges of the nine non-canonical bases, parameterized for parmbsc1 are given in Table S5, Supplementary Materials, with literature values [29–34] where available. Parenthetically, an inspection of Table S5 shows the RMS deviations between our charges and those from the literature to be on the same order of magnitude, and in the same direction, as the differences between charges derived from QM-based models for the canonical residues in parm94 (c.f., Figure 1A).

Is global fitting essential for parameterizing new residues? In principle, self-consistency is most retained if all residues are fitted simultaneously. The RESP implementation in AMBER provides this, and we fitted all nine non-canonical residues together with the four canonical residues. To render new residues transparent to the canonical bases, we fixed (as others have carried out) the deoxyribose and phosphate charges to the parm94 values. In global fitting, the optimization of each parameter set is not independent but subject to influence by the full data set. To evaluate the sensitivity of RESP-fitted charges to the statistics of global fitting, we compared the RESP charges of non-canonical residues that were fitted globally with charges derived from fitting the same residues individually with the canonical ff94 bases. Using inosine, 2-AP, and 5-MC as examples, we found negligible differences, no more than 10^{-5} in RMSD between the two approaches (Table S6, Supplementary Materials). This robustness appeared to reflect the greatly reduced number of statistical degrees of freedom when the phosphate and most of the deoxyribose charges were fixed to the parm94 values. The practical implication is that future residues could be conveniently added on their own, even one at a time, suggesting a good degree of "future-proofing" in the procedure.

Demonstration of newly incorporated DNA residues. The chemical shifts of imino protons are sensitive probes of nucleic acid conformation in solution. Chemical shifts depend on the nature of the nucleobase, the exchange behavior with bulk water for rapidly exchanging (broad) residues, and the local environment. We therefore tested several of the newly parameterized residues in Figure 2B by characterizing their experimental and computed chemical shift in a mixed-sequence duplex DNA construct. We designed a DNA hairpin in which a probe base **X** was part of an internal sequence 5′-d(CG**X**AA)-3′ (Figure 3A). To maximize the duplex structure of the cassette, the stack was flanked by a standard cassette consisting of an extra base pair from the terminus on one side and a T_4 hairpin on the other. All bases in the cassette were canonical DNA. A reference hairpin harboring 5′-d(CG**G**AA)-3′ serves as a control.

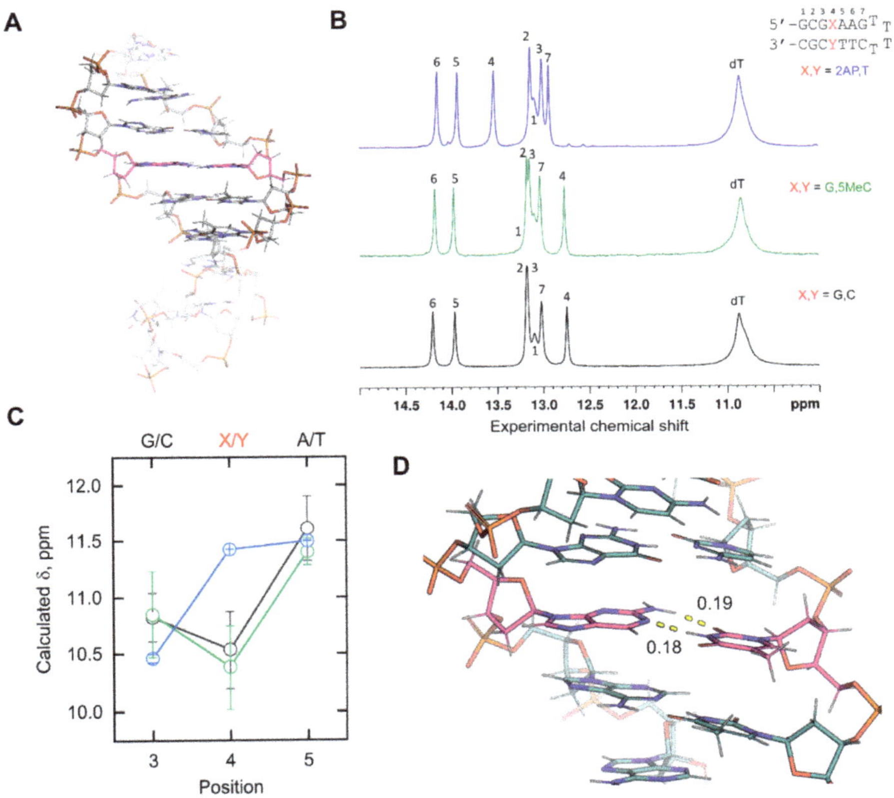

Figure 3. Comparison of experimental ^{1}H chemical shifts with predictions using simulated models of DNA duplexes harboring non-canonical residues. (**A**) Hairpin cassette designed for this study. The entire hairpin was simulated as an unrestrained *NPT* ensemble, and an internal 5 bp stack, rendered in opaque colors, was used for chemical shift calculations. The central probe residue position in the duplex is indicated with magenta carbon. (**B**) Experimental imino ^{1}H spectra of three test sequences, referenced against DSS and optimally resolved at 288 K. Peaks were assigned by ^{1}H-^{1}H NOESY experiments (not shown). (**C**) Calculated chemical shifts for test sequences, referenced against the averaged computed methyl ^{1}H of an optimized DSS structure. Colors follow the spectra in Panel B. Points represent the means ± standard deviation of the triplicate averaged structures. (**D**) Illustrative averaged conformation of the 2AP:T base pair. Watson–Crick bonds are shown with units in nm.

Experimental ^{1}H chemical shifts were measured by NMR spectroscopy in NaH_2PO_4/Na_2HPO_4 buffer (20 mM, pH 6.4) containing 10% D_2O and 50 mM NaCl (Figure 3B). For MD simulations, the unrestrained *NPT* ensemble was sampled in explicit TIP3P water with nominally 50 mM Na^+/Cl^- counterions at 298 K (see Supplementary Materials). Supported by the strongly convergent trajectory (over 200 ns) based on the RMSD of atomic coordinates, we generated averaged structures of the 5 bp stack from non-overlapping 50 ns segments of trajectory (Figure S1, Supplementary Materials). Given the substantial atom count of the stack, we optimized the averaged structures by energy minimization against parmbsc1 with a complement of TIP3P water and Na^+/Cl^- ions at the same nominal concentration as in the NMR samples. For each minimized structure, chemical shifts were computed at the B3LYP/6-31G* level in implicit water. Following this procedure, we compared the imino proton chemical shifts for the inner three positions of the stack, i.e., d(G**X**A), with the experimental values. Chemical shifts for distal base pairs of the stack, which were fully exposed to solvent in the calculations, were expected to diverge strongly from the experimental values.

For the canonical-only control, the imino protons of the inner triplet from the simulation (Figure 3C) showed the same rank order as the experimental chemical shifts. A lack of quantitative agreement with the experimental values might be expected given the structure minimization (which was needed but also perturbative), forgoing the computational demands of larger basis sets, and limitations in capturing hydration effects with implicit water [35]. To test the consistency of the results, we examined a hairpin in which the probe dG was paired with 5-methyl-dC. The experimental imino ^{1}H chemical shifts varied at all positions by ± 0.03 ppm or less between the reference and methyl-substituted hairpin. We were encouraged to find that the computed imino proton shifts of the inner triplet overlapped closely within their standard deviations between the two simulated hairpins. This was not a trivial result, as a comparison of D5MC (Table S5) shows charge redistribution relative to DC (Table S4) over multiple atoms.

We next tested d(2AP) paired with dT, whose imino peak (contributed by T) was significantly downshifted, by 0.8 ppm, relative to the other two base pairs probes. We found that the computed probe chemical shift was also downshifted, inverting its position with position 3 as observed experimentally. While imino proton shifts from T are generally downfield from G counterparts, the order of the computed chemical shifts for the mixed purine/pyrimidine 5-d(G-2AP-A)-3′ triplet remained in agreement with the experimental one, lending further credence to the physical relevance of the simulated DNA models harboring novel residues. Examination of the trajectory showed that the 2AP paired with the T via expected Watson–Crick interactions (Figure 3D). Similar to active efforts for RNA [36,37], the semi-quantitative agreement with experimental chemical shifts suggests that ns-timescale simulations represent reasonable starting points for chemical shift prediction in the case of duplex DNA structures. Currently, classical forcefields admit a tradeoff in neglecting nuclear quantum effects [9,11] in their treatment of H-bonding. One may envision that, as the computational burden of AIMD becomes more tractable for macromolecules, prediction of the conformational and hydration contributions to experimental spectroscopic observables would be further improved.

4. Conclusions

Benchmarked against the parm94 charge set of canonical DNA residues, RESP-fitted charges derived from QM-optimized models exhibit RMS deviation on the order of 10%, while those from MM-minimized (ff86-based) structures deviate no more than 1%. The higher self-consistency in MM-minimized structures over QM models for charge derivation of DNA residues for current AMBER nucleic acids forcefields thus motivates the use of MM-minimized structures for the parameterization of new residues. To bridge the gaps left by the approach [18] and legacy software used in the development of the parm94 DNA charge set, which is used also in the most current updates (parmbsc1 and OL15), we have devised a workflow for generating appropriate structural models from ff86 using contemporary computational methodologies. We presented a panel of nine non-canonical residues, some of which have been previously derived using ab initio structural models. While we do not suggest that RESP-fitted charges of residues derived from QM-optimized structures are invalid for their intended purposes, they are demonstrably less self-consistent with the charge set of the original canonical residues and can be readily improved through a well-defined change in structure preparation (Figure 2A). As high-level optimization of whole nucleosides, particularly of large and complex bases (e.g., dye-conjugation or heavy metal substitution), remains computationally demanding, we suggest that the proposed protocol is compelling in providing higher-quality charges at reduced computational effort.

Supplementary Materials: The following are available online at https://www.mdpi.com/article/10.3390/life12050666/s1, Supplemental Methods. Table S1. RESP-fitted atomic charges of QM- and MM-optimized canonical DNA nucleosides. Table S2. Reproducibility of legacy minimization results by contemporary versions of AMBER. Table S3. Comparison of fits to charge models between experimental and optimized ab initio structures. Table S4. QM-optimized structures of nucleobase fragment analogs reproduce ff86 atomic charges. Table S5. Final RESP-fitted atomic charges and comparison with literature. Table S6. Perturbation of global fitting on RESP-fitted charges. Figure S1. Equilibration of the DNA hairpin in unrestrained MD simulations.

Author Contributions: Conceptualization, G.M.K.P.; data curation, A.L.S., A.V.A., K.H., M.W.G. and G.M.K.P.; formal analysis, A.V.A., K.H., M.W.G. and G.M.K.P.; funding acquisition, M.W.G. and G.M.K.P.; investigation, A.L.S., M.W.G. and G.M.K.P.; methodology, K.H., M.W.G. and G.M.K.P.; project administration, A.L.S., G.M.K.P.; resources, G.M.K.P., Markus Germann; supervision, G.M.K.P.; validation, A.L.S., G.M.K.P.; visualization, G.M.K.P.; writing—original draft, A.L.S., G.M.K.P.; writing—review and editing, A.L.S., M.W.G. and G.M.K.P. All authors have read and agreed to the published version of the manuscript.

Funding: This is work is supported by NSF grant MCB 2028902 to M.W.G. and G.M.K.P., and NIH grants HL155178 and GM137160 to G.M.K.P.

Informed Consent Statement: Not applicable.

Data Availability Statement: The data presented in this study are available on reasonable request from the corresponding authors.

Acknowledgments: We are grateful to Hector Baldoni for sharing a copy of AMBER5, as well as cited colleagues who generously shared their charge sets. We also thank W. David Wilson, Samer Gozem, and Ivaylo Ivanov for insightful suggestions. This work is dedicated to Thomas Bloom (University of Toronto).

Conflicts of Interest: The authors declare no conflict of interest.

References

1. Cornell, W.D.; Cieplak, P.; Bayly, C.I.; Gould, I.R.; Merz, K.M.; Ferguson, D.M.; Spellmeyer, D.C.; Fox, T.; Caldwell, J.W.; Kollman, P.A. A Second Generation Force Field for the Simulation of Proteins, Nucleic Acids, and Organic Molecules. *J. Am. Chem. Soc.* **1995**, *117*, 5179–5197. [CrossRef]
2. Galindo-Murillo, R.; Robertson, J.C.; Zgarbova, M.; Sponer, J.; Otyepka, M.; Jurecka, P.; Cheatham, T.E., 3rd. Assessing the Current State of Amber Force Field Modifications for DNA. *J. Chem. Theory Comput.* **2016**, *12*, 4114–4127. [CrossRef] [PubMed]
3. Dans, P.D.; Danilane, L.; Ivani, I.; Drsata, T.; Lankas, F.; Hospital, A.; Walther, J.; Pujagut, R.I.; Battistini, F.; Gelpi, J.L.; et al. Long-timescale dynamics of the Drew-Dickerson dodecamer. *Nucleic. Acids Res.* **2016**, *44*, 4052–4066. [CrossRef]
4. Dans, P.D.; Ivani, I.; Hospital, A.; Portella, G.; Gonzalez, C.; Orozco, M. How accurate are accurate force-fields for B-DNA? *Nuclei. Acids Res.* **2017**, *45*, 4217–4230. [CrossRef]
5. Perez, A.; Luque, F.J.; Orozco, M. Dynamics of B-DNA on the microsecond time scale. *J. Am. Chem. Soc.* **2007**, *129*, 14739–14745. [CrossRef] [PubMed]
6. Zgarbova, M.; Otyepka, M.; Sponer, J.; Mladek, A.; Banas, P.; Cheatham, T.E., 3rd; Jurecka, P. Refinement of the Cornell. Nucleic Acids Force Field Based on Reference Quantum Chemical Calculations of Glycosidic Torsion Profiles. *J. Chem. Theory Comput.* **2011**, *7*, 2886–2902. [CrossRef] [PubMed]
7. Ivani, I.; Dans, P.D.; Noy, A.; Perez, A.; Faustino, I.; Hospital, A.; Walther, J.; Andrio, P.; Goni, R.; Balaceanu, A.; et al. Parmbsc1: A refined force field for DNA simulations. *Nat. Methods* **2016**, *13*, 55–58. [CrossRef]
8. Duan, Y.; Wu, C.; Chowdhury, S.; Lee, M.C.; Xiong, G.; Zhang, W.; Yang, R.; Cieplak, P.; Luo, R.; Lee, T.; et al. A point-charge force field for molecular mechanics simulations of proteins based on condensed-phase quantum mechanical calculations. *J. Comput. Chem.* **2003**, *24*, 1999–2012. [CrossRef]
9. Ceriotti, M.; Cuny, J.; Parrinello, M.; Manolopoulos, D.E. Nuclear quantum effects and hydrogen bond fluctuations in water. *Proc. Natl. Acad. Sci. USA* **2013**, *110*, 15591–15596. [CrossRef]
10. Cassone, G.; Kruse, H.; Sponer, J. Interactions between cyclic nucleotides and common cations: An ab initio molecular dynamics study. *Phys. Chem. Chem. Phys.* **2019**, *21*, 8121–8132. [CrossRef]
11. Cassone, G. Nuclear Quantum Effects Largely Influence Molecular Dissociation and Proton Transfer in Liquid Water under an Electric Field. *J. Phys. Chem. Lett.* **2020**, *11*, 8983–8988. [CrossRef] [PubMed]
12. Markland, T.E.; Ceriotti, M. Nuclear quantum effects enter the mainstream. *Nat. Rev. Chem.* **2018**, *2*, 109. [CrossRef]
13. Singh, U.C.; Kollman, P.A. An approach to computing electrostatic charges for molecules. *J. Comput. Chem.* **1984**, *5*, 129–145. [CrossRef]

14. Kuyper, L.F.; Hunter, R.N.; Ashton, D. Free energy calculations on the relative solvation free energies of benzene, anisole, and 1,2,3-trimethoxybenzene: Theoretical and experimental analysis of aromatic methoxy solvation. *J. Phys. Chem.* **1991**, *95*, 6661–6666. [CrossRef]
15. Bayly, C.I.; Cieplak, P.; Cornell, W.; Kollman, P.A. A well-behaved electrostatic potential based method using charge restraints for deriving atomic charges: The RESP model. *J. Phys. Chem.* **1993**, *97*, 10269–10280. [CrossRef]
16. Wang, J.; Cieplak, P.; Kollman, P.A. How well does a restrained electrostatic potential (RESP) model perform in calculating conformational energies of organic and biological molecules? *J. Comput. Chem.* **2000**, *21*, 1049–1074. [CrossRef]
17. Woods, R.J.; Chappelle, R. Restrained electrostatic potential atomic partial charges for condensed-phase simulations of carbohydrates. *J. Mol. Struct.* **2000**, *527*, 149–156. [CrossRef]
18. Cieplak, P.; Cornell, W.D.; Bayly, C.; Kollman, P.A. Application of the multimolecule and multiconformational RESP methodology to biopolymers: Charge derivation for DNA, RNA, and proteins. *J. Comput. Chem.* **1995**, *16*, 1357–1377. [CrossRef]
19. Cornell, W.D.; Cieplak, P.; Bayly, C.I.; Kollman, P.A. Application of RESP charges to calculate conformational energies, hydrogen bond energies, and free energies of solvation. *J. Am. Chem. Soc.* **1993**, *115*, 9620–9631. [CrossRef]
20. Dupradeau, F.Y.; Pigache, A.; Zaffran, T.; Savineau, C.; Lelong, R.; Grivel, N.; Lelong, D.; Rosanski, W.; Cieplak, P. The R.E.D. tools: Advances in RESP and ESP charge derivation and force field library building. *Phys. Chem. Chem. Phys.* **2010**, *12*, 7821–7839. [CrossRef]
21. Weiner, S.J.; Kollman, P.A.; Case, D.A.; Singh, U.C.; Ghio, C.; Alagona, G.; Profeta, S.; Weiner, P. A new force field for molecular mechanical simulation of nucleic acids and proteins. *J. Am. Chem. Soc.* **1984**, *106*, 765–784. [CrossRef]
22. Weiner, S.J.; Kollman, P.A.; Nguyen, D.T.; Case, D.A. An all atom force field for simulations of proteins and nucleic acids. *J. Comput. Chem.* **1986**, *7*, 230–252. [CrossRef] [PubMed]
23. Aduri, R.; Psciuk, B.T.; Saro, P.; Taniga, H.; Schlegel, H.B.; SantaLucia, J. AMBER Force Field Parameters for the Naturally Occurring Modified Nucleosides in RNA. *J. Chem. Theory Comput.* **2007**, *3*, 1464–1475. [CrossRef] [PubMed]
24. Ballenegger, V.; Cerda, J.J.; Holm, C. How to Convert SPME to P3M: Influence Functions and Error Estimates. *J. Chem. Theory Comput.* **2012**, *8*, 936–947. [CrossRef]
25. Bussi, G.; Donadio, D.; Parrinello, M. Canonical sampling through velocity rescaling. *J. Chem. Phys.* **2007**, *126*, 014101. [CrossRef]
26. Bernetti, M.; Bussi, G. Pressure control using stochastic cell rescaling. *J. Chem. Phys.* **2020**, *153*, 114107. [CrossRef]
27. Cremer, D.; Pople, J.A. General definition of ring puckering coordinates. *J. Am. Chem. Soc.* **1975**, *97*, 1354–1358. [CrossRef]
28. Chan, L.; Hutchison, G.R.; Morris, G.M. Understanding Ring Puckering in Small Molecules and Cyclic Peptides. *J. Chem. Inf. Model.* **2021**, *61*, 743–755. [CrossRef]
29. Carvalho, A.T.P.; Gouveia, L.; Kanna, C.R.; Warmlander, S.K.T.S.; Platts, J.A.; Kamerlin, S.C.L. Understanding the structural and dynamic consequences of DNA epigenetic modifications: Computational insights into cytosine methylation and hydroxymethylation. *Epigenetics* **2014**, *9*, 1604–1612. [CrossRef]
30. Bachmann, J.; Schonrath, I.; Muller, J.; Doltsinis, N.L. Dynamic Structure and Stability of DNA Duplexes Bearing a Dinuclear Hg(II)-Mediated Base Pair. *Molecules* **2020**, *25*, 4942. [CrossRef]
31. Schneider, M.; Trummer, C.; Stengl, A.; Zhang, P.; Szwagierczak, A.; Cardoso, M.C.; Leonhardt, H.; Bauer, C.; Antes, I. Systematic analysis of the binding behaviour of UHRF1 towards different methyl- and carboxylcytosine modification patterns at CpG dyads. *PLoS ONE* **2020**, *15*, e0229144. [CrossRef] [PubMed]
32. Marco, E.; Negri, A.; Luque, F.J.; Gago, F. Role of stacking interactions in the binding sequence preferences of DNA bis-intercalators: Insight from thermodynamic integration free energy simulations. *Nucleic Acids Res.* **2005**, *33*, 6214–6224. [CrossRef] [PubMed]
33. Remington, J.M.; McCullagh, M.; Kohler, B. Molecular Dynamics Simulations of 2-Aminopurine-Labeled Dinucleoside Monophosphates Reveal Multiscale Stacking Kinetics. *J. Phys. Chem. B* **2019**, *123*, 2291–2304. [CrossRef]
34. Lankas, F.; Cheatham, T.E., 3rd; Spacková, N.; Hobza, P.; Langowski, J.; Sponer, J. Critical effect of the N2 amino group on structure, dynamics, and elasticity of DNA polypurine tracts. *Biophys. J.* **2002**, *82*, 2592–2609. [CrossRef]
35. Victora, A.; Moller, H.M.; Exner, T.E. Accurate ab initio prediction of NMR chemical shifts of nucleic acids and nucleic acids/protein complexes. *Nucleic Acids Res.* **2014**, *42*, e173. [CrossRef] [PubMed]
36. Frank, A.T.; Law, S.M.; Brooks, C.L., 3rd. A simple and fast approach for predicting (1)H and (13)C chemical shifts: Toward chemical shift-guided simulations of RNA. *J. Phys. Chem. B* **2014**, *118*, 12168–12175. [CrossRef]
37. Shi, H.; Rangadurai, A.; Abou Assi, H.; Roy, R.; Case, D.A.; Herschlag, D.; Yesselman, J.D.; Al-Hashimi, H.M. Rapid and accurate determination of atomistic RNA dynamic ensemble models using NMR and structure prediction. *Nat. Commun.* **2020**, *11*, 5531. [CrossRef]

life

Article

Stabilization of G-Quadruplex-Duplex Hybrid Structures Induced by Minor Groove-Binding Drugs

Lily Scott and Tigran V. Chalikian *

Department of Pharmaceutical Sciences, Leslie Dan Faculty of Pharmacy, University of Toronto, 144 College Street, Toronto, ON M5S 3M2, Canada; lilyf.scott@mail.utoronto.ca
* Correspondence: t.chalikian@utoronto.ca; Tel.: +1-416-946-3715

Abstract: Once it had been realized that G-quadruplexes exist in the cell and are involved in regulation of genomic processes, the quest for ligands recognizing these noncanonical structures was underway. Many organic compounds that tightly associate with G-quadruplexes have been identified. However, the specificity of G-quadruplex-binding ligands towards individual structures remains problematic, as the common recognition element of these ligands is the G-tetrad. In this paper, we focus on G-quadruplex-duplex hybrids (QDH) containing a hairpin duplex incorporated as a stem-loop into the G-quadruplex core. The presence of a stem-loop renders QDH amenable to sequence-specific recognition by duplex-binding drugs. Should the thermodynamic crosstalk between the stem-loop and the tetraplex core be sufficiently strong, the drug binding to the loop would lead to the stabilization of the entire structure. We studied the stabilizing influence of the minor groove-binders netropsin and Hoechst 33258 on a family of QDH structures, as well as a G-quadruplex and a hairpin modeling the G-quadruplex core and the stem-loop of the QDH's. We found that the binding of either drug results in an enhancement of the thermal stability of all DNA structures, as expressed by increases in the melting temperature, T_M. Analysis of the hierarchical order of increases in T_M revealed that the drug-induced stabilization arises from drug binding to the G-quadruplex domain of a QDH and the stem-loop, if the latter contains an all-AT binding site. This result attests to the thermodynamic crosstalk between the stem-loop and the tetraplex core of a QDH. Given the existing library of minor groove-binding drugs recognizing mixed A·T and G·C DNA sequences, our results point to an untapped avenue for sequence-specific recognition of QDH structures in vitro and, possibly, in vivo; thereby, opening the way for selective stabilization of four-stranded DNA structures at predetermined genomic loci, with implications for the control of genomic events.

Keywords: quadruplex-duplex hybrids; minor groove binding drugs; thermodynamics; stability; optical spectroscopy

Citation: Scott, L.; Chalikian, T.V. Stabilization of G-Quadruplex-Duplex Hybrid Structures Induced by Minor Groove-Binding Drugs. *Life* **2022**, *12*, 597. https://doi.org/10.3390/life12040597

Academic Editor: Paolo Mariani

Received: 25 March 2022
Accepted: 13 April 2022
Published: 18 April 2022

Publisher's Note: MDPI stays neutral with regard to jurisdictional claims in published maps and institutional affiliations.

1. Introduction

G-quadruplexes have entered the fields of biophysics and the structural biology of nucleic acids, as well as those of molecular and cell biology and nanotechnology, as previously unappreciated secondary structures that may act as regulatory elements in genomic events, as well as cation-responsive structural components in DNA-based nanotechnological devices [1–10]. Once the regulatory properties of G-quadruplexes were recognized, the quest for low molecular weight ligands (drugs) that can recognize and stabilize individual genomic G-quadruplexes has been underway [3,10–15]. These studies have identified a number of high-affinity ligands that bind to G-quadruplexes [13–15]. For all such drugs, the main recognition element is the G-tetrad, which is the common structural building block for any G-quadruplex. The ensuing challenge facing research groups working on the rational design of G-quadruplex-binding drugs is to discriminate between G-quadruplex structures. One strategy to meet this challenge is to equip the drug with functional groups that sample, in addition to G-tetrads, grooves and loops that would render the drug selective with respect to a particular G-quadruplex structure [14,16]. An alternative and potentially propitious approach is to concentrate on G-quadruplex-duplex hybrids (QDH), a sub-class of four-stranded DNA structures.

A QDH incorporates a stem-loop duplex linked to the G-quadruplex core via a duplex-quadruplex junction [17–19]. More than 80,000 DNA sequences capable of folding into QDH structures with stem-loops of 20 nucleotides or less have been identified across important regulatory sites in the human genome, including transcription/mutagenesis hotspots and cancer-associated genes [19]. The folding topologies of various QDH structures have been characterized in vitro by NMR [19–22]. The stem-loop in QDH may be positioned in a coaxial or orthogonal orientation relative to the G-quadruplex core, and the topology of the core is sensitive to the orientation of the loop [17]. QDH structures are thermodynamically stable and melt via two-state transitions, as revealed in the careful calorimetric and spectroscopic studies by Giancola and colleagues [23].

The presence of a double-stranded hairpin loop renders a QDH amenable to recognition by minor groove-binding drugs, which can be specifically targeted to the sequence of the loop [24]. Drug binding to double stranded DNA leads to an increase in the stability of the host duplex [25–28]. Depending on the extent of the cooperative link between the double-stranded stem-loop and the tetraplex core of QDH, the drug binding to the stem-loop may or may not cause a global increase in the stability of the hybrid structure. Should the cooperativity be sufficiently strong to allow the drug-induced stabilization of the loop to propagate to the tetraplex core, the sequence-specific association of a minor groove-binding drug with the stem-loop may suggest a new approach to selective tetraplex stabilization.

Our results, described herein, reveal a significant increase in the stability of two QDH structures following the binding of netropsin or Hoechst 33258 to the AATT/AATT sequence of the stem-loop. It follows that, at least for the QDH structures studied in this work, there is a strong thermodynamic coupling between the duplex and tetraplex domains; the coupling allows the binding of a drug to the stem-loop to stabilize the entire structure. Taken together, our results point to a novel, allosteric mechanism for selective stabilization of four-stranded DNA structures with associated structural domains.

2. Materials and Methods

2.1. Materials

The DNA oligonucleotides d($T_2G_3TG_3\underline{T_3CA_2T_2G_2CACA_2T_2GT_3}G_3TG_3T$) (G4HP), d($\underline{CA_2T_2G_2CACA_2T_2G}$) (HP), d($T_2G_3TG_3\underline{T_3CGCGA_2GCAT_2CGCGT_3}G_3TG_3T$) (K6bp6T), d($T_2G_3TG_3TG_3TG_3T$) (G4), and d($T_2G_3TG_3\underline{T_2CA_2T_2GTGCATCA_2T_2GT_2}G_3TG_3T$) (QDH5L) were purchased from Integrated DNA Technologies (Coralville, IA, USA). DNA samples were dissolved in 50 mM CsCl, dialyzed exhaustively against distilled water in Spectra/Por (2000-Da cut-off, Repligen, Waltham, MA, USA), and lyophilized. The lyophilized DNA samples, with the exception of G4, were dissolved in a pH 6.7 buffer, consisting of 10 mM cacodylic acid/cesium cacodylate, 5 mM KCl, and 0.1 mM EDTA. In the presence of 5 mM KCl, G4 and its complexes with drugs are too stable and melt at temperatures which are experimentally unattainable. Therefore, G4 was dissolved in a pH 6.7 buffer, consisting of 10 mM cacodylic acid/cesium cacodylate, 0.2 mM KCl, and 0.1 mM EDTA. Before measurements, all DNA samples were annealed to ensure that, at room temperature, they exist not in a kinetically trapped but in the thermodynamically stable structural state.

Hoechst 33258·3HCl and netropsin·2HCl were obtained from Sigma-Aldrich Canada (Oakville, ON, Canada). The DNA and drug concentrations were determined spectrophotometrically with a Cary 300 Bio spectrophotometer (Varian Canada, Inc., Mississauga, ON, Canada). To determine the concentrations of G4HP, HP, K6bp6T, G4, and QDH5L, we used molar extinction coefficients ε_{260} of 360,200, 146,400, 354,000, 172,800, and 361,800 $M^{-1}cm^{-1}$, respectively. These values were computed for the unfolded states at 25 °C from the nearest-neighbor procedure described by Owczarzy [29]. The extinction coefficients of netropsin and Hoechst 33258 are ε_{296} = 21,500 $M^{-1}cm^{-1}$ and

ε_{338} = 42,000 $M^{-1}cm^{-1}$, respectively [30,31]. In CD spectropolarimetric measurements, the concentrations of the DNA were ~30 µM, while, in fluorometric measurements, they were ~50 nM.

2.2. Optical Spectroscopy

CD spectral measurements were conducted in a 1-mm path-length cuvette using a JASCO J-1100 Circular Dichroism Spectrophotometer (JASCO, Easton, MD, USA). Fluorescence intensity in DNA samples in the absence and presence of Hoechst 33258 were measured using an Aviv model ATF 105 spectrofluorometer (Aviv Associates, Lakewood, NJ, USA) with a bandwidth adjusted to 2 nm. CD and fluorescence titration profiles were measured at 25 °C by incremental addition of aliquots of a drug solution to the initial amount of DNA solution. The initial volume of DNA was delivered to the cuvette using a 1-mL Hamilton syringe, while aliquots of the drug were added using a 10-µL Hamilton syringe (Hamilton Co., Reno, NV, USA). The syringes were equipped with a Chaney adapter with a relative delivery accuracy of ±0.1%. For fluorescence measurements, the samples were excited at 359 nm, and the intensity of emission light was recorded through a monochromator at 470 nm. In the CD melting experiments, the temperature was changed at a rate of 1 °C per minute. The transitions temperatures, T_M, were evaluated from analysis of experimental melting profiles using standard procedures for a two-state helix-to-coil transition [32–34].

3. Results

3.1. DNA Structures

The QDH-forming DNA sequences studied in this work were the NMR-characterized sequence $d(T_2G_3TG_3T_3CGCGA_2GCAT_2CGCGT_3G_3TG_3T)$, referred to as K6bp6T, and its two derivatives, G4HP $[d(T_2G_3TG_3\underline{T_3CA_2T_2G_2CACA_2T_2GT_3}G_3TG_3T)]$ and QDH5L $[d(T_2G_3TG_3\underline{T_2CA_2T_2GTGCATCA_2T_2GT_2}G_3TG_3T)]$. K6bp6T has been shown to form a QHD with a parallel G-quadruplex core and an orthogonally oriented duplex stem [21,22]. This structure is presented schematically in Figure 1a. In G4HP, the six base pairs in the stem-loop of the original construct (CGCGAA/TTCGCG) are replaced with (CAATTG/CAATTG) that contains the (AATT/AATT) binding site for the AT-selective minor groove-binding drugs netropsin and Hoechst 33258. QDH5L contains an AT-containing stem-loop, which is compositionally similar to that of G4HP. However, QDH5L contains a hairpin with a five-nucleotide loop (-TGCAT-) in contrast to the three-nucleotide hairpin loop of G4HP (-GCA-); the stem-loop in the former is more relaxed and should be more B-like than that of the latter. Therefore, one could expect a stronger binding of netropsin or Hoechst 33258 to QDH5L relative to G4HP. The loop sequences of K6bp6T, G4HP, and QDH5L and the sequence of the hairpin (HP) are aligned for comparison in Figure 1b.

To ensure that the mutations introduced to K6bp6T do not change the topology of the G-quadruplex core or the orientation of the stem-loop in the modified QDH structures (G4HP and QDH5L), we compared the CD spectra of G4HP and QDH5L with the spectrum of K6bp6T [22]. Figure 2 shows the three CD spectra; the similarity of the overall shapes of the spectra testifies to the fact that the mutations did not alter the topology of the QDH structures. The differences in the amplitude of the CD bands between G4HP, QDH5L, and K6bp6T may be related to the differential nearest neighbor interactions between the bases of the stem loops.

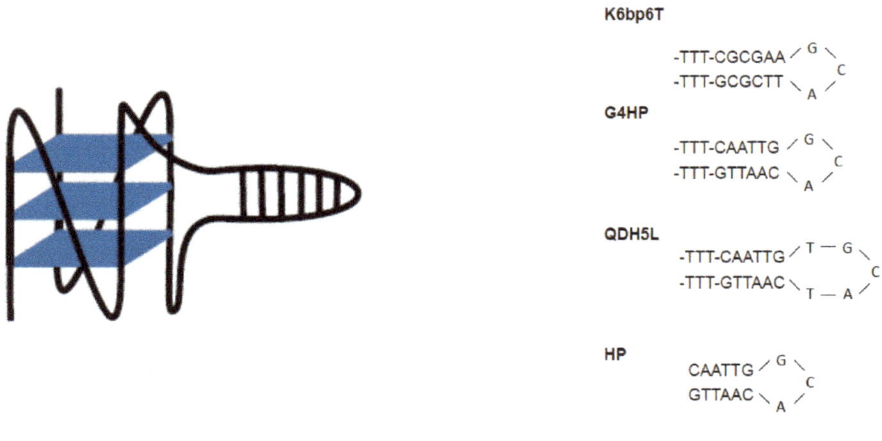

Figure 1. (**a**) Schematic representation of the QDH structures studied in this work; (**b**) the stem-loops of K6bb6T, G4HP, and QDH5L and the hairpin HP.

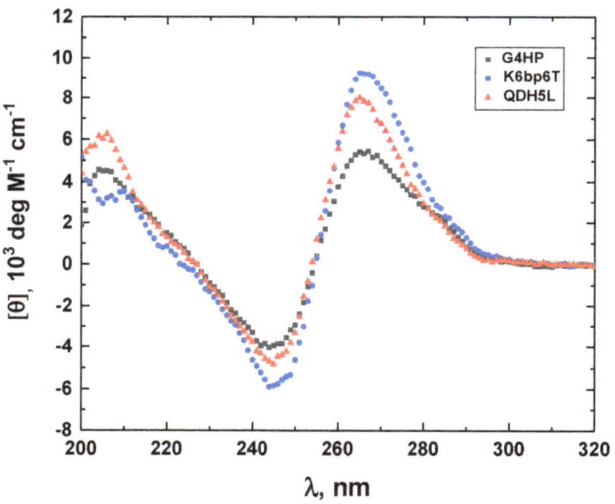

Figure 2. The CD spectra of the G4HP (black), QDH5L (red), and K6bp6T (blue) G-quadruplex–duplex hybrid structures.

The G-quadruplex G4 with a sequence of d($T_2G_3TG_3TG_3TG_3T$) and the hairpin HP with a sequence of d($CA_2T_2G_2CACA_2T_2G$) are the constituent parts of G4HP. We measured the CD spectra of these DNA structures to confirm that, under the experimental conditions of our study, G4 indeed forms a G-quadruplex, while HP folds into a hairpin. The monomolecular nature of the hairpin was additionally verified by UV melting at 260 nm, which revealed no dependence of the melting temperature, T_M, on DNA concentration. These CD spectra are shown in Figure 3. The CD spectrum of G4 with a negative minimum at 244 nm and a positive maximum at 264 nm is consistent with that of a parallel G-quadruplex.

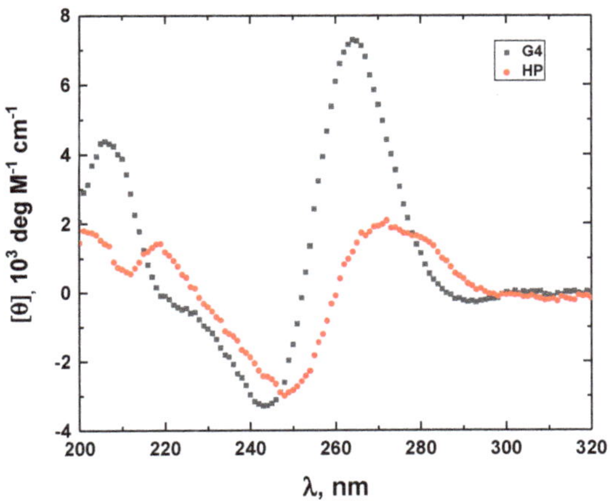

Figure 3. The CD spectra of the G4 G-quadruplex (black) and the HP hairpin (red).

3.2. Drug Binding

To confirm that the two minor groove-binding drugs bind to the host structures under the experimental conditions of this study, we measured the CD spectra of each DNA in the absence and presence of netropsin and Hoechst 33258. Additionally, the binding of Hoechst 33258 to the DNA was monitored by fluorescence. The drug-induced CD spectral variations of G4HP, QDH5L, HP, K6bp6T, and G4 are shown in Figures 4–8, respectively. In each figure, panel A shows the data on netropsin, while panel B shows the data on Hoechst 33258.

The CD spectra of G4HP, QDH5L, and HP depicted in Figures 4–6 exhibit characteristic binding-induced bands at ~315 (panel A) and ~350 (panel B) nm, which are the CD spectroscopic signatures of the association of duplex DNA with netropsin and Hoechst 33258, respectively [35,36]. Thus, both netropsin and Hoechst 33258 bind to the all-AT binding sites of the stem-loops of the G4HP and QDH5L G-quadruplex–duplex hybrids and the hairpin HP. In contrast, the CD spectra of K6bp6T and G4, shown in Figures 7 and 8, either lack or exhibit very weak CD bands above 300 nm. This observation suggests that the two minor groove binders do not bind, or bind only weakly, to the minor groove of a duplex; this is an expected result, given that the G-quadruplex (G4) lacks the duplex portion, while the stem-loop of K6bp6T does not contain the cognate all-AT binding site.

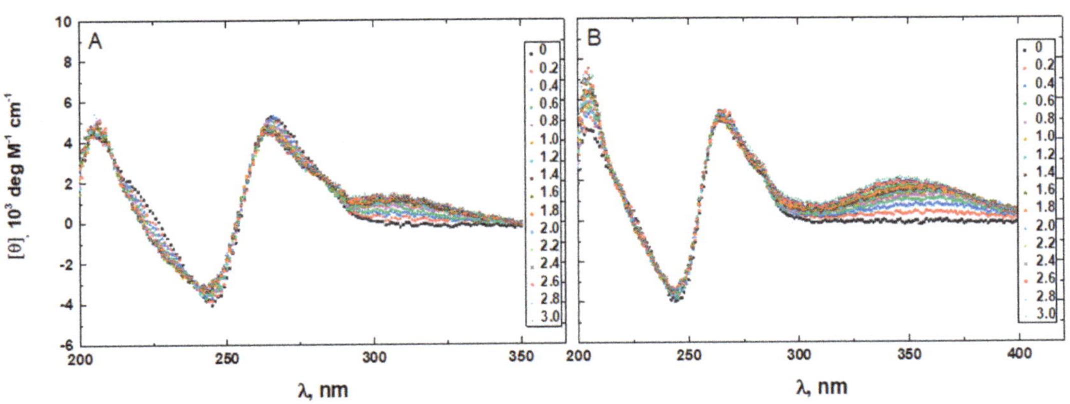

Figure 4. The CD spectra of G4HP in the absence and presence of netropsin (**A**) and Hoechst 33258 (**B**) at various drug-to-DNA ratios, r (shown in the insets), at 25 °C. The DNA concentration is ~20 μM.

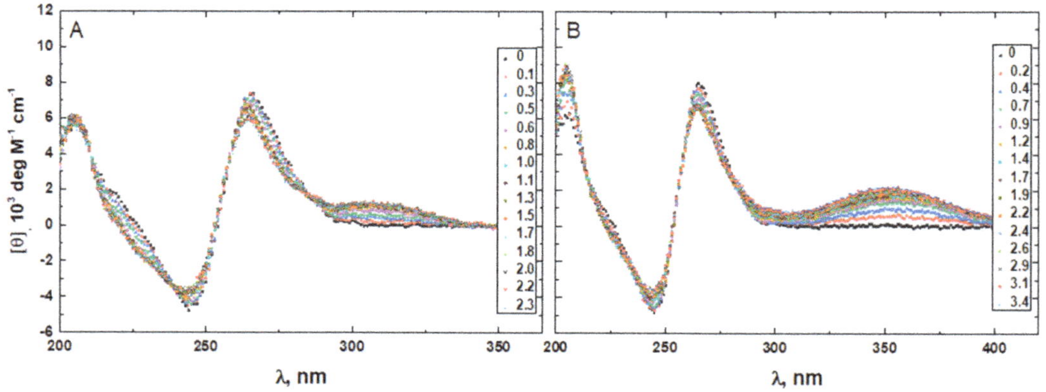

Figure 5. The CD spectra of QDH5L in the absence and presence of netropsin (**A**) and Hoechst 33258 (**B**) at various drug-to-DNA ratios, r (shown in the insets), at 25 °C. The DNA concentration is ~20 μM.

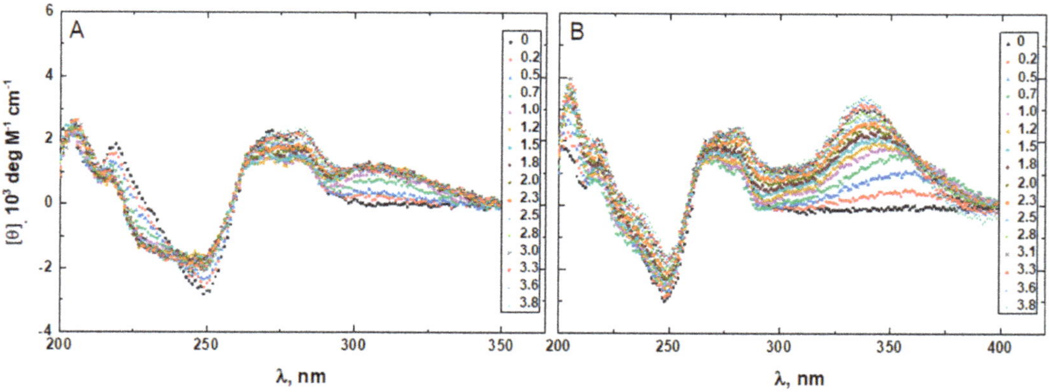

Figure 6. The CD spectra of HP in the absence and presence of netropsin (**A**) and Hoechst 33258 (**B**) at various drug-to-DNA ratios, r (shown in the insets), at 25 °C. The DNA concentration is ~20 μM.

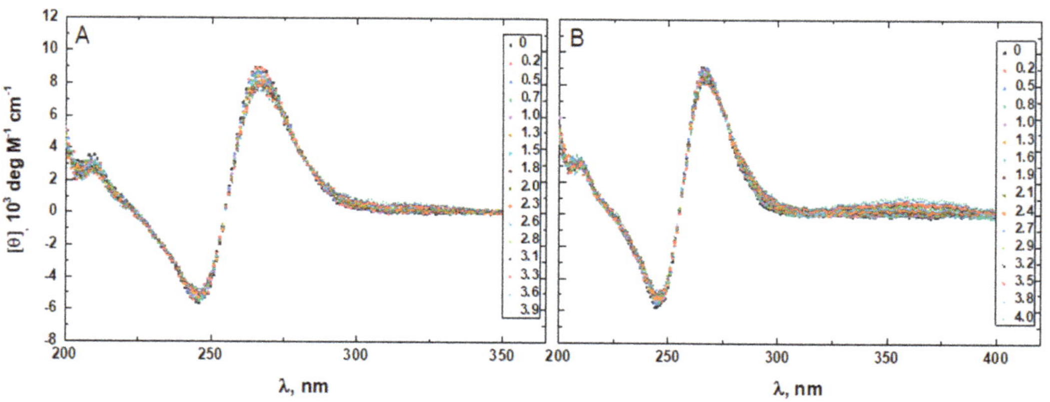

Figure 7. The CD spectra of K6bp6T in the absence and presence of netropsin (**A**) and Hoechst 33258 (**B**) at various drug-to-DNA ratios, r (shown in the insets), at 25 °C. The DNA concentration is ~20 μM.

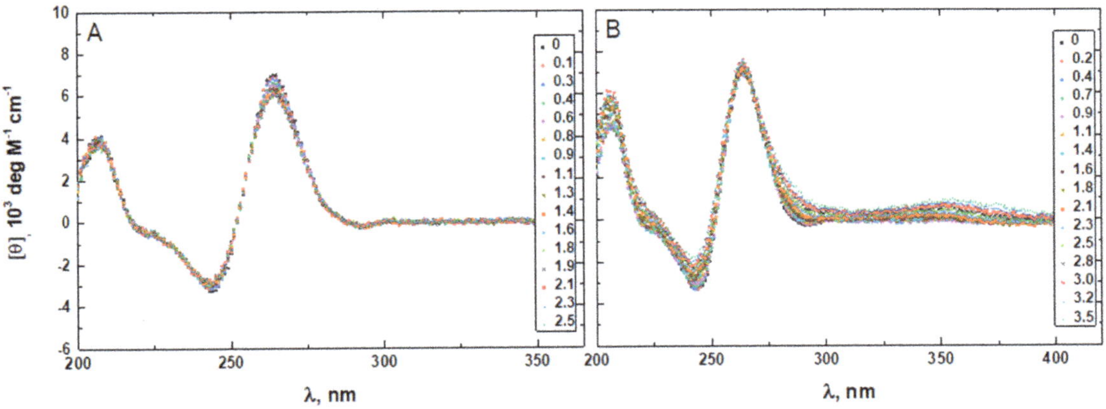

Figure 8. The CD spectra of G4 in the absence and presence of netropsin (**A**) and Hoechst 33258 (**B**) at various drug-to-DNA ratios, r (shown in the insets), at 25 °C. The DNA concentration is ~20 μM.

Figure 9 presents the fluorescence spectra of G4HP (panel A), QDH5L (panel B), HP (panel C), K6bp6T (panel D), and G4 (panel E) in the absence and presence of Hoechst 33258. The drug-dependent spectral variations shown in Figure 9A–E suggest strong binding of Hoechst 33258 to all DNA structures studied here. This result agrees with the results of previous studies that reported the association of minor groove-binding drugs with G-quadruplex structures [37–42].

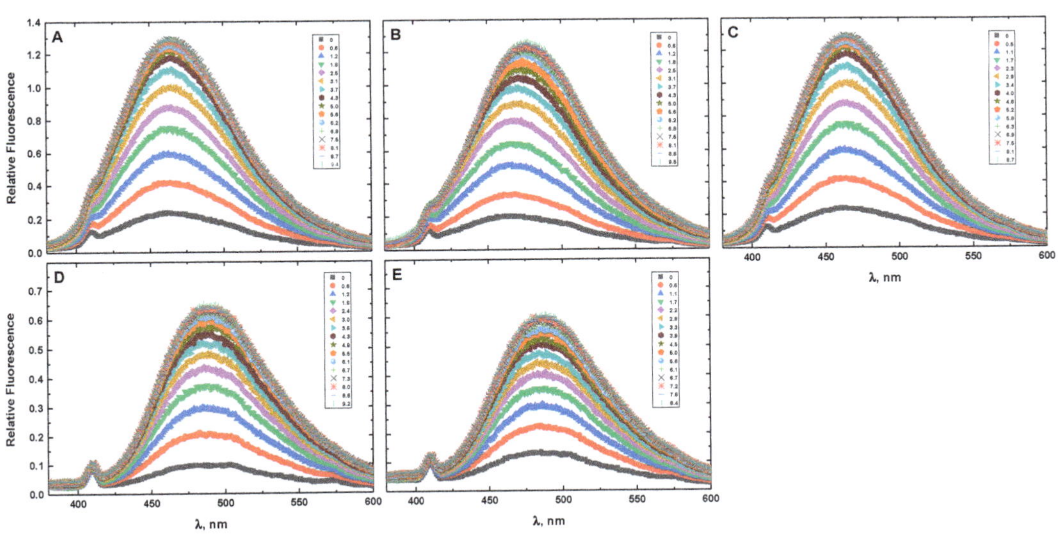

Figure 9. The emission spectra of G4HP (**A**), QDH5L (**B**), HP (**C**), K6bp6T (**D**), and G4 (**E**) in the absence and presence of Hoechst 33258 at various drug-to-DNA ratios, r (shown in the inset), at 25 °C. The excitation wavelength is 359 nm. The DNA concentration is ~50 nM.

Drug-induced spectral changes, such as those shown in Figures 4–13, can be used to construct CD- or fluorescence-detected binding profiles at specific wavelengths. Those profiles can then be fitted within the framework of an appropriate binding model, to determine the number of drug-binding sites and their respective affinities. Any such model must accommodate the mutual depletion of DNA and either ligand upon formation of the complex. In case of Hoechst 33258, the model must also accommodate its propensity to aggregate [43]. The effects of mutual depletion and aggregation should be greater at the concentrations of DNA required for measurements of CD (~20 μM) than at those required for measurements of fluorescence (~50 nM). Studies along these lines are in progress. In the present paper, the spectral changes depicted in Figures 4–9 are treated as a manifestation of ligand binding rather than a source of quantitative thermodynamic information. We concentrate instead on the binding-induced increases in melting temperatures, ΔT_M, which represent a global measure of drug-induced stabilization of DNA structures.

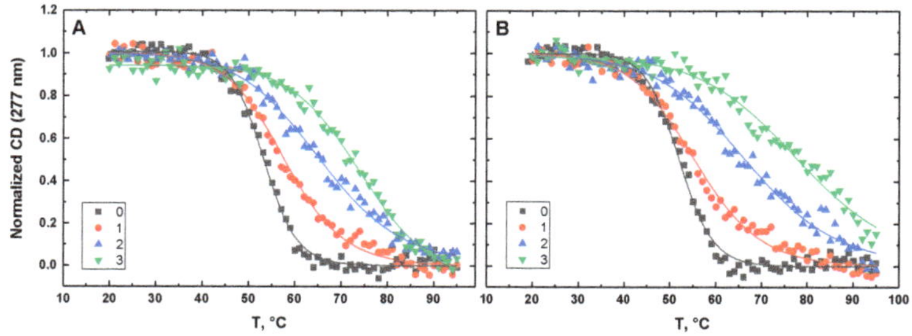

Figure 10. Normalized CD-melting profiles at 277 nm of G4HP in the absence and presence of netropsin (**A**) and Hoechst 33258 (**B**) at various drug-to-DNA ratios, r (shown in the insets). The DNA concentration is ~30 μM. Experimental points were fitted analytically based on the two-state transition thermodynamic formalism [32] (solid lines).

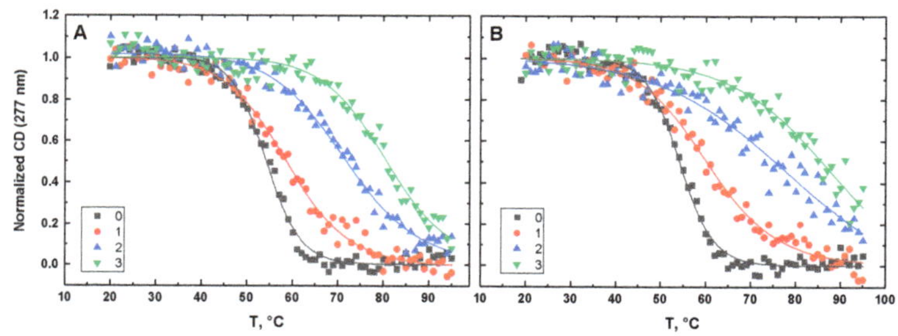

Figure 11. Normalized CD-melting profiles at 277 nm of QDH5L in the absence and presence of netropsin (**A**) and Hoechst 33258 (**B**) at various drug-to-DNA ratios, r (shown in the insets). The DNA concentration is ~30 μM. Experimental points were fitted analytically based on the two-state transition thermodynamic formalism [32] (solid lines).

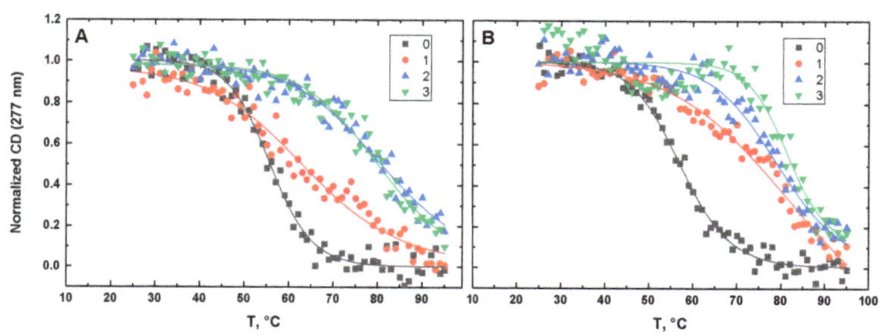

Figure 12. Normalized CD-melting profiles at 277 nm of HP in the absence and presence of netropsin (**A**) and Hoechst 33258 (**B**) at various drug-to-DNA ratios, r (shown in the insets). The DNA concentration is ~30 μM. Experimental points were fitted analytically based on the two-state transition thermodynamic formalism [32] (solid lines).

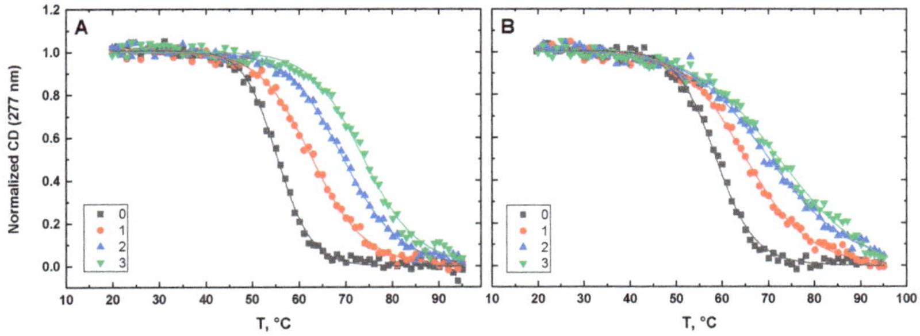

Figure 13. Normalized CD-melting profiles at 277 nm of K6bp6T in the absence and presence of netropsin (**A**) and Hoechst 33258 (**B**) at various drug-to-DNA ratios, r (shown in the insets). The DNA concentration is ~30 μM. Experimental points were fitted analytically based on the two-state transition thermodynamic formalism [32] (solid lines).

3.3. Thermal Stability

Figures 10–14 show the CD-detected melting profiles of G4HP, QDH5L, HP, K6bp6T, and G4, respectively, at drug-to-DNA ratios from 0 to 3. In each figure, panel A presents melting profiles in the presence of netropsin, while panel B shows the thermal data in the presence of Hoechst 33258. Inspection of Figures 10–14 reveals that the binding of either drug to each DNA structure studied here results in its stabilization, as can be judged by significant increases in T_M at increasing drug-to-DNA ratios.

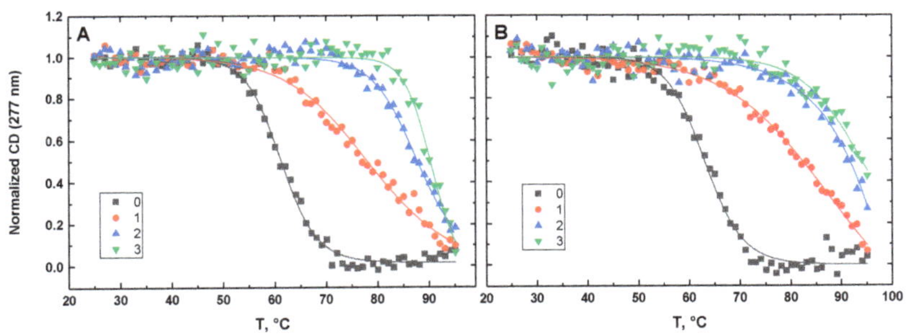

Figure 14. Normalized CD-melting profiles at 277 nm of G4 in the absence and presence of netropsin (**A**) and Hoechst 33258 (**B**) at various drug-to-DNA ratios, r (shown in the insets). The DNA concentration is ~30 μM. Experimental points were fitted analytically based on the two-state transition thermodynamic formalism [32] (solid lines).

Below, we derive the equation for the ligand-induced increase in T_M for a biopolymer with multiple ligand-binding sites. We assume that the biopolymer may exist in either the native, *N*, or denatured, *D*, state and that the ligand binds only to the native state. The equilibrium between the native state with n ligand-binding sites and the denatured state (which does not bind ligands) is given by the relationship:

$$K = \frac{[D]}{[N]} = \frac{[D]}{[N_0]\prod_{i=1}^{n}(1+k_i[L])} = \frac{K_0}{\prod_{i=1}^{n}(1+k_i[L])} \tag{1}$$

where $K_0 = \frac{[D]}{[N_0]}$ is the equilibrium constant between the native and denatured states of the biopolymer in the absence of the ligand; k_i is the microscopic affinity of the ligand for the *i*-th binding site; $[L]$ is the concentration of free ligand; and $\prod_{i=1}^{n}(1+k_i[L])$ is the binding polynomial [44].

The stability of the ligand–biopolymer complex is given by the equation:

$$\Delta G = -RTlnK = \Delta G_0 - \Delta G_b = \Delta H_0 - T\Delta S_0 + RTln\left[\prod_{i=1}^{n}(1 + k_i[L])\right] \quad (2)$$

where $\Delta G_0 = -RTlnK_0 = \Delta H_0 - T\Delta S_0$ is the stability of unligated biopolymer; ΔH_0 and ΔS_0 are the respective changes in enthalpy and entropy; $\Delta G_b = -RT\prod_{i=1}^{n}(1 + k_i[L])$ is the binding free energy.

At the melting temperature, T_M, the value of $\Delta G = 0$. Thus, one obtains the following relationship by equating Equation (2) to zero:

$$\Delta H_0 - T_M\Delta S_0 + RT_M \ln\left[\prod_{i=1}^{n}(1 + k_i[L])\right] = 0 \quad (3)$$

From Equation (3), T_M is given by the relationship:

$$T_M = \Delta H_0 / \left(\Delta S_0 - Rln\left[\prod_{i=1}^{n}(1 + k_i[L])\right]\right) \quad (4)$$

By dividing the numerator and the dominator of RHS of Equation (4) by ΔH_0 and taking into account that the melting temperature of unligated biopolymer, T_{M0}, equals $\Delta H_0/\Delta S_0$, one obtains the following:

$$\Delta T_M = T_M - TM_0 = (RTM_0T_M/\Delta H_0)\ \ln\left[\prod_{i=1}^{n}(1 + k_i[L])\right] \quad (5)$$

where T_{M0} and T_M are the melting temperatures of the biopolymer at free ligand concentrations of 0 and $[L]$, respectively.

For $n = 1$, Equation (5) converts into the well-known equation derived by Crothers in 1971 [25]. To the best of our knowledge, Equation (5), relating the melting temperature of a biopolymer with multiple binding sites to the concentration of a ligand and its affinity for the sites, has not been reported previously. Inspection of Equation (5) reveals that ΔT_M represents an unambiguous global measure of the net drug-induced stabilization of DNA structures.

We fitted the melting profiles presented in Figures 10–14 by a two-state transition model [32]. The values of T_M were determined from the fits as the helix-to-coil transition midpoints. Table 1 presents increases in T_M for each DNA construct studied here in the presence of netropsin and Hoechst 33258 at a drug-to-DNA ratio, r, of 3.

Table 1. Ligand-induced changes in melting temperatures, ΔT_M (°C), of DNA structures at a drug-to-DNA ratio, r, equal to 3.

	G4HP	HP	K6bp6T	G4	QDH5L
ΔT_M (Netropsin)	21.5 ± 0.4	24.7 ± 0.4	18.5 ± 0.2	29.3 ± 0.5	26.6 ± 0.7
ΔT_M (Hoechst 33258)	26.4 ± 0.3	24.4 ± 0.8	14.8 ± 0.4	31.8 ± 3.8	35.9 ± 0.4

4. Discussion

The minor-groove binding drugs studied here associate with the G-quadruplex domain of a G-quadruplex–duplex hybrid structure and its stem-loop if the latter contains the all-AT binding site. As alluded to above, a change in T_M represents a quantitative measure of the cumulative effect of drug binding on the structural stability of DNA. Inspection of the data on T_M presented in Table 1 reveals a hierarchical order of drug-induced stabilization, which is consistent with the following picture. Both netropsin and Hoechst 33258 bind to the G-quadruplex G4 and strongly stabilize it, as suggested by increases in T_M of 29.3 ± 0.5 and 31.8 ± 3.8 °C, respectively. It follows, by extension, that the two minor groove-binding drugs also bind to the G-quadruplex domain of the G4HP, QDH5L, and K6bp6T hybrids. Comparison of the stability data on K6bp6T and G4 reveals that the presence of the stem-loop in the hybrid causes a large reduction in the extent of drug-induced stabilization; the values of ΔT_M diminish to 18.5 $\pm$ 0.2 and 14.8 $\pm$ 0.4 °C for netropsin and Hoechst 33258, respectively. The stem-loop in K6bp6T lacks the all-AT binding site and, therefore, cannot associate with either of the two drugs. The observed decrease in the extent of drug-induced stabilization of K6bp6T suggests that the stem-loop either abolishes or sterically hinders the access of the drugs to one or more of the binding sites on its G-quadruplex domain.

In contrast to K6bp6T, the stem-loop of G4HP contains an all-AT binding site. Hence, netropsin and Hoechst 33258 bind, in addition to the G-quadruplex domain, to the stem-loop of G4HP, which is reflected in the CD spectra in Figure 4A,B. Comparison of the drug-induced increases in T_M for G4HP and K6bp6T reveals that complexation of the minor groove-binding drugs with the stem-loop of G4HP leads to its stabilization, which significantly exceeds that afforded by the drug binding to K6bp6T. The values of ΔT_M for G4HP are 21.5 $\pm$ 0.4 and 26.4 $\pm$ 0.3 °C for netropsin and Hoechst 33258, respectively. Coincidentally, these numbers are close to those observed for the HP hairpin, where the values of ΔT_M are 24.7 $\pm$ 0.4 and 24.4 $\pm$ 0.8 °C for netropsin and Hoechst 33258, respectively.

These results are consistent with the notion that the drug-induced stabilization of G4HP reflects the concerted effect of drug binding to both the stem-loop and the G-quadruplex domain of the hybrid. Comparison of the stability data for QDH5L and G4HP provides further evidence that drug binding to the stem-loop contributes significantly to the stabilization of a hybrid structure. QDH5L contains a five-loop hairpin in contrast to the more constrained three-loop hairpin of G4HP. The unconstrained stem-loop duplex in QDH5L should be more B-like relative to G4HP; hence, the binding of the minor groove binders to the stem-loop of QDH5L should be tighter than that to the stem-loop of G4HP. One can, therefore, expect stronger drug-induced stabilization of QDH5L than of G4HP. In agreement with this expectation, the values of ΔT_M for QDH5L were 26.6 $\pm$ 0.7 and 35.9 $\pm$ 0.4 °C for netropsin and Hoechst 33258, respectively, significantly exceeding those for G4HP. Therefore, an increase in the number of the nucleotides in the loop of the hairpin from three to five leads to substantially enhanced drug-induced stabilization of the QDH structure.

Our results suggest that minor groove-binding drugs stabilize QDH structures by binding to both the stem-loop and the G-quadruplex domains. While the contribution of drug binding to the G-quadruplex domain is large, the contribution of the binding to the stem-loop is also substantial. The two contributions are difficult to separate, and the combined effect of drug binding to the stem-loop and the G-quadruplex domain may be synergistic.

Our results may seem to contradict those of a study that suggested that netropsin binding to the hairpin stem-loop portion of one particular QDH does not lead to any appreciable increase in stability, despite enhancing the stability of the isolated hairpin [24]. The observed disparity reflects the structural differences in the QDH structures studied in this work and that studied by Nguyen et al. [24]. In the QDH studied in ref. [24], a nick is introduced to the G-tract bordering the quadruplex–duplex junction. Thus, in that structure, the stem-loop hangs from the 3′-end of the tetraplex core and is anchored to its terminal G-tetrad, which serves as the foothold for the loose end of the hairpin. The latter is not a loop in the tetraplex core of the hybrid structure, in the sense that it does not link G-runs; it is rather an appendix which is structurally and thermodynamically separate from the core.

The data described in the present work outline an untapped avenue for sequence-selective recognition and stabilization of stem-loop-containing QDH structures in vitro and, possibly and more importantly, in vivo. Given the existing library of minor groove-binding drugs recognizing mixed AT- and GC-rich DNA sequences [45–49], our results outline a potential method for stabilization, or even induction, of QDH structures with stem-loops containing cognate sequences, with clear extensions to, and implications for, the control of genomic events.

There is a subtle point that merits discussion. The thermodynamic data presented in this work leave no doubt that drug binding to the stem-loop contributes significantly to an increase in the stability of an isolated QDH. It is less clear if the same or a similar extent of drug-induced stabilization would be afforded within the context of duplex DNA, when the QDH is exposed to its complementary C-rich strand. In this scenario, the QDH-plus-single strand conformation with a single drug-binding site should compete with the duplex conformation with two drug-binding sites (given the palindromic nature of the stem-loop). The equilibrium between the QDH and duplex species would, clearly, depend on the specific DNA sequence and the identity of the drug; the QDH conformation may or may not outcompete the duplex conformation. Further studies are needed to explore the effect of drug binding on duplex–QDH equilibrium.

5. Conclusions

We characterized the effect of the minor groove binders netropsin and Hoechst 33258 on the stability of a family of QDH structures, as well as a G-quadruplex and a hairpin modeling the G-quadruplex core and stem-loop of the QDH's. The binding of either drug results in an increase in thermal stability for all DNA structures, as evidenced by increases in melting temperature, T_M. Analysis of the hierarchical order of increases in T_M reveals that the drug-induced stabilization arises from drug binding to both the G-quadruplex domain of a QDH and its stem-loop if the latter contains an all-AT binding site. This result attests to the thermodynamic crosstalk between the stem-loop and tetraplex core of the QDH. Our results allude to an untapped avenue for sequence-specific recognition of QDH structures.

Author Contributions: L.S. and T.V.C. performed research, analyzed data, and wrote the paper. All authors have read and agreed to the published version of the manuscript.

Funding: This work was supported by a grant from the Natural Sciences and Engineering Research Council of Canada to TVC (RGPIN-2017-06574).

Institutional Review Board Statement: Not applicable.

Informed Consent Statement: Not applicable.

Data Availability Statement: Not applicable.

Acknowledgments: The authors gratefully acknowledge many useful discussions with James W. Wells.

Conflicts of Interest: The authors declare no conflict of interests.

References

1. Huppert, J.L. Four-stranded nucleic acids: Structure, function and targeting of G-quadruplexes. *Chem. Soc. Rev.* **2008**, *37*, 1375–1384. [CrossRef] [PubMed]
2. Lane, A.N.; Chaires, J.B.; Gray, R.D.; Trent, J.O. Stability and kinetics of G-quadruplex structures. *Nucleic Acids Res.* **2008**, *36*, 5482–5515. [CrossRef] [PubMed]
3. Balasubramanian, S.; Hurley, L.H.; Neidle, S. Targeting G-quadruplexes in gene promoters: A novel anticancer strategy? *Nat. Rev. Drug Disc.* **2011**, *10*, 261–275. [CrossRef] [PubMed]
4. Varshney, D.; Spiegel, J.; Zyner, K.; Tannahill, D.; Balasubramanian, S. The regulation and functions of DNA and RNA G-quadruplexes. *Nat. Rev. Mol. Cell Biol.* **2020**, *21*, 459–474. [CrossRef]
5. Spiegel, J.; Adhikari, S.; Balasubramanian, S. The structure and function of DNA G-quadruplexes. *Trends Chem.* **2020**, *2*, 123–136. [CrossRef]
6. Tateishi-Karimata, H.; Sugimoto, N. Chemical biology of non-canonical structures of nucleic acids for therapeutic applications. *Chem. Commun.* **2020**, *56*, 2379–2390. [CrossRef]
7. Mergny, J.L.; Sen, D. DNA quadruple helices in nanotechnology. *Chem. Rev.* **2019**, *119*, 6290–6325. [CrossRef] [PubMed]
8. Burge, S.; Parkinson, G.N.; Hazel, P.; Todd, A.K.; Neidle, S. Quadruplex DNA: Sequence, topology and structure. *Nucleic Acids Res.* **2006**, *34*, 5402–5415. [CrossRef]
9. Gudanis, D.; Kaniowski, D.; Kulik, K.; Baranowski, D.; Gdaniec, Z.; Nawrot, B. Formation of an RNA quadruplex-duplex hybrid in living cells between mRNA of the epidermal growth factor receptor (EGFR) and a G-rich antisense oligoribonucleotide. *Cells* **2020**, *9*, 2375. [CrossRef]
10. Pandya, N.; Bhagwat, S.R.; Kumar, A. Regulatory role of non-canonical DNA polymorphisms in human genome and their relevance in cancer. *Biochim. Biophys. Acta Rev. Cancer* **2021**, *1876*, 188594. [CrossRef]

11. Husby, J.; Todd, A.K.; Platts, J.A.; Neidle, S. Small-molecule G-quadruplex interactions: Systematic exploration of conformational space using multiple molecular dynamics. *Biopolymers* **2013**, *99*, 989–1005. [CrossRef]
12. Neidle, S. Human telomeric G-quadruplex: The current status of telomeric G-quadruplexes as therapeutic targets in human cancer. *FEBS J.* **2010**, *277*, 1118–1125. [CrossRef] [PubMed]
13. Neidle, S. Quadruplex nucleic acids as novel therapeutic targets. *J. Med. Chem.* **2016**, *59*, 5987–6011. [CrossRef] [PubMed]
14. Savva, L.; Georgiades, S.N. Recent developments in small-molecule ligands of medicinal relevance for harnessing the anticancer potential of G-quadruplexes. *Molecules* **2021**, *26*, 841. [CrossRef] [PubMed]
15. Luedtke, N.W. Targeting G-quadruplex DNA with small molecules. *Chimia* **2009**, *63*, 134–139. [CrossRef]
16. Collie, G.W.; Parkinson, G.N. The application of DNA and RNA G-quadruplexes to therapeutic medicines. *Chem. Soc. Rev.* **2011**, *40*, 5867–5892. [CrossRef]
17. Lim, K.W.; Phan, A.T. Structural basis of DNA quadruplex-duplex junction formation. *Angew. Chem. Int. Ed. Engl.* **2013**, *52*, 8566–8569. [CrossRef]
18. Risitano, A.; Fox, K.R. The stability of intramolecular DNA quadruplexes with extended loops forming inter- and intra-loop duplexes. *Org. Biomol. Chem.* **2003**, *1*, 1852–1855. [CrossRef]
19. Lim, K.W.; Jenjaroenpun, P.; Low, Z.J.; Khong, Z.J.; Ng, Y.S.; Kuznetsov, V.A.; Phan, A.T. Duplex stem-loop-containing quadruplex motifs in the human genome: A combined genomic and structural study. *Nucleic Acids Res.* **2015**, *43*, 5630–5646. [CrossRef]
20. Tan, D.J.Y.; Winnerdy, F.R.; Lim, K.W.; Phan, A.T. Coexistence of two quadruplex-duplex hybrids in the PIM1 gene. *Nucleic Acids Res.* **2020**, *48*, 11162–11171. [CrossRef]
21. Lim, K.W.; Khong, Z.J.; Phan, A.T. Thermal stability of DNA quadruplex-duplex hybrids. *Biochemistry* **2014**, *53*, 247–257. [CrossRef] [PubMed]
22. Nguyen, T.Q.N.; Lim, K.W.; Phan, A.T. Folding kinetics of G-quadruplexes: Duplex stem loops drive and accelerate G-quadruplex folding. *J. Phys. Chem. B* **2020**, *124*, 5122–5130. [CrossRef] [PubMed]
23. Fotticchia, I.; Amato, J.; Pagano, B.; Novellino, E.; Petraccone, L.; Giancola, C. How are thermodynamically stable G-quadruplex-duplex hybrids? *J. Them. Anal. Calorim.* **2015**, *121*, 1121–1127. [CrossRef]
24. Nguyen, T.Q.N.; Lim, K.W.; Phan, A.T. A Dual-specific targeting approach based on the simultaneous recognition of duplex and quadruplex motifs. *Sci. Rep.* **2017**, *7*, 11969. [CrossRef] [PubMed]
25. Crothers, D.M. Statistical thermodynamics of nucleic acid melting transitions with coupled binding equilibria. *Biopolymers* **1971**, *10*, 2147–2160. [CrossRef] [PubMed]
26. McFedries, A.; Schwaid, A.; Saghatelian, A. Methods for the elucidation of protein-small molecule interactions. *Chem. Biol.* **2013**, *20*, 667–673. [CrossRef] [PubMed]
27. Carvalho, J.; Queiroz, J.A.; Cruz, C. Circular dichroism of G-quadruplex: A laboratory experiment for the study of topology and ligand binding. *J. Chem. Ed.* **2017**, *94*, 1547–1551. [CrossRef]
28. Redhead, M.; Satchell, R.; McCarthy, C.; Pollack, S.; Unitt, J. Thermal shift as an entropy-driven effect. *Biochemistry* **2017**, *56*, 6187–6199. [CrossRef]
29. Tataurov, A.V.; You, Y.; Owczarzy, R. Predicting ultraviolet spectrum of single stranded and double stranded deoxyribonucleic acids. *Biophys. Chem.* **2008**, *133*, 66–70. [CrossRef]
30. Loontiens, F.G.; Regenfuss, P.; Zechel, A.; Dumortier, L.; Clegg, R.M. Binding characteristics of Hoechst 33258 with calf thymus DNA, poly[d(A-T):, and d(CCGGAATTCCGG): Multiple stoiciometries and determination of tight binding with a wide spectrum site affinities. *Biochemistry* **1990**, *29*, 9029–9039. [CrossRef]
31. Chalikian, T.V.; Plum, G.E.; Sarvazyan, A.P.; Breslauer, K.J. Influence of drug binding on DNA hydration: Acoustic and densimetric characterizations of netropsin binding to the poly(dAdT)·poly(dAdT) and poly(dA)·poly(dT) duplexes and the poly(dT)·poly(dA)·poly(dT) triplex at 25 °C. *Biochemistry* **1994**, *33*, 8629–8640. [CrossRef] [PubMed]
32. Marky, L.A.; Breslauer, K.J. Calculating thermodynamic data for transitions of any molecularity from equilibrium melting curves. *Biopolymers* **1987**, *26*, 1601–1620. [CrossRef] [PubMed]
33. Rachwal, P.A.; Fox, K.R. Quadruplex melting. *Methods* **2007**, *43*, 291–301. [CrossRef] [PubMed]
34. Mergny, J.L.; Lacroix, L. Analysis of thermal melting curves. *Oligonucleotides* **2003**, *13*, 515–537. [CrossRef]
35. Park, Y.W.; Breslauer, K.J. Drug binding to higher ordered DNA structures: Netropsin complexation with a nucleic acid triple helix. *Proc. Natl. Acad. Sci. USA* **1992**, *89*, 6653–6657. [CrossRef]
36. Han, F.; Taulier, N.; Chalikian, T.V. Association of the minor groove binding drug Hoechst 33258 with d(CGCGAATTCGCG)$_2$: Volumetric, calorimetric, and spectroscopic characterizations. *Biochemistry* **2005**, *44*, 9785–9794. [CrossRef]
37. Martino, L.; Virno, A.; Pagano, B.; Virgilio, A.; Di Micco, S.; Galeone, A.; Giancola, C.; Bifulco, G.; Mayol, L.; Randazzo, A. Structural and thermodynamic studies of the interaction of distamycin A with the parallel quadruplex structure [d(TGGGGT)]$_4$. *J. Am. Chem. Soc.* **2007**, *129*, 16048–16056. [CrossRef]
38. Maiti, S.; Chaudhury, N.K.; Chowdhury, S. Hoechst 33258 binds to G-quadruplex in the promoter region of human c-myc. *Biochem. Biophys. Res. Commun.* **2003**, *310*, 505–512. [CrossRef]
39. Jain, A.K.; Bhattacharya, S. Interaction of G-quadruplexes with nonintercalating duplex-DNA minor groove binding ligands. *Bioconjug. Chem.* **2011**, *22*, 2355–2368. [CrossRef]

40. Zhou, J.; Le, V.; Kalia, D.; Nakayama, S.; Mikek, C.; Lewis, E.A.; Sintim, H.O. Diminazene or berenil, a classic duplex minor groove binder, binds to G-quadruplexes with low nanomolar dissociation constants and the amidine groups are also critical for G-quadruplex binding. *Mol. Biosyst.* **2014**, *10*, 2724–2734. [CrossRef]
41. Boncina, M.; Podlipnik, C.; Piantanida, I.; Eilmes, J.; Teulade-Fichou, M.P.; Vesnaver, G.; Lah, J. Thermodynamic fingerprints of ligand binding to human telomeric G-quadruplexes. *Nucleic Acids Res.* **2015**, *43*, 10376–10386. [CrossRef] [PubMed]
42. Pagano, B.; Fotticchia, I.; De Tito, S.; Mattia, C.A.; Mayol, L.; Novellino, E.; Petraccone, L.; Giancola, C. Selective Binding of Distamycin A Derivative to G-Quadruplex Structure [d(TGGGGT)]$_4$. *J. Nucleic Acids* **2010**, *2010*, 247137. [CrossRef] [PubMed]
43. Busto, N.; Cano, B.; Tejido, R.; Biver, T.; Leal, J.M.; Venturini, M.; Secco, F.; Garcia, B. Aggregation features and fluorescence of Hoechst 33258. *J. Phys. Chem. B* **2015**, *119*, 4575–4581. [CrossRef] [PubMed]
44. Schellman, J.A. Macromolecular binding. *Biopolymers* **1975**, *14*, 999–1018. [CrossRef]
45. Paul, A.; Guo, P.; Boykin, D.W.; Wilson, W.D. A new generation of minor-groove-binding-heterocyclic diamidines that recognize G·C base pairs in an AT sequence context. *Molecules* **2019**, *24*, 946. [CrossRef]
46. Wemmer, D.E.; Dervan, P.B. Targeting the minor groove of DNA. *Curr. Opin. Struct. Biol.* **1997**, *7*, 355–361. [CrossRef]
47. Finn, P.B.; Bhimsaria, D.; Ali, A.; Eguchi, A.; Ansari, A.Z.; Dervan, P.B. Single position substitution of hairpin pyrrole-imidazole polyamides imparts distinct DNA-binding profiles across the human genome. *PLoS ONE* **2020**, *15*, e0243905. [CrossRef]
48. Kang, J.S.; Meier, J.L.; Dervan, P.B. Design of sequence-specific DNA binding molecules for DNA methyltransferase inhibition. *J. Am. Chem. Soc.* **2014**, *136*, 3687–3694. [CrossRef]
49. Harika, N.K.; Germann, M.W.; Wilson, W.D. First structure of a designed minor groove binding heterocyclic cation that specifically recognizes mixed DNA base pair sequences. *Chemistry* **2017**, *23*, 17612–17620. [CrossRef]

life

Article

Replication Control of Human Telomere G-Quadruplex DNA by G-Quadruplex Ligands Dependent on Solution Environment

Shuntaro Takahashi [1], Sudipta Bhowmik [2], Shinobu Sato [3], Shigeori Takenaka [3] and Naoki Sugimoto [1,4,*]

1 FIBER (Frontier Institute for Biomolecular Engineering Research), Konan University, Kobe 650-0047, Japan; shtakaha@konan-u.ac.jp
2 Department of Biophysics, Molecular Biology & Bioinformatics, University of Calcutta, Kolkata 700009, India; sbbmbg@caluniv.ac.in
3 Department of Applied Chemistry, Kyushu Institute of Technology, Fukuoka 804-8550, Japan; shinobu@che.kyutech.ac.jp (S.S.); shige@che.kyutech.ac.jp (S.T.)
4 FIRST (Graduate School of Frontiers of Innovative Research in Science and Technology), Konan University, Kobe 650-0047, Japan
* Correspondence: sugimoto@konan-u.ac.jp

Abstract: The human telomere region is known to contain guanine-rich repeats and form a guanine-quadruplex (G4) structure. As telomeres play a role in the regulation of cancer progression, ligands that specifically bind and stabilize G4 have potential therapeutic applications. However, as the human telomere sequence can form G4 with various topologies due to direct interaction by ligands and indirect interaction by the solution environment, it is of great interest to study the topology-dependent control of replication by ligands. In the present study, a DNA replication assay of a template with a human telomere G4 sequence in the presence of various ligands was performed. Cyclic naphthalene diimides (cNDI1 and cNDI2) efficiently increased the replication stall of the template DNA at G4 with an anti-parallel topology. This inhibition was stability-dependent and topology-selective, as the replication of templates with hybrid or parallel G4 structures was not affected by the cNDI and cNDI2. Moreover, the G4 ligand fisetin repressed replication with selectivity for anti-parallel and hybrid G4 structures without stabilization. Finally, the method used, referred to as quantitative study of topology-dependent replication (QSTR), was adopted to evaluate the correlation between the replication kinetics and the stability of G4. Compared to previous results obtained using a modified human telomere sequence, the relationship between the stability of G4 and the effect on the topology-dependent replication varied. Our results suggest that native human telomere G4 is more flexible than the modified sequence for interacting with ligands. These findings indicate that the modification of the human telomeric sequence forces G4 to rigidly form a specific structure of G4, which can restrict the change in topology-dependent replication by some ligands.

Keywords: G-quadruplex; replication; thermodynamics; ligand; topology

Citation: Takahashi, S.; Bhowmik, S.; Sato, S.; Takenaka, S.; Sugimoto, N. Replication Control of Human Telomere G-Quadruplex DNA by G-Quadruplex Ligands Dependent on Solution Environment. *Life* **2022**, *12*, 553. https://doi.org/10.3390/life12040553

Academic Editors: Tigran Chalikian and Jens Völker

Received: 22 March 2022
Accepted: 5 April 2022
Published: 7 April 2022

1. Introduction

DNAs containing tandem repeats of guanine form guanine-quadruplexes (G4s) comprising guanine quartets linked via Hoogsteen hydrogen bonds. Telomeres contain approximately 100 repeats of T_2AG_3, which can form G4 structures. As telomerase binds to these regions to elongate telomeres, G4 formation in telomeres can inhibit the binding and catalysis of telomerase [1]. G4 structures perturb replication [2,3] and transcription, both in vitro and in vivo [4,5]. Previous studies have suggested that the efficiency of the inhibition of these reactions by G4 formation depends not only on the structural stability but also on the topology of G4 formed in the template DNA and RNA [2,4]. As there are more than 300,000 G4-forming potential regions in the human genome [6], the development of ligands that selectively tune the formation of a specific G4 and regulate replication is highly valuable for therapeutic treatment [3,7].

Although G4 topologies with antiparallel, hybrid, and parallel strands (Figure 1a) have been characterized, only certain G4 structures can be targeted to regulate gene expression. The major interaction between G4 and the ligand is the stacking interaction on the G-quartet [8]. As the G-quartet is a common motif in any type of G4 topology, such G4 ligands cannot distinguish a particular G4 structure. G4 structures differ in their sequence, length, and the configuration of the loop regions. In cells, G4 binding proteins possibly recognize differences in G4 structures [9], but only a few ligands are available to discriminate G4 topologies. These ligands can interact with the loops or grooves of G4 and G-quartets [10–12]. Replication of the telomeric region should efficiently occur to maintain the biological roles of telomeres in cells [13], which represents a promising target for cancer therapy [14]. As replication can be influenced by the G4 topology on the DNA template [2], an index for ligand development that shows topology-specific binding and function is highly required.

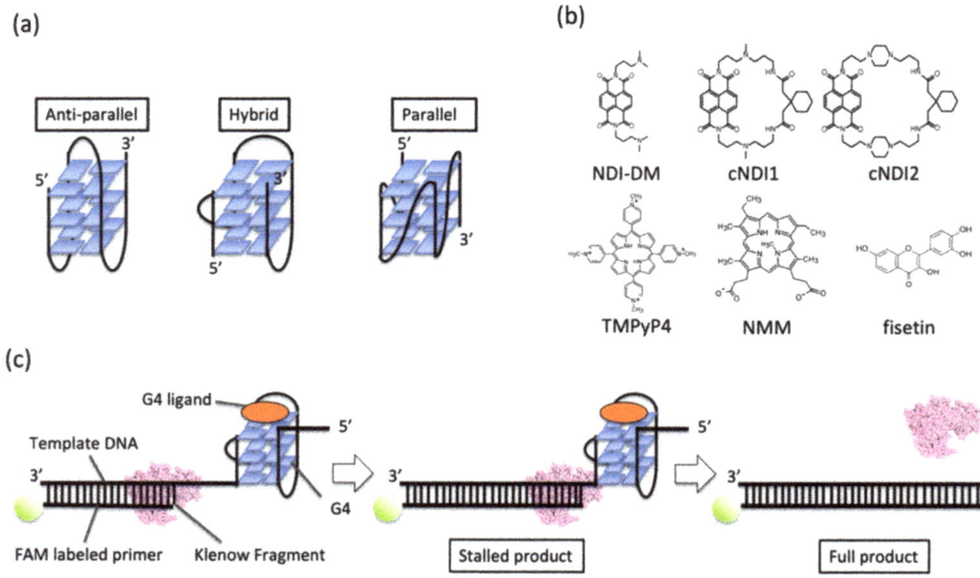

Figure 1. (**a**) Images of G4 topologies. (**b**) Ligands tested in this study. NDI-DM is a non-cyclic naphthalene diimide (NDI) derivative. cNDI1 and cNDI2 are based on NDI-DM and modified with 2,2′-(cyclohexane-1,1-diyl)diacetic acid with linkers of different lengths. Fisetin is a natural product. *Meso*-tetrakis-(*N*-methyl-4-pyridyl)-porphyrin TMPyP4 and *N*-methyl mesoporphyrin IX (NMM) are conventional G4 ligands. (**c**) Schematic illustration of the replication assay in this study.

DNA energetics is fundamental to understanding the stability of DNA structures as well as reaction-related DNA metabolism, such as DNA damage recognition, repair, and replication. Prof Kenneth J. Breslauer, who is the pioneer of this research area, demonstrated a significant relationship between the energetic values of DNA stability and replication efficiency and fidelity in a predictable sequence-dependent manner [15]. Therefore, quantitative analysis of the stability of DNA structures and replication efficiency is fundamental. We previously studied G4 replication in the presence of G4 ligands along a template DNA including a modified sequence of the human telomeric G4-forming sequence (5′-TTGGGTTAGGGTTAGGGTTAGGG-3′), which was different from the native sequence (5′-TTAGGGTTAGGGTTAGGGTTAGGGA-3′) [16]. The modified sequence was developed for NMR measurements to obtain clear spectra. As sequence modification can cause changes in the stability and topology of the G4 structure of human telomeres, the G4-topology-dependent replication of human telomeres should also be affected by sequence modification, although there has been no demonstration. Here, replication of DNA containing a native G4 sequence derived from human telomeres was investigated in the presence of various G4 ligands under different buffer conditions [17]. As shown here, it was found that the replication stall on G4 depended not only on the stability of the G4-ligand complex but also on the manner of interaction of the ligand depending on the G4 topology. The results of this study provide quantitative data showing that the interaction between ligands and G4 controls the energetic barrier of the replicating intermediate, resulting in different unwinding kinetics of G4. Our quantitative study of topology-dependent replication will facilitate drug design to efficiently control gene expression when a mutation occurs in the G4 sequence.

2. Materials and Methods

Materials. The deoxynucleoside triphosphates (dNTPs) were purchased from Takara Bio (Shiga, Japan). NDI-DM, cNDI1, and cNDI2 previously described [18]. All other ligands are purchased and prepared as described [16].

Oligonucleotides. Short DNA for the thermodynamics assay, primer DNA labeled with fluorescein amidite, and template DNA purified using high-performance liquid chromatography (HPLC) were purchased from Japan Bio Service. All of the DNA sequences used in this study are listed in Table S1.

Replication assay. For the polymerase, the Klenow fragment exo- (KF exo-) was prepared and used as previously reported [16]. Primer and template DNA were mixed and annealed in the buffer used in the replication reaction: 10 mM Tris-hydrochloric acid (HCl) (pH 7.5), 8 mM magnesium chloride ($MgCl_2$), 1 μM DNA, 10 μM each ligand (the concentration of fisetin was 50 μM to facilitate efficient replication stall), and 250 μM dNTPs. Potassium chloride (KCl) and poly(ethylene glycol200) (PEG200) were added as indicated. After preparing the solution, the mixtures were incubated at 37 °C for 5 min followed by the addition of 1 μM KF exo- to initiate the reaction. The reactants were collected at each time point and added to a solution containing 10 mM EDTA and 80 wt% formamide. After running urea-polyacrylamide gel electrophoresis (PAGE) containing 8 M urea at 70 °C for 1 h at 200 V in TBE buffer, the gel images were captured using a Fluoreimager FLA-5100 (Fujifilm, Tokyo, Japan) before and after staining with SYBR Gold (Thermo Fisher Scientific, Tokyo, Japan). The intensities of the bands were analyzed using NIH ImageJ software. The amount of full-length product (P) was quantified by calculating the ratio of the intensity of the full-length product band to that of all bands. A kinetic model was applied to a two-step sequential model.

$$P_0 \overset{k_s}{\rightarrow} P_S \overset{k_f}{\rightarrow} P_F \tag{1}$$

where P_0 is the starting state of the reaction, P_s is the state immediately after dissolving the stall, P_F is the state after replication of the full-length product is completed, k_s (min^{-1}) is the rate constant from the start of the reaction to dissolving the stall, and k_f (min^{-1}) is the rate constant from dissolving the stall to complete synthesis of the full-length product. The kinetic analyses were performed as previously described [16].

UV melting assay. The solution used for the ultra violet (UV) melting assay included 5 μM $(T_2AG_3)_4$ and 10 μM of each ligand. For fisetin, a 50 μM of the solution was used. All samples were dissolved in the solution containing the buffer used in the replication assay except for dNTPs. Melting and thermodynamic analyses were performed as previously described [16].

CD Spectroscopy Measurements. The sample preparation for CD measurement was same as the UV melting assay, followed by sample annealing. CD measurement were performed using a JASCO J-1500 instrument at 37 °C as previously [16].

3. Results

3.1. Structure and Stability of Human Telomeric G4 in the Presence of Ligands in a Low K^+ Ion Concentration

The replication of G4 DNA was studied using the Klenow fragment of DNA polymerase I from *Escherichia coli* that lacks exonuclease activity (KF exo-) in the presence of non-cyclic (NDI-DM) and cyclic naphthalene diimide (NDI) derivatives (cNDI1 and cNDI2) [18], and conventional G4 porphyrin ligands (*meso*-tetrakis-(*N*-methyl-4-pyridyl)-porphyrin (TMPyP4) and *N*-methyl mesoporphyrin IX (NMM) (Figure 1b,c). Fisetin, which binds to the loop regions of G4 [19], was assayed. The template DNA contained a G4-forming sequence (four repeats of the human telomeric DNA sequence ($T_2AG_3)_4$) downstream of the region complementary to the primer-binding site (Table S1). This sequence can form various topologies, depending on the solution conditions [17].

The topology of G4 DNA from human telomeric sequence is dependent on the concentration of K^+ ion; for example, the oligonucleotide $(T_2AG_3)_4$ forms an anti-parallel structure at a low concentration of K^+ ions [20]. In the presence of 1 mM KCl in a buffer containing 10 mM Tris HCl (pH 7.5) and 8 mM $MgCl_2$, the circular dichroism (CD) spectra of 5 μM $(T_2AG_3)_4$ in the absence of ligands at 37 °C showed a large positive peak around 295 nm (Figure 2a), indicative of an anti-parallel conformation [11]. As the stability of G4 is low in this buffer condition, the spectrum shows the mixture of the spectrum with the coil form, which showed peaks at 220 nm and 255 nm. We measured the CD spectra of the DNA with 10 μM NDI ligands. Except for the spectrum in the presence of cNDI1 showing a hybrid topology with peaks at 265 and 295 nm, all indicated that G4s in these conditions adopted an anti-parallel conformation (Figure 2a) [11]. The spectrum in the presence of fisetin also indicated an antiparallel topology with the mixture of coil form due to low stability of G4.

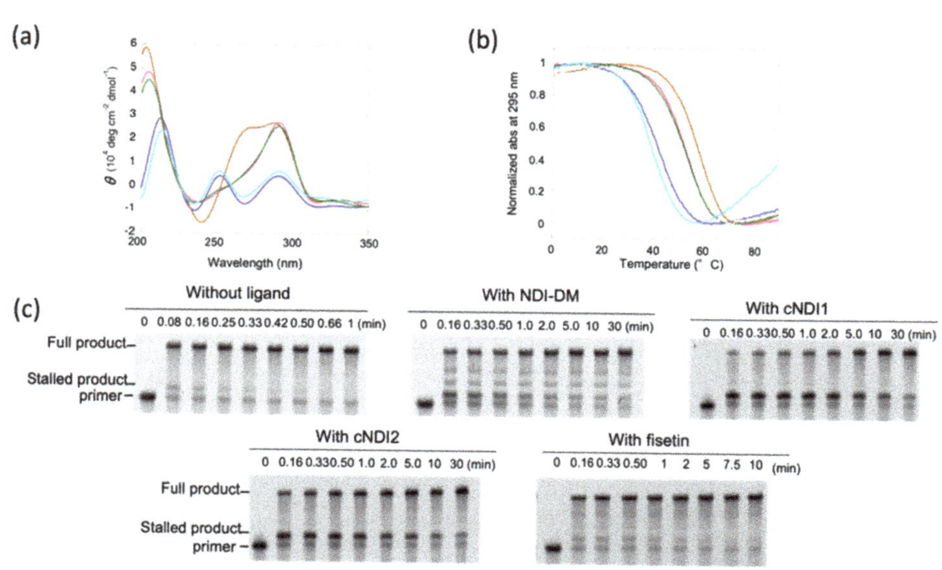

Figure 2. (**a**) CD spectra at 37 °C and (**b**) ultra violet (UV) melting curves of 5 μM $(T_2AG_3)_4$ in the absence (blue) and presence of 10 μM non-cyclic naphthalene diimide (NDI)-DM (pink), cyclic (**c**) NDI1 (brown), cNDI2 (green), or 50 μM fisetin (light blue). The buffer condition was 10 mM Tris-hydrochloric acid (HCl) (pH 7.5), 1 mM potassium chloride (KCl), and 8 mM magnesium chloride ($MgCl_2$). (**c**) Polyacrylamide gel electrophoresis (PAGE) images of aliquots of replication reactions of 1 μM $(T_2AG_3)_4$ template DNA in the absence or presence of 10 μM NDI-DM, cNDI1, cNDI2, or 50 μM fisetin. The reaction was carried out with 1 μM KF exo- and 250 μM deoxynucleoside triphosphates (dNTPs) at 37 °C. Aliquots were taken at time intervals that depended on the ligand.

Next, UV melting measurements were performed to quantitatively analyze the thermodynamic stability of G4 structures in the absence and presence of ligands (Figure 2b). In the absence of ligands, the melting temperature was 42.7 °C, and the thermodynamic stability at 37 °C ($-\Delta G°_{37}$) of the G4 structures, which was used to quantitatively compare the stability, was 0.6 kcal mol^{-1} (Table 1). On the other hand, $-\Delta G°_{37}$ in the presence of NDI-DM was 1.9 kcal mol^{-1}, indicating that the structure was stabilized by the ligand. For the cases of cyclic NDI ligands, $-\Delta G°_{37}$ was 2.5 (cNDI1) and 1.8 (cNDI2) kcal mol^{-1}, respectively, which were also larger than those in the absence of ligands. Fisetin slightly destabilized the G4 with a $-\Delta G°_{37}$ of 0.4 kcal mol^{-1} (Table 1).

Table 1. Thermodynamic parameters for the formation of human native G4 structure and kinetic parameters for G4 replication at 37 °C.

DNA [a] with Ligand	Condition [b]	T_m (°C) (°C)	$-\Delta G°_{37}$ (kcal mol^{-1})	k_s (min^{-1}) (min^{-1})	k_f (min^{-1}) (min^{-1})
$(T_2AG_3)_4$	1 mM KCl, 0 wt% PEG200	42.7 ± 0.7	0.6 ± 0.1	4.4 ± 2.2	71 ± 40
$(T_2AG_3)_4$ + NDI-DM		52.9 ± 1.0	1.9 ± 0.1	0.68 ± 0.01	10 ± 1.5
$(T_2AG_3)_4$ + cNDI1		58.7 ± 1.0	2.5 ± 0.2	0.24 ± 0.11	8.5 ± 3.0
$(T_2AG_3)_4$ + cNDI2		52.3 ± 1.8	1.8 ± 0.3	0.14 ± 0.06	9.6 ± 1.3
$(T_2AG_3)_4$ + fisetin		41.4 ± 0.2	0.4 ± 0.2	0.60 ± 0.04	9.6 ± 2.8
$(T_2AG_3)_4$	100 mM KCl, 0 wt% PEG200	73.7 ± 0.2	4.0 ± 0.1	0.36 ± 0.08	7.6 ± 1.6
$(T_2AG_3)_4$ + NDI-DM		76.6 ± 0.2	5.0 ± 0.2	0.042 ± 0.019	1.0 ± 0.3
$(T_2AG_3)_4$ + cNDI1		79.3 ± 0.9	5.4 ± 0.1	0.023 ± 0.064	0.8 ± 0.1
$(T_2AG_3)_4$ + cNDI2		76.0 ± 0.2	4.6 ± 0.1	0.047 ± 0.025	1.6 ± 0.4
$(T_2AG_3)_4$ + TMPyP4		74.8 ± 0.3	5.4 ± 0.3	0.056 ± 0.005	1.3 ± 0.1
$(T_2AG_3)_4$ + NMM		72.4 ± 0.4	4.1 ± 0.1	0.23 ± 0.03	4.9 ± 0.9
$(T_2AG_3)_4$ + fisetin		70.6 ± 0.1	3.4 ± 0.1	0.028 ± 0.003	1.3 ± 0.1
$(T_2AG_3)_4$	1 mM KCl, 20 wt% PEG200	45.4 ± 0.1	1.1 ± 0.1	0.49 ± 0.19	6.3 ± 1.4
$(T_2AG_3)_4$ + NDI-DM		53.3 ± 0.9	2.1 ± 0.1	0.042 ± 0.001	2.3 ± 0.14
$(T_2AG_3)_4$ + cNDI1		57.0 ± 1.2	2.5 ± 0.2	0.039 ± 0.008	1.9 ± 0.03
$(T_2AG_3)_4$ + cNDI2		51.7 ± 1.5	2.0 ± 0.2	0.039 ± 0.009	3.3 ± 0.29
$(T_2AG_3)_4$ + TMPyP4		47.5 ± 0.4	1.2 ± 0.1	0.069 ± 0.009	2.7 ± 0.1
$(T_2AG_3)_4$ + NMM		42.7 ± 0.1	0.6 ± 0.1	0.073 ± 0.006	3.7 ± 0.1
$(T_2AG_3)_4$ + fisetin		40.1 ± 0.2	0.5 ± 0.1	1.4 ± 0.4	14 ± 2.4

[a] The concentration of DNA was 5 µM. [b] The sample solution contained 10 mM Tris-hydrochloric acid (HCl) (pH 7.5), 1 or 100 mM potassium chloride (KCl), and 8 mM magnesium chloride ($MgCl_2$) with or without poly(ethylene glycol200) (PEG200).

3.2. Relationship between Thermodynamic Stability of the G4 Structure and Its Replication Efficiency in a Low K$^+$ Concentration

The replication reaction was performed in the same buffer in the presence of 1 µM KF exo- and 250 µM dNTPs. After 0.08 min (=5 s), the two products were detected by denaturing polyacrylamide gel electrophoresis (PAGE; Figure 2c). The product migrating faster was confirmed to be the fully extended primer after staining in comparison with the mobility of the template DNA (Figure S1). The product migrating slower, which extended only a few nucleotides from the primer (Figures 2c and S1), indicated a stalled product due to its G4 structure. This stalled product disappeared after 0.33 min (=20 s), indicating that replication was stalled to some extent by the G4 structure.

In the presence of NDI-DM, these two replication products were mainly as observed in the absence of the ligand. However, the stalled product was much accumulated more than that without the ligand (Figure 2c), which indicates that NDI-DM stabilized G4 and thus increased the replication stall compared to that without the ligand. The minor bands observed between the bands of the stalled product and full-length product may be derived from replication stall by the nonspecific binding of NDI-DM due to the relatively low specificity for G4 [18]. More stalled products were detected in the presence of cNDI1 and cNDI2 than in the presence of NDI-DM (Figure 2c). Contrary to our expectations based on the melting results, which showed that the $-\Delta G^{\circ}_{37}$ values were not significantly different in the presence of NDI-DM and cNDI2, although cNDI1 slightly stabilized the $(T_2AG_3)_4$ structure relative to NDI-DM, more significant stalling of the polymerase enzyme was induced by cNDI1 and cNDI2 than NDI-DM. Replication was also analyzed in the presence of TMPyP4 and NMM as conventional G4 ligands, and found almost no polymerase stalling (Figure S2a,b), likely due to the fact that TMPyP4 and NMM do not bind well to anti-parallel G4 [21]. Therefore, cNDIs may stall replication effectively due to the interaction between the cyclic moieties and G4 loop regions. Moreover, in the presence of fisetin, a small amount of the stalled product was observed.

To compare the replication kinetics and thermodynamic stability of $(T_2AG_3)_4$, we determined the kinetics of the replication of the full-length products, as previously reported (Figure S3) [2]. A two-step sequential model consisting of the stall in replication was assumed, which included the prior replication to the stall position, and generating a full-length product after resolving the replication stall. The rate constants (k_s) at 37 °C before releasing the replication stall of KF exo- were 4.4 min^{-1} (without ligand), 0.68 min^{-1} (with NDI-DM), 0.24 min^{-1} (with cNDI1), and 0.14 min^{-1} (with cNDI2) (Table 1). The rate-limiting step of these reactions was the resolution of G4 since the rate constants k_f for replication after dissolving the stall was much larger than k_s (Table 1). For fisetin, the k_s values (0.60 min^{-1}) were also smaller than the k_f values. Therefore, stall of replication is the rate-limiting step of the reaction. The k_f values were similar for all of the ligands, and the relatively large difference between the k_f without ligand and that with ligand indicated a decrease in the processivity of the single-stranded region downstream of the G4 structure owing to the effect of nonspecific binding of ligands to the template DNA.

The k_s values indicated that the cNDI ligands inhibited the replication of the $(T_2AG_3)_4$ template more effectively than the parent compound NDI-DM. Each of these NDI compounds has a naphthalene moiety that presumably interacts in a similar manner to a G-quartet; thus, the cyclic moieties of cNDI1 and cNDI2 influence binding to $(T_2AG_3)_4$. Interestingly, NDI-DM and cNDI2 similarly stabilized the G4 conformation of $(T_2AG_3)_4$, but the efficiency of replication repression differed. Furthermore, the k_s of fisetin was much smaller than that in the absence of ligands, although the stability of the $(T_2AG_3)_4$ G4 did not increase. These effects on the replication rate that do not correlate with G4 stability imply that the binding mode of the ligand to G4 influences how efficiently the structure is resolved by polymerase.

3.3. Quantitative Analysis of G4 Stability and G4 Replication Affected by G4 Ligands

To quantitatively analyze the effects of the ligand on replication through G4s of different topologies, a method was developed for novel categorization of G4 ligands named quantitative study of topology-dependent replication (QSTR) to analyze the phase diagram of the stability of G4 vs. activation energy of G4 replication, which we established previously (Figure 3) [2,16,22]. Briefly, QSTR categorizes and quantitatively characterizes the effect of ligands on stalling at G4. In the analysis, it was demonstrated that the plot between the $-\Delta\Delta G^{\circ}_{37}$ values ($-\Delta G^{\circ}_{37}$ in the presence of ligand—($-\Delta G^{\circ}_{37}$ in the absence of ligand)) and $\Delta\Delta G^{\ddagger}_{37}$ values ($\Delta G^{\ddagger}_{37}$ is the activation free energy of replication of G4 at 37 °C). Furthermore, the slope ($\Delta\Delta G^{\ddagger}_{37}/(-\Delta\Delta G^{\circ}_{37})$) of the x-y surface of the ligand of interest, S_i, was divided by that of the standard ligand S. A comparison of the slopes represents the obtained value α, that is, S_i/S. Here, the standard is NDI-DM.

To visualize the relationship between the stability of G4 and replication kinetics, the plots were categorized into upper, lower, and right or left quadrants in different colors (groups I, II, III, and IV are yellow, green, light blue, and purple, respectively) (Figure 3A). Ligands in groups I and III changed the unfolding kinetics of G4, which resulted in replication control. However, ligands in groups II and IV exhibit control over replication that is independent of the effect on the stability of G4. The categorized ligands were also arranged in in category i ($\alpha \approx 1$), ii ($\alpha > 1$ indicating that the reaction is $\Delta\Delta G^{\ddagger}_{37}$ dependent), and iii ($\alpha < 1$ indicating that the reaction is $\Delta\Delta G^{\circ}_{37}$ dependent) in addition to iv ($\alpha < 0$). The QSTR of the data for 1 mM KCl without PEG200 (Figure 3B and Table 2) indicated that cNDI1 was group I-i, which means that cNDI1 has the same mechanism of replication stall ($\alpha \approx 1$) as that of NDI-DM. The cNDI2 data corresponded to relationship (I-ii), suggesting that these ligands stabilize G4 and inhibit the replicating polymerase differently and more efficiently than NDI-DM. The fisetin data also deviated and represented relationship (II-iv), indicating that the ligand repressed the replication of G4 with a smaller $\Delta G^{\ddagger}$ than that of NDI-DM.

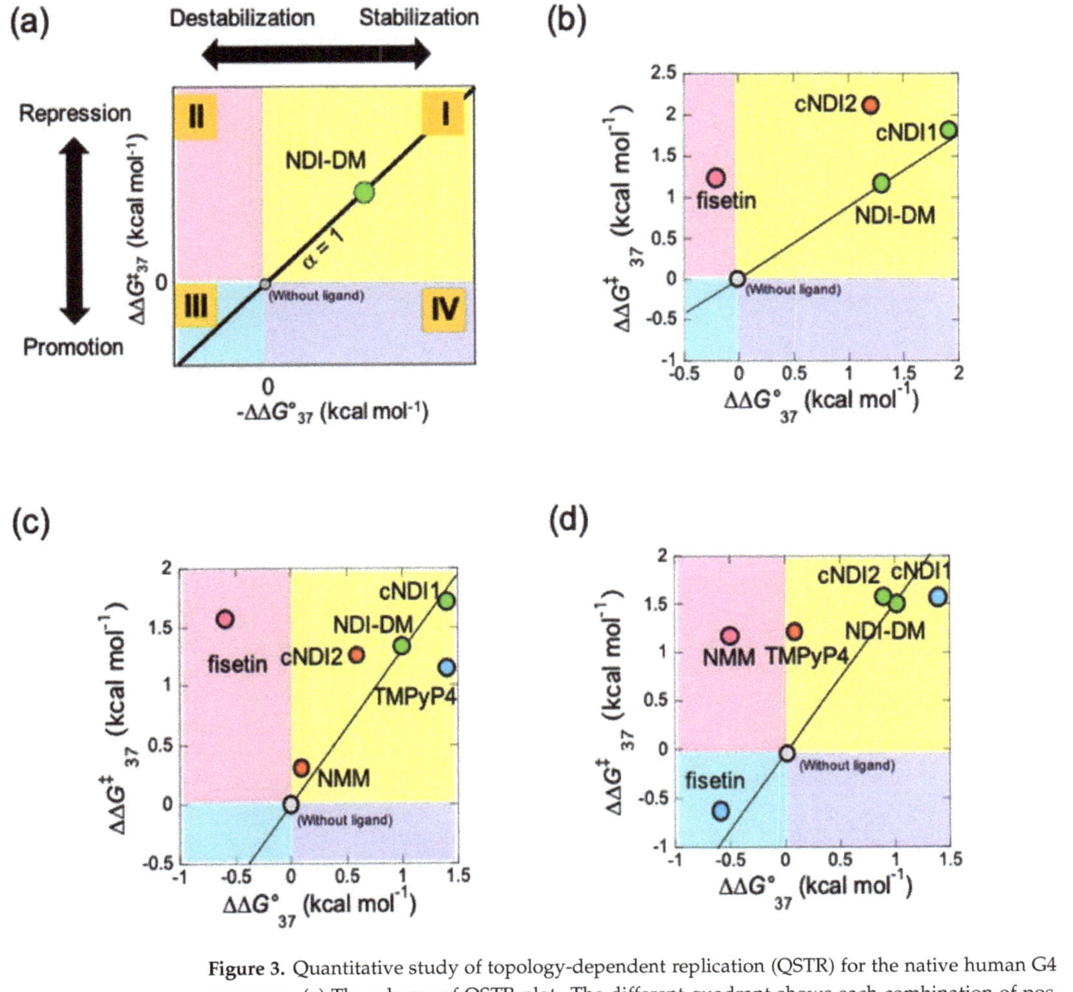

Figure 3. Quantitative study of topology-dependent replication (QSTR) for the native human G4 sequence. (**a**) The scheme of QSTR plot. The different quadrant shows each combination of positive and negative value of $-\Delta\Delta G°_{37}$ and $\Delta\Delta G^{\ddagger}_{37}$ (I–IV). The value α, which indicates the ratio of $\Delta\Delta G^{\ddagger}_{37}/(-\Delta\Delta G°_{37})$ of the ligand of interest to that of the NDI-DM. Reprinted with permission from [16]. Copyright 2021 American Chemical Society. (**b–d**) Obtained QSTR plots in (**b**) 1 mM potassium chloride (KCl) without poly(ethylene glycol200) (PEG200), (**c**) in the presence of 100 mM KCl without PEG200, and (**d**) 1 mM KCl with 20 wt% PEG200. The degree of α was shown in different colors in the plot ($\alpha \approx 1$; green, $\alpha > 1$; red, $0 < \alpha < 1$; blue, and $\alpha < 0$; pink).

Table 2. G4 topologies of $(T_2AG_3)_4$ and categories of QSTR.

Condition	Ligand	G4 Topology $(T_2AG_3)_4$	QSTR Category $(T_2AG_3)_4$	G4 Topology $T_2(G_3T_2A)_3A$ [b]	QSTR Category $T_2(G_3T_2A)_3A_4$ [b]
1 mM KCl without PEG200	NDI-DM	Anti-parallel	I-i [a]	Anti-parallel	I-i [a]
	cNDI1	Hybrid	I-i	Hybrid	I-ii
	cNDI2	Anti-parallel	I-ii	Anti-parallel	I-ii
	fisetin	Anti-parallel	II-iv	n.d.	n.d.
100 mM KCl without PEG200	NDI-DM	Anti-parallel	I-i [a]	Anti-parallel	I-i [a]
	cNDI1	Hybrid	I-i	Hybrid	I-ii
	cNDI2	Anti-parallel	I-ii	Anti-parallel	I-ii
	TMPyP4	Parallel	I-iii	n.d.	n.d.
	NMM	Parallel	I-ii	n.d.	n.d.
	fisetin	Anti-parallel	II-iv	Hybrid	III-iii
1 mM KCl and 20 wt% PEG200	NDI-DM	Anti-parallel	I-i [a]	Anti-parallel	I-i [a]
	cNDI1	Hybrid	I-i	Hybrid	I-ii
	cNDI2	Hybrid	I-i	Anti-parallel	I-ii
	TMPyP4	Parallel	I-ii	n.d.	n.d.
	NMM	Parallel	II-iv	Hybrid	II-iv
	fisetin	Hybrid	III-iii	n.d.	n.d.

[a] NDI-DM is a standard ligand. [b] Results obtained from our previous study.

3.4. Categorization of Ligands by Quantitative Study of Topology-Dependent Replication (QSTR) Analysis

Regarding 100 mM KCl without PEG200 condition, from the CD spectrum at 37 °C, the native human G4 sequence showed a hybrid topology, as previously reported [23], whereas all NDI ligands induced an anti-parallel topology, except for cNDI1 (Figure S4a). The spectrum in the presence of cNDI1 differed slightly from that of the others, suggesting a slightly different structural change forming the hybrid topology. For TMPyP4 and NMM, the spectra indicated a parallel topology, whereas fisetin had an antiparallel topology (Figures S2g and S4a).

At 100 mM KCl, stability and replication kinetics parameters were obtained (Figures S2i and S4b,c, Table 1). $-\Delta\Delta G^{\circ}_{37}$ vs. $\Delta\Delta G^{\ddagger}_{37}$ for NDI-DM, and cNDI1 fell in the QSTR categories I-i (Figure 3c and Table 2). Under these conditions, the G4 of $(T_2AG_3)_4$ formed a hybrid topology in the presence of these ligands. Thus, QSTR indicates that these ligands affected the replication stall through the same mechanism as the standard. cNDI2 and NMM were categorized as I-ii, but the difference in the relative activation energy suggested that NMM repressed replication more efficiently than cNDI2. TMPyP4 was categorized as I-iii, indicating that replication was repressed less efficiently than in the case of other ligands due to a small effect on the intermediate formed during replication. Fisetin was categorized as into II-vi, as was the case with 1 mM KCl without PEG200, suggesting that the effect of fisetin on replication stalling was not influenced by the topology of G4 but that the inhibition mechanism may be shared in each case.

As molecular crowding can change G4 topology [17], the QSTR analysis was also performed in the presence of 20 wt% poly(ethylene glycol) 200 (PEG200; average molecular weight 200). In 20 wt% PEG200, 10 mM Tris-HCl (pH 7.5), 8 mM $MgCl_2$, and 1 mM KCl at 37 °C, the CD spectrum of $(T_2AG_3)_4$ indicated a hybrid topology (Figure S4d). Despite the low concentration of K^+, PEG200 partially induced a structural change from the antiparallel to the hybrid form. The spectra with cNDI1, cNDI2, and fisetin were indicative of similar CD patterns, although NDI-DM showed an anti-parallel topology (Figure S4d). TMPyP4 and NMM also exhibited spectra with parallel topologies (Figure S2h).

Based on the thermodynamic and replication assays in 1 mM KCl and 20 wt% PEG200 (Figures S2j and S4e,f, Table 1), QSTR analysis indicated that α the of NDI-DM, cNDI1, and cNDI2 data were almost 1, resulting in QSTR categories I-i (Figure 3d and Table 2). Fisetin was categorized as III-iii, indicating that fisetin destabilized G4 and promoted replication due to the acceleration of the kinetics of G4 unfolding. TMPyP4 was categorized into I-ii, which may depend on how the ligand induces the parallel topology. The QSTR category for NMM was II-iv, suggesting that the energy barrier for the replicating intermediate influences replication kinetics. All of the plots of NDI ligands may fall into category I-i with the use of TMPyP4 or NMM if all NDI ligands and fisetin bind to the perfectly formed parallel topology. In summary, the interaction of G4 ligands with G4 structures can vary under different conditions, which causes the regulation of G4 replication in a topology-dependent manner.

4. Discussion

In this study, QSTR analysis of G4 replication along the template DNA was performed, including a native human telomere sequence (5′-TTAGGGTTAGGGTTAGGGTTAGGG-3′). As shown in Table 2, the results obtained in this study were compared with those obtained for the modified sequence (5′-TTGGGTTAGGGTTAGGGTTAGGGA-3′). Some QSTR categories of the native sequences were the same as those of the modified sequences. For example, cNDI2 in 1 or 100 mM KCl without PEG200 showed the same QSTR categories. Contrastingly, other QSTR categories differed from those of the modified sequence. In the case of cNDI1, the QSTR categories for the native sequence were I-i in all solution conditions and I-ii for the modified sequence. In a previous study, an interaction between the loop region of the G4 and ligand moiety of cNDI1 and cNDI2 was identified, which caused different QSTR categories from the standard ligand NDI-DM [16]. As NDI-DM has no cyclic moiety, the interaction between the ligand and the loop region should be small. The QSTR category I-i of cNDI1 indicates that the unwinding mechanism of replication is similar to NDI-DM. Therefore, the results of this study suggest that the interaction of cNDI1 with the loop region of G4 was not strong and had no significant effect on the efficient replication stall. For cNDI2, the QSTR category in the solution containing 1 mM KCl and 20 wt% PEG200 was I-i, which was different from that for the modified sequence. This could be due to topology change, which lowered the interaction between the loop region of G4 and cNDI2. A similar change was observed in the case of fisetin in the solution of 100 mM KCl without PEG200, which showed different QSTR categories compared with the native and modified sequences owing to the topology change of G4. The authors hypothesize that these behaviors of the native sequence are derived from the flexibility of the structure. Considering the availability of structural analysis, the modified sequence has only a small amount of structural polymorphism and is beneficial for structural analyses such as NMR measurements [16]. Thus, the rigid structure provides a specific interaction geometry for the ligands and causes a topology-dependent replication stall. On the other hand, the G4 structure from the native sequence was flexible and formed a different structure from the ligand-G4 complex. Therefore, the effect of topology-dependent replication was different from that of the ligand structure. In truth, we could assign NMM and TMPyP4 to the QSTR categories in the case of the native sequence, whereas NMM and TMPyP4 were not enough to stabilize the G4 of the modified sequence in most cases. QSTR provides information on the potential for replication control along different template sequences. From the view point of the specific interaction derived from the difference of the sequence, there are two possible interactions by the additional adenine in the sequence of the native telomere sequence. One is the stack on the G-quartet, which can differently regulate the topology of G-quadruplex compared with the modified sequence. Another one is the adenine base in the native sequence may differently interact with the ligands compared with the case of the modified sequence. In the anti-parallel and hybrid structure, these interactions happened affect QSTR categories of the NDI ligands as you suggested, since NDI ligands interact each G-quartet of the G-quadruplex and loop regions as well.

In conclusion, it was identified that the regulation of topology-dependent replication by G4 ligands depends on the structural flexibility of G4 owing to sequence modification. These results suggest that naturally occurring modifications, such as methylation and oxidation, can cause perturbation of G4 flexibility and thus, changes in topology-dependent replication of the telomere region, which would be useful for designing ligands under many conditions for treating diseases.

Supplementary Materials: The following are available online at https://www.mdpi.com/article/10.3390/life12040553/s1. Figure S1: Images of gels shown in Figure 2b stained with SYBR Gold reagent, Figure S2: Effect of TMPyP4 and NMM on replication and structure of G4; Figure S3: Time course of full-length product of the replication reaction; Figure S4: Spectroscopic data and replication results in the presence of 100 mM KCl or in the presence of 1 mM KCl and 20 wt% PEG200; Table S1: DNA sequences used in this study; Table S2: QSTR data calculated using the parameters listed in Table 2.

Author Contributions: Conceptualization, S.T. (Shuntaro Takahashi) and N.S.; methodology, S.T. (Shuntaro Takahashi); investigation, S.T. (Shuntaro Takahashi); resources, S.B., S.S. and S.T. (Shigeori Takenaka); data curation, S.T. (Shuntaro Takahashi) and N.S.; writing—original draft preparation, S.T. (Shuntaro Takahashi); writing—review and editing, all authors. All authors have read and agreed to the published version of the manuscript.

Funding: This work was supported by Grants-in-Aid for Scientific Research from the Ministry of Education, Culture, Sports, Science and Technology (MEXT) and the Japan Society for the Promotion of Science (JSPS), especially a Grant-in-Aid for Scientific Research on Innovative Areas "Chemistry for Multimolecular Crowding Biosystems"(JSPS KAKENHI Grants JP17H06351, 18KK0164, and 19K05723), JSPS Core-to-Core Program (grant number: JPJSCCA20220005), the Hirao Taro Foundation of Konan Gakuen for Academic Research, the Nakatani Foundation for Advancement of Measurement in Biomedical Engineering, the Chubei Itoh Foundation, and DST, Govt. of India.

Institutional Review Board Statement: Not applicable.

Informed Consent Statement: Not applicable.

Acknowledgments: We thank A. Matsuyama and J. Inoue for their assistance with the experiments.

Conflicts of Interest: The authors declare no conflict of interest.

References

1. Zahler, A.; Williamson, J.R.; Cech, T.R.; Prescott, D.M. Inhibition of telomerase by G-quartet DMA structures. *Nature* **1991**, *350*, 718–720. [CrossRef] [PubMed]
2. Takahashi, S.; Brazier, J.A.; Sugimoto, N. Topological impact of noncanonical DNA structures on Klenow fragment of DNA polymerase. *Proc. Natl. Acad. Sci. USA* **2017**, *114*, 9605–9610. [CrossRef] [PubMed]
3. Lopes, J.; Piazza, A.; Bermejo, R.; Kriegsman, B.; Colosio, A.; Teulade-Fichou, M.-P.; Foiani, M.; Nicolas, A. G-quadruplex-induced instability during leading-strand replication. *EMBO J.* **2011**, *30*, 4033–4046. [CrossRef]
4. Tateishi-Karimata, H.; Isono, N.; Sugimoto, N. New Insights into Transcription Fidelity: Thermal Stability of Non-Canonical Structures in Template DNA Regulates Transcriptional Arrest, Pause, and Slippage. *PLoS ONE* **2014**, *9*, e90580. [CrossRef] [PubMed]
5. Broxson, C.; Beckett, J.; Tornaletti, S. Transcription Arrest by a G Quadruplex Forming-Trinucleotide Repeat Sequence from the Human c-myb Gene. *Biochemistry* **2011**, *50*, 4162–4172. [CrossRef] [PubMed]
6. Huppert, J.L.; Balasubramanian, S. Prevalence of quadruplexes in the human genome. *Nucleic Acids Res.* **2005**, *33*, 2908–2916. [CrossRef] [PubMed]
7. Neidle, S. Quadruplex Nucleic Acids as Novel Therapeutic Targets. *J. Med. Chem.* **2016**, *59*, 5987–6011. [CrossRef]
8. Parkinson, G.N.; Ghosh, R.; Neidle, S. Structural Basis for Binding of Porphyrin to Human Telomeres. *Biochemistry* **2007**, *46*, 2390–2397. [CrossRef]
9. Zaug, A.J.; Podell, E.R.; Cech, T.R. Human POT1 disrupts telomeric G-quadruplexes allowing telomerase extension in vitro. *Proc. Natl. Acad. Sci. USA* **2005**, *102*, 10864–10869. [CrossRef]
10. Dhamodharan, V.; Harikrishna, S.; Bhasikuttan, A.C.; Pradeepkumar, P.I. Topology Specific Stabilization of Promoter over Telomeric G-Quadruplex DNAs by Bisbenzimidazole Carboxamide Derivatives. *ACS Chem. Biol.* **2015**, *10*, 821–833. [CrossRef]
11. Diveshkumar, K.V.; Sakrikar, S.; Harikrishna, S.; Dhamodharan, V.; Pradeepkumar, P.I. Targeting Promoter G-Quadruplex DNAs by Indenopyrimidine-Based Ligands. *ChemMedChem* **2014**, *9*, 2754–2765. [CrossRef] [PubMed]
12. Sato, N.; Takahashi, S.; Tateishi-Karimata, H.; Hazemi, M.E.; Chikuni, T.; Onizuka, K.; Sugimoto, N.; Nagatsugi, F. Alkylating probes for the G-quadruplex structure and evaluation of the properties of the alkylated G-quadruplex DNA. *Org. Biomol. Chem.* **2018**, *16*, 1436–1441. [CrossRef] [PubMed]
13. Gilson, E.; Géli, V. How telomeres are replicated. *Nat. Rev. Mol. Cell Biol.* **2007**, *8*, 825–838. [CrossRef]
14. Jafri, M.A.; Ansari, S.A.; Alqahtani, M.H.; Shay, J.W. Roles of telomeres and telomerase in cancer, and advances in telomerase-targeted therapies. *Genome Med.* **2016**, *8*, 69. [CrossRef] [PubMed]

15. Breslauer, K.J. The shaping of a molecular linguist: How a career studying DNA energetics revealed the language of molecular communication. *J. Biol. Chem.* **2021**, *296*, 100522. [CrossRef]
16. Takahashi, S.; Kotar, A.; Tateishi-Karimata, H.; Bhowmik, S.; Wang, Z.-F.; Chang, T.-C.; Sato, S.; Takenaka, S.; Plavec, J.; Sugimoto, N. Chemical Modulation of DNA Replication along G-Quadruplex Based on Topology-Dependent Ligand Binding. *J. Am. Chem. Soc.* **2021**, *143*, 16458–16469. [CrossRef]
17. Miyoshi, D.; Karimata, H.; Sugimoto, N. Hydration Regulates Thermodynamics of G-Quadruplex Formation under Molecular Crowding Conditions. *J. Am. Chem. Soc.* **2006**, *128*, 7957–7963. [CrossRef] [PubMed]
18. Esaki, Y.; Islam, M.; Fujii, S.; Sato, S.; Takenaka, S. Design of tetraplex specific ligands: Cyclic naphthalene diimide. *Chem. Commun.* **2014**, *50*, 5967–5969. [CrossRef]
19. Sengupta, B.; Pahari, B.; Blackmon, L.; Sengupta, P.K. Prospect of Bioflavonoid Fisetin as a Quadruplex DNA Ligand: A Biophysical Approach. *PLoS ONE* **2013**, *8*, e65383. [CrossRef]
20. Marchand, A.; Gabelica, V. Folding and misfolding pathways of G-quadruplex DNA. *Nucleic Acids Res.* **2016**, *44*, 10999–11012. [CrossRef]
21. Kreig, A.; Calvert, J.; Sanoica, J.; Cullum, E.; Tipanna, R.; Myong, S. G-quadruplex formation in double strand DNA probed by NMM and CV fluorescence. *Nucleic Acids Res.* **2015**, *43*, 7961–7970. [CrossRef] [PubMed]
22. Takahashi, S.; Kim, K.T.; Podbevšek, P.; Plavec, J.; Kim, B.H.; Sugimoto, N. Recovery of the Formation and Function of Oxidized G-Quadruplexes by a Pyrene-Modified Guanine Tract. *J. Am. Chem. Soc.* **2018**, *140*, 5774–5783. [CrossRef] [PubMed]
23. Ambrus, A.; Chen, D.; Dai, J.; Bialis, T.; Jones, R.A.; Yang, D. Human telomeric sequence forms a hybrid-type intramolecular G-quadruplex structure with mixed parallel/antiparallel strands in potassium solution. *Nucleic Acids Res.* **2006**, *34*, 2723–2735. [CrossRef] [PubMed]

life

Review

Forces Driving a Magic Bullet to Its Target: Revisiting the Role of Thermodynamics in Drug Design, Development, and Optimization

Conceição A. Minetti * and David P. Remeta *

Department of Chemistry and Chemical Biology, Rutgers—The State University of New Jersey, Piscataway, NJ 08854, USA
* Correspondence: cminetti@chem.rutgers.edu (C.A.M.); dpremeta@chem.rutgers.edu (D.P.R.)

Abstract: Drug discovery strategies have advanced significantly towards prioritizing target selectivity to achieve the longstanding goal of identifying "magic bullets" amongst thousands of chemical molecules screened for therapeutic efficacy. A myriad of emerging and existing health threats, including the SARS-CoV-2 pandemic, alarming increase in bacterial resistance, and potentially fatal chronic ailments, such as cancer, cardiovascular disease, and neurodegeneration, have incentivized the discovery of novel therapeutics in treatment regimens. The design, development, and optimization of lead compounds represent an arduous and time-consuming process that necessitates the assessment of specific criteria and metrics derived via multidisciplinary approaches incorporating functional, structural, and energetic properties. The present review focuses on specific methodologies and technologies aimed at advancing drug development with particular emphasis on the role of thermodynamics in elucidating the underlying forces governing ligand–target interaction selectivity and specificity. In the pursuit of novel therapeutics, isothermal titration calorimetry (ITC) has been utilized extensively over the past two decades to bolster drug discovery efforts, yielding information-rich thermodynamic binding signatures. A wealth of studies recognizes the need for mining thermodynamic databases to critically examine and evaluate prospective drug candidates on the basis of available metrics. The ultimate power and utility of thermodynamics within drug discovery strategies reside in the characterization and comparison of intrinsic binding signatures that facilitate the elucidation of structural–energetic correlations which assist in lead compound identification and optimization to improve overall therapeutic efficacy.

Keywords: drug discovery; drug-target interactions; thermodynamic binding signatures; isothermal titration calorimetry; lead optimization; fragment-based drug discovery; drug-like properties; enthalpic efficiency; group efficiency; lipophilic efficiency

Citation: Minetti, C.A.; Remeta, D.P. Forces Driving a Magic Bullet to Its Target: Revisiting the Role of Thermodynamics in Drug Design, Development, and Optimization. *Life* **2022**, *12*, 1438. https://doi.org/10.3390/life12091438

Academic Editors: Tigran Chalikian and Jens Völker

Received: 25 July 2022
Accepted: 3 September 2022
Published: 15 September 2022

1. Introduction

Tracing the historical origins of drug discovery, scientists have long pursued the paradigm of *one-drug-for-one-disease* as illustrated by the pioneering studies of Paul Ehrlich, who inspired generations of medicinal chemists in the quest of identifying, modifying, and optimizing lead compounds to achieve the desired potency, selectivity, and specificity (as reviewed in [1]). For over a century, research has advanced towards prioritizing drug-target selectivity by seeking "magic bullets" amongst thousands of prospective candidates. This scenario has proven extremely challenging as the search for novel therapeutics to treat existing and emerging diseases often relies on multi-targeting, repurposing, and/or refining an approved drug [2]. In fact, growing awareness of the cellular interactome and increasing complexity of networks controlling biological processes has recently invoked a broader selectivity (i.e., shotgun) approach [3,4]. The latter explores complex diseases that often implicate multiple targets as intervening therapies. Moreover, seemingly undruggable targets that involve functionally important protein-protein interactions (PPIs) have been unraveled and the hotspots identified therein represent novel avenues for drug discovery campaigns [5]. Irrespective of the approach, nature, and/or type of validated target, drug discovery efforts continue to accelerate with the arsenal of available tools expanding dramatically to facilitate improvements in drug efficacy while minimizing off-target effects. Considering the strategies to tackle diseases of simple or complex etiology, one principle remains unvaried, namely a drug must be designed to interact with the desired target(s) specifically and elicit the appropriate response.

The studies described and reviewed herein are intended to rekindle the interest of biochemists, biophysicists and molecular biologists on the importance of employing thermodynamic information in drug design and development by assisting hit-to-lead progression and optimization while facilitating identification and/or improvement of repurposed drugs. Major health threats related to cancer, cardiovascular diseases, and neurodegeneration remain the focus of ongoing research in immunology, virology, and a host of medical fields. The recent emergence of SARS-CoV-2 as a global pandemic warrants development of effective therapeutics as a parallel path to assist immunization efforts and mitigate the impact of vaccine resistant variants. An alarming increase in bacterial resistance has challenged the research community and pharmaceutical industry to undertake immediate action by developing the next generation of novel antibiotics and strategic approaches to obviate an imminent risk with global repercussions. This review focuses on revisiting specific accomplishments achieved in drug discovery during the past two decades particularly with respect to the role of thermodynamics as an integral tool in providing unique insights that assist decision making processes in drug design, development, and optimization [6]. The acquisition and critical assessment of thermodynamic binding signatures [7,8], in conjunction with the elucidation of structural features at the molecular level, enhance our understanding of ligand-target interactions and set the stage for improvement of predictive capabilities. A combination of biological, structural, thermodynamic, and computational approaches integrated within a multidisciplinary drug discovery strategy provides the most effective and efficient route for developing future generations of improved therapeutics.

1.1. Scope of the Review

This review examines the utility of employing specific criteria and metrics to guide decisions regarding the screening and selection of prospective lead compounds in the drug design and development process. While biophysical approaches, such as isothermal titration calorimetry (ITC), have been utilized to bolster drug discovery efforts, there are instances in which the power of this technique has been underutilized, serving solely as an alternate tool to derive binding affinities with the information-rich thermodynamic signatures ostensibly overlooked. In the present review, we highlight studies that have specifically explored and exploited the use of energetic binding signatures to augment drug discovery strategies and discuss the wealth of information that has been gleaned towards advancing the overall design and development process. We examine and evaluate the fundamental role of thermodynamic-based metrics to assess the fate of a compound along the drug discovery pathway. The latter is accomplished in part by seeking empirical correlations between specific physicochemical properties and experimentally derived thermodynamic profiles deduced via analysis of biomolecular interactions deposited in binding databases. Towards this end, we have scrutinized the biophysical properties and binding energetics of over 800 ligand–target complexes curated from BindingDB (http://www.bindingdb.org/bind/index.jsp, accessed on 24 April 2022) [9,10], Scorpio (http://scorpio.biophysics.ismb.lon.ac.uk/scorpio.html, accessed on 24 April 2022) [11], and PDBcal [12] tabulated within a single compiled master file [13]. The utility of employing thermodynamics-based metrics benefits from the availability of such binding data, as one can evaluate general trends and/or specific correlations that may assist in drug discovery. This review concludes by acknowledging the power of energetics in elucidating fundamental aspects of biomolecular interactions and the need for accelerating development of high-throughput technologies that integrate biological, biophysical, computational, and structural approaches in a multidisciplinary strategy to advance the design and optimization of lead compounds as novel therapeutics.

1.2. Acknowledging Professor Breslauer's Contributions to Drug Discovery

This review celebrates Professor Breslauer's seventy-fifth birthday by reflecting on his pioneering contributions towards characterizing the thermodynamic properties of biological systems and their fundamental role in accelerating drug discovery efforts. During the past four decades, Professor Breslauer and his colleagues have focused their attention on timely topics of biological relevance by conducting biophysical research studies that assisted parallel biomedical efforts to identify effective treatment strategies for a number of devastating diseases. Such efforts have culminated in characterizing the energetics of drug–DNA interactions, particularly those directed towards the treatment of cancer and related ailments [14–19], revealing insightful trends in terms of thermodynamic binding signatures [18–22] and yielding invaluable information content for advancing drug discovery strategies [23–25]. In a series of seminal studies, Professor Breslauer established a database of nearest neighbor thermodynamics for canonical DNA duplexes [26], advanced the concept of enthalpy-entropy compensation in ligand–target interactions [15] which represents a formidable challenge in drug design efforts, explored the relevance of heat capacity changes on DNA duplex energetics [27,28] and macromolecular processes [29], and characterized the thermodynamics of template-directed polymerase synthesis [30]. In collaboration with Professors Grollman and Johnson at Stony Brook University, the Breslauer research group has investigated oxidative DNA damage, which causes a myriad of diseases including cancer and neurodegeneration. These efforts have complemented biophysical studies of canonical DNA with detailed thermodynamic analysis of lesion impacts on duplex energetics [31–42] as well as lesion recognition and repair by DNA glycosylases [23,29,43,44]. While establishing the foundation required for understanding the energetic basis of complex biological systems, the Breslauer research group has focused on gene transcription regulation, repair, and replication [29,30,39,45,46]. Professor Breslauer's impressive academic career spanning five decades is chronicled in a recent retrospective [23] as his research group continues to pursue effective therapeutic interventions for combating and treating infectious diseases such as the SAR-CoV-2 pandemic [47]. In summary, the indelible impact of Professor Breslauer's pioneering contributions to the field of biothermodynamics and their consequential impact on advancing drug discovery strategies is gratefully acknowledged and recognized by a legion of peers.

2. Drug Discovery Strategies

The design, development, and optimization of prospective compounds as effective therapeutic agents require broad-based drug discovery strategies that incorporate high throughput screenings (HTS) in conjunction with multidisciplinary characterization protocols. HTS aims at generating selected hits from millions of compounds that are further optimized into chemical leads. The identification of an ideal chemical lead and its overall quality is dictated by a host of properties that extend well beyond inhibitory activities and binding affinities. These include specific pharmacokinetic properties, such as absorption, distribution, metabolism, excretion, and toxicity that are collectively termed ADMET. While deemed essential to ascertain a sustained therapeutic level upon oral administration, ADMET properties are generally more difficult to optimize than primary biological activity [48]. Considering their fundamental role in evaluating drug efficacy, complementary approaches to HTS are instrumental for monitoring the optimal predicted performance of a compound following in vivo administration. Drug discovery strategies are continually expanding capabilities in terms of both restructuring facilities to experimentally access and accelerate the production of hits [48] while intensifying the use of complementary computational approaches to expedite compound screening. During the screening process, several rules are imposed on the basis of selected criteria and/or metrics to categorize the drug-likeness of a compound. Such criteria rely on a host of properties that assist in evaluating a compound based on its desired qualities as a pharmacophore in addition to specific biological and ADMET characteristics. The criteria employed for screening potential therapeutic agents are presented and reviewed within the context of information gleaned from several publically accessible databases.

2.1. Fragment-Based Drug Discovery

The strategy of fragment-based drug discovery (FBDD) originated as a natural progression from a concept initially proposed by Jencks to characterize the binding energetics of protein–ligand complexes [49]. Invoking this empirical approach, the binding free energy of a ligand might be visualized as the sum of its component fragment interactions with specific protein sub-sites. A subsequent study employed these principles to examine the differential impact of tethered small molecules *versus* their isolated counterparts on protein-inhibitor binding energetics [50]. Significantly, the tethered ligand exhibited a dramatic enhancement in binding affinity approaching three orders of magnitude (i.e., nanomolar vs. micromolar). This proof-of-principle demonstrated the potential utility of incorporating additivity in target-directed research and effectively served as the predecessor of FBDD. A follow-up study applied this methodology towards the screening of prospective papillomavirus E2 protein inhibitors and identified ligands that bind weakly to adjacent sites with the prospect of tethering these molecules within a higher affinity compound [51]. Given the collective success of such protocols in contrast with the limitations of prior approaches, FBDD has advanced into a multidisciplinary endeavor (as reviewed in [52]) and is considered one of the preferred strategies in current drug discovery programs [53]. FBDD essentially consists of screening and identifying small molecules that bind proximal subsites in the target protein, which are subsequently optimized and the resultant fragments linked to produce higher affinity ligands with the desired biological activity/potency (as reviewed in [54]). A schematic overview of FBDD presented in Figure 1 illustrates how low affinity molecules that interact with subpockets of the binding site are subjected to biophysical screening and the optimized fragments assembled into high affinity ligands [48].

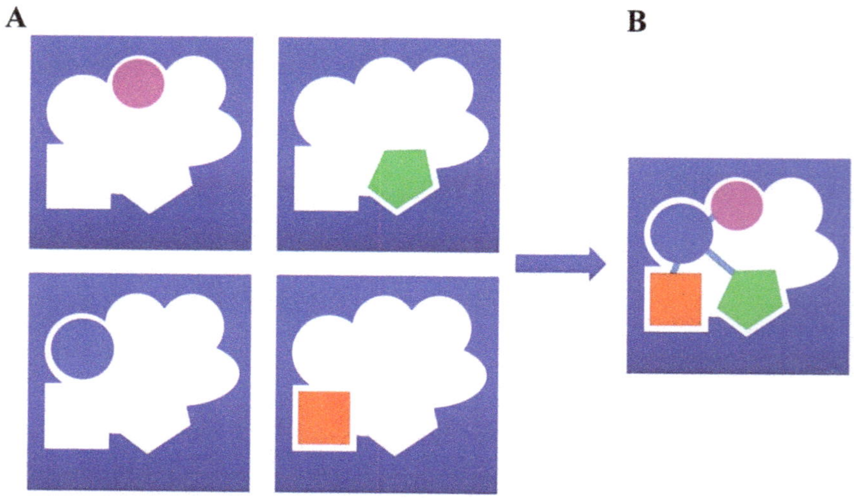

Figure 1. Representative linking approach commonly employed in FBDD to generate an optimized lead compound. (**A**) Several small fragments bind to the receptor within specific subpockets (as represented by blue and magenta spheres, a green pentagon, and red square). (**B**) The fragments are linked according to their spatial distribution within the binding site, which allows the resultant lead molecule to interact with a larger surface and therefore exhibit a significantly higher affinity relative to its constituent fragments [48].

FBDD has been introduced as a powerful alternative and complementary technique to traditional HTS strategies for the identification of lead molecules. Several new compounds derived via FBDD have reached the clinical development stage with progressively more attention devoted towards applying this method in drug discovery protocols [55]. Fragment libraries of relatively small molecular weight compounds (~100–300 Da) are screened via a combination of structural, functional, and biophysical techniques. The initial screening process in FBDD employs X-ray crystallography and/or nuclear magnetic resonance (NMR) spectrometry to identify prospective target-fragment complexes. Subsequent characterization via biophysical and functional approaches yields the requisite fragment binding affinities and inhibition activities. Optimization of the initial hits generates a higher molecular weight ligand comprised of multiple fragments which exhibits both enhanced affinity and potency. In most successful examples of fragment optimization, discrete contacts are responsible for securing the anchoring fragments and these binding modes are preserved throughout the optimization process. Specific details regarding the various steps encompassing *hit-to-lead-to-drug*, the strategies utilized in FBDD, and optimization of the final molecules are described and reviewed elsewhere [56]. Some of the most successful applications of fragment-based methods have involved deconstructing known leads and reassembling these to generate a new chemical series with improved biological properties [57].

Despite the overall success achieved in identifying prospective lead compounds, FBDD screening methodologies often encounter challenges when the binding mode of assembled fragments within adjacent pockets are disrupted as a consequence of geometric constraints imposed by the linker [58]. This represents a potential shortcoming of FBDD as the anticipated gain in potency may be impaired by the difficulty of linking fragments to form an active compound [59]. Moreover, potential undesired physicochemical properties may compromise a drug's performance in vivo, despite its ability to bind the target and elicit a desired effect in vitro. An experimental strategy that incorporates binding thermodynamics and prediction metrics during the early stages of FBDD may therefore prove extremely useful in terms of ensuring a successful outcome. Specifically, a rigorous thermodynamic analysis may furnish invaluable information regarding biophysical properties that are often overlooked when a molecule is evaluated solely in terms of binding affinities and inhibition constants. The utility of thermodynamics in drug design, development, and optimization is the focus of this review and discussed in subsequent sections.

2.2. Computational and Experimental Methods to Assess Drug Potential: Criteria and Metrics

A multitude of parameters for evaluating the quality of a molecule in terms of specific "drug-likeness" attributes have been adopted over the years and are generally considered as useful metrics to assess whether a compound fulfills basic physicochemical properties that serve as a prerequisite for subsequent development into an effective therapeutic agent. Several of the proposed metrics are better suited for a particular class of compounds, route of administration, target location, or nature of the screening protocols (as reviewed in [60]). These empirical criteria continue to evolve on the basis of advancements achieved in the design, development, and optimization of successful lead compounds. In the following sections, we discuss specific criteria and metrics that have been proposed and utilized as cut-off filters in drug discovery strategies.

2.2.1. Criteria for Selecting Prospective Lead Compounds

Among commonly adopted criteria to assess the overall fitness of a molecule as a prospective drug, Lipinski's *rule of 5* (Ro5) [61] has been widely employed to predict permeability and solubility characteristics, particularly in the case of compounds intended for *oral* administration. Computational data have assumed a significant role in providing general criteria and guidelines to determine whether a compound satisfies Ro5 stringency. In specific terms, the limits for reasonable absorption and/or permeation require a molecular weight (MW) no greater than 500, a maximum of 5 H-bond donors (HBD) and

10 H-bond acceptors (HBA), a calculated LogP (CLogP) that does not exceed 5, and the absence of toxic groups. Considered a useful parameter to analyze small molecules, LogP corresponds to the logarithm of its partition coefficient between octanol and water (i.e., Log $[C_o]/[C_w]$), and therefore represents a well-known measure of molecular hydrophobicity and/or lipophilicity [61,62]. LogP is determined to assess biological properties that are relevant to drug performance, including lipid solubility, tissue distribution, receptor binding, cellular uptake, metabolism, and bioavailability. LogP is commonly estimated via parameterization methods and the calculated values (i.e., CLogP or AlogP) are routinely used in the prediction of LogP with relatively comparable performance [63].

Although predictions are useful for evaluating related analog series, experimentally derived binding energetics and solubility measurements are deemed essential to unequivocally ascertain whether a molecule will perform as planned. Ro5 is insightful and serves as a useful guideline or extra filter in screening prospective compounds. Nevertheless, the fluidity of emerging data requires constant vigilance and updates to monitor novel classes of compounds and targets identified in the drug discovery process. The ultimate goal of achieving high potency and selectivity via application of specific metrics is paralleled by the expectation that an ideal compound operates within a reasonable therapeutic window. In this respect, basic drug-like properties must be viewed within the prism of achieving efficacy while precluding toxicity. The latter is accomplished by incorporating easily conjugated (e.g., hydroxyl, amino, carbonyl), metabolically cleaved (e.g., ester, amide), oxidizable, and excretable (e.g., methyl) groups/metabolites, all of which ensure maximum efficacy and minimal side effects (as reviewed in [64]).

The need for standardized metrics to predict/evaluate compound potency is justified due to the realization that screening processes are generally biased towards selecting higher MW compounds [65] given their propensity to exhibit enhanced binding affinities. An observable trend is that lead developments have progressively resulted in the selection of compounds with increased molecular weight. In fact, this screening bias is evident as ligand potency often tracks with molecular size as noted by inspection of pK_d distribution tendencies as a function of heavy atom (HA) number in Figure 2. Such qualitative trends are quite visible in larger datasets and particularly striking when higher potency compounds are selected for analysis [66]. On the basis of Ro5 guidelines and in view of anticipated problems associated with drug solubility and biodistribution, lower molecular weight compounds are typically the most effective therapeutic agents with an optimal MW below 400 Da.

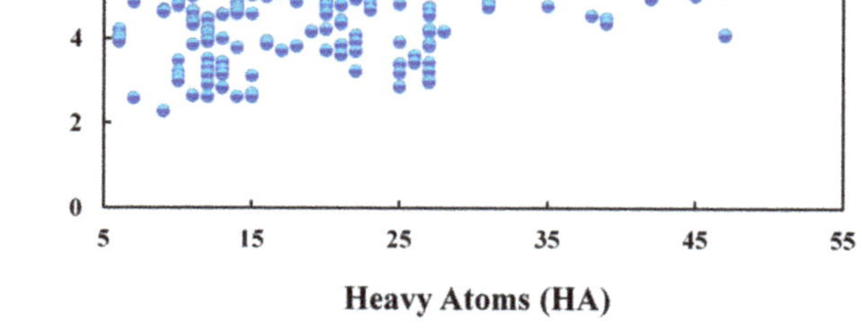

Figure 2. Correlation of ligand potency and molecular size deduced via pK_d distribution as a function of ligand heavy atom (HA) number [13].

A significant observation is that drug candidates and novel molecular entities (NME) tend to exhibit increased hydrophobicity, generating concerns regarding their overall biophysical properties and effectiveness [61]. Assessment of available data in terms of its conformance with Ro5 criteria can be gleaned via inspection of LogP values as a function of molecular weight. Viewing the graphical depiction in Figure 3, most of the approved oral drugs are clustered within an area represented by the black box and thereby fulfill this general rule (i.e., LogP $\leq$ 5). As with all metrics employed in the evaluation of an ideal drug, there are successful outcomes that do not strictly comply with stringent rules [67]. In such cases, the overall distribution extends beyond Ro5 [68] as evidenced by compounds residing in the dashed blue box (Figure 3) which still appears densely populated. In the specific example of HIV-protease inhibitors represented as light green diamonds (Figure 3), the overall molecular size is in fact greater than 500 Da and LogP values are border line Ro5 limits.

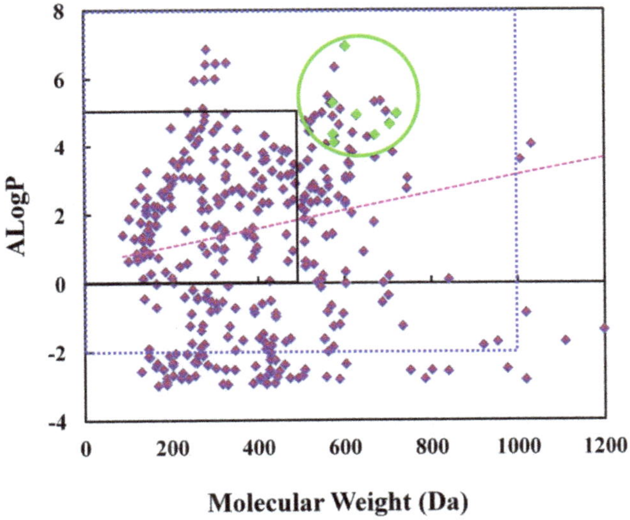

Figure 3. LogP (calculated as AlogP) values [13] plotted as a function of compound molecular weight (MW). The black box delineates an area within which the majority of drugs approved for oral administration are located. The blue dashed-box extends this area to higher molecular weight compounds distributed over an expanded range of LogP which contains exceptions to the rules (Log P $\leq$ 5 and/or MW < 500) yet results in approved drugs. The encircled green diamonds represent some of the compounds employed successfully in anti-HIV therapeutics.

Another critical issue concerning the application of Ro5 in drug design pertains to HB donors (HBD) and acceptors (HBA), the former of which has been scrutinized extensively in lead optimizations. Inspection of available data suggests that most compounds appear to abide within the limit of HBD < 5 as illustrated in Figure 4. Despite this evidence, there has been a recent upsurge in the number of drugs approved for therapeutic use that tend to violate HBD and pertinent Ro5 criteria [69–72]. In view of these developments, suggestions to expand such limits have been considered and a modified version designated as *beyond Ro5* (bRo5) proposed [69]. In retrospect, compounds harboring bRo5 chemical space already existed amongst natural products (NP), and the advent of semi-synthetic as well as total synthesis protocols has expedited the availability of such molecules [73]. Inspired by nature, an increasing number of compounds isolated from plant or marine sources have been investigated and their potential applications explored. These include biologically active defense molecules and antimicrobial peptides that have been studied and elaborated as future templates to combat the alarming rise in bacterial resistance [74,75].

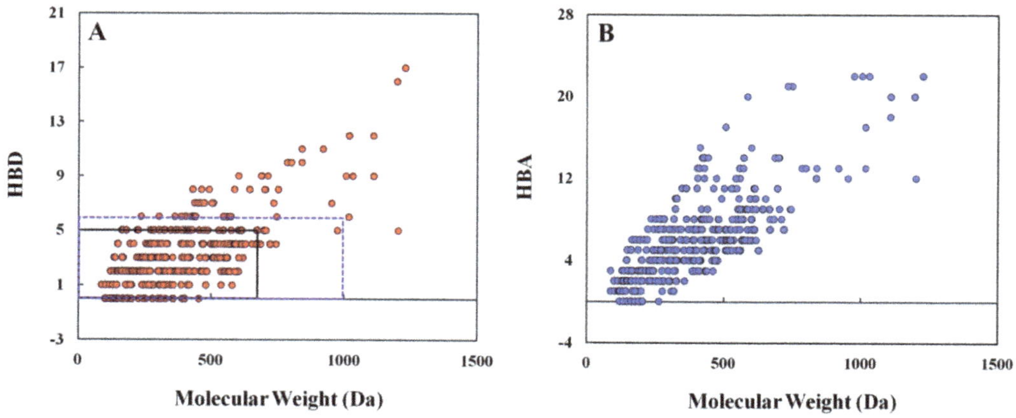

Figure 4. Molecular size distribution of compounds deposited in public databases [13] according to the number of (**A**) hydrogen-bond donors (HBD) and (**B**) hydrogen-bond acceptors (HBA).

Given the recent emergence of SARS-CoV-2, chemists have resorted to evaluating a myriad of novel, repurposed, synthetic, and/or NPs as prospective antiviral molecules [76]. A large number of compounds are now viewed beneath the umbrella of expanded criteria and include targets with "difficult" binding sites that are commonly designated as difficult-to-drug targets. The latter comprise large, highly lipophilic or polar, and/or flexible binding sites [77]. This category includes protein–protein interaction (PPI) inhibitors, a rapidly expanding class of compounds that are designed to disrupt the large protein-protein interface and as a consequence do not fulfill basic Ro5 criteria. Moreover, recent advances in drug discovery directed towards targets that comprise intrinsically disordered proteins (IDP), which are implicated in a host of disorders including cancer [2] and neurodegenerative diseases [78], have resulted in candidates evading traditional metrics. Unusual target sites cannot accommodate conventional therapeutics, and thus require higher molecular weight, lipophilic, or polar compounds that are more amenable to the bRo5 concept [68] or evade any set of criteria/rules that define a drug-like molecule [67].

The drug-like properties of therapeutic compounds approved for oral administration have been evaluated in terms of their molecular and physical properties [79]. The overall findings suggest that reduced flexibility, expressed in terms of rotatable bonds (i.e., ROTB $\leq$ 10), low polar surface area (i.e., PSA $\leq$ 140 Å^2), or total hydrogen bond count (i.e., sum of HBD and HBA $\leq$ 12) are reasonable predictors of oral bioavailability, irrespective of molecular weight. These observations underscore the need to critically assess and balance pros versus cons when evaluating the overall degree of success exhibited by a drug, as concurrent factors such as target affinity, selectivity, permeation, and clearance may benefit from distinct or even opposing molecular properties. In this respect, the unanticipated positive impact of molecular rigidity (i.e., reduced ROTBs) has to be considered in terms of a shape that retains the ability to interact with a target, thereby enhancing activity and selectivity to bind carrier/clearance proteins and optimally permeate membranes.

The optimization process in FBDD arises from a small fragment (e.g., MW~150 Da; millimolar binding affinity) in which most atoms are involved in the desired target–ligand interaction. Therefore, the size, complexity, and physical properties of the molecule are more easily controlled relative to a higher-affinity compound containing groups that are not essential for the desired binding (e.g., MW~400 Da; K_d~nM). Less potent fragments may undergo high-quality interactions [48] that form the basis for optimization into larger drugs exhibiting enhanced potency [80,81] (refer to Figure 1). Advancement of drug optimization protocols led to the realization that Ro5 and/or bRo5 did not adequately address the needs of FBDD, and therefore, other metrics have been introduced to fill this void. A newly proposed *rule of 3* (Ro3) has been implemented to assist decision making processes involving FBDD [82]. Analogous to Ro5 and based on prior successes, ideal fragments should adhere to the following Ro3: MW < 300 Da; ClogP < 3; HBD < 3; HDA < 3; and, ROTB < 3. FBDD employs fragment-sized compounds that usually comply with the Ro3 for initial screening against a biomolecular target. In this respect, FBDD has better chances of hit identification due to its more efficient sampling of the chemical space. An added advantage is that smaller sized fragment hits are more amenable to structural optimization [82]. Challenges associated with this strategy include the sensitivity of detection methods as discussed in a subsequent section. A significant number of criteria and rules have been proposed as metrics for evaluating prospective compounds in terms of achieving ideal drug-like properties. Representative examples are summarized in Scheme 1 and a comprehensive list is described elsewhere [83].

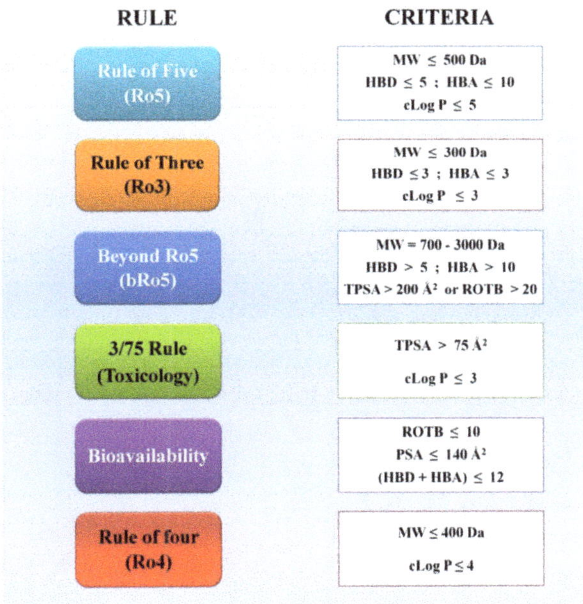

Scheme 1. Several commonly applied criteria/rules to identify and characterize the drug-like properties of therapeutic compounds. A comprehensive and extensive list appears in [83] while specific details of each rule are described accordingly: Ro5 [61]; Ro3 [84]; bRo5 [69]; 3/75 [85]; Bioavailability [79]; Ro4 [85].

2.2.2. Ligand Efficiency (LE)

The concept of ligand efficiency (LE) has been proposed [86] in accordance with the observations of Kuntz and colleagues [87] to counter the common misperception that larger molecules generally represent superior compounds. Ligand efficiency is a parameter introduced to avoid the biases of molecular weight differences and normalize the binding affinity on a MW basis [65]. LE is considered a useful indicator of compound quality [83] since the measured binding affinity is normalized to the number of heavy atoms (HA). The popularity of this metric is directly associated with the fact that FBDD has expanded steadily, thereby necessitating emphasis on the optimization of low molecular weight compounds. The binding energy per heavy atom or binding 'efficiency' of a ligand has been deemed a useful parameter in the selection and development of a lead compound or fragment irrespective of its molecular weight. The binding efficiency index (BEI) is another metric based on a molecular weight scale [88] that provides a facile and effective ranking en route to compound optimization. The overall utility of these parameters in the drug discovery process are visualized by plotting LE or BEI derived from available binding databases as a function of molecular size. A graphical presentation of the resultant data in Figure 5 reveals hyperbolic behavior with a tendency to reach a plateau at ~25 heavy atoms and MW~400 [66]. Comparing lead compounds on the basis of LE/BEI might prove useful to assess the potential for further optimization of particular 'hits' and chemical scaffolds, providing LE is not employed as a *definitive cut-off* in the filtering process. The latter incurs a risk of inadvertently overlooking and potentially discarding useful leads in the drug development process.

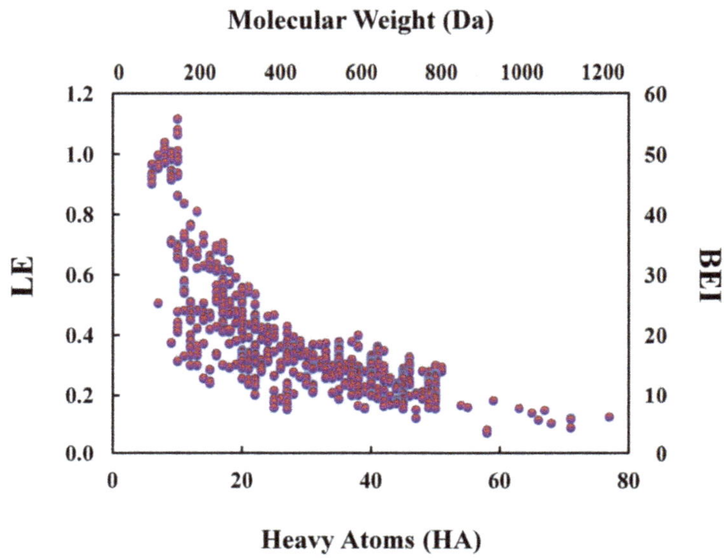

Figure 5. Graphical representation of calculated ligand efficiency (LE) and binding efficiency index (BEI) as a function of Molecular Weight (Da) and Heavy Atoms (HA) [66] curated from public databases [13] and converted as follows: LE = $\Delta G/HA$; BEI = IC_{50}/MW (Da) [83,88].

A critical and fundamental observation is that these metrics rely on ΔG, K_d, K_i, or IC_{50} and thereby render the resultant parameters dependent on a concentration used to define the standard state. Accordingly, a number of investigators have questioned their validity [89–91], and several critics have concluded that LE should not be used in absolute terms for individual compounds nor considered as a cut-off criterion [90]. In the midst of this ongoing debate [92] and irrespective of warnings against overusing such metrics as an absolute filter in drug development [91,93], several investigators have adopted a conciliatory approach by reiterating the utility of this metric for specific applications [83,94–96]. Alternate metrics such as group efficiency (GE) utilizes differential values (i.e., $\Delta\Delta G/\Delta HA$ or $\Delta pK_d/\Delta HA$) to assess the impact of a fragment on compound affinity [93] as described below.

2.2.3. Group Efficiency (GE)

Group efficiency (GE) is a measure of the relative contribution that a specific group confers to ligand potency and is evaluated systematically by differential ($\Delta\Delta G$) analysis [97,98] involving the comparison of compounds harboring each successive substituent. The latter is related to Free-Wilson relationships [99] that mathematically assess the contribution of group substitutions via QSAR. Considered a more sensitive metric, GE bears the anchoring principle in which $\Delta\Delta G$ refers to the fragment atoms (i.e., ΔHA) that form each succeeding derivative via the following relation: GE = $\Delta\Delta G/\Delta HA$ or $\Delta pK_d/\Delta HA$. The interactions between fragment and target should exhibit high ligand efficiency and favorable binding thermodynamics to offset the rotational and translational freedom lost during complex association. Accordingly, fragment binding is generally enthalpy-driven to overcome the loss of rigid-body entropy [57].

The evaluation of fragment binding efficiency to overall molecular performance is deduced via the relation: $\Delta G = \Delta G_{int} + \Delta G_{rigid}$, where ΔG_{int} is the intrinsic binding free energy and ΔG_{rigid} is the free energy associated with loss of rigid body entropy upon interaction with the target [97]. The latter has been estimated as 4.2 kcal·mol^{-1} [81,97] and is incorporated in the initial scaffold fragment binding free energy as illustrated in Figure 6. This particular GE analysis systematically evaluates and improves the potency of pantothenate synthetase inhibitors against Mycobacterium tuberculosis [100] as drug candidates in tuberculosis therapy. Inspection of the free energy landscape reveals that compound 5 is dissected into its component fragments and the binding contributions ($\Delta\Delta G$) calculated from these building blocks. The resultant $\Delta\Delta G$ (−8.2 kcal·mol^{-1}) for the indole group (i.e., compound 1) includes an unfavorable entropy loss upon binding [i.e., $\Delta G = \Delta G_{int}$ (−8.2 kcal·mol^{-1}) − ΔG_{rigid} (+ 4.2 kcal·mol^{-1}) = −4.0 kcal·mol^{-1}]. The addition of successive sulfoamyl, pyridine-methyl, and acetic acid fragments significantly enhances ligand binding free energy. Fragment linking approaches utilizing GE as a metric therefore facilitate the optimization of these compounds, leading to improved inhibitor potency against pantothenate synthetase.

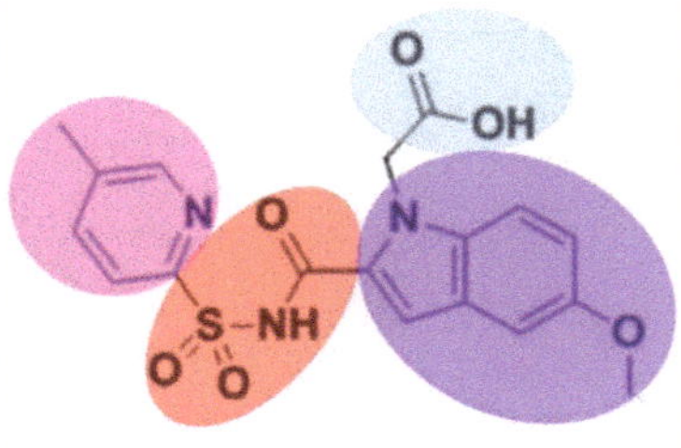

Compound 5

Compound	Indole	Sulfoamyl	Pyr-Met	Acetic Acid	ΔG
1	X				-4.0
3	X	X			-5.0
4	X	X	X		-6.2
5	X	X	X	X	-7.9
ΔΔG	-8.2 *	-1.0	-1.2	-1.7	
HA	11	6	7	4	
GE	0.75	0.17	0.17	0.43	

Figure 6. Utility of employing group efficiency (GE) as a metric in lead optimizations [100]. The binding affinity acquired via ITC measurements yields the corresponding free energy (ΔG) for each derivative. The resultant $\Delta\Delta G$ and heavy atom (HA) numbers facilitate calculation of group efficiency (GE) via the relation $GE = \Delta\Delta G/HA$. Compound **5** is deconstructed into its constituent fragments (i.e., **1** = indol (purple); **3** = **1** + sulfamoyl (red); **4** = **3** + pyridine-methyl (magenta); **5** = **4** + Acetic acid moiety (blue). * The $\Delta\Delta G$ of −8.2 kcal·mol^{-1} includes an initial ΔG_{rigid} (i.e., 4.2 kcal·mol^{-1}) that is added to the measured ΔG of −4.0 kcal·mol^{-1}.

2.2.4. Lipohilic Ligand Efficiency (LLE)

Based on the original hypothesis advanced nearly 35 years ago [101] that advocated for enhancing drug hydrophilicity while preserving overall efficacy, the LLE concept has withstood the test of time as a molecular descriptor in drug design and development [86]. Ligand lipophilic efficiency (LLE) or LipE has received significant attention recently, as this metric combines potency and lipophilicity in such a manner that drug efficiency is normalized to its degree of lipophilicity (as reviewed in [102]). LipE is considered a metric that predicts the quality of a compound and encapsulates both lipophilicity and potency. This parameter is defined according to the relation: LipE (LLE) = pKd (pIC50) − LogP (aLogP, cLogP), where the terms in parentheses are interchangeable with those in the equation. LipE therefore provides an estimate of ligand affinity that is primarily derived from polar interactions as opposed to its hydrophobicity characteristics. In contrast with LE, which has been the subject of significant scrutiny for a number of aforementioned reasons [91,93], LipE is broadly accepted and considered a reliable metric in the overall drug screening and optimization process [102].

The utility of estimating LLE (LipE) may be illustrated by the hypothetical representation in Figure 7 that depicts a compound undergoing optimizations to acquire potencies greater than 8 (e.g., 10 nM affinity) yet maintaining low LogP values (e.g., <3) with a consequent enhancement in overall lipophilic efficiency [103]. While the LLE metric represents an excellent criterion to guide compound optimizations, there are competing factors, such as membrane permeability, that must be considered when assessing overall drug quality. This realization infers that there is a limit to which optimizations should focus solely on potency and low LogP, as there is a delicate balance between lipophilicity and solubility to achieve both potency and suitable ADMET properties [104].

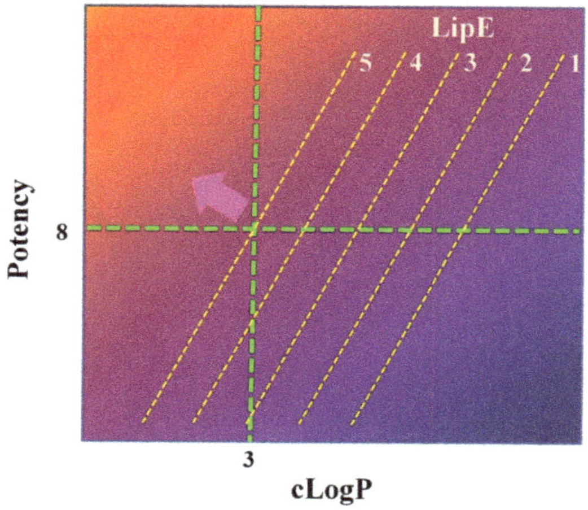

Figure 7. Hypothetical representation of a compound exhibiting a potency (pK_d) of 8 and octanol-water partition coefficient (cLogP) of 3 resulting in a LipE of 5 (i.e., LipE = pK_d − cLogP). The horizontal green dotted line projects outcomes where LogP increases without changes in affinity yielding a significant reduction in lipophilic efficiency. Alternatively, the vertical green dotted line demonstrates how potency increases at constant cLogP resulting in enhanced lipophilic efficiency. A magenta arrow represents the direction for improvement of compound properties.

2.2.5. Enthalpic Efficiency (EE)

Thermodynamics has been employed extensively to assist decision making processes in drug design and development. Considering the relevance of employing thermodynamic signatures in lead optimizations [6,105–112], the metric of *enthalpic efficiency* (EE) has been proposed as a primary decision criterion in drug discovery [113]. This metric is calculated using the measured binding enthalpy (i.e., ΔH) that is normalized to the number of non-hydrogen (N_{NH}) or heavy atoms (HA) (i.e., EE = $\Delta H/N_{NH}$) in the molecule. The foundation of this approach is based on the seminal findings published by Freire and colleagues [114], who demonstrated that drug performance may be optimized more effectively via detailed knowledge of the binding thermodynamics. Based on these findings, the energetic signatures of ligand–target interactions represent a core metric for evaluating lead compound quality and accelerating drug design efforts [110,115,116].

Several of the metrics described in previous sections incorporate affinity (K_a) and/or inhibition (K_i, IC_{50}) data derived from a host of methodologies that are subsequently converted to the binding free energy (ΔG). In contrast, the EE metric requires rigorous determination of complete thermodynamic binding parameters [i.e., free energy (ΔG), enthalpy (ΔH), and entropy (TΔS)] via model-independent analytical techniques, such as isothermal titration calorimetry (ITC). An alternative approach is to employ model-dependent van't Hoff optical measurements or robust computational methods [117] as substitutes for the more precise calorimetric instrumentation.

The proliferation of thermodynamic binding parameters has facilitated curation of energetic databases that are useful for observing specific correlations and/or trends that may be evaluated to identify commonalities and develop further predictive capabilities. As an example, inspection of published databases reveals a qualitative trend in which the binding enthalpy (ΔH) decreases with MW as illustrated in Figure 8A. Conversely, the entropic (−TΔS) term depicted in Figure 8B exhibits a favorable correlation with increasing size, an observation that is consistent with the burial of larger surface areas upon binding. As a direct consequence, binding-induced apolar surface area desolvation is characterized by a slight increase of ΔG as observed in Figure 8C. While the binding enthalpies tend to gradually decrease as a function of MW, one observes a sharp molecular weight-dependent reduction when the ΔH values are converted to EE, as visualized in Figure 9.

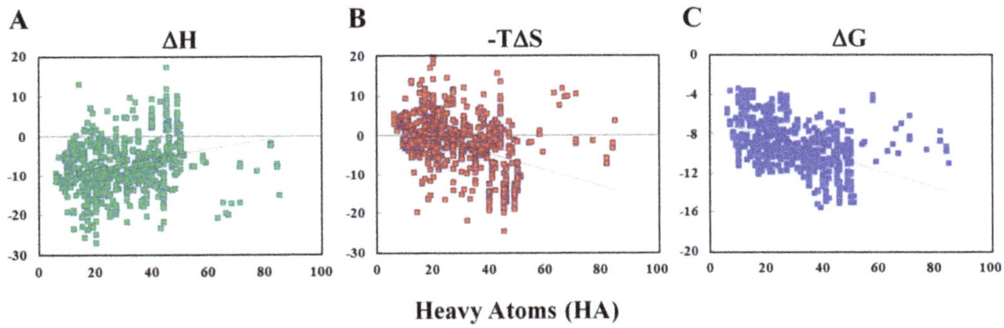

Figure 8. Correlation of ligand-target thermodynamic binding parameters in Panels (**A**) Enthalpy, (**B**) Entropy, and (**C**) Free Energy as a function of Heavy Atom (HA) number curated from public databases [13]. Inspection of the respective plots reveals qualitative trends including a slight reduction in favorable enthalpy and enhancement in both favorable entropy and free energy with increasing number of heavy atoms.

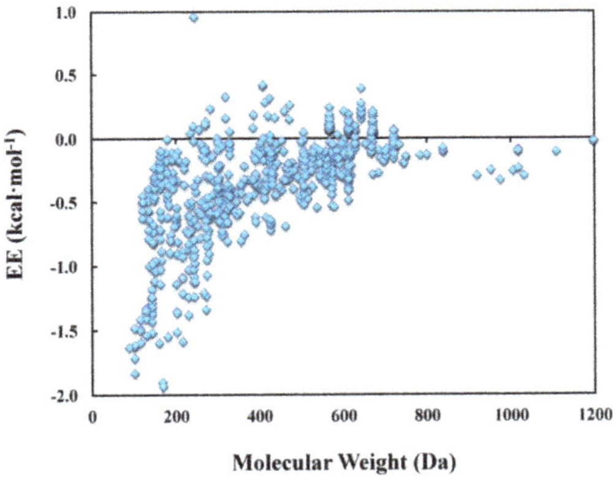

Figure 9. Enthalpic efficiency (EE) calculated using the measured binding enthalpy (ΔH) normalized to number of non-hydrogen (N_{NH}) or heavy atoms (HA) (i.e., EE = $\Delta H/N_{NH}$). This approach is based on the finding that drug performance may be optimized more effectively via knowledge of binding thermodynamics [114] in which the energetic signatures of ligand-target interactions assist drug design efforts.

Subsequent efforts to guarantee an unbiased hit selection and further hit-to-lead progression have culminated in the development of a *size independent enthalpic efficiency* (SIHE) metric [13], which normalizes EE by the number of heavy atoms as illustrated in Figure 10. SIHE is defined as $pK_H/40{\cdot}HA^{0.3}$, where $pK_H = \Delta H/(2.303{\cdot}RT)$ and HA is the number of heavy atoms [13]. This metric has been proposed on the basis of its relevance to monitor enthalpic gains/losses during lead optimization while simultaneously normalizing the data to molecular size. SIHE represents an informative metric for analyzing MW-independent trends in binding enthalpies following normalization for molecular size.

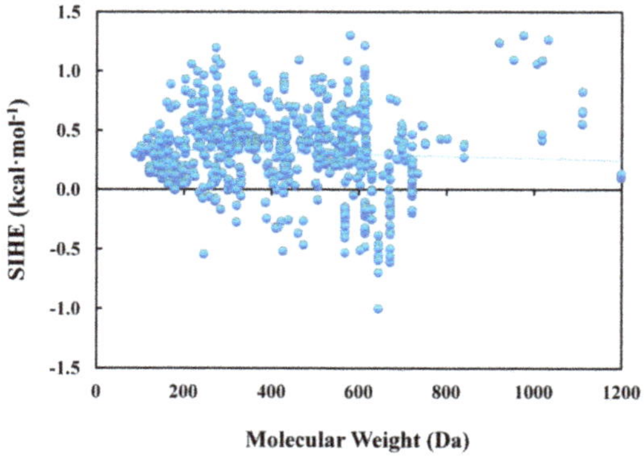

Figure 10. Size-independent enthalpic efficiency (SIHE) defined as $pK_H/40 \cdot HA^{0.3}$ where $pK_H = \Delta H/(2.303 \cdot RT)$ and HA is the number of heavy atoms [13].

Since binding affinities and inhibition constants may miss important physical properties of a lead compound, complementary approaches that exploit thermodynamic binding properties provide an additional level of critical information which has proven enlightening in the drug discovery process. The compilation of thermodynamic databases has facilitated evaluation of binding energetics as a core metric in lead compound development, resulting in a number of interesting qualitative observations (as reviewed in [13]), namely: (a) drug binding affinities tend to increase with MW; and, (b) enhanced binding free energies are generally accompanied by a reduction in enthalpy and increase in entropy. These findings suggest that EE has a fundamental role in FBDD optimization approaches as the initial screening process should focus on identifying enthalpy-dominated binding fragments [53,118,119] to overcome the natural tendency for a size-dependent reduction in enthalpic efficiency. The early stages of lead development and optimization therefore benefit from a rigorous assessment and characterization of thermodynamic binding signatures.

2.2.6. Exploiting LipE and EE as Core Metrics to Achieve Optimal Success

Considering the strengths and weaknesses of specific metrics discussed in the preceding sections, the number of available analytical tools to assist decision making in the drug discovery process allows a robust lead evaluation. The combined efforts of structural, computational, and biophysical methods, in conjunction with desired biological properties, provide assurance that drug discovery protocols achieve the ultimate goal of hitting a specified target. Following two decades of establishing rules and metrics as guidelines in the drug discovery arena, several criteria stand the test of time for evaluating lead compound quality during the early stages of development and subsequent optimization into a successful therapeutic agent. Although the use of a metric does not necessarily guarantee or increase the probability of success, a wealth of evidence suggests that LipE might be viewed as a core metric that best describes compound quality (as reviewed in [67]). Remarkably, considering its emphasis on polarity and lipophilicity control, this metric serves as a gauge of ADMET properties and therefore tracks with the EE of a ligand [120]. An example of their collective utility is depicted in Figure 11 for HIV protease inhibitors whereby one observes a reasonable correlation between LipE and EE. Significantly, both of these metrics track with drug development, thereby corroborating the notion that LipE and EE reflect ligand properties to a similar degree [67].

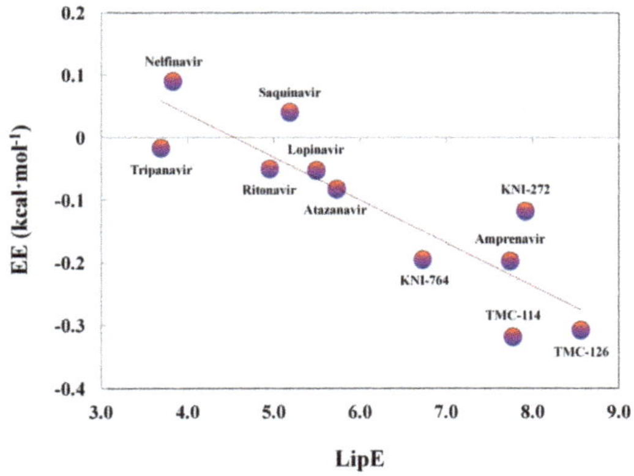

Figure 11. Correlation of enthalpic (EE) and lipophilic (LipE) efficiencies for HIV protease inhibitors retrieved from the PDcal database. The overall distribution reveals that ligands characterized by higher LipE generally exhibit a more favorable enthalpic contribution upon association with the target.

Enthalpic optimization has proven invaluable at early stages of drug discovery, given its unique power and versatility in discriminating compounds on the basis of selectivity and improved ADMET properties. Collectively, the LipE and EE metrics exhibit reasonable correlation and generally represent efficient criteria for discriminating between selective versus promiscuous ligands. The application of these core metrics at the outset of lead optimization may therefore assist in achieving successful outcomes. A detailed review [120] using available binding databases critically evaluates correlations between LipE and other metric indices, including enthalpic efficiency of ligand–target systems, and documents the advantages and caveats of these approaches. Scheme 2 summarizes the core metrics that are commonly employed in drug design and lead optimization protocols. The reader is referred to a detailed review that contains a comprehensive and extensive listing of drug discovery metrics [83].

METRIC	FORMULATION
LE	$\Delta G/HA$
BEI	IC_{50}/MW (Da)
GE	$\Delta\Delta G/\Delta HA$ or $\Delta pK_d/\Delta HA$
ΔG	$\Delta G_{int} + \Delta G_{rigid}$
ΔΔG	$\Delta G_{frag\ added} - \Delta G_{prior\ to\ addition}$
SIHE	$(pK_H/40) \times HA^{0.3}$ where $pK_H = \Delta H/(2.303\ RT)$
SILE	$(pK_i$ or pK_d or $pIC_{50})/HA^{0.3}$
EE	$\Delta H/MW$ or $\Delta H/HA$
ADMET Efficiency Index	$(pK_i$ or pK_d or $pIC_{50} - \lvert LogP\rvert / PSA) \times 100$
LipE (LLE)	$(pK_i$ or pK_d or $pIC_{50}) - LogP$
LogP	$Log([C_o]/[C_w])$

Scheme 2. Several commonly employed drug design metrics applied to prioritize hit/lead compounds and their respective formulations. Specific details of each metric are reported elsewhere: LE [65]; BEI [88]; GE and related calculations [98]; SIHE [13]; SILE [121]: EE [113]; ADMET Efficiency Index [122]; LipE (or LLE) [123]; LogP [61].

2.2.7. Pre-Screening Library Fragments Conserves Resources/Time and Is "PAIN-Less"

Initial fragment libraries contain a myriad of potential therapeutic compounds that must be evaluated in accordance with specific criteria and metrics designed to identify prospective leads. The resultant screening process is resource/time intensive and often compromised by "bad actors" that may ultimately prolong drug discovery protocols [57]. These "misbehaving" molecules may include nonspecific ligands, reactive modifiers, chelators, and/or aggregating compounds that represent undesirable false positive hits [124]. An initial pre-screening inspection of prospective fragments can ascertain whether a specific library contains molecules that are formally included in the PAINS (Pan-Assay Interference Compounds) database [125]. There are a number of recommendations for pre-cleaning fragment libraries prior to initiating compound screenings [57] and several PAINS pre-filters have been proposed as described elsewhere [126]. PAINS filters may assist the drug screening process by eliminating promiscuous and reactive compounds that are "frequent hitters" and unlikely to represent useful leads. Orthogonal approaches are recommended as caution should be exercised whenever implementing cut-off filters to ensure that an otherwise potentially innovative drug candidate bearing possible PAINS alerts is inadvertently overlooked.

3. Insights Gleaned from Thermodynamic Binding Signatures

3.1. Enthalpy-Entropy Plots

Protein–ligand binding databases provide valuable insights regarding the thermodynamic properties of drug-target interactions. Critical analysis of available binding data furnishes information on the utility of obtaining thermodynamic parameters in drug discovery efforts. An instructive example of this strategy appears in Figure 12A which depicts a subset of binding data cast in the form of ΔH/TΔS linear plots divided into four quadrants. Based on this graphical representation, it is possible to visualize the density of interactions that are either entropy-driven (top left quadrant) or enthalpy-driven (bottom right quadrant), versus those that are both enthalpically and entropically favorable (bottom left quadrant). As proposed by Freire [110], an ideal scenario is to optimize enthalpic contributions to the interaction such that a compound achieves a greater selectivity towards its target, with the added advantage of preferred physical properties vis-à-vis solubility characteristics.

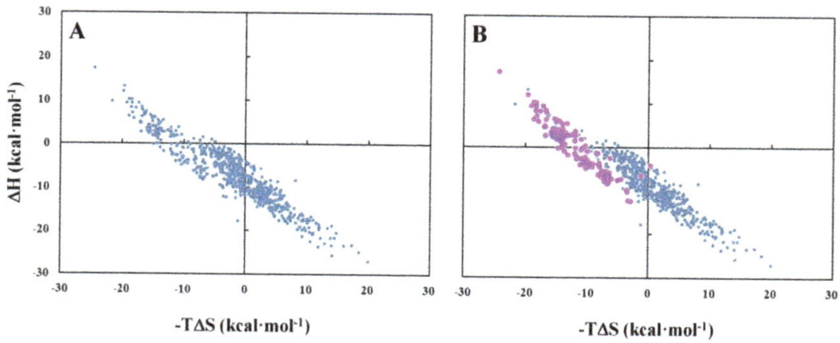

Figure 12. Distribution of compounds retrieved from databases of protein-ligand complexes according to their corresponding enthalpic (ΔH) and entropic (−TΔS) signatures. (**A**) Linear correlation of ΔH and −TΔS reflecting interactions that are enthalpically unfavorable and entropically favorable (top-left quadrant); enthalpically and entropically favorable (bottom-left quadrant); enthalpically favorable and entropically unfavorable (bottom-right quadrant). (**B**) Linear correlation of ΔH and −TΔS highlighting HIV-protease inhibitor interactions as magenta spheres spanning the top- and bottom-left quadrants. The strategy of enthalpic optimization is signified by a gradual shift in distribution of these compounds towards the lower quadrant.

An example of applying enthalpy-entropy plots to track drug development is illustrated in Figure 12B for a series of HIV protease inhibitors against wild type and mutant variants (highlighted in magenta). This representation of the data reveals that binding interactions are distributed within two major groups, namely those that are entropy-driven ($\Delta H > 0$; $-T\Delta S < 0$) and those that are both enthalpy- and entropy-driven ($\Delta H < 0$; $-T\Delta S < 0$). In a retrospective analysis spanning 12 years since the commercialization of these drugs, it is interesting to note that binding data corresponding to the introduction of newer compounds progressively populated the lower quadrant, which comprises enthalpically favorable interactions, as noted in Figure 12B. This finding is consistent with enthalpic optimization [24,109], a strategy that has gained momentum since the original proposition over two decades ago [109,114]. Freire and colleagues [115] have further illuminated the roadblocks to drug optimization (as reviewed in [24]) visualized through the lens of an optimization funnel as discussed below.

3.2. The Optimization Funnel

In the pursuit of novel therapeutic agents, a common goal is to identify high affinity potent compounds, and the design of optimization filters can represent an informative metric for monitoring the drug development process [115]. An observation gleaned over the years is that the optimization of a chemical scaffold results in high affinity/high potency compounds. Generally speaking, micromolar binding affinities (K_d) are enhanced three to six orders of magnitude (i.e., nano or picomolar) with a corresponding binding free energy increase of $\Delta\Delta G \sim 4$–7 kcal·mol^{-1} [115]. A practical demonstration of ligand optimization is easily visualized by inspection of the enthalpic (ΔH) contribution to net binding free energy (ΔG) as the affinity of a compound is optimized [24]. In the absence of optimization, low affinity compounds can exhibit a broad range of $\Delta H/T\Delta S$ combinations [115] as a consequence of intrinsic hydrophobic and polar interactions. As the binding affinity increases due to compound optimization, the range of available combinations narrow and converge, resembling a funnel as illustrated in Figure 13A. Significantly, optimization funnels are observed in databases containing a wide range of compounds and unrelated targets as the number of enthalpic-entropic combinations is reduced and the average compound affinity increases. By overlaying specific examples within the global data, one observes sub-funnels, such as that of trypsin inhibitors in Figure 13B (green dots) as well as different generations of HIV protease inhibitors complexed to protease variants in Figure 13C (magenta dots). While the average $\Delta H/\Delta G$ ratio changes with each system studied, a characteristic and common feature is conversion of higher affinity complexes to a narrower distribution within the optimization funnel.

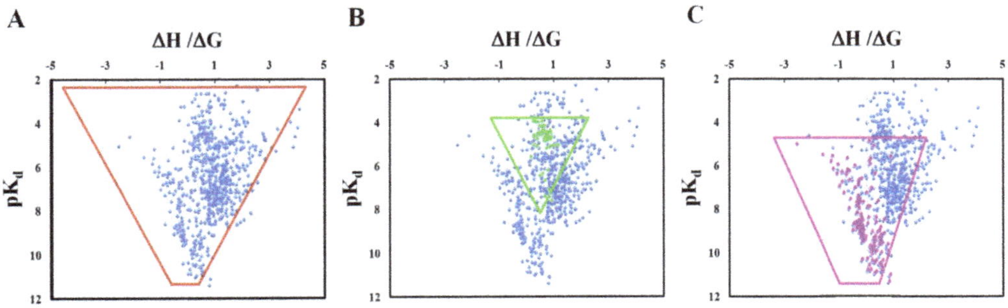

Figure 13. Optimization funnels generated by plotting pK_d as a function of $\Delta H/\Delta G$ ratios for (**A**) protein-ligand interactions (red) derived from thermodynamic databases [13] highlighting (**B**) trypsin inhibitors (green) and (**C**) HIV protease inhibitors (magenta).

4. Biophysical Methods Employed in Drug Discovery

Valuable information can be gained from the judicious inspection of molecular interactions that document outcomes based on compound hits derived from library screenings. There are a multitude of experimental methodologies and techniques available for initial screenings of fragment libraries [48,53,127]. These include nuclear magnetic resonance (NMR), surface plasmon resonance (SPR), thermal shift assay (TSA), capillary electrophoresis (CE), microscale thermophoresis (MST), biolayer interferometry (BLI), weak affinity chromatography (WAC), grating-coupled interferometry (GCI), cryo-electron microscopy (CEM), photoaffinity probes, fluorescence-based techniques, and isothermal titration calorimetry (ITC). This review specifically focuses on the use of ITC both in terms of its unique advantages and experimental challenges.

4.1. Characterization of Ligand-Target Interactions via ITC

ITC is widely regarded as an essential tool in the repertoire of biophysicists for characterizing the energetics of macromolecular interactions. Since its introduction to the scientific community approximately three decades ago, ITC has gained critical acclaim as the experimental technique of choice for the quantitative assessment of association processes. ITC is routinely applied in drug design and therapeutic strategies for the discovery of lead compounds that target specific macromolecules. In drug discovery protocols, ITC assays are typically performed employing millimolar and micromolar concentrations of lead compounds and targets, respectively. The most significant advantage afforded by ITC is that a single well-designed experiment facilitates the precise determination of the association constant (K_a), Gibbs free energy (ΔG), enthalpy (ΔH), entropy (ΔS), and stoichiometry (n) of a binding interaction.

The simultaneous assessment of binding affinity (K_a) and energetic driving forces (ΔH, ΔS) renders ITC an indispensable technique for characterizing ligand-target interactions [6]. A complete thermodynamic binding profile assists in elucidating the interrelationships between ligand affinity and overall biophysical properties. Following data acquisition and analysis via specialized methods and programs [128–130], the resultant thermodynamic signatures are employed to evaluate the efficacy of prospective compounds on the basis of binding energetics. In addition to the intrinsic value afforded by another layer of information, ITC-derived thermodynamic signatures have been recognized as an invaluable metric in drug design, development, and optimization given their integral role for assessing enthalpic efficiency (refer to the discussion on EE and SIHE in Section 2.2.5).

The thermodynamic parameters describing a particular ligand-target association process are a function of specific contributions that can be attributed to molecular interactions and driving forces. Parsing the Gibbs free energy into its enthalpic and entropic components permits identification of favorable/unfavorable interactions and discrimination of primary driving forces. Specifically, the binding enthalpy is comprised of favorable (e.g., hydrogen bonding and van der Waals) and unfavorable (e.g., polar group desolvation) ligand–target interactions. Conversely, the binding entropy includes favorable (e.g., desolvation and release of water molecules to bulk solvent) and unfavorable (e.g., conformational and/or motion restrictions) contributions arising from both the ligand and target (as reviewed in [8]). Acquisition of the requisite thermodynamic binding profiles as a function of temperature affords evaluation of heat capacity changes (ΔC_p) accompanying the binding process [131,132]. This extra-thermodynamic information is essential for elucidating binding processes that are coupled with desolvation and/or folding.

4.2. Overview of ITC Methodology

The basic principle of operation in an ITC experiment involves the use of a thermostatted titration syringe to dispense precise aliquots of a ligand into the sample cell containing a prospective receptor. The addition of titrant is accompanied by a measurable reaction heat due to ligand dilution and potential interactions with the target. Each ligand injection is characterized by the absorption or evolution of heat, triggering a difference in temperature between the sample and reference cells as illustrated in Figure 14A. The resultant temperature differential causes a feedback system to either lower or raise the thermal power as a means of compensating for this temperature imbalance. The experimental protocol is designed to allow sufficient time between successive injections to restore the temperature balance and thereby ensure that the system achieves equilibrium.

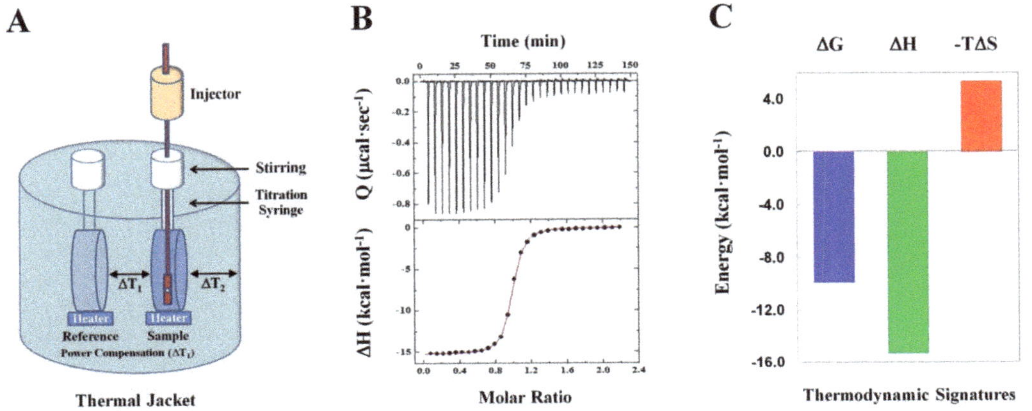

Figure 14. (**A**) ITC schematic depicting basic instrument components housed within an adiabatic jacket that includes a titration syringe delivering precise aliquots of ligand into a sample compartment containing the target solution under constant stirring conditions with the reaction signal detected via thermopiles strategically positioned on the exterior faces of sample and reference cells. (**B**) A representative ITC profile depicting exothermic reaction heats (top panel) integrated and normalized into enthalpy as a function of molar ratio (bottom panel). Non-linear least squares analysis (red line) yields the binding affinity (K_a), enthalpy (ΔH), and stoichiometry (n). (**C**) Resultant thermodynamic binding signatures derived via analysis of the ITC reaction profile: ΔG (blue), ΔH (green), and $-T\Delta S$ (red).

The thermal signal is detected via thermopiles that are strategically positioned on the exterior faces of the sample and reference cells. An electrical impulse is transmitted to the computer and the heat deflection is registered in the form of an integratable peak per injection, yielding a titration profile that is fit to a binding isotherm (red line) as depicted in Figure 14B. Detailed reviews containing complete method descriptions are available elsewhere [128]. The ITC method is sufficiently robust to accommodate a diverse range of interacting species within biological processes, irrespective of their biochemical and physicochemical properties. ITC analysis affords the evaluation of small molecule interactions with a target macromolecule, as the resultant binding parameters provide an energetic description of the association process in the form of characteristic thermodynamic signatures (i.e., ΔG, ΔH, and $T\Delta S$), as illustrated in Figure 14C.

5. Thermodynamic Binding Signatures as a Metric of Drug Potency, Selectivity, and Adaptability

5.1. Achieving Superior Lead Compound Selectivity

In addition to several lines of evidence that suggest favorable enthalpic interactions optimize compound efficacy, a potential link between thermodynamic signatures and drug selectivity has been proposed [115] and reviewed extensively [133]. These findings infer that improved shape complementarity is directly correlated with enthalpically-driven interactions, which create a bias towards specific targets versus off-targets, thereby enhancing overall selectivity [115]. As a case in point, lead compounds targeting distinct aldose-reductases exhibit high selectivity towards a particular species with the interaction accompanied by a significant enthalpic advantage [134]. In searching for compounds that selectively interact with and inhibit a specific aldose reductase, small molecules harboring a common 3-benzyluracil-1-acetic acid scaffold containing a chloronitrobenzyl group substituent selectively inhibit an aldose reductase (AR), which is implicated in diabetes. In contrast, a compound containing bulkier ortho/meta substitutions targets the cognate enzyme AKR1B10, an aldose reductase associated with cancer. The latter occurs conceivably via displacement of disordered water that is trapped in the enzyme hydrophobic pocket. As a consequence, each of the enzyme–inhibitor complexes exhibits optimal selectivity and their interactions are characterized by greater enthalpic contributions relative to the remaining ligands. These differential preferences afforded selectivity in cell cultures as the first inhibitor can potentially prevent sorbitol accumulation in retinal cells, whereas the second blocks the proliferation of cancer cells.

5.2. *Achieving In Vivo Efficacy*

While binding affinities toward a target tend to correlate with the desired biological property of a ligand, there are cases in which such correspondence is not readily apparent. In fact, there are examples documenting the utility of acquiring thermodynamic binding signatures in addition to seeking improved efficiency, selectivity, and ADMET properties. A study on CD4/gp120 inhibitors [116] has reported correspondence between HIV-1 cell infectivity inhibition and $\Delta H/T\Delta S$ ratios for various compounds, despite modest differences in the overall binding free energies. The rationale for these results resides in the realization that unwanted conformational effects contribute to modulate the $\Delta H/T\Delta S$ balance, the latter serving as a reporter in the selection of compounds that do not elicit such undesired effects. Therefore, thermodynamic optimization has proven of significant value for the selection of compounds that form complexes with the target protein in a "pre-organized" conformation, a finding that can be confirmed via thermodynamic signatures.

In cases where structural information is neither available nor sufficiently reliable to determine whether conformational changes might impact an outcome for a series of inhibitors, the $\Delta H/T\Delta S$ balance may serve as an indicator/predictor of structural alterations. It is interesting to evaluate the applicability of such a hypothesis on a selected class of peptidyl-nitrile compounds against *T. cruzi* strains [135] for which cruzain binding thermodynamics have been derived [136]. Despite the lack of correlation between cruzain affinity/inhibition and trypanocidal activity, the binding enthalpies (and corresponding enthalpic efficiencies) vary linearly with the PEC_{50} values ($r^2 \sim 0.98$) for a series of compounds studied. These findings suggest that a certain degree of conformational constraint is predicted based on the $\Delta H/T\Delta S$ balance measured for this congeneric series of compounds, whereby comparable affinity ligands may exhibit conformational preferences leading to a differential outcome in situ.

5.3. *Achieving Adaptability to Drug Resistance Mutations*

In general, engineering effective therapeutic agents harboring potent antiviral properties requires identification of specific factors that dictate selectivity and minimize susceptibility to mutations via adaptability (as reviewed in [8]). In drug discovery strategies, an ideal molecule is evaluated in terms of its potency, selectivity, specificity, selectivity, and adaptability to mutations, thereby preventing drug resistance [137]. Collectively, these characteristic properties can be achieved via knowledge of the enthalpic and entropic contributions to inhibitor-target interactions. In fact, such lead optimizations require an inhibitor to maintain interactions with conserved residues that do not normally undergo mutations, while simultaneously acquiring some flexibility to allow interactions with less stable, variable regions in the target that commonly undergo mutations. Although the design of adaptive ligands may incur an enthalpic penalty as a consequence of suboptimal complementarity, such interactions retain an enthalpic character with compensating favorable entropy, thereby maintaining the desired binding affinity to variant targets and exhibiting effective anti-viral activity.

6. ITC in Drug Discovery, Development, and Optimization

6.1. ITC in FBDD: Case Studies

During the past decade, several reports have documented the use of ITC as a strategic approach in the initial hit-to-lead phases of fragment-based drug design. There is ample evidence to suggest that selection of fragments on the basis of enthalpic efficiency irrespective of binding affinity leads to superior molecules [118]. As a validation technique in drug development and optimization, ITC has proven invaluable as it furnishes novel information on lead compounds for which thermodynamic properties have not been assessed from the outset. Specific examples on the use of ITC in fragment optimization include the identification of ligands for acetylcholine-binding protein (AChBP) [138] and the characterization of small molecule carriers for siRNA [139]. ITC has been utilized in the validation of fragment hits derived from screenings of ligands targeting the anti-apoptotic protein target Bcl-x(L) via automated mass spectrometry [140]. In a subsequent study, investigators employed ITC to evaluate the binding of compounds generated during FBDD campaigns against two functionally distinct proteins, the X-linked inhibitor of apoptosis protein (XIAP) and cyclin-dependent kinase 2 (CDK2) [141].

Alternate drug discovery strategies have been proposed including one developed in an academic setting yet suitable for industrial scale applications [142]. This protocol consists of a fragment screening cascade to identify hits employing a combination of differential scanning fluorimetry (DSF), validation by NMR spectroscopy, and final characterization of binding fragments via ITC and X-ray crystallography. Along these lines, ITC has been utilized in the hit validation of halogen-enriched fragments that exhibit low micromolar affinities and high ligand efficiencies [143]. FBDD strategies have been employed to identify suitable antibiotic candidates targeting tRNA (m1G37) methyltransferase (TrmD) from *Mycobacterium abscessus* (Mab) [144], a rapidly growing multidrug resistant mycobacteria. The screening for compound prioritization involved DSF followed by ITC validation, yielding a lead compound comprised of two merged fragments bound to the active site.

In a quest to identify therapeutic agents against the opportunistic pathogen *Pseudomonas aeruginosa*, investigators recently designed and optimized a series of compounds to interact with pseudomonas quinolone signal receptor (PqsR) [145], a key transcription factor that controls bacterial pathogenicity. Hit optimizations monitored via SPR and ITC evaluated the corresponding enthalpic efficiency (EE) of introducing flexible in lieu of rigid linkers in these compounds. While apparently counterintuitive, the finding that flexible linkers boost PqsR activity and enhance anti-virulence potency can be rationalized on the basis of their respective thermodynamic binding profiles [145].

6.2. ITC in FBDD: Experimental Challenges

Upon combining the selected fragments discovered via FBDD approaches, the ultimate objective is to derive a lead compound with high ligand efficiency. Pursuing this strategy, the initial fragments are expected to exhibit affinities spanning the range of 100 μM–10 mM to generate a final product with affinities on the order of 10 nM [48]. Specialized analytical techniques are therefore required to screen the low affinity fragments. Biophysical methods commonly employed in the fragment screening process include NMR spectroscopy, X-ray diffraction analysis, mass spectrometry, surface plasmon resonance, fluorescence-based techniques, and isothermal titration calorimetry. Since FBDD requires detection of low-potency hits, a caveat on the use of most biophysical techniques is to obtain sufficient amounts of purified protein target (>10 mg) and fragments that are soluble at the concentrations needed (mM) for screening and optimization. Standard operating practice is to acquire ITC measurements on systems that exhibit a moderate range of affinities (i.e., $10^3 > K_a > 10^7\ M^{-1}$) with target solution concentrations at least tenfold greater than the ligand dissociation constant (K_d). Despite these recommended guidelines, a well-designed ITC experiment may yield informative results with significantly less material and/or lower affinities.

While the thermodynamic data acquired via ITC is invaluable in FBDD, the tendency is to employ this technique later in the drug development process for compound optimization given its lower sample throughput and higher target protein requirements relative to other biophysical techniques. Considering these experimental caveats, there are success stories in which ITC has assumed a lead role in the entire drug discovery process. Under specific conditions, ITC has furnished valuable information in FBDD decision making processes [138–140,142,145–148]. Although ITC might pose an experimental challenge in initial FBDD screenings, this technique represents a powerful method during the secondary screening stage and might retrospectively provide important clues that assist further improvements and optimization of the starting fragment candidates. Given the material and time constraints associated with ITC characterization of ligand–target interactions, continuing efforts are devoted towards enhancing overall sample throughput and sensitivity. Recent studies have focused on overcoming such experimental limitations by developing enthalpy screening methods with the goal of accelerating data acquisition [149,150] and expanding the high-throughput capabilities of calorimetric instrumentation.

6.3. Protein-Protein Interactions (PPI) as Targets in Drug Discovery

Drug discovery programs focusing on protein-protein interaction (PPI) inhibitors are challenging and relatively few FBDD approaches have tackled these projects to identify active small molecules against a number of chemotherapeutic targets (as reviewed in [151]). Considering the fundamental role of PPIs in disease, conventional wisdom has postulated that these interactions might be 'undruggable', thereby rendering drug discovery efforts targeting these macromolecular systems particularly cumbersome [152]. Recent developments have challenged this notion, as the physicochemical properties of small-molecule PPI modulators undergoing clinical trial progressively demonstrate typical characteristics of drug-like molecules. These findings suggest that future drug discovery campaigns aimed at targeting PPIs may follow traditional design parameters, albeit along the lines of bRo5 space described in Section 2.2.

One successful example of a PPI inhibition by small molecules involves interactions between tumor suppressor BRCA2 and recombination enzyme RAD51 [153]. In this case, fragment hits capable of interacting with the PPI interface have been validated by ITC, NMR, and X-ray approaches. Another relevant approach in drug design is to identify PPI inhibitors that are essential for a particular cellular function/dysfunction and responsible for a number of pathological conditions. In order to develop treatments for oxidative stress-related diseases, a recent study [154] focused on inhibiting the interaction of transcription factor Nrf2 with its negative regulator Keap1, thereby upregulating Nrf2 transcriptional activity. Seeking to identify inhibitors of these PPI interactions, a number of lead compounds inspired by a natural molecule that interacts with Keap1 have been developed and evaluated via multiple approaches including detailed structural and thermodynamic analysis.

This combined structural-energetic strategy led to improvement of the compound series by introducing chemical modifications that impact solvation and residue flexibility. One of the binding modes involved displacement of a coordinated water molecule, which resulted in an additional entropic gain, complementing the favorable enthalpic contributions of productive key residue interactions. This series of compounds is clearly distinct in terms of respective energetic signatures, as manifest in a striking reduction of the entropic penalty incurred by binding-induced desolvation of the interaction site, as visualized in Figure 15A where entropically favored compounds are designated with a red line (i.e., entries **60–78**). The accompanying $\Delta H/T\Delta S$ plot in Figure 15B reveals that these compounds (magenta dots) are clustered in the region of lower $-T\Delta S$ values ($-2.5 < -T\Delta S < 2.5$ kcal·mol^{-1}) and variable levels of favorable enthalpic contributions ($-12 < \Delta H < -7$ kcal·mol^{-1}). Significantly, the broad variation in thermodynamic signatures occurs via a simple substitution in the ligand as depicted in Figure 15C that conceivably dislodges a coordinated water molecule bridging residues S508 and R415 of Keap1.

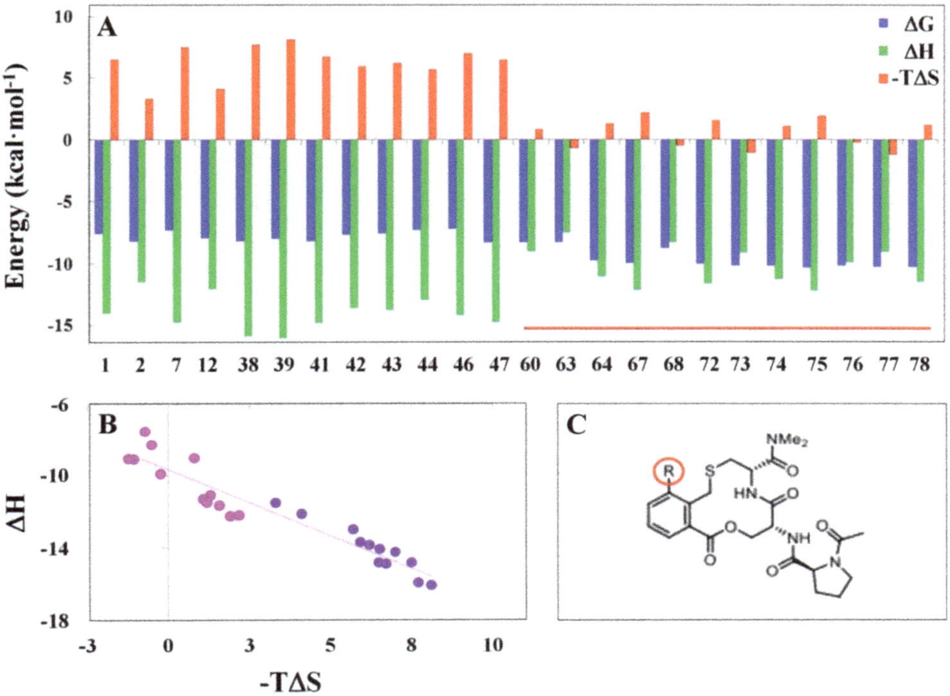

Figure 15. Impact of binding-induced desolvation on the energetic signatures of PPI inhibitors. (**A**) Selected compounds reported in [154] are sorted according to original ID number (**1–78**) and comprise two distinct ligand classes on the basis of their respective thermodynamic signatures. Inhibitors **1–47** exhibit a higher entropy penalty (red histograms) relative to compounds **60–78**, (underlined by a **solid red line**) that are characterized by enhanced affinity as reflected in the higher binding free energy (blue histograms) due to reduced entropic penalty despite the lower binding enthalpy (green histograms). (**B**) An enthalpy-entropy plot discriminates between higher affinity compounds exhibiting lower unfavorable entropy (magenta circles) versus lower affinity compounds (purple circles) characterized by higher unfavorable binding entropies ($-T\Delta S \geq 3$ kcal·mol^{-1}). (**C**) Structure of compounds with a substituent R in the ortho-position that exhibit improved binding energetics. Inhibitors **67**, **68**, and **72**–**78** contain either a carboxyl or tetrazole group on an aromatic moiety attached at the ortho-position of the phenylene ring. The entropy gain observed for compounds **60–78** may arise from dislodging a coordinated water molecule that bridges residues S508 and R415 of Keap1.

Along the lines of modulating PPI interactions as therapeutic interventions, strategies have been developed with the assistance of computational approaches, affording the design of enzyme-inhibiting peptides that transiently convert an active enzyme into a proenzyme. The latter is co-delivered with a pro-drug substrate to specific target sites for subsequent activation by local proteases [155]. This approach eliminates the risks of systemic toxicity as the pro-drug is harmless to the body in its latent form and only converted to an active toxic agent by the accompanying pro-enzyme in local target tissues. Applying this strategy to the therapeutically relevant protein carboxypeptidase G2 (CPG2), a combination of biochemical, computational, structural, and thermodynamic methods has assisted in identifying optimal peptide candidates to fulfill the role of a transient enzyme inhibitor.

6.4. Emerging Infectious Diseases: SARS-CoV-2 Therapeutic Interventions

The emergence of highly infectious diseases such as the SARS-CoV-2 pandemic represents a global health threat that requires development of effective therapeutic treatment regimens. While efforts have intensified to immunize the world population through administration of vaccines, SARS-CoV-2 likely represents a long-lasting endemic that will continue to evolve with the appearance of novel variants evading immunization protocols. In an effort to address this void, therapeutic interventions are urgently needed and recent developments towards this goal have resulted in the approval of several drugs and anti-inflammatory agents. Despite limited progress on this front, specific targets of viral or host origins have been identified for further exploration in the development of prospective therapeutic agents. Current drug discovery strategies have employed a diverse array of biophysical approaches including ITC to assist in validating hits from HTS and characterizing the efficacy of new and repurposed drugs as antiviral therapeutics against SARS-CoV-2 targets.

The coronavirus designated SARS-CoV-2 encodes two proteases, namely 3CL^{pro} (or M^{pro}) and PL^{pro}, both of which are considered as primary antiviral research targets in SBDD and drug repurposing strategies (as reviewed in [156]). CL-Pro is a class of proteases for which a host of promising molecules have been developed against other coronaviruses. A representative example of these therapeutic agents is GC376 that has already been used to treat feline coronavirus and subsequently evaluated against several SARS-CoV variants. In a recent study, the efficacy of this compound has been validated by ITC and proven specific against SARS-CoV-2 M^{Pro} [157]. A separate study reported on the design and synthesis of dipeptidyl inhibitors with novel P_3 scaffolds exhibiting potent inhibitory activity against SARS-CoV 3CL^{pro} and nanomolar affinities measured via ITC [158]. The SARS-CoV-2 target PL^{pro} is a papain-like protease with deubiquitinating and deISGylating activities [156]. By removing the ubiquitin-like ISG15 (interferon-stimulated gene 15 protein) modifications from host proteins, this viral protease causes suppression of the innate immune response and promotes viral replication. PL^{pro} is effectively inhibited by GRL0617, a non-covalent inhibitor that interacts with and blocks the association of PL^{pro} with ISG15 [159]. As a consequence, this small molecule inhibits viral replication while potentially promoting anti-viral immunity [160].

Several of the SARS-CoV viral nonstructural proteins including nsp12 (i.e., RdRp) and nsp7-8 (i.e., auxiliary proteins) assemble to form an active replication transcription complex. Employing a combination of biochemical and biophysical approaches, investigators are currently evaluating a number of repurposed compounds against RdRp. Inspired by studies conducted on protein synthesis inhibitors over five decades ago [161], these lead compounds have demonstrated the ability to interact with SARS-CoV-2 RdRp and inhibit ribosomal protein synthesis [47]. The screening and characterization of prospective SARS-CoV-2 therapeutics have been documented including a HTS using the nsp10-nsp16 complex. Repurposed drugs have been assessed for emergency use to treat SARS-CoV-2 infection [162] and the positive hits validated by ITC.

One of the mechanisms underlying SARS-CoV-2 mediated suppression of interferon responses occurs due to orf9b interactions with host cell components. This includes the Hsp90/TOM70 complex for which interactions have been explored structurally and thermodynamically [163,164]. A recent study employing ITC reported on the allosteric inhibition of macromolecular interactions by orf9b [165]. Remarkably, the binding affinity of Hsp90 EEVD motif to TOM70 NTD is reduced by ~29-fold when orf9b occupies the TOM70 CTD pocket, supporting the proposition that orf9b allosterically inhibits Hsp90/TOM70 interactions. This finding sheds light on the mechanism underlying SARS-CoV-2 orf9b mediated suppression of interferon responses [166], thereby providing opportunities for therapeutic interventions.

Another potential SARS-CoV-2 target that has recently received attention is the DC-SIGN (dendritic cell-specific intercellular adhesion molecule 3 grabbing nonintegrin), a C-type lectin receptor that mediates infection and dissemination of numerous viruses [167].

Since the SARS-CoV-2 spike protein is heavily glycosylated, its interaction with DC-SIGN represents a potential ACE2-independent infection route of innate immune cells. Inhibition of virus binding to DC-SIGN therefore represents an attractive host-directed strategy to attenuate overshooting innate immune responses and prevent disease progression. In a recent study, ligand optimization has been monitored via biophysical approaches and calorimetric determination of the resultant thermodynamic signatures for DC-SIGN–glycopolymer interactions revealed a correlation of potency with ligand binding enthalpies [168].

An alternate strategy has been pursued to identify aptamers that efficiently bind DNA-susceptible peptide structures in SARS-CoV-2 proteins critical for infectivity such as the receptor binding domain (RBD) of spike protein and SARS-CoV-2 RdRp. By repurposing existing aptamers, investigators have identified positive hits validated by a number of techniques including ITC [169]. The number of studies documenting new anti SARS-CoV-2 candidates published on a daily basis suggest that the use of existing/repurposed drugs, natural chemicals, and/or novel entities has progressively increased since the pandemic onset in 2019. Multiple targets of viral or host origin have already been selected for treatment, as drug repurposing offers an attractive prospect in terms of accelerated therapeutic development. In retrospect, significant progress has been achieved during an abbreviated timeframe given the identification of protease inhibitors and selection of prospective candidates undergoing clinical trials to evaluate their antiviral efficacy against SARS-CoV-2. The arsenal of potential leads and repurposed drugs has advanced dramatically [170–172] with the promise of developing more effective SARS-CoV-2 therapeutics to counter an ever-expanding array of novel viral variants.

7. Parsing Thermodynamic Binding Signatures

7.1. Role of Solvation on Binding Energetics

Numerous studies have succeeded in elucidating the forces driving protein–ligand interactions with particular emphasis on the role of hydration, which is often obscured during static structural analysis or overlooked via computational techniques that do not implicitly or explicitly incorporate solvation models (as reviewed in [173]. The power of computational methods is enhanced by rigorous evaluation of theoretical predictions based on experimental data. In fact, armed with experimental thermochemical and thermophysical data (e.g., ThermoML) [174], molecular simulations can predict specific physicochemical properties of compounds such as infinite dilution activity coefficients (IDAC), which essentially reflect interactions between a single solute molecule surrounded by solvent and may therefore provide valuable insights regarding the solvation energies of compounds [175]. The powerful combination of structural [153,176,177], computational [178,179], and experimental biophysical methods, including calorimetric techniques [180], can furnish invaluable information on the molecular forces driving ligand–target interactions, including the role of solvation in binding energetics.

In principle, ITC measurements yield global thermodynamic signatures that cannot resolve the contributions of each water molecule participating or displaced upon protein–ligand complex formation, and sometimes presents a challenge in overall data interpretation [181]. Examples include certain thrombin–ligand complexes that exhibit nearly identical binding enthalpies despite distinct HB patterns and water networks. This is presumably due to associated structural changes within the interaction site that mask differential binding modes of these compounds [182]. The role of water molecules in ligand recognition is multifactorial, a primary reason why predictions of binding energetics represent a formidable challenge even when structural and thermodynamic information is available [183]. A well-designed multiparametric strategy incorporating systematic group substitutions in the ligand [184,185] and/or amino acid replacements in the target protein [186] may resolve hydration contributions to a ligand-target interaction at the molecular level.

In conjunction with this approach, the acquisition of ITC measurements under rigorous experimental conditions should provide clarification on the origins of net thermodynamic binding signatures. Examples include the use of multiple buffers with distinct ionization enthalpies at different pH to determine intrinsic ligand-target binding enthalpies [187–189], and multiple assay temperatures to assess heat capacity changes [132,190]. Some molecular binding events that are detectable at the enthalpy/entropy level are not necessarily reflected in the binding free energy due to enthalpy/entropy compensations as illustrated by oligosaccharide interactions with a number of mutant proteins [186]. The energetics of protein–ligand complexes may be modulated by introducing specific functional groups resulting in enhancement of either the ligand binding enthalpy or entropy. As an example, target–ligand complexes can be stabilized by inserting H-bonding functional groups that interact with or replace interfacial water molecules resulting in a favorable contribution to binding enthalpy. Conversely, the introduction of certain functional groups within a ligand may promote expulsion of surface waters into bulk solvent thereby increasing binding entropy.

The relevance of solvation in molecular recognition events is evident based on numerous observations reported over the years [46,183,191–199]. The critical role of solvation is clearly illustrated within the context of drug design and development by citing a classic example of the serine protease family, which includes a number of disease-associated enzymes that have been used as targets in drug discovery campaigns. Klebe and colleagues [183] have provided an illustrative example of individual water molecules eliciting dramatic impacts on the thermodynamic binding signatures of compounds interacting within the enzyme S_1 pocket. Specifically, the binding of a compound in which the benzamidine anchor has been removed is accompanied by expulsion of one water molecule, a favorable enthalpic event that enhances the overall binding affinity. These results set the stage for development of several orally available anticoagulants designed on the basis of important thermodynamic findings [183]. In summary, there is considerable interest regarding the fundamental role of solvent in modulating thermodynamic signatures of ligand–target interactions, information that is not readily gleaned from structural studies. Solvent molecules may play a significant role in binding energetics and therefore represent a major force in driving a ligand to its desired target.

7.2. Conformational Impacts: Ligand Preorganization

The combined efforts of structural and energetics studies may facilitate the design of ligands with improved affinities by lowering their entropic cost of association due to immobilization. In this respect, thermodynamic characterization of ligand-target interactions assists in the overall design process by monitoring a decrease in the entropic penalty to identify compounds with enhanced binding energetics. Ligands that adopt bioactive conformations during late-stage optimization by introducing structural constraints may exhibit greater affinity/potency over their flexible counterparts [200]. The rationale for this improvement relates to entropic advantages assigned to rigid versus flexible ligands. Such techniques have been used to design peptide ligands by conformationally constraining their unbound structural freedom [201]. There are numerous approaches to achieve this goal, which are often performed chemically or enzymatically [202,203].

In terms of drug discovery strategies, thermodynamic analysis at late-stage lead optimizations have yielded significant insights into the role of ligand preorganization on interaction affinity and potency of small molecules [200,204]. The ultimate goal is to identify a preferred bound geometry of ligand within the binding site and simulate its conformation in the solution state via chemical modifications. The latter may include intramolecular hydrogen bonds, cyclization, and other means to shift the ensemble of ligand conformations towards a bioactive form. One particular challenge resides in ensuring that the free ligand-populated conformation faithfully reproduces the active bound state in such a manner that both the conformational strain and entropic penalty are alleviated. An example in which such approaches have been applied to small molecules is the class of BACE inhibitors involving introduction of cyclopropane moieties in the ligand molecular structure [205]. The cyclopropane substituted aminopyrimidone-type BACE-1 inhibitors exhibit improved activity relative to a non-restricted ethylene linker compound. These data correlate well with the resultant entropic gain derived from imposition of conformational constraints due to incorporation of the cyclopropane ring.

7.3. Impact of Cooperativity on Binding Energetics

Systematic group substitutions in the ligand [184,185], and/or amino acid replacements in the target protein [186] provide insights into the origins of net thermodynamic binding signatures at the molecular level. The strategy underlying FBDD involves assembling several low affinity fragments with the appropriate geometry and attachment pattern to generate a high affinity ligand that specifically recognizes the target. Deconstruction of the resultant ligand into its respective fragments may or may not yield a binding affinity that reflects the sum of its constituents, which represents a measure of cooperativity [206]. The impact of cooperativity or nonadditivity on binding energetics has been assessed by methyl group substitutions, a popular approach that aims at improving ligand geometry and complementarity in the bound state [207]. The concept of nonadditivity can be appreciated when evaluating congeneric series of compounds [184], where hydrophobic contacts and hydrogen-bond formation between various substituents elicit cooperative effects. A typical example of nonadditivity involving double functional group replacement is presented in Figure 16 for thrombin inhibitors [208]. Inspection of the schematic reveals that single site substitutions of an amino group or benzyl side chain result in binding free energy enhancements (ΔΔG) of -1.2 and -2.1 kcal·mol^{-1}, respectively. The simultaneous introduction of both functional groups yields a binding free energy enhancement of -4.3 kcal·mol^{-1}, representing a positive cooperativity (i.e., nonadditivity) of ΔΔG = -1.0 kcal·mol^{-1} relative to the sum of individual substitutions (ΔΔG = -3.3 kcal·mol^{-1}). It is interesting to note that the order in which group replacements are introduced may play a significant role in the eventual outcome [185].

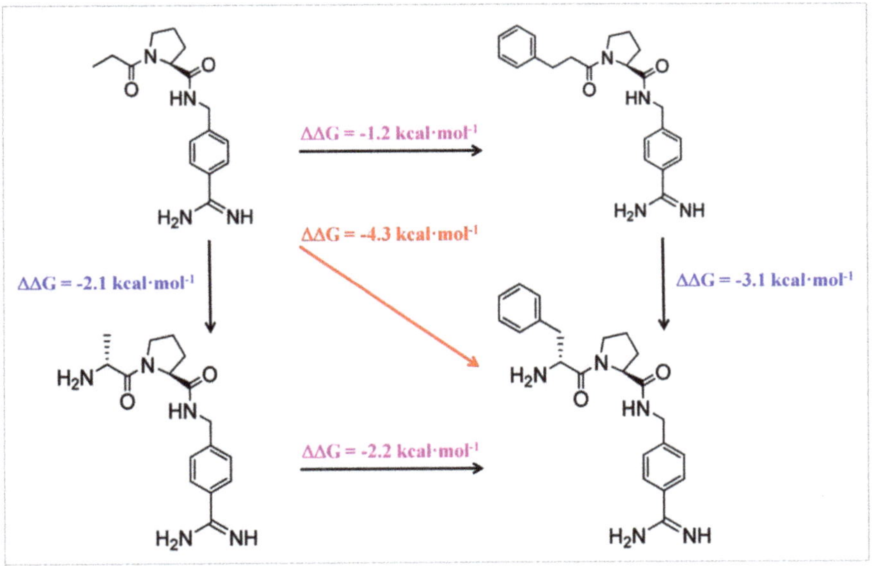

Figure 16. Cooperativity (nonadditivity) of hydrogen bond formation and hydrophobic contacts in a set of thrombin inhibitors analyzed via double replacement cycle. Introduction of a lipophilic side chain (upper horizontal arrow) increases the affinity by 1.2 kcal·mol^{-1}. Addition of an amino group (left vertical arrow) increases the affinity by 2.1 kcal·mol^{-1}. The combined chemical group substitutions yield a binding free energy enhancement of 4.3 kcal·mol^{-1} (diagonal red arrow) representing a positive cooperativity of ΔΔG = 4.3 − 2.1 − 1.2 = 1.0 kcal·mol^{-1}. (Data derived from [208]).

8. Challenges Associated with Interpretation of Thermodynamic Data

The concept of incorporating thermodynamic measurements within the drug discovery arena initially elicited an enthusiastic response that has been tempered by guarded optimism. A notable milestone is the proposal of employing thermodynamic binding signatures as an additional metric in lead optimizations [109], which has been embraced with the expectation that real-time experimental observables might potentially drive the drug development process more effectively. Indeed, such measurements represent a valuable tool for drug development and design, based on the premise that enthalpically driven binders are considered superior ligands in terms of aqueous solubility, decreased toxicity, and higher selectivity [7]. A successful case history that proves this point is illustrated by the various generations of anti-HIV drugs introduced commercially over time, as discussed in Sections 3.1 and 3.2. Significantly, one observes a general trend in which the newer drugs acquire a thermodynamic signature that gradually shifts from entropy-driven to enthalpically favorable (see Figure 12B).

The most compelling evidence for dissemination of this concept is the recognition that enthalpically driven ligands with polar characteristics retain an added advantage of exhibiting superior pharmacokinetic properties and thereby score additional points in terms of ADMET criteria. As reviewed herein, this paradigm in drug discovery and optimization remains a matter of prioritization in current design strategies yet requires judicious planning and analysis, which integrates rigorous buffer-dependent thermodynamic assessments with high resolution structural capabilities to define binding energetics at the molecular level (refer to additional relevant reviews on this topic [8,209]). Given these constraints and guidelines, contributions arising from solvation, ligand preorganization, cooperativity, and linked processes must be considered explicitly when interpreting thermodynamic signatures within the context of structural data. A clearly defined structural–energetic correlation finally emerges following successful resolution of potential competing events and extrinsic factors that may mask the intrinsic thermodynamic parameters. In this section, we present several caveats associated with the use of thermodynamics in drug discovery strategies and offer suggestions to rationalize some of the unanticipated findings.

8.1. Resolving Paradoxes in Thermodynamic Characterizations

Given the increasing use of ITC as an indispensable biophysical tool in drug discovery, coupled with the availability of comprehensive thermodynamic databases, a number of questions have arisen as a consequence of several studies reporting unanticipated outcomes. This prompted serious re-evaluation of available data to corroborate the assumption that *enthalpy-driven binders are favored in drug development* [7]. Additional concerns and queries regarding method protocols, the role of solvation, and impact of ligand conformational constraints [182,210], must be addressed and reassessed [117,211]. Following a period of relative euphoria with the prospect of achieving a successful structural-energetics consensus, a certain level of skepticism has surfaced as drug design laboratories seeking rapid and straightforward decision criteria often face the realization that critical data are lacking for a complete interpretation [7]. One of the caveats associated with employing ITC as a tool in drug development originated as a consequence of the variability observed when conducting measurements under disparate solution conditions, thereby hampering comparison of calorimetric data on an absolute scale. The implementation of rigorous and stringent experimental protocols reduced overall variability and ensured systematic evaluations to resolve intrinsic thermodynamic binding parameters for ligand-target interactions [138,212–214].

While ITC is a model-independent technique that provides a direct measure of the heat absorbed/released upon ligand–target association, the resultant enthalpy may be the result of multiple events that occur concomitantly to the binding process. The latter may include and is not restricted to coupled protonation/deprotonation reactions [105,215], and binding-induced conformational changes amongst a host of heat absorbing/releasing events [216]. These linked processes require additional measurements to resolve intrinsic thermodynamic parameters [189,217]. In this respect, ITC protocols must be designed to derive the requisite intrinsic thermodynamic data by conducting measurements in an array of buffers with distinct ionization enthalpies at various pH. The calorimetric experiments should be conducted over a sufficiently broad temperature range to account for heat capacity changes involved in the association process (as reviewed in [131]).

The finding that binding parameters routinely measured in drug discovery campaigns generally do not represent intrinsic thermodynamic signatures has created some reluctance in applying ITC for time-sensitive lead optimizations and drug development [182]. A range of experimental parameters, including temperature and solution conditions (e.g., buffer, solute effects), can significantly influence the observed thermodynamic signatures and must be explicitly considered to derive intrinsic binding enthalpies. Moreover, structural flexibility and allosteric mechanisms may lead to obfuscation in the intrinsic thermodynamic parameters, rendering the latter experimentally inaccessible. The utility of thermodynamics as a core metric to evaluate the potential for success and accelerate drug discovery efforts has been enthusiastically embraced. Nevertheless, a number of unforeseen challenges have hampered the desire to adopt this technique as a routine strategy. A critical review of specific intricacies associated with the use of ITC in drug discovery [182] suggests that while enthalpy and entropy should not be viewed as direct end points, the latter may significantly enhance our understanding of ligand–target interactions when employed in conjunction with structural and/or computational approaches.

8.2. Caveats Associated with the Design of a Constrained Ligand

Ligand preorganization may represent a useful strategy in thermodynamic approaches designed to improve binding affinities by reducing entropic penalties associated with conformational strain (Refer to Section 7.2). Considering its potential utility, several unanticipated experimental outcomes led to the realization that additional factors must be considered when designing a "bioactive conformation". As a case in point, studies on Grb2 SH2 domain-peptide interactions [201] reveal that designed ligands adopting a bound-like geometry are characterized by unexpected improvements in binding enthalpies yet exhibit unfavorable entropic contributions relative to the original flexible conformers [200,211]. These findings suggest a counterproductive outcome when attempting to improve ligand potency by purposely avoiding entropic penalties due to binding-induced conformational strain. A plausible explanation for this apparent discrepancy has been offered by assessing the conformational energetics computationally [117]. Specifically, the conformationally-constrained ligands might further reduce binding site residue entropies, thereby offering a counterargument to the expected improvement in overall binding entropy upon ligand pre-organization.

In the final analysis, thermodynamic binding signatures must explicitly consider multiple factors and parameters in order to enable reliable predictions. Along these lines, the inventory of water molecules involved in a ligand-target interaction is an important consideration [200,211], as solvation plays a fundamental role in the binding energetics and overall enthalpy/entropy balance. The general consensus is to introduce some constraints in the small molecule that create a *pre-organized state*, which is poised to interact with the binding site while introducing no major strain. In principle, this represents a reasonable approach yet requires a number of important assumptions, namely: (a) the constrained ligand conformation is equivalent to the final bound state; and, (b) hydration and hydrogen bond donor/acceptor capabilities are intact. Under these conditions, the resultant energetic signatures are characterized by an entropy gain and enhanced binding affinity. In practice, there are systems in which the entropy gain is balanced by an enthalpy loss resulting in marginal improvement of the binding affinity. Conceivable origins for such unexpected outcomes associated with conformationally constrained ligands may reside in violation of specific conditions related to the integrity of hydration and/or hydrogen bond donor/acceptor capabilities as a consequence of ligand design.

8.3. *Origins of Enthalpy-Driven Hydrophobic Interactions*

The well described hydrophobic effect suggests that the burial of non-polar surfaces represents an entropically favorable process, implying that addition of non-polar residues to small molecules enhances the binding affinity via entropy gain. Although typically viewed as an entropy-driven process, examples of hydrophobic interactions that are enthalpic in nature have been documented in a comprehensive review on this topic [218]. There are several plausible explanations for enthalpy-driven hydrophobic effects including binding-induced release of water molecules from the hydrophobic environment to bulk water [219]. Conversely, an opposing entropic component arises due to the elimination of solvent fluctuations inside the binding pocket. Such contributions override the favorable entropic effects of extracting a small non-polar ligand from bulk solvent.

Water molecules that form non-optimal hydrogen-bonds within a hydrophobic surface are released upon ligand binding and optimally hydrogen-bonded in the bulk solvent which is an enthalpically favorable process. Another possibility arises when the binding pocket of an apo protein is suboptimally hydrated. In such situations, solute-solvent dispersion interactions in the hydrated complex might not be completely offset by dispersion interactions for the hydrated protein and ligand. The concomitant increase in dispersive interactions upon complexation may therefore result in a favorable enthalpic contribution. An excellent example is the interaction between major urinary protein (MUP-1) and primary alcohols of various chain length in which a combination of calorimetric, structural, and computational studies reveal that the association process is enthalpy-driven, despite the hydrophobicity of interacting species [220]. The authors rationalize these findings as conceivably arising from favorable solute-solute dispersion interactions following protein–ligand complexation. Indeed, the apo-MUP-I pocket is suboptimally hydrated [221], which implies an inequality between solute–solvent dispersion interactions prior to protein–ligand binding versus solute–solute dispersion interactions in the associated complex.

Several studies involving congeneric ligands highlight some of the challenges associated with understanding how adding nonpolar surface area to small molecules affects their protein-binding energetics. An intriguing example in which increased non-polar surface area modulates the thermodynamic binding signature is illustrated in Figure 17 for a series of thermolysin inhibitors [222]. These enzyme inhibitors have been designed via $P_{2'}$ substitutions, as noted in the inset of Figure 17A. Inspection of the enthalpy-entropy plot in Figure 17A and resultant thermodynamic signatures sorted by molecular size in Figure 17B reveals that while compounds **1–3** exhibit an increased enthalpic contribution as a function of size, ligands **4–5** are virtually indistinguishable thermodynamically, and compounds **6–9** exhibit the expected size-dependent enhancement in binding entropy as deduced from increased surface area burial. In the latter group, entropy gains are balanced by reduced enthalpic interactions, which is consistent with a structural loss of definition and increased motion in some complexes. An overall evaluation of the enthalpy/entropy balance reveals that whereas enthalpy prevails for smaller substituents, the entropic component dominates ΔG for larger substituents. It is interesting to note that ligands with medium-sized $P_{2'}$-substituents (i.e., compounds **4** and **5**) exhibit the highest affinities. In an effort to rationalize these observations, inspection of the water geometries adjacent to $P_{2'}$ via X-ray crystallography [222] suggests correlation between an optimal solvation network and the resultant thermodynamic profiles. These findings underscore the importance of performing systematic evaluations within congeneric compounds to establish correlations between structural and energetic properties, as useful information can be gleaned from such comparisons [223], including the paradoxical nature of enthalpically driven hydrophobic interactions.

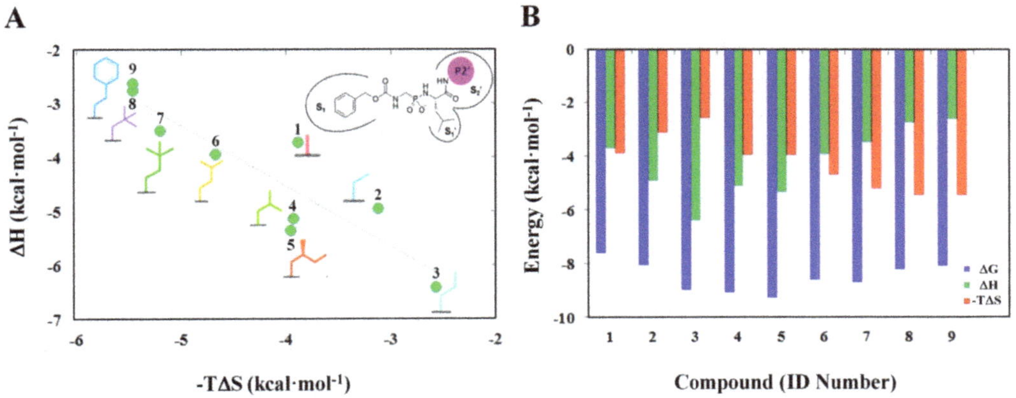

Figure 17. (**A**) Enthalpy-entropy plot for a congeneric series of thermolysin inhibitors harboring distinct substitutions at the $P_{2'}$ position with each compound identified according to entry number (**1–9**) and respective group substituent; (**B**) Thermodynamic binding signatures [i.e., ΔG (blue), ΔH (green), and −TΔS (red)] for the thermolysin inhibitors presented in Panel A. Ligands with medium-sized P_2' substituents (i.e., compounds **4** and **5**) exhibit the highest affinities as a consequence of favorable enthalpy and entropy contributions. (Data derived from [222]).

9. Potential of Structure-Energetic Correlations in Accelerating Drug Design Predictions

Elucidation of biomolecular interactions at the atomic level is extremely complex as these involve formation and/or disruption of multiple non-covalent bonds between the interacting molecules as well as solvent [113]. Despite several decades of experimental studies and extensive analyses, a complete understanding of the thermodynamic driving forces governing biomolecular interactions remains elusive. While ITC provides the most accurate and reliable experimental technique to achieve a complete thermodynamic characterization, the utility of such information in drug design and development still represents a challenge. Correlations between thermodynamic data and structural features yield invaluable insights on biomolecular interactions and computational tools have empowered such correlations at the molecular level. Incorporation of multiparametric approaches in drug discovery strategies represents a powerful infrastructure for the development of future treatment regimens exhibiting enhanced efficacy. Thermodynamics can still provide the requisite input into decision-making processes as the goal of rational drug design is to identify small molecule substitutions that increase compound efficiency, potency, and specificity while minimizing overall toxicity. Thermodynamics may assist in this endeavor by gauging the progress achieved or drawbacks encountered when certain molecular manipulations tend to elicit an unfavorable response. Specifically, characterization of the binding energetics associated with chemical modifications/substitutions facilitates assessment of group-group interactions, conformational constraints, and accommodation within the binding site. The resultant thermodynamic signature in conjunction with a high-resolution structure furnishes a complete molecular description of biomolecular interactions within the ligand-target complex.

There is a fundamental need to expand predictive capabilities based on experimental structural-energetics data that inform decisions regarding the selection of hits worth pursuing at the next level of drug design and development. In order to identify the most promising candidates for lead optimization, predictive biophysical parameters are required, and thermodynamic data can furnish valuable insights to achieve this goal. Approaches towards this direction have been pursued and score functions derived from structural and calorimetric data [224]. ITC experiments provide direct access to ΔG, ΔH, TΔS and ΔH/TΔS compensation as a function of specific experimental variables including buffer, temperature, osmolytes, and pH [188]. Resolving intrinsic thermodynamic parameters is therefore essential for an unbiased structural-energetic analysis. While the current review mines available thermodynamic databases as an illustration of overall trends and provides insights on the interplay between experimental observables and design metrics, it is highly advisable to evaluate differential binding profiles across specific data sets comprising congeneric ligand series for precise interpretations [223]. Such systematic evaluations have proven particularly illuminating by resolving apparent paradoxes and gaining a more detailed understanding of the forces driving specific ligand target interactions at both the structural and thermodynamic levels.

A considerable body of evidence has demonstrated that ligand preorganization abrogates the entropic penalty associated with structural rearrangements. Conversely, entropically favorable interactions arising from residual mobility of ligand or target in the bound state is often compensated by an unfavorable enthalpic term. In most cases, desolvation is an entropically favorable process, yet there are instances in which such interactions are enthalpically favored. An unexpected observation is that the role of local water structure may dramatically alter the thermodynamic binding signatures without any appreciable impact on ligand affinity. These findings underscore the need to evaluate biomolecular interactions thermodynamically, as characterization of binding affinities alone may miss important molecular events that are only distinguished in the enthalpic and/or entropic terms. Intrinsic thermodynamic binding parameters should be correlated with high resolution crystallographic data, as low resolution structures lack fundamental aspects related to water inventory, thereby precluding optimal structural-energetic correlations at atomic resolution [7].

Despite the challenges of correlating specific fragment substitutions and design changes in small molecules with their respective thermodynamic binding profiles, the latter provide additional layers of valuable information to elucidate the forces governing specific ligand-target interactions. Computational studies aimed at complementing these assessments rely on experimental databases that are incomplete yet necessary to correlate specific biophysical properties at a molecular level with the net thermodynamic consequences of chemical modifications/substitutions. A critical review of progress achieved spanning three decades suggests that there is a fundamental role for thermodynamics in drug discovery strategies. Although thermodynamic signatures might not necessarily constitute endpoints for lead optimizations such as cases in which a multitude of complex events hamper resolution of the driving forces [225], thermodynamic profiling combined with high resolution structural data represent an enormous asset to accelerate drug development and optimization [182]. The collective efforts of dedicated thermodynamicists, structural biologists, and computational scientists armed with the requisite biomolecular data and biophysical parameters should improve overall predictive capabilities whereby structure informs biology and energetics provides the foundation for decision making on hit-to-lead optimizations in drug design and development.

10. Concluding Remarks

This manuscript presents an overview of experimental approaches employed in drug discovery strategies to identify and develop prospective molecules for further optimization as lead compounds in treatment regimens. Particular emphasis is focused on the fundamental role of thermodynamics in drug design and optimization by applying calorimetric methods to characterize the forces driving ligand-target association processes in solution. Considering the relevance of utilizing direct model-independent calorimetric data to elucidate macromolecular binding energetics, ongoing efforts are concentrated on the development of high-throughput methodologies aimed at deriving rapid yet accurate thermodynamic parameters while employing minimal quantities of biomaterials. The necessity and urgency of devising novel technological approaches to evaluate the biological/biophysical properties of lead compounds for the express purpose of optimizing their efficacy in terms of bioavailability, potency, and specificity is readily apparent. Drug discovery technologies rely on high throughput screenings whereby hit-to-lead decisions and compound optimization inevitably requires validations by combining structural and functional methodologies. Energetics-based approaches successfully bridge the gap between structure and function as calorimetric data fill the void by furnishing a complete description of biomolecular interactions via the elucidation of thermodynamic binding signatures and driving forces.

11. Dedication

The authors express their grateful and heartfelt appreciation to Professor Kenneth J. Breslauer, a distinguished mentor and lifelong friend, who has served as an exemplary role model throughout our scientific careers. In recognition of his seventh-fifth birthday celebration, we dedicate this manuscript as a personal tribute with our deepest admiration, affection, gratitude, and respect.

Funding: This research received no external funding.

Institutional Review Board Statement: Not Applicable.

Informed Consent Statement: Not Applicable.

Data Availability Statement: Not Applicable.

Conflicts of Interest: The authors declare no conflict of interest.

Abbreviations

ADMET	Absorption, Distribution, Metabolism, Excretion, Toxicity
AlogP	Computationally Derived LogP
BEI	Binding Efficiency Index
BLI	Biolayer Interferometry
bRo5	Beyond Rule of Five
CE	Capilary Electrophoresis
ClogP	Computationally Derived LogP
CryoEM	Cryo-Electron Microscopy
DSF	Differential Scanning Fluorimetry
EE	Enthalpic Efficiency
FBDD	Fragment-Based Drug Discovery
GCI	Grating-Coupled Interferometry
GE	Group Efficiency
HA	Heavy Atoms
HBA	Hydrogen Bond Acceptors
HBD	Hydrogen Bond Donors
HTS	High Throughput Screening
IDP	Intrinsically Disordered Protein
ITC	Isothermal Titration Calorimetry
LE	Ligand Efficiency
LipE	Lipophilic Efficiency
LLE	Ligand Lipophilic Efficiency
LogP	Logarithm of Octanol/Water Partition Coefficient
MST	Microscale Thermophoresis
NME	New Molecular Rntities
NMR	Nuclear Magnetic Resonance
NNH	Number of non-hydrogen atoms
NP	Natural Products
PAINS	Pan-Assay Interference Compounds
PSA	Polar Surface Area
QSAR	Quantitative Structure-Activity Relationships
Ro3	Rule of Three
Ro5	Rule of Five

ROTB	Rotatable Bonds
SBDD	Structure-Based Drug Design
SIHE	Size Independent Enthalpic Efficiency
SPR	Surface Plasmon Resonance
TPSA	Topological Polar Surface Area
TSA	Thermal Shift Assay
WAC	Weak Affinity Chromatography

References

1. Strebhardt, K.; Ullrich, A. Paul Ehrlich's magic bullet concept: 100 years of progress. *Nat. Rev. Cancer* **2008**, *8*, 473–480. [CrossRef] [PubMed]
2. Rizzuti, B.; Lan, W.; Santofimia-Castaño, P.; Zhou, Z.; Velázquez-Campoy, A.; Abián, O.; Peng, L.; Neira, J.L.; Xia, Y.; Iovanna, J.L. Design of Inhibitors of the Intrinsically Disordered Protein NUPR1: Balance between Drug Affinity and Target Function. *Biomolecules* **2021**, *11*, 1453. [CrossRef] [PubMed]
3. Erickson, R.P. From "magic bullet" to "specially engineered shotgun loads": The new genetics and the need for individualized pharmacotherapy. *Bioessays* **1998**, *20*, 683–685. [CrossRef]
4. Saenz-Méndez, P.; Eriksson, L.A. Exploring Polypharmacology in Drug Design. In *Rational Drug Design*; Humana Press: New York, NY, USA, 2018; Volume 1824, pp. 229–243. [CrossRef]
5. Lu, H.; Zhou, Q.; He, J.; Jiang, Z.; Peng, C.; Tong, R.; Shi, J. Recent advances in the development of protein–protein interactions modulators: Mechanisms and clinical trials. *Signal Transduct. Target. Ther.* **2020**, *5*, 1–23. [CrossRef] [PubMed]
6. Campoy, A.V.; Freire, E. ITC in the post-genomic era. . . ? Priceless. *Biophys. Chem.* **2005**, *115*, 115–124. [CrossRef] [PubMed]
7. Klebe, G. Broad-scale analysis of thermodynamic signatures in medicinal chemistry: Are enthalpy-favored binders the better development option? *Drug Discov. Today* **2019**, *24*, 943–948. [CrossRef]
8. Claveria-Gimeno, R.; Vega, S.; Abian, O.; Velazquez-Campoy, A. A Look at Ligand Binding Thermodynamics in Drug Discovery. *Expert Opin. Drug Discov.* **2017**, *12*, 363–377. [CrossRef]
9. Gilson, M.K.; Liu, T.; Baitaluk, M.; Nicola, G.; Hwang, L.; Chong, J. BindingDB in 2015: A public database for medicinal chemistry, computational chemistry and systems pharmacology. *Nucleic Acids Res.* **2015**, *44*, D1045–D1053. [CrossRef]
10. Liu, T.; Lin, Y.; Wen, X.; Jorissen, R.N.; Gilson, M.K. BindingDB: A web-accessible database of experimentally determined protein-ligand binding affinities. *Nucleic Acids Res.* **2007**, *35*, D198–D201. [CrossRef]
11. Olsson, T.S.; Williams, M.A.; Pitt, W.R.; Ladbury, J.E. The Thermodynamics of Protein–Ligand Interaction and Solvation: Insights for Ligand Design. *J. Mol. Biol.* **2008**, *384*, 1002–1017. [CrossRef]
12. Li, L.; Dantzer, J.J.; Nowacki, J.; O'Callaghan, B.J.; Meroueh, S.O. PDBcal: A Comprehensive Dataset for Receptor–Ligand Interactions with Three-dimensional Structures and Binding Thermodynamics from Isothermal Titration Calorimetry. *Chem. Biol. Drug Des.* **2008**, *71*, 529–532. [CrossRef] [PubMed]
13. Ferenczy, G.G.; Keserű, G.M. Enthalpic Efficiency of Ligand Binding. *J. Chem. Inf. Model.* **2010**, *50*, 1536–1541. [CrossRef]
14. Marky, L.A.; Snyder, J.G.; Breslauer, K.J. Calorimetric and spectroscopic investigation of drug-DNA interactions: II. Dipyrandlam binding to poly d(AT). *Nucleic Acids Res.* **1983**, *11*, 5701–5715. [CrossRef] [PubMed]
15. Breslauer, K.J.; Remeta, D.P.; Chou, W.Y.; Ferrante, R.; Curry, J.; Zaunczkowski, D.; Snyder, J.G.; Marky, L.A. Enthalpy-entropy compensations in drug-DNA binding studies. *Proc. Natl. Acad. Sci. USA* **1987**, *84*, 8922–8926. [CrossRef] [PubMed]
16. Snyder, J.G.; Hartman, N.G.; D'Estantoit, B.L.; Kennard, O.; Remeta, D.P.; Breslauer, K.J. Binding of actinomycin D to DNA: Evidence for a nonclassical high-affinity binding mode that does not require GpC sites. *Proc. Natl. Acad. Sci. USA* **1989**, *86*, 3968–3972. [CrossRef]
17. Lee, M.; Shea, R.G.; Hartley, J.A.; Kissinger, K.; Pon, R.T.; Vesnaver, G.; Breslauer, K.J.; Dabrowiak, J.C.; Lown, J.W. Molecular Recognition between Oligopeptides and Nucleic-Acids-Sequence-Specific Binding of the Naturally-Occurring Antibiotic (4s)-(+)-Anthelvencin-a and Its (4r)-(-) Enantiomer to Deoxyribonucleic Acids Deduced from H-1-NMR, Footprinting, and Thermodynamic Datat. *J. Am. Chem. Soc.* **1989**, *111*, 345–354.
18. Remeta, D.P.; Mudd, C.P.; Berger, R.L.; Breslauer, K.J. Thermodynamic characterization of daunomycin-DNA interactions: Microcalorimetric measurements of daunomycin-DNA binding enthalpies. *Biochemistry* **1991**, *30*, 9799–9809. [CrossRef]
19. Remeta, D.P.; Mudd, C.P.; Berger, R.L.; Breslauer, K.J. Thermodynamic characterization of daunomycin-DNA interactions: Comparison of complete binding profiles for a series of DNA host duplexes. *Biochemistry* **1993**, *32*, 5064–5073. [CrossRef]
20. Marky, L.A.; Breslauer, K.J. Origins of netropsin binding affinity and specificity: Correlations of thermodynamic and structural data. *Proc. Natl. Acad. Sci. USA* **1987**, *84*, 4359–4363. [CrossRef]
21. Pilch, D.S.; Poklar, N.; Gelfand, C.A.; Law, S.M.; Breslauer, K.J.; Baird, E.E.; Dervan, P.B. Binding of a hairpin polyamide in the minor groove of DNA: Sequence-specific enthalpic discrimination. *Proc. Natl. Acad. Sci. USA* **1996**, *93*, 8306–8311. [CrossRef]
22. Xu, Z.T.; Pilch, D.S.; Srinivasan, A.R.; Olson, W.K.; Geacintov, N.E.; Breslauer, K.J. Modulation of nucleic acid structure by ligand binding: Induction of a DNA center dot RNA center dot DNA hybrid triplex by DAPI intercalation. *Bioorg. Med. Chem.* **1997**, *5*, 1137–1147. [CrossRef]
23. Breslauer, K.J. The shaping of a molecular linguist: How a career studying DNA energetics revealed the language of molecular communication. *J. Biol. Chem.* **2021**, *296*, 100522. [CrossRef] [PubMed]

24. Garbett, N.C.; Chaires, J.B. Thermodynamic studies for drug design and screening. *Expert Opin. Drug Discov.* **2012**, *7*, 299–314. [CrossRef] [PubMed]
25. Holt, P.A.; Buscaglia, R.; Trent, J.O.; Chaires, J.B. A discovery funnel for nucleic acid binding drug candidates. *Drug Dev. Res.* **2010**, *72*, 178–186. [CrossRef]
26. Breslauer, K.J.; Frank, R.; Blocker, H.; Marky, L.A. Predicting DNA duplex stability from the base sequence. *Proc. Natl. Acad. Sci. USA* **1986**, *83*, 3746–3750. [CrossRef]
27. Volker, J.; Plum, G.E.; Breslauer, K.J. Heat Capacity Changes (Delta C-p) for Interconversions between Differentially-Ordered DNA States within Physiological Temperature Domains: Implications for Biological Regulatory Switches. *J. Phys. Chem. B* **2020**, *124*, 5614–5625. [CrossRef]
28. Chalikian, T.V.; Völker, J.; Plum, G.E.; Breslauer, K.J. A more unified picture for the thermodynamics of nucleic acid duplex melting: A characterization by calorimetric and volumetric techniques. *Proc. Natl. Acad. Sci. USA* **1999**, *96*, 7853–7858. [CrossRef]
29. Minetti, C.A.; Remeta, D.P.; Zharkov, D.O.; Plum, G.E.; Johnson, F.; Grollman, A.P.; Breslauer, K.J. Energetics of Lesion Recognition by a DNA Repair Protein: Thermodynamic Characterization of Formamidopyrimidine-glycosylase (Fpg) Interactions with Damaged DNA Duplexes. *J. Mol. Biol.* **2003**, *328*, 1047–1060. [CrossRef]
30. Minetti, C.A.S.A.; Remeta, D.P.; Miller, H.; Gelfand, C.A.; Plum, G.E.; Grollman, A.P.; Breslauer, K.J. The thermodynamics of template-directed DNA synthesis: Base insertion and extension enthalpies. *Proc. Natl. Acad. Sci. USA* **2003**, *100*, 14719–14724. [CrossRef]
31. Plum, G.E.; Grollman, A.P.; Johnson, F.; Breslauer, K.J. Influence of an Exocyclic Guanine Adduct on the Thermal-Stability, Conformation, and Melting Thermodynamics of a DNA Duplex. *Biochemistry* **1992**, *31*, 12096–12102. [CrossRef]
32. Plum, G.E.; Breslauer, K.J. DNA Lesions: A Thermodynamic Perspective. *Ann. N. Y. Acad. Sci.* **1994**, *726*, 45–56. [CrossRef]
33. Pilch, D.; Plum, G.E.; Breslauer, K.J. The thermodynamics of DNA structures that contain lesions of guanine tetrads. *Curr. Opin. Struct. Biol.* **1995**, *5*, 334–342. [CrossRef]
34. Plum, G.E.; Grollman, A.P.; Johnson, F.; Breslauer, K.J. Influence of the Oxidatively Damaged Adduct 8-Oxodeoxyguanosine on the Conformation, Energetics, and Thermodynamic Stability of a DNA Duplex. *Biochemistry* **1995**, *34*, 16148–16160. [CrossRef] [PubMed]
35. Gelfand, C.A.; Plum, G.E.; Grollman, A.P.; Johnson, F.; Breslauer, K.J. The impact of a bistrand abasic lesion on DNA duplex properties. *Biopolymers* **1996**, *38*, 439–445. [CrossRef]
36. Gelfand, C.A.; Plum, G.E.; Grollman, A.P.; Johnson, F.; Breslauer, K.J. The Impact of an Exocyclic Cytosine Adduct on DNA Duplex Properties: Significant Thermodynamic Consequences Despite Modest Lesion-Induced Structural Alterations. *Biochemistry* **1998**, *37*, 12507–12512. [CrossRef]
37. Gelfand, C.A.; Plum, G.E.; Grollman, A.P.; Johnson, F.; Breslauer, K.J. Thermodynamic Consequences of an Abasic Lesion in Duplex DNA Are Strongly Dependent on Base Sequence. *Biochemistry* **1998**, *37*, 7321–7327. [CrossRef]
38. Gelfand, C.A.; Plum, G.E.; Mielewczyk, S.; Remeta, D.P.; Breslauer, K.J. A quantitative method for evaluating the stabilities of nucleic acids. *Proc. Natl. Acad. Sci. USA* **1999**, *96*, 6113–6118. [CrossRef]
39. Minetti, C.; Remeta, D.; Johnson, F.; Iden, C.R.; Breslauer, K.J. Impact of Î±-Hydroxy-Propanodeoxyguanine adducts on DNA duplex energetics: Opposite base modulation and implications for mutagenicity and genotoxicity. *Biopolymers* **2009**, *93*, 370–382. [CrossRef]
40. Lukin, M.; Minetti, C.A.S.A.; Remeta, D.P.; Attaluri, S.; Johnson, F.; Breslauer, K.J.; de los Santos, C. Novel post-synthetic generation, isomeric resolution, and characterization of Fapy-dG within oligodeoxynucleotides: Differential anomeric impacts on DNA duplex properties. *Nucleic Acids Res.* **2011**, *39*, 5776–5789. [CrossRef]
41. Minetti, C.A.S.A.; Remeta, D.P.; Iden, C.R.; Johnson, F.; Grollman, A.P.; Breslauer, K.J. Impact of thymine glycol damage on DNA duplex energetics: Correlations with lesion-induced biochemical and structural consequences. *Biopolymers* **2015**, *103*, 491–508. [CrossRef]
42. Minetti, C.A.; Sun, J.Y.; Jacobs, D.P.; Kang, I.; Remeta, D.P.; Breslauer, K.J. Impact of bistrand abasic sites and proximate orientation on DNA global structure and duplex energetics. *Biopolymers* **2018**, *109*, e23098. [CrossRef]
43. Minetti, C.A.S.A.; Remeta, D.P.; Breslauer, K.J. A continuous hyperchromicity assay to characterize the kinetics and thermodynamics of DNA lesion recognition and base excision. *Proc. Natl. Acad. Sci. USA* **2008**, *105*, 70–75. [CrossRef] [PubMed]
44. Völker, J.; Breslauer, K.J. Differential repair enzyme-substrate selection within dynamic DNA energy landscapes. *Q. Rev. Biophys.* **2021**, *55*, 1–56. [CrossRef]
45. Minetti, C.A.S.A.; Remeta, D.P.; Dickstein, R.; Breslauer, K.J. Energetic signatures of single base bulges: Thermodynamic consequences and biological implications. *Nucleic Acids Res.* **2009**, *38*, 97–116. [CrossRef] [PubMed]
46. Privalov, P.L.; Dragan, A.I.; Crane-Robinson, C.; Breslauer, K.J.; Remeta, D.P.; Minetti, C.A. What Drives Proteins into the Major or Minor Grooves of DNA? *J. Mol. Biol.* **2007**, *365*, 1–9. [CrossRef]
47. Minetti, C.A.; Remeta, D.P.; Hashimoto, K.; Bonala, B.; Chennamshetti, R.; Yin, X.; Garcia-Diaz, M.; Grollman, A.P.; Johnson, F.; Sidorenko, V.S. Characterization of Aurintricarboxylic Acid (ATA) Interactions with Plasma Transporter Protein and SARS-CoV-2 Viral Targets: Correlation of Functional Activity and Binding Energetics. *Life* **2022**, *22*, 872. [CrossRef]
48. Murray, C.; Rees, D.C. The rise of fragment-based drug discovery. *Nat. Chem.* **2009**, *1*, 187–192. [CrossRef]
49. Jencks, W.P. On the attribution and additivity of binding energies. *Proc. Natl. Acad. Sci. USA* **1981**, *78*, 4046–4050. [CrossRef] [PubMed]

50. Shuker, S.B.; Hajduk, P.J.; Meadows, R.P.; Fesik, S.W. Discovering High-Affinity Ligands for Proteins: SAR by NMR. *Science* **1996**, *274*, 1531–1534. [CrossRef]
51. Hajduk, P.J.; Dinges, J.; Miknis, G.F.; Merlock, M.; Middleton, T.; Kempf, D.J.; Egan, D.A.; Walter, K.A.; Robins, T.S.; Shuker, S.B.; et al. NMR-Based Discovery of Lead Inhibitors That Block DNA Binding of the Human Papillomavirus E2 Protein. *J. Med. Chem.* **1997**, *40*, 3144–3150. [CrossRef] [PubMed]
52. Romasanta, A.K.S.; van der Sijde, P.; Hellsten, I.; Hubbard, R.E.; Keseru, G.M.; van Muijlwijk-Koezen, J.; de Esch, I.J.P. When fragments link: A bibliometric perspective on the development of fragment-based drug discovery. *Drug Discov. Today* **2018**, *23*, 1596–1609. [CrossRef] [PubMed]
53. Konteatis, Z. What makes a good fragment in fragment-based drug discovery? *Expert Opin. Drug Discov.* **2021**, *16*, 723–726. [CrossRef] [PubMed]
54. Kirsch, P.; Hartman, A.M.; Hirsch, A.K.H.; Empting, M. Concepts and Core Principles of Fragment-Based Drug Design. *Molecules* **2019**, *24*, 4309. [CrossRef]
55. Davis, B.J.; Roughley, S.D. Fragment-based lead discovery. In *Annual Reports in Medicinal Chemistry*; Academic Press: Cambridge, MA, USA, 2017; pp. 371–439.
56. Rees, D.C.; Congreve, M.; Murray, C.; Carr, R. Fragment-based lead discovery. *Nat. Rev. Drug Discov.* **2004**, *3*, 660–672. [CrossRef] [PubMed]
57. Erlanson, D.A.; Fesik, S.W.; Hubbard, R.E.; Jahnke, W.; Jhoti, H. Twenty years on: The impact of fragments on drug discovery. *Nat. Rev. Drug Discov.* **2016**, *15*, 605–619. [CrossRef] [PubMed]
58. Ichihara, O.; Barker, J.; Law, R.J.; Whittaker, M. Compound Design by Fragment-Linking. *Mol. Inform.* **2011**, *30*, 298–306. [CrossRef]
59. Chung, S.; Parker, J.B.; Bianchet, M.; Amzel, L.M.; Stivers, J.T. Impact of linker strain and flexibility in the design of a fragment-based inhibitor. *Nat. Chem. Biol.* **2009**, *5*, 407–413. [CrossRef]
60. Mignani, S.; Rodrigues, J.; Tomas, H.; Jalal, R.; Singh, P.P.; Majoral, J.-P.; Vishwakarma, R.A. Present drug-likeness filters in medicinal chemistry during the hit and lead optimization process: How far can they be simplified? *Drug Discov. Today* **2018**, *23*, 605–615. [CrossRef]
61. Lipinski, C.A.; Lombardo, F.; Dominy, B.W.; Feeney, P.J. Experimental and computational approaches to estimate solubility and permeability in drug discovery and development settings. *Adv. Drug Deliv. Rev.* **1997**, *23*, 3–25. [CrossRef]
62. Kenny, P.W.; Montanari, C.A.; Prokopczyk, I.M. ClogP(alk): A method for predicting alkane/water partition coefficient. *J. Comput.-Aided Mol. Des.* **2013**, *27*, 389–402. [CrossRef]
63. Ghose, A.K.; Viswanadhan, V.N.; Wendoloski, J.J. Prediction of Hydrophobic (Lipophilic) Properties of Small Organic Molecules Using Fragmental Methods: An Analysis of ALOGP and CLOGP Methods. *J. Phys. Chem. A* **1998**, *102*, 3762–3772. [CrossRef]
64. Klebe, G. *Drug Design-Methodology, Concepts, and Mode-of-Action*; Springer: Berlin/Heidelberg, Germany, 2013.
65. Hopkins, A.L.; Groom, C.R.; Alex, A. Ligand efficiency: A useful metric for lead selection. *Drug Discov. Today* **2004**, *9*, 430–431. [CrossRef]
66. Reynolds, C.H.; Bembenek, S.D.; Tounge, B.A. The role of molecular size in ligand efficiency. *Bioorg. Med. Chem. Lett.* **2007**, *17*, 4258–4261. [CrossRef]
67. Shultz, M.D. Two Decades under the Influence of the Rule of Five and the Changing Properties of Approved Oral Drugs. *J. Med. Chem.* **2018**, *62*, 1701–1714. [CrossRef]
68. Lipinski, C.A. Rule of five in 2015 and beyond: Target and ligand structural limitations, ligand chemistry structure and drug discovery project decisions. *Adv. Drug Deliv. Rev.* **2016**, *101*, 34–41. [CrossRef]
69. Doak, B.C.; Over, B.; Giordanetto, F.; Kihlberg, J. Oral Druggable Space beyond the Rule of 5: Insights from Drugs and Clinical Candidates. *Chem. Biol.* **2014**, *21*, 1115–1142. [CrossRef]
70. Doak, B.; Zheng, J.; Dobritzsch, D.; Kihlberg, J. How Beyond Rule of 5 Drugs and Clinical Candidates Bind to Their Targets. *J. Med. Chem.* **2015**, *59*, 2312–2327. [CrossRef]
71. Selwood, D.L. Macrocycles, the edge of drug-likeness chemical space or Goldilocks zone? *Chem. Biol. Drug Des.* **2017**, *89*, 164–168. [CrossRef]
72. Poongavanam, V.; Doak, B.; Kihlberg, J. Opportunities and guidelines for discovery of orally absorbed drugs in beyond rule of 5 space. *Curr. Opin. Chem. Biol.* **2018**, *44*, 23–29. [CrossRef]
73. Tyagi, M.; Begnini, F.; Poongavanam, V.; Doak, B.C.; Kihlberg, J. Drug Syntheses Beyond the Rule of 5. *Chem. Eur. J.* **2020**, *26*, 49–88. [CrossRef]
74. Liu, T.Y.; Minetti, C.A.; Fortes-Dias, C.L.; Liu, T.; Lin, L.; Lin, Y. C-reactive proteins, limunectin, lipopolysaccharide-binding protein, and coagulin. Molecules with lectin and agglutinin activities from Limulus Polyphemus. *Ann. N. Y. Acad. Sci.* **1994**, *712*, 146–154. [CrossRef] [PubMed]
75. Zanjani, N.T.; Miranda-Saksena, M.; Cunningham, A.L.; Dehghani, F. Antimicrobial Peptides of Marine Crustaceans: The Potential and Challenges of Developing Therapeutic Agents. *Curr. Med. Chem.* **2018**, *25*, 2245–2259. [CrossRef] [PubMed]
76. Rizzuti, B.; Grande, F.; Conforti, F.; Jimenez-Alesanco, A.; Ceballos-Laita, L.; Ortega-Alarcon, D.; Vega, S.; Reyburn, H.T.; Abian, O.; Velazquez-Campoy, A. Rutin Is a Low Micromolar Inhibitor of SARS-CoV-2 Main Protease 3CLpro: Implications for Drug Design of Quercetin Analogs. *Biomedicines* **2021**, *9*, 375. [CrossRef] [PubMed]

77. Begnini, F.; Poongavanam, V.; Over, B.; Castaldo, M.; Geschwindner, S.; Johansson, P.; Tyagi, M.; Tyrchan, C.; Wissler, L.; Sjö, P.; et al. Mining Natural Products for Macrocycles to Drug Difficult Targets. *J. Med. Chem.* **2020**, *64*, 1054–1072. [CrossRef]
78. Liu, A.Y.; Minetti, C.A.; Remeta, D.P.; Breslauer, K.J.; Chen, K.Y. HSF1, Aging and Neurodegeneration. In *Advances in Experimental Medicine and Biology*; Turksen, K., Ed.; Springer International Publishing: Cham, Switzerland, 2022.
79. Veber, D.F.; Johnson, S.R.; Cheng, H.-Y.; Smith, B.R.; Ward, K.W.; Kopple, K.D. Molecular Properties That Influence the Oral Bioavailability of Drug Candidates. *J. Med. Chem.* **2002**, *45*, 2615–2623. [CrossRef]
80. Page, M.I.; Jencks, W.P. Entropic Contributions to Rate Accelerations in Enzymic and Intramolecular Reactions and the Chelate Effect. *Proc. Natl. Acad. Sci. USA* **1971**, *68*, 1678–1683. [CrossRef]
81. Murray, C.W.; Verdonk, M.L. The consequences of translational and rotational entropy lost by small molecules on binding to proteins. *J. Comput. Mol. Des.* **2002**, *16*, 741–753. [CrossRef]
82. Shi, L.; Zhang, N. Applications of Solution NMR in Drug Discovery. *Molecules* **2021**, *26*, 576. [CrossRef]
83. Cavalluzzi, M.M.; Mangiatordi, G.F.; Nicolotti, O.; Lentini, G. Ligand efficiency metrics in drug discovery: The pros and cons from a practical perspective. *Expert Opin. Drug Discov.* **2017**, *12*, 1087–1104. [CrossRef]
84. Congreve, M.; Carr, R.; Murray, C.; Jhoti, H. A 'rule of three' for fragment-based lead discovery? *Drug Discov. Today* **2003**, *8*, 876–877. [CrossRef]
85. Hughes, J.D.; Blagg, J.; Price, D.A.; Bailey, S.; DeCrescenzo, G.A.; Devraj, R.V.; Ellsworth, E.; Fobian, Y.M.; Gibbs, M.E.; Gilles, R.W.; et al. Physiochemical drug properties associated with in vivo toxicological outcomes. *Bioorg. Med. Chem. Lett.* **2008**, *18*, 4872–4875. [CrossRef]
86. Hopkins, A.L.; Keserü, G.M.; Leeson, P.D.; Rees, D.C.; Reynolds, C.H. The role of ligand efficiency metrics in drug discovery. *Nat. Rev. Drug Discov.* **2014**, *13*, 105–121. [CrossRef]
87. Kuntz, I.D.; Chen, K.; Sharp, K.A.; Kollman, P.A. The maximal affinity of ligands. *Proc. Natl. Acad. Sci. USA* **1999**, *96*, 9997–10002. [CrossRef] [PubMed]
88. Abad-Zapatero, C.; Metz, J.T. Ligand efficiency indices as guideposts for drug discovery. *Drug Discov. Today* **2005**, *10*, 464–469. [CrossRef]
89. Zhou, H.-X.; Gilson, M.K. Theory of Free Energy and Entropy in Noncovalent Binding. *Chem. Rev.* **2009**, *109*, 4092–4107. [CrossRef]
90. Kenny, P.W.; Leitao, A.; Montanari, C. Ligand efficiency metrics considered harmful. *J. Comput. Mol. Des.* **2014**, *28*, 699–710. [CrossRef] [PubMed]
91. Shultz, M.D. Improving the Plausibility of Success with Inefficient Metrics. *ACS Med. Chem. Lett.* **2013**, *5*, 2–5. [CrossRef] [PubMed]
92. Murray, C.W.; Erlanson, D.A.; Hopkins, A.L.; Keserü, G.M.; Leeson, P.D.; Rees, D.C.; Reynolds, C.H.; Richmond, N.J. Validity of Ligand Efficiency Metrics. *ACS Med. Chem. Lett.* **2014**, *5*, 616–618. [CrossRef]
93. Kenny, P.W. The nature of ligand efficiency. *J. Chemin.* **2019**, *11*, 1–18. [CrossRef]
94. Reynolds, C.H. Ligand efficiency metrics: Why all the fuss? *Futur. Med. Chem.* **2015**, *7*, 1363–1365. [CrossRef]
95. Sheridan, R.P. Debunking the Idea that Ligand Efficiency Indices Are Superior to pIC50 as QSAR Activities. *J. Chem. Inf. Model.* **2016**, *56*, 2253–2262. [CrossRef]
96. Polanski, J.; Tkocz, A.; Kucia, U. Beware of ligand efficiency (LE): Understanding LE data in modeling structure-activity and structure-economy relationships. *J. Chemin.* **2017**, *9*, 1–8. [CrossRef] [PubMed]
97. Saxty, G.; Woodhead, S.J.; Berdini, V.; Davies, T.G.; Verdonk, M.L.; Wyatt, P.G.; Boyle, R.G.; Barford, D.; Downham, R.; Garrett, A.M.D.; et al. Identification of Inhibitors of Protein Kinase B Using Fragment-Based Lead Discovery. *J. Med. Chem.* **2007**, *50*, 2293–2296. [CrossRef]
98. Verdonk, M.L.; Rees, D.C. Group Efficiency: A Guideline for Hits-to-Leads Chemistry. *ChemMedChem* **2008**, *3*, 1179–1180. [CrossRef]
99. Free, S.M.; Wilson, J.W. A Mathematical Contribution to Structure-Activity Studies. *J. Med. Chem.* **1964**, *7*, 395–399. [CrossRef] [PubMed]
100. Hung, A.W.; Silvestre, H.L.; Wen, S.; George, G.P.C.; Boland, J.; Blundell, T.L.; Ciulli, A.; Abell, C. Optimization of Inhibitors of *Mycobacterium tuberculosis* Pantothenate Synthetase Based on Group Efficiency Analysis. *ChemMedChem* **2015**, *11*, 38–42. [CrossRef] [PubMed]
101. Hansch, C.; Björkroth, J.; Leo, A. Hydrophobicity and Central Nervous System Agents: On the Principle of Minimal Hydrophobicity in Drug Design. *J. Pharm. Sci.* **1987**, *76*, 663–687. [CrossRef]
102. Freeman-Cook, K.D.; Hoffman, R.L.; Johnson, T.W. Lipophilic efficiency: The most important efficiency metric in medicinal chemistry. *Futur. Med. Chem.* **2013**, *5*, 113–115. [CrossRef]
103. Young, R.J.; Leeson, P.D. Mapping the Efficiency and Physicochemical Trajectories of Successful Optimizations. *J. Med. Chem.* **2018**, *61*, 6421–6467. [CrossRef]
104. Johnson, T.W.; Gallego, R.A.; Edwards, M.P. Lipophilic Efficiency as an Important Metric in Drug Design. *J. Med. Chem.* **2018**, *61*, 6401–6420. [CrossRef]
105. Velazquez-Campoy, A.; Luque, I.; Todd, M.J.; Milutinovich, M.; Kiso, Y.; Freire, E. Thermodynamic dissection of the binding energetics of KNI-272, a potent HIV-1 protease inhibitor. *Protein Sci.* **2000**, *9*, 1801–1809. [CrossRef] [PubMed]

106. Velazquez-Campoy, A.; Kisob, Y.; Freire, E. The Binding Energetics of First- and Second-Generation HIV-1 Protease Inhibitors: Implications for Drug Design. *Arch. Biochem. Biophys.* **2001**, *390*, 169–175. [CrossRef] [PubMed]
107. Velazquez-Campoy, A.; Luque, I.; Freire, E. The application of thermodynamic methods in drug design. *Thermochim. Acta* **2001**, *380*, 217–227. [CrossRef]
108. Muzammil, S.; Ross, P.; Freire, E. A Major Role for a Set of Non-Active Site Mutations in the Development of HIV-1 Protease Drug Resistance. *Biochemistry* **2003**, *42*, 631–638. [CrossRef]
109. Freire, E. Do enthalpy and entropy distinguish first in class from best in class? *Drug Discov. Today* **2008**, *13*, 869–874. [CrossRef] [PubMed]
110. Freire, E. A Thermodynamic Approach to the Affinity Optimization of Drug Candidates. *Chem. Biol. Drug Des.* **2009**, *74*, 468–472. [CrossRef]
111. Ladbury, J.E.; Klebe, G.; Freire, E. Adding calorimetric data to decision making in lead discovery: A hot tip. *Nat. Rev. Drug Discov.* **2010**, *9*, 23–27. [CrossRef]
112. Kawasaki, Y.; Freire, E. Finding a better path to drug selectivity. *Drug Discov. Today* **2011**, *16*, 985–990. [CrossRef]
113. Ladbury, J.E. Calorimetry as a tool for understanding biomolecular interactions and an aid to drug design. *Biochem. Soc. Trans.* **2010**, *38*, 888–893. [CrossRef]
114. Velazquez-Campoy, A.; Todd, M.J.; Freire, E. HIV-1 Protease Inhibitors: Enthalpic versus Entropic Optimization of the Binding Affinity. *Biochemistry* **2000**, *39*, 2201–2207. [CrossRef]
115. Ruben, A.J.; Kiso, Y.; Freire, E. Overcoming Roadblocks in Lead Optimization: A Thermodynamic Perspective. *Chem. Biol. Drug Des.* **2005**, *67*, 2–4. [CrossRef] [PubMed]
116. Schön, A.; Madani, N.; Smith, A.B.; LaLonde, J.M.; Freire, E. Some Binding-Related Drug Properties are Dependent on Thermodynamic Signature. *Chem. Biol. Drug Des.* **2010**, *77*, 161–165. [CrossRef] [PubMed]
117. Li, A.; Gilson, M.K. Protein-ligand binding enthalpies from near-millisecond simulations: Analysis of a preorganization paradox. *J. Chem. Phys.* **2018**, *149*, 072311. [CrossRef]
118. Ferenczy, G.G.; Keserű, G.M. On the enthalpic preference of fragment binding. *MedChemComm* **2015**, *7*, 332–337. [CrossRef]
119. Keserű, G.M.; Erlanson, D.A.; Ferenczy, G.; Hann, M.M.; Murray, C.; Pickett, S. Design Principles for Fragment Libraries: Maximizing the Value of Learnings from Pharma Fragment-Based Drug Discovery (FBDD) Programs for Use in Academia. *J. Med. Chem.* **2016**, *59*, 8189–8206. [CrossRef]
120. Shultz, M.D. The thermodynamic basis for the use of lipophilic efficiency (LipE) in enthalpic optimizations. *Bioorganic Med. Chem. Lett.* **2013**, *23*, 5992–6000. [CrossRef]
121. Nissink, J.W.M. Simple Size-Independent Measure of Ligand Efficiency. *J. Chem. Inf. Model.* **2009**, *49*, 1617–1622. [CrossRef]
122. Barton, P.; Riley, R.J. A new paradigm for navigating compound property related drug attrition. *Drug Discov. Today* **2016**, *21*, 72–81. [CrossRef]
123. Tarcsay, Á.; Nyíri, K.; Keserű, G.M. Impact of Lipophilic Efficiency on Compound Quality. *J. Med. Chem.* **2012**, *55*, 1252–1260. [CrossRef]
124. Baell, J.; Walters, M.A. Chemistry: Chemical con artists foil drug discovery. *Nature* **2014**, *513*, 481–483. [CrossRef]
125. Baell, J.B. Feeling Nature's PAINS: Natural Products, Natural Product Drugs, and Pan Assay Interference Compounds (PAINS). *J. Nat. Prod.* **2016**, *79*, 616–628. [CrossRef] [PubMed]
126. Baell, J.B.; Nissink, J.W.M. Seven Year Itch: Pan-Assay Interference Compounds (PAINS) in 2017—Utility and Limitations. *ACS Chem. Biol.* **2017**, *13*, 36–44. [CrossRef]
127. Murray, C.W.; Verdonk, M.L.; Rees, D.C. Experiences in fragment-based drug discovery. *Trends Pharmacol. Sci.* **2012**, *33*, 224–232. [CrossRef] [PubMed]
128. Brautigam, C.; Zhao, H.; Vargas, C.; Keller, S.; Schuck, P. Integration and global analysis of isothermal titration calorimetry data for studying macromolecular interactions. *Nat. Protoc.* **2016**, *11*, 882–894. [CrossRef] [PubMed]
129. Keller, S.; Vargas, C.; Zhao, H.; Piszczek, G.; Brautigam, C.A.; Schuck, P. High-precision isothermal titration calorimetry with automated peak-shape analysis. *Anal. Chem.* **2012**, *84*, 5066–5073. [CrossRef]
130. Piñeiro, Á.; Muñoz, E.; Sabín, J.; Costas, M.; Bastos, M.; Velázquez-Campoy, A.; Garrido, P.F.; Dumas, P.; Ennifar, E.; García-Río, L.; et al. AFFINImeter: A software to analyze molecular recognition processes from experimental data. *Anal. Biochem.* **2019**, *577*, 117–134. [CrossRef]
131. Minetti, C.A.S.A.; Privalov, P.L.; Remeta, D.P. Calorimetric Methods to Characterize the Forces Driving Macromolecular Association and Folding Processes. In *Proteins in Solution and at Interfaces: Methods and Applications in Biotechnology and Materials Science*; Ruso, J.M., Piñeiro, Á., Eds.; John Wiley & Sons, Inc.: New York, NY, USA, 2013.
132. Vega, S.; Abian, O.; Velazquez-Campoy, A. On the link between conformational changes, ligand binding and heat capacity. *Biochim. Biophys. Acta-Gen. Subj.* **2016**, *1860*, 868–878. [CrossRef]
133. Tarcsay, Á.; Keserű, G.M. Is there a link between selectivity and binding thermodynamics profiles? *Drug Discov. Today* **2015**, *20*, 86–94. [CrossRef]
134. Ruiz, F.X.; Cousido-Siah, A.; Porté, S.; Domínguez, M.; Crespo, I.; Rechlin, C.; Mitschler, A.; de Lera, R.; Martín, M.J.; de la Fuente, J.; et al. Structural Determinants of the Selectivity of 3-Benzyluracil-1-acetic Acids toward Human Enzymes Aldose Reductase and AKR1B10. *ChemMedChem* **2015**, *10*, 1989–2003. [CrossRef]

135. Avelar, L.A.A.; Camilo, C.D.; De Albuquerque, S.; Fernandes, W.B.; Gonçalez, C.; Kenny, P.W.; Leitão, A.; McKerrow, J.H.; Montanari, C.A.; Orozco, E.V.; et al. Molecular Design, Synthesis and Trypanocidal Activity of Dipeptidyl Nitriles as Cruzain Inhibitors. *PLOS Negl. Trop. Dis.* **2015**, *9*, e0003916. [CrossRef]
136. Minetti, C.A.S.A.; Prokopczyk, I.M.; Orozco, E.V.M.; Rosini, F.; Remeta, D.P.; Montanari, C.A. Energetic basis for optimization of cysteine protease inhibitors. *Protein Sci.* **2016**, *25*, 168.
137. Ohtaka, H.; Muzammil, S.; Schon, A.; Velazquez-Campoy, A.; Vega, S.; Freire, E. Thermodynamic rules for the design of high affinity HIV-1 protease inhibitors with adaptability to mutations and high selectivity towards unwanted targets. *Int. J. Biochem. Cell Biol.* **2004**, *36*, 1787–1799. [CrossRef] [PubMed]
138. Edink, E.; Rucktooa, P.; Retra, K.; Akdemir, A.; Nahar, T.; Zuiderveld, O.; van Elk, R.; Janssen, E.; van Nierop, P.; van Muijlwijk-Koezen, J.; et al. Fragment Growing Induces Conformational Changes in Acetylcholine-Binding Protein: A Structural and Thermodynamic Analysis. *J. Am. Chem. Soc.* **2011**, *133*, 5363–5371. [CrossRef]
139. Gooding, M.; Tudzarova, S.; Worthington, R.J.; Kingsbury, S.R.; Rebstock, A.-S.; Dube, H.; Simone, M.I.; Visintin, C.; Lagos, D.; Quesada, J.-M.F.; et al. Exploring the Interaction Between siRNA and the SMoC Biomolecule Transporters: Implications for Small Molecule-Mediated Delivery of siRNA. *Chem. Biol. Drug Des.* **2011**, *79*, 9–21. [CrossRef] [PubMed]
140. Maple, H.J.; Garlish, R.A.; Rigau-Roca, L.; Porter, J.; Whitcombe, I.; Prosser, C.E.; Kennedy, J.; Henry, A.J.; Taylor, R.J.; Crump, M.P.; et al. Automated Protein–Ligand Interaction Screening by Mass Spectrometry. *J. Med. Chem.* **2011**, *55*, 837–851. [CrossRef]
141. Gavriilidou, A.F.M.; Holding, F.P.; Coyle, J.E.; Zenobi, R. Application of Native ESI-MS to Characterize Interactions between Compounds Derived from Fragment-Based Discovery Campaigns and Two Pharmaceutically Relevant Proteins. *SLAS Discov. Adv. Sci. Drug Discov.* **2018**, *23*, 951–959. [CrossRef]
142. Mashalidis, E.H.; Śledź, P.; Lang, S.; Abell, C. A three-stage biophysical screening cascade for fragment-based drug discovery. *Nat. Protoc.* **2013**, *8*, 2309–2324. [CrossRef] [PubMed]
143. Dammann, M.; Kramer, M.; Zimmermann, M.O.; Boeckler, F.M. Quadruple Target Evaluation of Diversity-Optimized Halogen-Enriched Fragments (HEFLibs) Reveals Substantial Ligand Efficiency for AP2-Associated Protein Kinase 1 (AAK1). *Front. Chem.* **2022**, *9*, 815567. [CrossRef]
144. Whitehouse, A.J.; Thomas, S.E.; Brown, K.P.; Fanourakis, A.; Chan, D.S.-H.; Libardo, M.D.J.; Mendes, V.; Boshoff, H.I.M.; Floto, R.A.; Abell, C.; et al. Development of Inhibitors against *Mycobacterium abscessus* tRNA (m^1G37) Methyltransferase (TrmD) Using Fragment-Based Approaches. *J. Med. Chem.* **2019**, *62*, 7210–7232. [CrossRef]
145. Zender, M.; Witzgall, F.; Kiefer, A.F.; Kirsch, B.; Maurer, C.K.; Kany, A.M.; Xu, N.; Schmelz, S.; Börger, C.; Blankenfeldt, W.; et al. Flexible Fragment Growing Boosts Potency of Quorum-Sensing Inhibitors against *Pseudomonas aeruginosa* Virulence. *ChemMedChem* **2019**, *15*, 188–194. [CrossRef]
146. Orgován, Z.; Ferenczy, G.G.; Keseru, G.M. Fragment-Based Approaches for Allosteric Metabotropic Glutamate Receptor (mGluR) Modulators. *Curr. Top. Med. Chem.* **2019**, *19*, 1768–1781. [CrossRef] [PubMed]
147. Scott, A.D.; Phillips, C.; Alex, A.; Flocco, M.; Bent, A.; Randall, A.; O'Brien, R.; Damian, L.; Jones, L.H. Thermodynamic Optimisation in Drug Discovery: A Case Study using Carbonic Anhydrase Inhibitors. *ChemMedChem* **2009**, *4*, 1985–1989. [CrossRef] [PubMed]
148. Zender, M.; Witzgall, F.; Drees, S.L.; Weidel, E.; Maurer, C.K.; Fetzner, S.; Blankenfeldt, W.; Empting, M.; Hartmann, R.W. Dissecting the Multiple Roles of PqsE in *Pseudomonas aeruginosa* Virulence by Discovery of Small Tool Compounds. *ACS Chem. Biol.* **2016**, *11*, 1755–1763. [CrossRef]
149. Schön, A.; Freire, E. Enthalpy screen of drug candidates. *Anal. Biochem.* **2016**, *513*, 1–6. [CrossRef]
150. Baggio, C.; Udompholkul, P.; Barile, E.; Pellecchia, M. Enthalpy-Based Screening of Focused Combinatorial Libraries for the Identification of Potent and Selective Ligands. *ACS Chem. Biol.* **2017**, *12*, 2981–2989. [CrossRef]
151. Scott, D.E.; Bayly, A.R.; Abell, C.; Skidmore, J. Small molecules, big targets: Drug discovery faces the protein-protein interaction challenge. *Nat. Rev. Drug Discov.* **2016**, *15*, 533–550. [CrossRef]
152. Truong, J.; George, A.; Holien, J.K. Analysis of physicochemical properties of protein–protein interaction modulators suggests stronger alignment with the "rule of five". *RSC Med. Chem.* **2021**, *12*, 1731–1749. [CrossRef] [PubMed]
153. Scott, D.E.; Ehebauer, M.T.; Pukala, T.; Marsh, M.; Blundell, S.T.L.; Venkitaraman, A.R.; Abell, C.; Hyvönen, M. Using a Fragment-Based Approach To Target Protein-Protein Interactions. *ChemBioChem* **2013**, *14*, 332–342. [CrossRef]
154. Begnini, F.; Geschwindner, S.; Johansson, P.; Wissler, L.; Lewis, R.J.; Danelius, E.; Luttens, A.; Matricon, P.; Carlsson, J.; Lenders, S.; et al. Importance of Binding Site Hydration and Flexibility Revealed When Optimizing a Macrocyclic Inhibitor of the Keap1–Nrf2 Protein–Protein Interaction. *J. Med. Chem.* **2022**, *65*, 3473–3517. [CrossRef]
155. Yachnin, B.J.; Azouz, L.R.; White, R.E.; Minetti, C.A.S.A.; Remeta, D.P.; Tan, V.M.; Drake, J.M.; Khare, S.D. Massively parallel, computationally guided design of a proenzyme. *Proc. Natl. Acad. Sci. USA* **2022**, *119*, e2116097119. [CrossRef]
156. Tan, H.; Hu, Y.; Jadhav, P.; Tan, B.; Wang, J. Progress and Challenges in Targeting the SARS-CoV-2 Papain-like Protease. *J. Med. Chem.* **2022**, *65*, 7561–7580. [CrossRef] [PubMed]
157. Wang, Y.C.; Yang, W.H.; Yang, C.S.; Hou, M.H.; Tsai, C.L.; Chou, Y.Z.; Hung, M.C.; Chen, Y. Structural basis of SARS-CoV-2 main protease inhibition by a broad-spectrum anti-coronaviral drug. *Am. J. Cancer Res.* **2020**, *10*, 2535. [PubMed]
158. Thanigaimalai, P.; Konno, S.; Yamamoto, T.; Koiwai, Y.; Taguchi, A.; Takayama, K.; Yakushiji, F.; Akaji, K.; Chen, S.E.; Naser-Tavakolian, A.; et al. Development of potent dipeptide-type SARS-CoV 3CL protease inhibitors with novel P3 scaffolds: Design, synthesis, biological evaluation, and docking studies. *Eur. J. Med. Chem.* **2013**, *68*, 372–384. [CrossRef]

159. Fu, Z.; Huang, B.; Tang, J.; Liu, S.; Liu, M.; Ye, Y.; Liu, Z.; Xiong, Y.; Zhu, W.; Cao, D.; et al. The complex structure of GRL0617 and SARS-CoV-2 PLpro reveals a hot spot for antiviral drug discovery. *Nat. Commun.* **2021**, *12*, 488. [CrossRef]
160. Shin, D.; Mukherjee, R.; Grewe, D.; Bojkova, D.; Baek, K.; Bhattacharya, A.; Schulz, L.; Widera, M.; Mehdipour, A.R.; Tascher, G.; et al. Papain-like protease regulates SARS-CoV-2 viral spread and innate immunity. *Nature* **2020**, *587*, 657–662. [CrossRef]
161. Grollman, A.P. Aurintricarboxylic Acid—Unique Inhibitor of Initiation of Protein Synthesis. *Pharmacologist* **1969**, *11*, 284.
162. Perveen, S.; Yazdi, A.K.; Devkota, K.; Li, F.L.; Ghiabi, P.; Hajian, T.; Loppnau, P.; Bolotokova, A.; Vedadi, M. A High-Throughput RNA Displacement Assay for Screening SARS-CoV-2 nsp10-nsp16 Complex toward Developing Therapeutics for COVID-19. *Slas Discov.* **2021**, *26*, 620–627. [CrossRef]
163. Young, J.C.; Hoogenraad, N.J.; Hartl, F. Molecular Chaperones Hsp90 and Hsp70 Deliver Preproteins to the Mitochondrial Import Receptor Tom70. *Cell* **2003**, *112*, 41–50. [CrossRef]
164. Zanphorlin, L.M.; de Lima, T.B.; Wong, M.J.; Balbuena, T.S.; Minetti, C.; Remeta, D.; Young, J.C.; Barbosa, L.; Gozzo, F.C.; Ramos, C.H.I. Heat Shock Protein 90 kDa (Hsp90) Has a Second Functional Interaction Site with the Mitochondrial Import Receptor Tom70. *J. Biol. Chem.* **2016**, *291*, 18620–18631. [CrossRef]
165. Gao, X.P.; Zhu, K.X.; Qin, B.; Olieric, V.; Wang, M.T.; Cui, S. Crystal structure of SARS-CoV-2 Orf9b in complex with human TOM70 suggests unusual virus-host interactions. *Nat. Commun.* **2021**, *12*, 2843. [CrossRef]
166. Ayinde, K.S.; Pinheiro, G.M.S.; Ramos, C.H.I. Binding of SARS-CoV-2 protein ORF9b to mitochondrial translocase TOM70 prevents its interaction with chaperone HSP90. *Biochimie* **2022**, *200*, 99–106. [CrossRef] [PubMed]
167. Cramer, J.; Aliu, B.; Jiang, X.; Sharpe, T.; Pang, L.; Hadorn, A.; Rabbani, S.; Ernst, B. Poly-l-lysine Glycoconjugates Inhibit DC-SIGN-mediated Attachment of Pandemic Viruses. *ChemMedChem* **2021**, *16*, 2345–2353. [CrossRef] [PubMed]
168. Cramer, J.; Lakkaichi, A.; Aliu, B.; Jakob, R.P.; Klein, S.; Cattaneo, I.; Jiang, X.; Rabbani, S.; Schwardt, O.; Zimmer, G.; et al. Sweet Drugs for Bad Bugs: A Glycomimetic Strategy against the DC-SIGN-Mediated Dissemination of SARS-CoV-2. *J. Am. Chem. Soc.* **2021**, *143*, 17465–17478. [CrossRef] [PubMed]
169. Weisshoff, H.; Krylova, O.; Nikolenko, H.; Dungen, H.D.; Dallmann, A.; Becker, S.; Gottel, P.; Muller, J.; Haberland, A. Aptamer BC 007—Efficient binder of spreading-crucial SARS-CoV-2 proteins. *Heliyon* **2020**, *6*, e05421. [CrossRef] [PubMed]
170. Ma, C.L.; Sacco, M.D.; Hurst, B.; Townsend, J.A.; Hu, Y.M.; Szeto, T.; Zhang, X.J.; Tarbet, B.; Marty, M.T.; Chen, Y.; et al. Boceprevir, GC-376, and calpain inhibitors II, XII inhibit SARS-CoV-2 viral replication by targeting the viral main protease. *Cell Res.* **2020**, *30*, 678–692. [CrossRef]
171. Shigdel, U.K.; Lee, S.-J.; Sowa, M.E.; Bowman, B.R.; Robison, K.; Zhou, M.; Pua, K.H.; Stiles, D.T.; Blodgett, J.A.V.; Udwary, D.W.; et al. Genomic discovery of an evolutionarily programmed modality for small-molecule targeting of an intractable protein surface. *Proc. Natl. Acad. Sci. USA* **2020**, *117*, 17195–17203. [CrossRef]
172. Douangamath, A.; Fearon, D.; Gehrtz, P.; Krojer, T.; Lukacik, P.; Owen, C.D.; Resnick, E.; Strain-Damerell, C.; Aimon, A.; Abranyi-Balogh, P.; et al. Crystallographic and electrophilic fragment screening of the SARS-CoV-2 main protease. *Nat. Commun.* **2020**, *11*, 5047. [CrossRef]
173. Gilson, M.K.; Zhou, H.-X. Calculation of Protein-Ligand Binding Affinities. *Annu. Rev. Biophys. Biomol. Struct.* **2007**, *36*, 21–42. [CrossRef]
174. Frenkel, M.; Chirico, R.D.; Diky, V.; Brown, P.L.; Dymond, J.H.; Goldberg, R.N.; Goodwin, A.R.H.; Heerklotz, H.; Königsberger, E.; Ladbury, J.E.; et al. Extension of ThermoML: The IUPAC standard for thermodynamic data communications (IUPAC Recommendations 2011). *Pure Appl. Chem.* **2011**, *83*, 1937–1969. [CrossRef]
175. Matos, G.D.R.; Calabrò, G.; Mobley, D.L. Infinite Dilution Activity Coefficients as Constraints for Force Field Parametrization and Method Development. *J. Chem. Theory Comput.* **2019**, *15*, 3066–3074. [CrossRef]
176. Dias, D.M.; Van Molle, I.; Baud, M.; Galdeano, C.; Geraldes, C.; Ciulli, A. Is NMR Fragment Screening Fine-Tuned to Assess Druggability of Protein–Protein Interactions? *ACS Med. Chem. Lett.* **2013**, *5*, 23–28. [CrossRef] [PubMed]
177. Silvestre, H.L.; Blundell, T.L.; Abell, C.; Ciulli, A. Integrated biophysical approach to fragment screening and validation for fragment-based lead discovery. *Proc. Natl. Acad. Sci. USA* **2013**, *110*, 12984–12989. [CrossRef]
178. Ben-Shalom, I.Y.; Lin, C.; Radak, B.K.; Sherman, W.; Gilson, M.K. Fast Equilibration of Water between Buried Sites and the Bulk by Molecular Dynamics with Parallel Monte Carlo Water Moves on Graphical Processing Units. *J. Chem. Theory Comput.* **2021**, *17*, 7366–7372. [CrossRef]
179. Heinzelmann, G.; Gilson, M.K. Automation of absolute protein-ligand binding free energy calculations for docking refinement and compound evaluation. *Sci. Rep.* **2021**, *11*, 1–18. [CrossRef]
180. Salillas, S.; Galano-Frutos, J.J.; Mahia, A.; Maity, R.; Conde-Gimenez, M.; Anoz-Carbonell, E.; Berlamont, H.; Velazquez-Campoy, A.; Touati, E.; Mamat, U.; et al. Selective Targeting of Human and Animal Pathogens of the Helicobacter Genus by Flavodoxin Inhibitors: Efficacy, Synergy, Resistance and Mechanistic Studies. *Int. J. Mol. Sci.* **2021**, *22*, 10137. [CrossRef] [PubMed]
181. Geschwindner, S.; Ulander, J. The current impact of water thermodynamics for small-molecule drug discovery. *Expert Opin. Drug Discov.* **2019**, *14*, 1221–1225. [CrossRef] [PubMed]
182. Geschwindner, S.; Ulander, J.; Johansson, P. Ligand Binding Thermodynamics in Drug Discovery: Still a Hot Tip? *J. Med. Chem.* **2015**, *58*, 6321–6335. [CrossRef]

183. Schiebel, J.; Gaspari, R.; Wulsdorf, T.; Ngo, K.; Sohn, C.; Schrader, T.E.; Cavalli, A.; Ostermann, A.; Heine, A.; Klebe, G. Intriguing role of water in protein-ligand binding studied by neutron crystallography on trypsin complexes. *Nat. Commun.* **2018**, *9*, 1–15. [CrossRef]
184. Baum, B.; Muley, L.; Smolinski, M.; Heine, A.; Hangauer, D.; Klebe, G. Non-additivity of Functional Group Contributions in Protein–Ligand Binding: A Comprehensive Study by Crystallography and Isothermal Titration Calorimetry. *J. Mol. Biol.* **2010**, *397*, 1042–1054. [CrossRef]
185. Biela, A.; Betz, M.; Heine, A.; Klebe, G. Water Makes the Difference: Rearrangement of Water Solvation Layer Triggers Non-additivity of Functional Group Contributions in Protein-Ligand Binding. *ChemMedChem* **2012**, *7*, 1423–1434. [CrossRef]
186. Kunstmann, S.; Gohlke, U.; Broeker, N.K.; Roske, Y.; Heinemann, U.; Santer, M.; Barbirz, S. Solvent Networks Tune Thermodynamics of Oligosaccharide Complex Formation in an Extended Protein Binding Site. *J. Am. Chem. Soc.* **2018**, *140*, 10447–10455. [CrossRef] [PubMed]
187. Krimmer, S.G.; Klebe, G. Thermodynamics of protein-ligand interactions as a reference for computational analysis: How to assess accuracy, reliability and relevance of experimental data. *J. Comput. Aided Mol. Des.* **2015**, *29*, 867–883. [CrossRef] [PubMed]
188. Zubrienė, A.; Smirnovienė, J.; Smirnov, A.; Morkūnaitė, V.; Michailovienė, V.; Jachno, J.; Juozapaitienė, V.; Norvaišas, P.; Manakova, E.; Gražulis, S.; et al. Intrinsic thermodynamics of 4-substituted-2,3,5,6-tetrafluorobenzenesulfonamide binding to carbonic anhydrases by isothermal titration calorimetry. *Biophys. Chem.* **2015**, *205*, 51–65. [CrossRef] [PubMed]
189. Linkuvienė, V.; Zubrienė, A.; Manakova, E.; Petrauskas, V.; Baranauskienė, L.; Zakšauskas, A.; Smirnov, A.; Gražulis, S.; Ladbury, J.E.; Matulis, D. Thermodynamic, kinetic, and structural parameterization of human carbonic anhydrase interactions toward enhanced inhibitor design. *Q. Rev. Biophys.* **2018**, *51*, e10. [CrossRef] [PubMed]
190. Freire, E. Thermodynamics of protein folding and molecular recognition. *Pure Appl. Chem.* **1997**, *69*, 2253–2262. [CrossRef]
191. Biela, A.; Sielaff, F.; Terwesten, F.; Heine, A.; Steinmetzer, T.; Klebe, G. Ligand Binding Stepwise Disrupts Water Network in Thrombin: Enthalpic and Entropic Changes Reveal Classical Hydrophobic Effect. *J. Med. Chem.* **2012**, *55*, 6094–6110. [CrossRef]
192. Morton, C.J.; Ladbury, J.E. Water mediated protein-DNA interactions: The relationship of thermodynamics to structural detail. *Protein Sci.* **1996**, *5*, 2115–2118. [CrossRef]
193. Cramer, J.; Krimmer, S.G.; Heine, A.; Klebe, G. Paying the Price of Desolvation in Solvent-Exposed Protein Pockets: Impact of Distal Solubilizing Groups on Affinity and Binding Thermodynamics in a Series of Thermolysin Inhibitors. *J. Med. Chem.* **2017**, *60*, 5791–5799. [CrossRef] [PubMed]
194. Vukovic, S.; Brennan, P.E.; Huggins, D.J. Exploring the role of water in molecular recognition: Predicting protein ligandability using a combinatorial search of surface hydration sites. *J. Physics: Condens. Matter* **2016**, *28*, 344007. [CrossRef]
195. Robinson, D.; Bertrand, T.; Carry, J.-C.; Halley, F.; Karlsson, A.; Mathieu, M.; Minoux, H.; Perrin, M.-A.; Robert, B.; Schio, L.; et al. Differential Water Thermodynamics Determine PI3K-Beta/Delta Selectivity for Solvent-Exposed Ligand Modifications. *J. Chem. Inf. Model.* **2016**, *56*, 886–894. [CrossRef]
196. Rühmann, E.; Betz, M.; Heine, A.; Klebe, G. Fragment Binding Can Be Either More Enthalpy-Driven or Entropy-Driven: Crystal Structures and Residual Hydration Patterns Suggest Why. *J. Med. Chem.* **2015**, *58*, 6960–6971. [CrossRef] [PubMed]
197. Yang, Y.; Lightstone, F.C.; Wong, S.E. Approaches to efficiently estimate solvation and explicit water energetics in ligand binding: The use of WaterMap. *Expert Opin. Drug Discov.* **2013**, *8*, 277–287. [CrossRef] [PubMed]
198. Dubins, D.N.; Filfil, R.; Macgregor, J.R.B.; Chalikian, T.V. Role of Water in Protein−Ligand Interactions: Volumetric Characterization of the Binding of 2′-CMP and 3′-CMP to Ribonuclease A. *J. Phys. Chem. B* **1999**, *104*, 390–401. [CrossRef]
199. Chalikian, T.V. Does the release of hydration water come with a Gibbs energy contribution? *J. Chem. Thermodyn.* **2021**, *158*, 106409. [CrossRef]
200. Sandner, A.; Hüfner-Wulsdorf, T.; Heine, A.; Steinmetzer, T.; Klebe, G. Strategies for Late-Stage Optimization: Profiling Thermodynamics by Preorganization and Salt Bridge Shielding. *J. Med. Chem.* **2019**, *62*, 9753–9771. [CrossRef] [PubMed]
201. Davidson, J.P.; Lubman, O.; Rose, T.; Waksman, G.; Martin, S.F. Calorimetric and Structural Studies of 1,2,3-Trisubstituted Cyclopropanes as Conformationally Constrained Peptide Inhibitors of Src SH2 Domain Binding. *J. Am. Chem. Soc.* **2001**, *124*, 205–215. [CrossRef]
202. Nguyen, G.K.T.; Kam, A.; Loo, S.; Jansson, A.E.; Pan, L.X.; Tam, J.P. Butelase 1: A Versatile Ligase for Peptide and Protein Macrocyclization. *J. Am. Chem. Soc.* **2015**, *137*, 15398–15401. [CrossRef]
203. Schmidt, M.; Toplak, A.; Quaedflieg, P.J.; Nuijens, T. Enzyme-mediated ligation technologies for peptides and proteins. *Curr. Opin. Chem. Biol.* **2017**, *38*, 1–7. [CrossRef]
204. Rühmann, E.H.; Rupp, M.; Betz, M.; Heine, A.; Klebe, G. Boosting Affinity by Correct Ligand Preorganization for the S2 Pocket of Thrombin: A Study by Isothermal Titration Calorimetry, Molecular Dynamics, and High-Resolution Crystal Structures. *ChemMedChem* **2016**, *11*, 309–319. [CrossRef]
205. Yonezawa, S.; Fujiwara, K.; Yamamoto, T.; Hattori, K.; Yamakawa, H.; Muto, C.; Hosono, M.; Tanaka, Y.; Nakano, T.; Takemoto, H.; et al. Conformational restriction approach to beta-secretase (BACE1) inhibitors III: Effective investigation of the binding mode by combinational use of X-ray analysis, isothermal titration calorimetry and theoretical calculations. *Bioorg. Med. Chem.* **2013**, *21*, 6506–6522. [CrossRef]
206. Babaoglu, K.; Shoichet, B.K. Deconstructing fragment-based inhibitor discovery. *Nat. Chem. Biol.* **2006**, *2*, 720–723. [CrossRef] [PubMed]

207. Wienen-Schmidt, B.; Schmidt, D.; Gerber, H.-D.; Heine, A.; Gohlke, H.; Klebe, G. Surprising Non-Additivity of Methyl Groups in Drug–Kinase Interaction. *ACS Chem. Biol.* **2019**, *14*, 2585–2594. [CrossRef]
208. Muley, L.; Baum, B.; Smolinski, M.; Freindorf, M.; Heine, A.; Klebe, G.; Hangauer, D.G. Enhancement of Hydrophobic Interactions and Hydrogen Bond Strength by Cooperativity: Synthesis, Modeling, and Molecular Dynamics Simulations of a Congeneric Series of Thrombin Inhibitors. *J. Med. Chem.* **2010**, *53*, 2126–2135. [CrossRef] [PubMed]
209. Klebe, G. Applying thermodynamic profiling in lead finding and optimization. *Nat. Rev. Drug Discov.* **2015**, *14*, 95–110. [CrossRef] [PubMed]
210. Klebe, G. The Use of Thermodynamic and Kinetic Data in Drug Discovery: Decisive Insight or Increasing the Puzzlement? *ChemMedChem* **2014**, *10*, 229–231. [CrossRef] [PubMed]
211. DeLorbe, J.E.; Clements, J.H.; Teresk, M.G.; Benfield, A.P.; Plake, H.R.; Millspaugh, L.E.; Martin, S.F. Thermodynamic and Structural Effects of Conformational Constraints in Protein-Ligand Interactions. Entropic Paradoxy Associated with Ligand Preorganization. *J. Am. Chem. Soc.* **2009**, *131*, 16758–16770. [CrossRef]
212. Wang, Y.; Edalji, R.P.; Panchal, S.C.; Sun, C.; Djuric, S.W.; Vasudevan, A. Are We There Yet? Applying Thermodynamic and Kinetic Profiling on Embryonic Ectoderm Development (EED) Hit-to-Lead Program. *J. Med. Chem.* **2017**, *60*, 8321–8335. [CrossRef]
213. Su, H.; Xu, Y. Application of ITC-Based Characterization of Thermodynamic and Kinetic Association of Ligands With Proteins in Drug Design. *Front. Pharmacol.* **2018**, *9*, 1133. [CrossRef]
214. Ushiyama, F.; Amada, H.; Takeuchi, T.; Tanaka-Yamamoto, N.; Kanazawa, H.; Nakano, K.; Mima, M.; Masuko, A.; Takata, I.; Hitaka, K.; et al. Lead Identification of 8-(Methylamino)-2-oxo-1,2-dihydroquinoline Derivatives as DNA Gyrase Inhibitors: Hit-to-Lead Generation Involving Thermodynamic Evaluation. *ACS Omega* **2020**, *5*, 10145–10159. [CrossRef]
215. Baker, B.; Murphy, K. Evaluation of linked protonation effects in protein binding reactions using isothermal titration calorimetry. *Biophys. J.* **1996**, *71*, 2049–2055. [CrossRef]
216. Luque, I.; Freire, E. Structural parameterization of the binding enthalpy of small ligands. *Proteins Struct. Funct. Bioinform.* **2002**, *49*, 181–190. [CrossRef] [PubMed]
217. Zubrienė, A.; Smirnov, A.; Dudutienė, V.; Timm, D.D.; Matulienė, J.; Michailovienė, V.; Zakšauskas, A.; Manakova, E.; Gražulis, S.; Matulis, D. Intrinsic Thermodynamics and Structures of 2,4- and 3,4-Substituted Fluorinated Benzenesulfonamides Binding to Carbonic Anhydrases. *ChemMedChem* **2016**, *12*, 161–176. [CrossRef]
218. Meyer, E.A.; Castellano, R.K.; Diederich, F. Interactions with Aromatic Rings in Chemical and Biological Recognition. *Angew. Chem. Int. Ed.* **2003**, *42*, 1210–1250. [CrossRef]
219. Setny, P.; Baron, R.; McCammon, J.A. How Can Hydrophobic Association Be Enthalpy Driven? *J. Chem. Theory Comput.* **2010**, *6*, 2866–2871. [CrossRef] [PubMed]
220. Malham, R.; Johnstone, S.; Bingham, R.J.; Barratt, E.; Phillips, S.E.; Laughton, C.A.; Homans, S.W. Strong solute-solute dispersive interactions in a protein-ligand complex. *J. Am. Chem. Soc.* **2005**, *127*, 17061–17067. [CrossRef]
221. Barratt, E.; Bingham, R.J.; Warner, D.J.; Laughton, C.A.; Phillips, A.S.E.V.; Homans, S.W. Van der Waals Interactions Dominate Ligand−Protein Association in a Protein Binding Site Occluded from Solvent Water. *J. Am. Chem. Soc.* **2005**, *127*, 11827–11834. [CrossRef] [PubMed]
222. Krimmer, S.G.; Betz, M.; Heine, A.; Klebe, G. Methyl, Ethyl, Propyl, Butyl: Futile But Not for Water, as the Correlation of Structure and Thermodynamic Signature Shows in a Congeneric Series of Thermolysin Inhibitors. *ChemMedChem* **2014**, *9*, 833–846. [CrossRef] [PubMed]
223. Klebe, G. Protein-Ligand Interactions as the Basis for Drug Action. In *Multifaceted Roles of Crystallography in Modern Drug Discovery*; Scapin, G., Patel, D., Arnold, E., Eds.; Springer: Dordrecht, The Netherlands, 2015; pp. 83–92.
224. Tang, Y.T.; Marshall, G.R. PHOENIX: A Scoring Function for Affinity Prediction Derived Using High-Resolution Crystal Structures and Calorimetry Measurements. *J. Chem. Inf. Model.* **2011**, *51*, 214–228. [CrossRef]
225. Ferenczy, G.G.; Keseru, G.M. Thermodynamic profiling for fragment-based lead discovery and optimization. *Expert Opin. Drug Discov.* **2019**, *15*, 117–129. [CrossRef]

life

Review

Nucleic Acid Thermodynamics Derived from Mechanical Unzipping Experiments

Paolo Rissone and **Felix Ritort** *

Small Biosystems Lab, Condensed Matter Physics Department, University of Barcelona, Carrer de Martí i Franqués 1, 08028 Barcelona, Spain; rissone.paolo@gmail.com
* Correspondence: fritort@gmail.com; Tel.: +34-934035869

Abstract: Force-spectroscopy techniques have led to significant progress in studying the physicochemical properties of biomolecules that are not accessible in bulk assays. The application of piconewton forces with laser optical tweezers to single nucleic acids has permitted the characterization of molecular thermodynamics and kinetics with unprecedented accuracy. Some examples are the hybridization reaction between complementary strands in DNA and the folding of secondary, tertiary, and other heterogeneous structures, such as intermediate and misfolded states in RNA. Here we review the results obtained in our lab on deriving the nearest-neighbor free energy parameters in DNA and RNA duplexes from mechanical unzipping experiments. Remarkable nonequilibrium effects are also observed, such as the large irreversibility of RNA unzipping and the formation of non-specific secondary structures in single-stranded DNA. These features originate from forming stem-loop structures along the single strands of the nucleic acid. The recently introduced barrier energy landscape model quantifies kinetic trapping effects due to stem-loops being applicable to both RNA and DNA. The barrier energy landscape model contains the essential features to explain the many behaviors observed in heterogeneous nucleic-acid folding.

Keywords: single-molecule biophysics; nucleic acid thermodynamics; statistical mechanics; fluctuation theorems; computational biophysics

Citation: Rissone, P.; Ritort, F. Nucleic Acid Thermodynamics Derived from Mechanical Unzipping Experiments. *Life* **2022**, *12*, 1089. https://doi.org/10.3390/life12071089

Academic Editors: Tigran Chalikian and Jens Völker

Received: 13 June 2022
Accepted: 19 July 2022
Published: 20 July 2022

1. Introduction

In molecular biophysics, the accurate knowledge of the thermodynamics of nucleic acids (NAs) and proteins is essential to obtain reliable predictions of protein folding [1–5], DNA/RNA hybridization [6], and their interactions with enzymes and ions [7]. Over the past decades, measurements of molecular free-energies, entropies, and enthalpies have been obtained by using bulk techniques such as UV absorbance, fluorescence, and calorimetry, among others [8]. DNA and RNA hybridization are key reactions in many biochemical processes, such as NA synthesis, RNA folding, and DNA amplification by PCR. The energy parameters used to model the hybridization reaction can be directly obtained from melting curves of oligos of varying sequence and length. Accurate knowledge of these parameters is crucial in many ways, such as optimizing heating–cooling protocols for PCR products, structural predictions in NAs folding and biosensor devices that use DNA hybridization for detection (e.g., surface-plasmon resonance), and sequencing technologies. Unified sets of energy parameters have been derived from DNA and RNA melting temperature data obtained by many laboratories worldwide [9,10]. They are currently used as reference values by all main prediction tools [11].

However, bulk methods yield results that are incoherent temporal averages over a large population of molecules that are in different states (Figure 1, left). The measured signal depends on the dominant species and reactions, limiting the capability of detecting fast events, rare non-native states, and reaction pathways [12]. For example, RNAs and proteins often become trapped in non-productive, misfolded structures [3,13–17]. Such structures have been related to the development of many phenotype diseases such as Huntington's disease, fragile X-associated tremor ataxia syndrome, myotonic dystrophies, and spinocerebellar ataxias, among others [18–20]. By monitoring one molecule at a time, single-molecule techniques allow for the characterization of these structures, which go undetected in bulk assays. Notice that by averaging results over many events, single-molecule measurements approach the bulk limit to which results can be compared.

By overcoming the intrinsic limitations of bulk measurements, techniques such as single-molecule fluorescence [21], single-molecule translocation across nanopores [22], atomic-force microscopy (AFM), magnetic tweezers (MT), and laser optical tweezers (LOT) [23] offer a fresh view in biomolecular sciences (Figure 1, right panel). In particular, LOTs [24] were revealed to be a powerful tool to investigate the properties of NAs. In fact, LOTs permit the direct measurement of the mechanical work (and therefore of the free-energy) needed to unfold DNA and RNA hairpins, rendering force spectroscopy a valuable tool for NA thermodynamics [25,26]. Moreover, single-molecule manipulation sets a new bar for resolving complex molecular reactions [27], such as NAs' elastic response [28,29] and non-specific secondary structure formation [30], as well as their protein and NA folding [31,32].

During the past decade, single-molecule techniques have become standard for investigating NAs at the structural, biochemical, and thermodynamical levels. Here, we briefly review the thermodynamics of DNA and RNA folding as obtained by developments in our lab over the past ten years that combine LOTs with suitable data analysis methods. In this way, it has been possible to measure DNA and RNA base-pair free-energies from mechanical unzipping experiments with high accuracy (0.1 kcal/mol) in different experimental conditions [33–35]. Unzipping consists of pulling apart the $3'$ and $5'$ ends on one side of a helical NA structure stabilized by complementary (Watson–Crick) base-pair interactions. The reverse process, called rezipping, consists of the reformation or assembly of the helical structure (hybridization) by approaching the $3'$ and $5'$ ends. Even though both DNA and RNA form double-stranded helices, important differences are found in unzipping experiments. In particular, unzipping is a fully reversible process for DNA in a broad range of salt conditions and loading rates. Instead, RNA unzipping turns out to be strongly irreversible in the same experimental conditions. In this case, transient off-pathway misfolded structures appear during the unzipping–rezipping process, slowing down the hybridization reaction [36–38]. The characterization of these off-pathway structures that compete with the native stem is a challenging problem that requires the extraction of the free energies of kinetic (non-native) states from irreversible work measurements [25,39,40]. In unzipping experiments, DNA and RNA molecules unravel into a single NA chain (single-strand) of a given molecular extension. In contrast, in bulk experiments, no forces are involved, and the unfolded state is a relaxed random coil. To relate unzipping experiments with bulk measurements, it is necessary to correct free energy differences by the so-called stretching energy contribution. This term equals the work needed to stretch the NA chain from the random coil state (zero extension) to the elongated state (finite extension). Therefore, thermodynamic measurements with force spectroscopy require accurate knowledge of the ideal elastic response of the NA chains, which can also be measured by pulling the individual single-strands [29,30]. Here, we discuss the experimental results obtained in our lab and interpret them in the framework of the barrier energy landscape (BEL) model recently introduced by us to explain the strong irreversibility observed in RNA unzipping experiments [35]. We also extend the BEL model to predict of the force–extension curves reported in studies of the non-specific secondary structure in single-stranded DNA [30].

The paper is organized as follows: In Section 2, we describe the LOT setup and the experimental protocol to unzip NAs. In Section 3, we introduce the popular nearest-neighbor model used to characterize the hybridization reaction in DNA and RNA duplexes. In Section 4 we review the results obtained by unzipping long DNA and RNA hairpins with LOTs. In Section 5, we introduce the BEL model, which can be used to explain the strong irreversibility observed in RNA unzipping and the formation of non-specific secondary structures in DNA [30]. Finally, in Section 6, we discuss future perspectives.

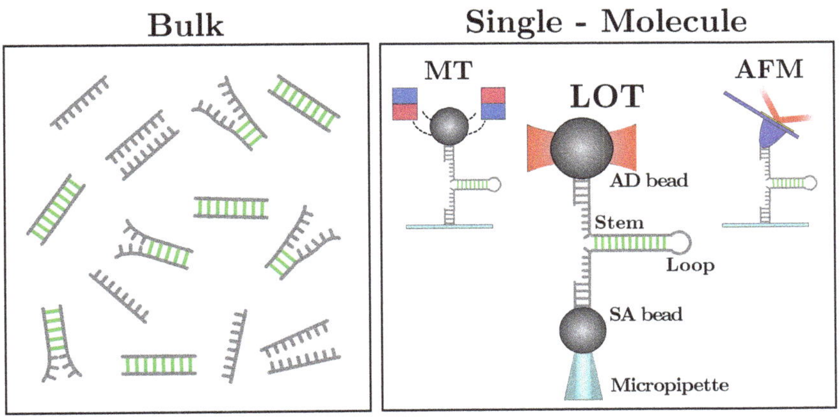

Figure 1. Information about biomolecules from bulk essays (**left panel**) is not a direct procedure and is obtained from the global behavior of a large number of molecules. On the contrary, single-molecule techniques (**right panel**) such as atomic-force microscopy (AFM), magnetic tweezers (MT), and laser optical tweezers (LOT), allow us to sample reactions one molecule at a time, characterizing the molecules at the microscopic level.

2. Mechanical Unzipping of Nucleic Acids (NAs)

To measure the NAs free-energy of formation at the single base-pair (bp) level, we unzip long DNA and RNA hairpins (a few thousand bases) consisting of a stem of fully complementary Watson–Crick base-pairs (Figure 2). The stem terminates in a tetra-loop GAAA. When fully unzipped, the hairpin converts to the single-stranded form. On the other side, the 3′ and 5′ ends of the hairpin are ligated to short (29 bp) double-stranded handles, one labeled with a digoxigenin tail (DIG) and the other with biotin (BIO). The two tags specifically bind to anti-DIG (AD)- and streptavidin-coated (SA) beads, respectively. The AD bead is optically trapped, while the SA one is held by air suction at the tip of a glass micro-pipette (Figure 1, right panel). In an unzipping experiment, the optical trap is moved with respect to the (fixed) micro-pipette at a constant speed. At the beginning of the protocol, the molecule is folded into its native double-stranded (ds) hairpin configuration (Figure 2, top-left). As the optical trap moves away from the pipette, the force applied to the hairpin increases until the intramolecular bonds at the beginning of the stem break open. As the unzipping progresses, groups of new bases open one after another in a sequential fashion. Unzipping is a stick–slip process consisting of the succession of an elastic deformation (stick) followed by the release of groups of bases that collectively unfold in a cooperative manner (slip), resulting in sudden force jumps. The unfolding protocol proceeds until the hairpin is fully unzipped and the single-strand (ss) form is fully stretched (Figure 2, bottom-left). At this point, the reverse process starts (rezipping), and the molecule refolds starting from the loop until the native ds hairpin has been reformed. Upon rezipping, groups of bases are cooperatively absorbed into the stem resulting in sudden increases in force. The force–distance curve (FDC) measured during unzipping and rezipping exhibits a saw-tooth pattern that depends on the sequence of the hairpin. In Figure 2 (right), we show the FDCs obtained by unzipping a 6.8 kbp DNA hairpin (top) in a 1 M NaCl solution and a 2 kbp RNA hairpin in a 500 mM NaCl solution (bottom). Notice that RNA FDCs exhibit many irreversible regions (insets) with large hysteresis and many long-lived intermediates. This does not happen in DNA, where unzipping experiments are carried out at equilibrium.

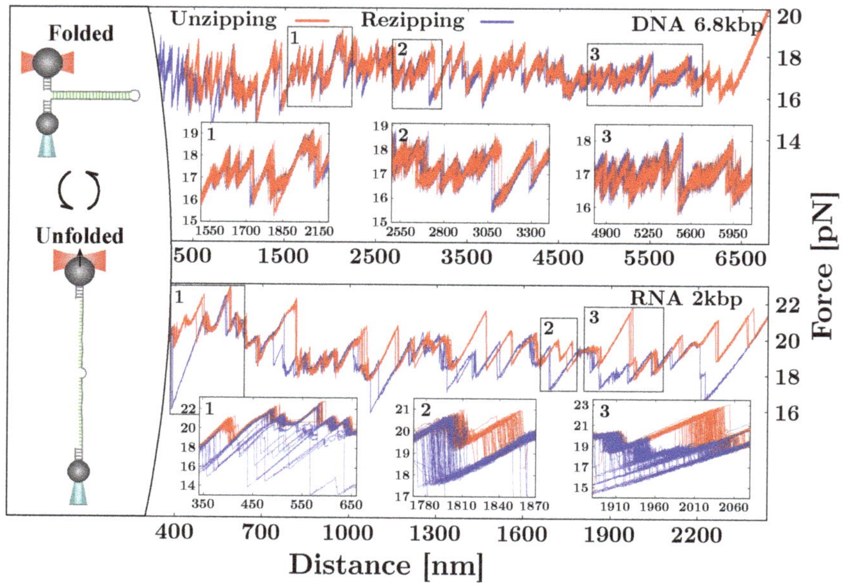

Figure 2. Unzipping experiment with LOTs. (**Left**) In a pulling experiment, the molecular construct is repeatedly unzipped and rezipped. By starting in the folded configuration (**top**), the trap–pipette distance is increased until the molecule is fully unfolded (**bottom**). Then, the process is reversed, and the hairpin refolds upon moving closer the optical trap to the micropipette. (**Right**) The typical FDCs obtained for a 6.8 kbp DNA hairpin (**top**) and a 2 kbp RNA hairpin (**bottom**) in a 1 M NaCl buffer and a 500 mM NaCl buffer, respectively. The insets magnify selected regions (gray squares) along the FDCs to point out the large irreversibility observed in RNA. In contrast, DNA unzipping is fully reversible.

3. Modeling the Unzipping Experiment

Single-molecule experiments can be used to measure the hybridization free-energy of NA, the interaction energy between hairpins and proteins [41–43], and ions [44]. They can also be used to monitor the folding pathway and detect intermediates and misfolded states [40]. As explained in the introduction, force spectroscopy measurements require the knowledge of the contributions of the experimental apparatus (optical trap) and the stretched molecular construct (handles and single-stranded NA). In fact, the standard free-energy of formation of a hairpin, ΔG_0, is defined as the free-energy difference between the native and the random coil state, the latter being the state (with zero extension) of the unfolded semi-flexible single-stranded NA when no external force is applied. However, in mechanical unzipping experiments, the unfolded molecule does not attain a random coil but a stretched semiflexible form, making the free energy difference $\Delta G_{\text{tot}} \neq \Delta G_0$. To obtain ΔG_0, we need to subtract the contribution by the experimental setup to the measured ΔG_{tot}. Therefore, a theoretical understanding of the experimental contributions to ΔG_{tot} is essential to derive reliable free energy values.

3.1. Bases Hybridization and Nearest-Neighbors Model

NAs are polymeric chains of monomers, the nucleotides that are linked together by covalent and non-covalent interactions [45] (see Figure 3A, left). The nucleotides are divided into purines, i.e., Adenine (A) and Guanine (G), and pyrimidines, i.e., Cytosine (C) and Thymine (T). For RNA, the nucleotide Uracil (U) substitutes Thymine. A nucleotide (see Figure 3A, right) is formed by one molecule of phosphoric acid (dark-yellow circle), one molecule of 2′-deoxyribose (light-yellow pentagon), and a nitrogenous base (which can be A,C,G,T/U). Two nucleotides concatenate through a bond between the phosphoric group of the first nucleotide and the third carbon of the second nucleotide. The concatenation of multiple nucleotides forms a phosphate–deoxyribose backbone of linked bases featuring a phosphate group on one terminus (5′) and a hydroxyl group on the other one (3′). The structure of NAs is conventionally given as a sequence of bases in the direction $5' \rightarrow 3'$.

The canonical interactions between nucleotides belonging to different strands are given by the Watson–Crick base-pairing rules [46] and account for purine–pyrimidine bonding: A links to T (U in the RNA case) and G to C with three and two hydrogen bonds, respectively. Although hydrogen bonding is responsible for the specificity of base interactions, much of the stability of the NA is due to base stacking [47]. In fact, nitrogenous planar bases are non-polar, and the hydrophobic stacking on top of each other maximizes the Van der Waals attraction. To account not only for the specific base-pairing but also for the stacking between adjacent base-pairs, the duplex energetics is described with the nearest-neighbor (NN) model [9,48–50].

In the NN model, the base-pairing energy of two complementary bases only depends on the base itself and on the first neighbor located in the same strand (in the $5' \rightarrow 3'$ direction). Therefore, the total free energy of formation of a duplex, ΔG_0, is given by the sum over all the nearest-neighbor base-pair (NNBP) motifs occurring along the sequence:

$$\Delta G_0(N) = \sum_{i=1}^{N} \Delta g_i \tag{1}$$

where Δg_i is the free energy of motif i. Notice that the NNBP energies are negative, as they are defined as the free-energy loss upon hybridizing a base-pair, i.e., $\Delta g_i = g_i^H - g_i^O < 0$, where $g_i^H(g_i^O)$ is the free-energy of the hybridized (open) motif. There are 16 different motifs accounting for all possible combinations of adjacent NNBPs (see Figure 3B). This number is reduced from 16 to 10 by considering the degeneracy of the free-energies due to the Watson–Crick complementarity. It is possible to further reduce this number from 10 to 8 independent parameters by considering the circular symmetry of the NN model [51,52]. This symmetry yields additional self-consistent relations so that out of the 10 NNBP energies, 2 can be expressed as linear combinations of the remaining 8 [34,52,53]. Motifs TA/AT (UA/AU) and GC/CG are usually expressed as a function of the others:

$$\begin{aligned} \Delta g_{\mathrm{TA(UA)}} &= \Delta g_{\mathrm{CG}} + \frac{1}{2}(\Delta g_{\mathrm{AC}} + \Delta g_{\mathrm{GA}} - \Delta g_{\mathrm{AG}} - \Delta g_{\mathrm{CA}}) \\ \Delta g_{\mathrm{GC}} &= \Delta g_{\mathrm{AT(AU)}} + \frac{1}{2}(\Delta g_{\mathrm{GA}} + \Delta g_{\mathrm{CA}} - \Delta g_{\mathrm{AG}} - \Delta g_{\mathrm{AC}}) \end{aligned} \tag{2}$$

where Δg_{XY} indicates the free energy of motif 5′-XY-3′ hybridized with its complementary sequence. The accurate measure of the NNBP free-energies is key for the correct estimation of the total free energy of formation of the duplex native state. The 10 independent parameters have been extracted from melting experiments of short duplexes of varying sequences and lengths [10,54–56] for both DNA and RNA and are accessible in the Mfold server [11]. Single-molecule techniques allow for much more accurate free-energy measurements with respect to bulk experiments. However, to derive the 10 (8 if circular symmetry is considered) NNBP parameters from unzipping experiments, it is fundamental to have a theoretical model of the unzipping process to predict the experimental FDC.

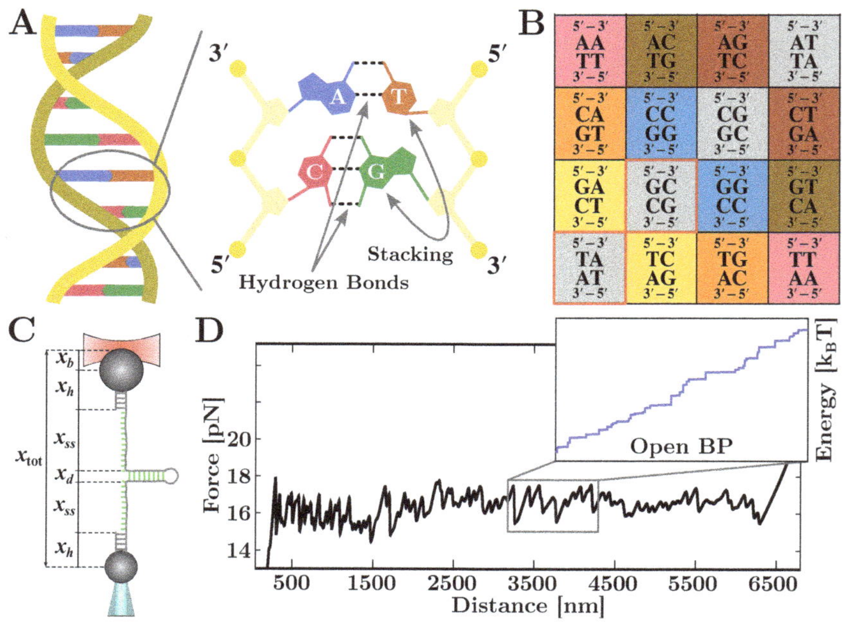

Figure 3. (**A**) Double-helix structure according the Watson–Crick base-pairing rules. (**B**) Matrix of the 16 NNBP motifs according to the NN model. Degenerate energies have the same cell color. Out of the 10 independent parameters, circular symmetry allows to express 2 NNBP energies (TA/AT and GC/CG—red-bordered cells) as a linear combination of the others (see Equation (2)). Notice that the same matrix is obtained for the RNA case by changing thymine (T) for uracil (U). (**C**) Different contributions of the molecular construct to the total extension. Each term (optical trap, dsDNA handles, ssDNA, and folded double-helix) has a different elastic response to the applied force. (**D**) Theoretical prediction of the FDC and of the minimum free-energy as a function of the open bp (inset).

3.2. *Computation of the System Free-Energy*

In unzipping experiments at controlled position, the trap–pipette distance, x_{tot}, is steadily increased (unfolding) or decreased (refolding) by moving the position of the optical trap. As the optical trap moves, the force f applied on the molecular construct changes and the number of open bases n varies accordingly. The total trap–pipette distance equals the sum of different contributions [57] (see Figure 3C), that is:

$$x_{\text{tot}}(f,n) = x_b(f) + 2x_h(f) + 2x_{ss}(f,n) + x_d(f) \tag{3}$$

where $x_b(f)$ is the displacement of the bead in relation to the center of the optical trap, $x_h(f)$ is the extension of the double-stranded handles, $x_{ss}(f,n)$ is the extension of the single-stranded (unfolded) molecule containing n bases, and $x_d(f)$ accounts for the extension of the diameter of the NA helix d (typically $d = 2$ nm for DNA and RNA hairpins [58]) projected along the force axis [59]. Notice that the latter term is zero when the hairpin is fully unzipped. The bead in the optical trap is modeled as a linear spring of stiffness k:

$$f(x_b) = kx_b. \tag{4}$$

The double-stranded handles and the single-stranded NA are modeled as elastic polymers. Typically, these terms are described by the inextensible or extensible Worm-Like Chain (WLC) model [60]. In the former case:

$$f(x) = \frac{k_B T}{4 l_p}\left[\left(1 - \frac{x}{n l_d}\right)^{-2} - 1 + 4\frac{x}{n l_d}\right] \tag{5}$$

where k_B is the Boltzmann constant, T is the temperature, l_p is the persistence length, and l_d is the interphosphate distance. Finally, the extension upon orienting the double helix is modeled as a dipole of length equal to the helix diameter, d, that aligns along the force axis:

$$x_d(f) = d\left[\coth\left(\frac{fd}{k_B T}\right) - \frac{k_B T}{fd}\right]. \tag{6}$$

Notice that while $x_b(f)$ and $x_d(f)$ are directly obtained from Equations (4) and (6), respectively, $x_{ss}(f,n)$ requires inverting Equation (5) [57].

For a given x_{tot} and n, the total free-energy of the NA hairpin is given by these elastic terms plus the hybridization free-energy in Equation (1) so that:

$$\Delta G_{\text{tot}}(x_{\text{tot}}, n) = \Delta G_b(x_b) + 2\Delta G_h(x_h) + 2\Delta G_{ss}(x_{ss}, n) + \Delta G_d(x_d) + \Delta G_0(N-n) \tag{7}$$

where N is the total number of base-pairs in the sequence and the distance constraint Equation (3). The elastic free-energy contributions are obtained by computing the following integral:

$$\Delta G_{el}(x) = \int_0^x f(x')dx' \tag{8}$$

of Equations (4)–(6). Therefore, it is straightforward to obtain:

$$\Delta G_b = \frac{1}{2}k x_b^2 \tag{9}$$

and

$$\Delta G_{\text{WLC}}(x) = \frac{k_B T}{4 l_p}\left[n l_d\left(1 - \frac{x}{n l_d}\right)^{-1} - x + 2\frac{x^2}{n l_d}\right] \tag{10}$$

from Equations (4) and (5), respectively. The computation of the free-energy needed to orient the dipole-like folded hairpin requires inverting Equation (6).

3.3. The Equilibrium FDC

Given the above model, it is possible to compute the equilibrium force of the system, f_{eq}, at each instant of the unzipping protocol. This ultimately allows us to obtain a theoretical prediction of the equilibrium FDC of the hairpin sequence. To do this, let us introduce the system partition function for a fixed x_{tot}, which is defined as the sum over all the possible states (all the possible values of n) so that:

$$Z(x_{\text{tot}}) = \sum_{n=0}^{N} \exp\left(-\frac{\Delta G_{\text{tot}}(x_{\text{tot}}, n)}{k_B T}\right) \tag{11}$$

where N is the total number of base-pairs of the sequence. By recalling that $\Delta G = -k_B T \ln Z$, the equilibrium force is then:

$$f_{\text{eq}}(x_{\text{tot}}) = -k_B T \frac{\partial \ln Z(x_{\text{tot}})}{\partial x_{\text{tot}}}. \tag{12}$$

Computing Equation (11) requires solving the transcendental equation (3) (that can be performed numerically) with respect to f and then computing Equation (7) for all $n \in [0, N]$. The value n^* minimizing the equilibrium free-energy $\Delta G_{\text{eq}} = \Delta G_{\text{tot}}(x_{\text{tot}}, n^*(x_{\text{tot}}))$ gives the most probable number of open base-pairs at a given x_{tot}. Eventually, the computation of the equilibrium force in Equation (12) gives a theoretical prediction for the unzipping curve of a given sequence (see Figure 3D). Notice that the total energy in Equation (7) is given by the balance between two independent contributions: the elastic terms and the hybridization energy. The latter term grows linearly with n (and does not depend by x_{tot}) while the elastic one is a non-linear function of (x_{tot}, n). This means that a variation of x_{tot} does not necessarily imply a variation of n^*. As a result, the equilibrium energy ΔG_{eq} is a rough function of x_{tot} exhibiting a sequence of steps (Figure 3D, inset) as n^* changes upon releasing groups of Δn^* bases. Each one of these jumps corresponds to a rip along the equilibrium FDC (Figure 3D).

4. Derivation of the NNBP Free-Energies

Unzipping data for long NA hairpins have been used to derive the formation free energies of the 10 NNBP motifs in DNA [33,34] and RNA [35]. To do this, we developed a Monte Carlo optimization algorithm that uses both the experimental data and the theoretical FDC prediction described in Section 3 [33]. By starting with an initial guess of the 10 independent parameters (or 8 if circular symmetry is applied), at each step of the optimization, a random increment in the energies is proposed and a prediction of the FDC is generated. The error made in approximating the experimental curve with the theoretical one, E, drives a Metropolis algorithm: a change in the energy parameters is accepted if the error difference with respect to the previous step is negative ($\Delta E < 0$). Otherwise ($\Delta E > 0$), the proposal is accepted if $\exp(\Delta E/T) < r$, with r being a random number uniformly distributed $r \in U(0,1)$. The algorithm continues until convergence is achieved, i.e., until ΔE is smaller than a given threshold. Let us note that because of the high number of parameters, only experimental data from the unzipping of long molecules (a few kbp) allow for an accurate estimation of the NNBP energies. In fact, the algorithm relies on the saw-tooth pattern characteristic of the sequence to accept or reject an energies proposal: the longer the sequence is, the more accurate the values of the NNBP energies are.

In Figure 4, we show the results obtained for the eight independent NNBP free-energies of DNA (circles) and RNA (squares) at different concentrations of sodium (blue) and magnesium (red). In particular, the DNA energies have been measured in a salt range from 10 mM to 1 M for Na^+ and from 0.01 mM to 10 mM for Mg^{++}, while the RNA energies have been measured at 500 mM Na^+ and 10 mM Mg^{++}. The last two parameters, GC/CG and TA/AT (UA/AU for RNA), were obtained by applying the circular symmetry relations. It is apparent that the (negative) energy parameters are lower for RNA than for DNA. Notice that this difference is more marked for motifs containing at least one purine (A, G), i.e., where stacking interactions are stronger, explaining why RNA tends to fold into more compact structures than DNA does [61].

Moreover, the NNBP RNA energies have been used to test the validity of the 1/100 salt equivalence rule of thumb. The rule states that the RNA energy parameters at a given divalent salt concentration equal that at a 100-fold monovalent salt concentration, i.e., $\Delta g_i[\mathrm{Div}^{++}] = \Delta g_i[\mathrm{Mon}^{+}]_{eq}$, where $[\mathrm{Mon}^{+}]_{eq} = 100 \times [\mathrm{Div}^{++}]$. Our results proved that 10 mM Mg^{++} corresponds to 800 mM Na^+ (approximately a 1/80 equivalence), which is compatible with the 1/100 rule within the experimental errors [35]. Although this result has been tested in a single-salt condition, its validity extends to the dilute salt regime where cooperative salt effects are negligible ($[Mg^{++}] < 0.05$ M) and competition effects with sodium are weak ($R = \sqrt{[Mg^{++}]}/[Na^{+}] > 0.22\mathrm{M}^{-1/2}$) [62,63]. The 1/100 salt equivalence rule of thumb has been disputed on the basis of experimental data obtained in bulk experiments using atomic emission spectroscopy in buffer-equilibrated samples [64]. Although this technique is capable of determining the fraction of cations that are dissociated and bound to the RNA, it does not provide a direct measurement of the free energies.

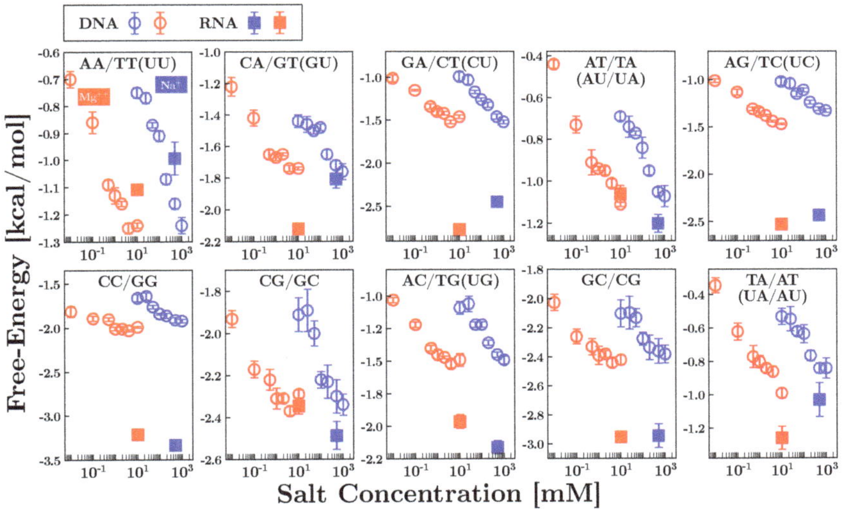

Figure 4. Experimentally measured NNBP free-energies of DNA (circles) and RNA (squares) at different salt concentrations of sodium (blue) and magnesium (red). The salt dependence of the NNBP energies in DNA has been studied in the ranges [10, 1000] mM NaCl and [0.01, 10] mM $MgCl_2$ [34]. In RNA, the energies have been measured at 500 mM NaCl and 10 mM $MgCl_2$ [35]. The energy of the last two motifs, GC/CG and TA/AT (UA/AU for RNA), has been computed by using the circular symmetry relations (see Equations (2)).

5. Out-of-Equilibrium Processes and Kinetics Effects

To characterize the irreversibility observed along the unzipping FDCs for RNA (see Figure 2), we introduced a many-valley barrier energy landscape (BEL) that accounts for the off-pathway competing stem-loop structures that can form along each single strand of the hairpin [35]. Upon unfolding (refolding) the hairpin, multiple stem-loops of varying lengths L can form along the single strands of the hairpin, i.e., in the vicinity of the hybridization junction separating the stem from the ssNA (Figure 5A, bottom right). Stem-loops slow down the hybridization of the native stem by stabilizing off-pathway conformations that kinetically trap the multiple apparent intermediates observed along the FDC (Figure 5B, grey circle snapshot). These complex structures are transiently stable; therefore, the stem-loops eventually unfold, and the native stem can form (Figure 5A).

We model the loop-BEL as follows (see Ref. [35] for details). Let us consider the set of all possible segments of L bases along a single-stranded sequence of N bases, $\mathcal{S}_L = \{[b_i, b_{i+L}]; 1 \leq i \leq N-L\}$, where b_i and b_{i+L} are the initial and the final bases of the segment, respectively. For a given L-segment, $[b_i, b_{i+L}]$, there are several competing folds, most of them stabilized by a few Watson–Crick base-pairs plus one or more loops of varying sizes (mostly from three to eight bases). By using the DINAmelt web application [65,66], we found the minimum free energy fold for a given segment. This gives the set of lowest free energy folds, $\{\epsilon^0_{L,i}\}$ for all L-segments in $\mathcal{S}_L$ at standard conditions. To construct the loop-BEL, one should consider all possible excitations (i.e., higher energy states) formed by multiple stem-loops folding along the two ssDNA strands at both sides of the junction. In principle, any number of stem-loops can form at arbitrary positions along the two strands. As counting all possible excitations is a daunting task, in the simplest approach, one can simplify the treatment by considering two stem-loops (one per strand) that are located at arbitrary and independent positions. However, stem-loops that are located far away from the junction, even if energetically favorable, cannot interfere with the unzipping–rezipping of the hairpin, a reaction taking place precisely at the junction. For a given stem-loop at a distance j from the junction and a given force f, we introduce an energy penalty equal to the mechanical work at constant f needed to bring that stem-loop from distance $j+L$ to the junction, $\int_0^f x_{j+L}(f')df'$ (Figure 5A, bottom left).

Given L, we define the loop-BEL at force f and junction position n as:

$$\Delta G_L(n,f) = -k_B T \log \sum_{j_1,j_2=0}^{N-n} \exp\left(-\frac{\Delta g_L^{(1)}(j_1,f) + \Delta g_L^{(2)}(j_2,f)}{k_B T}\right), \tag{13}$$

where $\Delta g_L^{(1,2)}(j,f)$ is the free-energy of forming a single stem-loop in strand $(1,2)$ of length L plus the work at force f to bring it from position j to the junction located at position n. $\Delta g_L^{(1,2)}(j,f)$ is then given by:

$$\Delta g_L^{(1,2)}(j,f) = \epsilon_{L,j}^{0(1,2)} + \int_0^f x_{L+j}(f')df'. \tag{14}$$

Here, $\epsilon_{L,j}^{0(1,2)}$ is the (negative) free energy of formation at zero force of the stem-loop $[b_j, b_{j+L}]$ in strand $(1,2)$. The integral term accounts for the free energy cost at force f to bring the base of the stem-loop located at the farthest end, $j+L$ bases away from the junction, towards the junction. As previously said, this term penalizes stem-loops that are formed far away from the junction because they cannot kinetically trap the native RNA hairpin. The elastic response term $x_{j+L}(f)$ has been modeled with the WLC in Equation (5).

We found that the minima of the loop-BEL (Figure 5B) directly correlate with the amount of hysteresis observed along the FDC [35]. This correlation is maximum for stem-loops of length $L \sim 20$ bases, indicating that L-segments of this size are the most likely to fold. Moreover, the analysis of the hairpin sequence showed that the regions of large irreversibility contain a high fraction of stacked (A, U) bases that are prone to Watson–Crick pairing within the same strand. On the contrary, regions of low hysteresis are characterized by lower stacking and intra-strand pairing interactions. The formation of stem-loops structures may contribute to explaining the broad phenomenology of heterogeneous RNA folding, from misfolding and multiplicity of native states to the formation of complex tertiary structures. In particular, the BEL model can be extended to the formation of non-specific secondary structures observed in pulling experiments of ssDNA and ssRNA. Although this phenomenon has not been investigated in RNA, it has been extensively studied in DNA [28,30,67–69]: upon stretching, the ssDNA's elastic response deviates from the expected ideal behavior of a polymeric chain (described for example with the WLC) forming a shoulder below $f \sim 10$–12 pN (see Figure 5C, bottom).

To model this phenomenon, let us consider an ssDNA of N bases of a random sequence, which, at difference with the previous case, cannot form a native hairpin (i.e., there is no hybridization junction). We consider the set of all possible excitations consisting of multiple stem-loops of a given length L along the sequence. The free energy of such an L-set of excitations equals:

$$\Delta G_L(f) = -k_B T \log \sum_{k=0}^{K} \exp\left(-\frac{\Delta g_L(k,f)}{k_B T}\right), \tag{15}$$

where $\Delta g_L(k,f)$ is the total free-energy contribution of $k \geq 0$ stem-loops and $K = \lfloor N/L \rfloor$ is the maximum number of stem-loops that can form along the single strand. This is given by:

$$\Delta g_L(k,f) = E_L(k) + \left[(N-kL)\Delta G_{ss}^1(f) + k\Delta G_d(f)\right]. \tag{16}$$

The term $E_L(k)$ accounts for the most energetically stable configuration of k stem-loops randomly positioned along the sequence. The term $(N-kL)\Delta G_{ss}^1(f)$ is the energy gain upon stretching the free $N-kL$ bases at force f corrected by the (smaller) energy contribution, $k\Delta G_d(f)$, of orienting k stem-loops along the force axis. Notice that $(N-kL)\Delta G_{ss}^1(f)$ is an extensive quantity, equal to the number of monomers, $N-kL$, times the energy cost to stretch a single monomer, $\Delta G^1_{ss}(f) = -\int_0^f x^1_{ss}(f')df'$, where $x^1_{ss}(f)$ has been modeled according to the WLC in Equation (5). The same consideration holds for the dipole contribution of k stem-loops, $\Delta G_d(f) = -\int_0^f x_d(f')df'$, where $x_d(f)$ is modeled according to Equation (6).

An exact computation of $E_L(k)$ in Equation (16) requires considering non-overlapping stem-loops: if a stem-loop of length L forms at position n along the sequence, the next stem-loop can only form outside the interval $[n - L : n + L]$. This is an unaffordable mathematical task that we simplified by considering overlapping stem-loops in a mean-field approximation. In this approximation, $E_L(k)$ is taken as the typical total energy of k stem-loops randomly chosen over the ensemble of C_k different realizations, without imposing any constraints on these loops (i.e., they can be overlapping or non-overlapping). In contrast, the stretching contribution $(N - kL)\Delta G^1_{ss}(f) + k\Delta G_d(f)$ is taken independent of the k stem-loops realization. Therefore, we have:

$$E_L(k) \approx \min_{C_k}\left\{\sum_k \epsilon^0_{L,k}\right\} - k_B T \log(C_k), \tag{17}$$

where, for the typical energy of k stem-loops ($\epsilon^0_{L,k} < 0, \forall k, L$), we took the most stable configuration (i.e., the one of lowest energy) within the ensemble C_k. For large N, the total number of configurations in C_k is enormous, so we restricted the sampling to a few hundreds of configurations (typically 500). The second term in the rhs of Equation (17) is an entropic contribution stabilizing the formation of stem-loops. The total number of configurations is given by the binomial coefficient, $C_k = \binom{K}{k}$, or the number of ways k objects (stem-loops) can be arranged into $K = \lfloor N/L \rfloor$ different positions.

From Equations (15) and (16), we can compute the average ssDNA extension for a given L, which is defined as:

$$x_{ss,L}(f) = -\frac{\partial \Delta G_L(f)}{\partial f} = \frac{1}{Z_L}\sum_{k=0}^{K} x_L(k,f)\exp\left(-\frac{\Delta g_L(k,f)}{k_B T}\right), \tag{18}$$

where Z_L is the system's partition function for a given L (c.f. Equation (15)):

$$Z_L = \exp\left(-\frac{\Delta G_L(f)}{k_B T}\right) = \sum_{k=0}^{K}\exp\left(-\frac{\Delta g_L(k,f)}{k_B T}\right) \tag{19}$$

and $x_L(k,f)$ is the ssDNA extension when k stem-loops are formed:

$$x_L(k,f) = (N - kL)x^1_{ss}(f) + kx_d(f)\,. \tag{20}$$

Finally, the thermodynamic free-energy and the ssDNA extension averaged over all L-segments are computed as:

$$\Delta G(f) = -k_B T \log \sum_L \exp\left(-\frac{\Delta G_L(f)}{k_B T}\right). \tag{21}$$

and

$$x_{ss}(f) = \frac{1}{Z}\sum_L x_{ss,L}(f)\exp\left(-\frac{\Delta G_L(f)}{k_B T}\right), \tag{22}$$

where $Z = \sum_L Z_L$ is the system's partition function.

In Figure 5C we show the ssDNA extension predicted by the BEL model for a random DNA sequence of $N = 2027$ bases at 10 mM (top panel) and 1 M (middle panel) NaCl salt concentrations. These results are compared with experimental data from pulling experiments of a long DNA hairpin in the same salt conditions (bottom panel). The expected ssDNA elastic response for $L \in [10, 100]$ computed with Equation (18) (solid lines) is shown along with $x_{ss}(f)$ in Equation (22) (dashed black line). It is apparent that the BEL model reproduces the deviation from the ideal WLC model (dashed gray line) experimentally observed below ~10 pN. This behavior results from the competition between the stem-loops of different sizes: the lower the force, the larger the contribution to Equation (22) by larger stem-loops. Therefore, as the ssDNA approaches the random coil state ($f = 0$ pN), the energetic gain to stretch large L-segments in Equation (16) tends to zero, while $E_L(k)$ remains constant, favoring the formation of large stem-loop structures. Remarkably, as the force approaches the dsDNA unzipping force ($f \approx 15$ pN), stem-loops of length $L \sim 30$–40 bases become the most likely folds. This number is not far from what has been reported for RNA, where $L \sim 20$ (see above and Ref. [35]). At higher forces, the elastic response collapses to the WLC as experimentally observed. Let us notice that at both 10 mM and 1 M NaCl, the predicted extension differs from the experimental data (solid dots in Figure 5C, bottom) when $f < 6$ pN. This is particularly evident at 10 mM NaCl as our model indicates that stem-loops still form at low force while no secondary structure is observed in the pulling trajectories. This and other potential inconsistencies come from the crude approximations made in Equation (17). In fact, this approximation only holds when $k \ll K$, i.e., when the typical distance between consecutive stem-loops is much larger than L, so that the overlapping is negligible. Despite the simplicity of the mean-field approximation, the BEL model is useful to study the complex behaviors observed in NA. A more rigorous analytical treatment may lead to a deeper understanding of heterogeneous folding in NA.

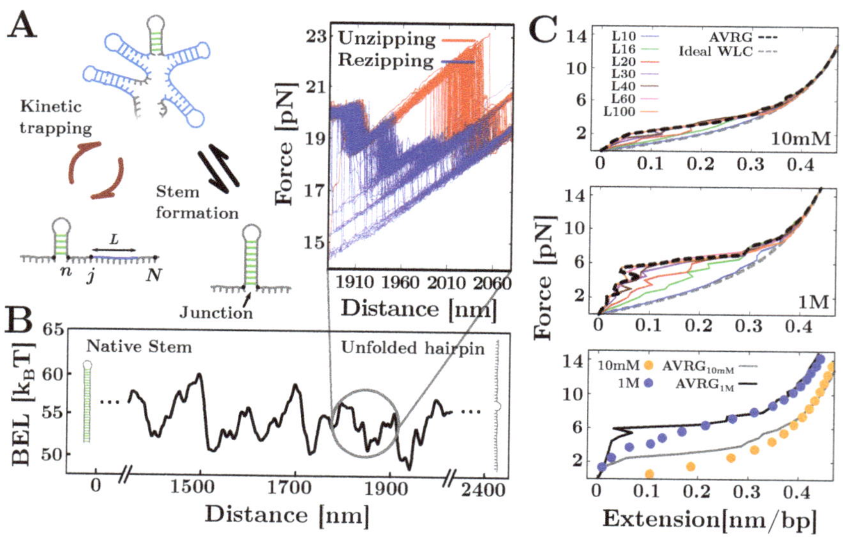

Figure 5. Stem-loops' formation along the single-stranded sequence. (**A**) Kinetic trapping during unfolding and refolding processes allows for the formation of stem-loops of different lengths folding along the unpaired strands of the molecule. As the state of highest stability is always the native state, the stem eventually forms. (**B**) Loop-BEL profile as a function of the trap–pipette distance. The zoom shows the FDC region corresponding to the circled loop-BEL. The minima of the landscape correlates with the measured hysteresis. (**C**) Extension of ssDNA predicted by the BEL model for different L-segments (solid lines) and over all L (dashed black lines) at 10 mM and 10 M NaCl (top and middle panels). The comparison between these predictions (solid lines) and experiments (solid dots) shows that the model reproduces the observed formation of non-specific secondary structure (bottom panel). The deviation form this behavior at low force ($f < 6$ pN) is due to the approximation made in computing the model.

6. Discussion

We discussed recent advances in single-molecule pulling experiments with laser optical tweezers (LOT) to investigate NAs' thermodynamics. Many of the results that we have presented cannot be obtained with bulk methods, which, by averaging over populations of molecules, do not permit sampling rare NAs folds such as misfolded configurations and intermediates. In addition to accurate experimental measurements, it is also key to have a reliable thermodynamic model for the unzipping reaction. In Section 3, we described the nearest-neighbor (NN) model that reproduces the experimental force–distance curves (FDCs). To date, the NN model provides the best theoretical description of the unzipping experiments. Conversely, unzipping experiments provide an elegant verification of the NN model. Our experimental-theoretical approach ultimately allows for predicting the unzipping FDC and the extraction of the nearest-neighbor base-pair (NNBP) energy parameters with 0.1 kcal/mol accuracy.

In Section 4, we summarized the NNBP energy parameters for DNA [34] and RNA [35]. Remarkably, the results are in agreement with the literature based on bulk assays [9,10]. The difference observed between the DNA and RNA energy parameters, where energies are lower for RNA, explains the higher stability of dsRNA duplexes in melting and unzipping experiments. It also explains the slower kinetics in RNA folding, where heterogeneous structures are often observed. This is confirmed by the multiple intermediates visible in the RNA FDCs that are not observed in DNA (Figure 2. Even though the irreversibility in RNA have been observed before in pulling experiments of short duplexes [70], a characterization of this general phenomenon in the unzipping of long RNA hairpins has been fundamental to measure the RNA NNBP free-energies. We do not exclude the possibility that similar intermediates also occur in DNA folding; however, this would require different experimental conditions, such as lower temperatures [5,71].

In Section 5, we have shown that the formation of stem-loops along the single strands explains the kinetic phenomena observed in unzipping experiments. To model this mechanism, we used the free-energy landscape formalism, useful to understand protein and NA folding into complex tertiary structures [5,72]. On the one hand, the barrier energy landscape (BEL) model introduced in Ref. [35] and defined in Equation (13) correlates with the hysteresis measured along the RNA FDCs. This shows that stem-loops are off-pathway structures that kinetically trap the RNA hairpin, slowing down the formation of the stem. On the other hand, the BEL model allows for predicting the formation of non-specific secondary structure in ssDNA pulling experiments [30]. This feature appears to be caused by the competition between multiple stem-loop folds of different sizes that make the extension to deviate from the ideal WLC behavior. Stem-loops formation appears as a general mechanism driving the folding and refolding processes of RNA hairpins as well as the elastic response of ssDNA sequences, and it may help understanding the broad phenomenology showed by NAs. However, the predictions obtained by the present model are limited by crude approximations introduced to simplify the enormous complexity of a complete stem-loops modeling. Firstly, in the computation of the stem-loops free-energy (Equations (16) and (17)), we disregarded overlapping effects between consecutive L-segments. This implies that partially overlapping L-segments can simultaneously fold into stem-loops. Moreover, we neglected the cooperativity effects, which favor the nucleation of contiguous stem-loops in an avalanche fashion reminiscent of phase transitions. Ultimately, a full BEL modeling would require allowing for the simultaneous formation of stem-loops of different lengths L, whereas the current model accounts for this effect through a mean-field approximation (Equations (21) and (22)). Although a comprehensive modeling of stem-loops formation is lacking, the phenomenon might explain many features of heterogeneous NA folding. The development of a more accurate description of the BEL model accounting for the complex phenomenology discussed above is left for future work.

Author Contributions: Conceptualization, P.R. and F.R.; methodology, P.R. and F.R.; formal analysis, P.R.; investigation, P.R. and F.R.; supervision, F.R.; writing—original draft preparation, P.R. and F.R.; writing—review and editing, P.R. and F.R.; funding acquisition, P.R. and F.R. All authors have read and agreed to the published version of the manuscript.

Funding: P.R. was supported by the Angelo della Riccia foundation. F.R. was supported by Spanish Research Council Grant PID2019-111148GB-I00 and the Institució Catalana de Recerca i Estudis Avançats Academia Prizes 2018.

Institutional Review Board Statement: Not applicable.

Informed Consent Statement: Not applicable.

Data Availability Statement: The results shown in Figures 4 and 5 have been published in [34,35], respectively. All the other results are original and will be shared upon request.

Conflicts of Interest: The authors declare no conflict of interest.

References

1. Bryngelson, J.D.; Onuchic, J.N.; Socci, N.D.; Wolynes, P.G. Funnels, pathways, and the energy landscape of protein folding: A synthesis. *Proteins* **1995**, *21*, 167–195. [CrossRef] [PubMed]
2. Aabert, B.; Johnson, A.L.J.; Raff, M.; Roberts, K.; Walter, P. *Molecular Biology of the Cell*, 4th ed.; Garland Science: New York, NY, USA, 2002.
3. Englander, S.W.; Mayne, L.; Krishna, M.M. Protein folding and misfolding: Mechanism and principles. *Q. Rev. Biophys.* **2007**, *40*, 1–41. [CrossRef] [PubMed]
4. Englander, S.W.; Mayne, L. The nature of protein folding pathways. *Proc. Natl. Acad. Sci. USA* **2014**, *111*, 15873–15880. [CrossRef] [PubMed]
5. Rico-Pasto, M.; Zaltron, A.; Davis, S.J.; Frutos, S.; Ritort, F. Molten globule–like transition state of protein barnase measured with calorimetric force spectroscopy. *Proc. Natl. Acad. Sci. USA* **2022**, *119*, e2112382119. [CrossRef] [PubMed]
6. Felsenfeld, G.; Miles, H.T. The physical and chemical properties of nucleic acids. *Annu. Rev. Biochem.* **1967**, *36*, 407–448. [CrossRef]
7. Shamsi, M.H.; Kraatz, H.B. Interactions of metal ions with DNA and some applications. *Inorg. Organomet. Polym. Mater.* **2013**, *23*, 4–23. [CrossRef]
8. Cantor, C.R.; Schimmel, P.R. *Biophysical Chemistry: Part II: Techniques for the Study of Biological Structure and Function*; Macmillan: New York, NY, USA, 1980.
9. SantaLucia, J. A unified view of polymer, dumbbell, and oligonucleotide DNA nearest-neighbor thermodynamics. *Proc. Natl. Acad. Sci. USA* **1998**, *95*, 1460–1465. [CrossRef]
10. Xia, T.; SantaLucia, J., Jr.; Burkard, M.E.; Kierzek, R.; Schroeder, S.J.; Jiao, X.; Cox, C.; Turner, D.H. Thermodynamic parameters for an expanded nearest-neighbor model for formation of RNA duplexes with Watson-Crick base-pairs. *Biochemistry* **1998**, *37*, 14719–14735. [CrossRef]
11. Zuker, M. Mfold web server for nucleic acid folding and hybridization prediction. *Nucleic Acids Res.* **2003**, *31*, 3406–3415. [CrossRef]
12. Neupane, K.; Foster, D.A.; Dee, D.R.; Yu, H.; Wang, F.; Woodside, M.T. Direct observation of transition paths during the folding of proteins and nucleic acids. *Science* **2016**, *352*, 239–242. [CrossRef]
13. Treiber, D.K.; Williamson, J.R. Exposing the kinetic traps in RNA folding. *Curr. Opin. Struct. Biol.* **1999**, *9*, 339–345. [CrossRef]
14. Johnson, S.M.; Wiseman, R.L.; Sekijima, Y.; Green, N.S.; Adamski-Werner, S.L.; Kelly, J.W. Native state kinetic stabilization as a strategy to ameliorate protein misfolding diseases: A focus on the transthyretin amyloidoses. *Acc. Chem. Res.* **2005**, *38*, 911–921. [CrossRef] [PubMed]
15. Manosas, M.; Junier, I.; Ritort, F. Force-induced misfolding in RNA. *Phys. Rev. E* **2008**, *78*, 061925. [CrossRef] [PubMed]
16. Russell, R. RNA misfolding and the action of chaperones. *Front. Biosci.* **2008**, *13*, 1–20. [CrossRef] [PubMed]
17. Rodgers, M.L.; Woodson, S.A. A roadmap for rRNA folding and assembly during transcription. *Trends Biochem. Sci.* **2021**, *46*, 889–901. [CrossRef]
18. Jain, A.; Vale, R.D. RNA phase transitions in repeat expansion disorders. *Nature* **2017**, *546*, 243–247. [CrossRef]
19. Błaszczyk, L.; Rypniewski, W.; Kiliszek, A. Structures of RNA repeats associated with neurological diseases. *Wiley Interdiscip. Rev. RNA* **2017**, *8*, e1412. [CrossRef]
20. Zhao, X.; Usdin, K. (Dys) function follows form: Nucleic acid structure, repeat expansion, and disease pathology in FMR1 disorders. *Int. J. Mol. Sci.* **2021**, *22*, 9167. [CrossRef]
21. Shashkova, S.; Leake, M.C. Single-molecule fluorescence microscopy review: Shedding new light on old problems. *Biosci. Rep.* **2017**, *37*, BSR20170031. [CrossRef]
22. Meller, A.; Nivon, L.; Branton, D. Voltage-driven DNA translocations through a nanopore. *Phys. Rev. Lett.* **2001**, *86*, 3435. [CrossRef]

23. Neuman, K.C.; Nagy, A. Single-molecule force spectroscopy: Optical tweezers, magnetic tweezers and atomic force microscopy. *Nat. Methods* **2008**, *5*, 491. [CrossRef] [PubMed]
24. Ashkin, A.; Mourou, G.; Strickland, D. The 2018 Nobel Prize in Physics: A gripping and extremely exciting tale of light. *Curr. Sci.* **2018**, *115*, 1844.
25. Junier, I.; Mossa, A.; Manosas, M.; Ritort, F. Recovery of free energy branches in single molecule experiments. *Phys. Rev. Lett.* **2009**, *102*, 070602. [CrossRef] [PubMed]
26. Tinoco, I.; Li, P.T.; Bustamante, C. Determination of thermodynamics and kinetics of RNA reactions by force. *Q. Rev. Biophys.* **2006**, *39*, 325–360. [CrossRef] [PubMed]
27. Bustamante, C.; Liphardt, J.; Ritort, F. The nonequilibrium thermodynamics of small systems. *Phys. Today* **2005**, *58*, 43. [CrossRef]
28. Bosco, A.; Camunas-Soler, J.; Ritort, F. Elastic properties and secondary structure formation of single-stranded DNA at monovalent and divalent salt conditions. *Nucleic Acids Res.* **2014**, *42*, 2064–2074. [CrossRef]
29. Camunas-Soler, J.; Ribezzi-Crivellari, M.; Ritort, F. Elastic properties of nucleic acids by single-molecule force spectroscopy. *Annu. Rev. Biophys.* **2016**, *45*, 65–84. [CrossRef]
30. Viader-Godoy, X.; Pulido, C.; Ibarra, B.; Manosas, M.; Ritort, F. Cooperativity-dependent folding of single-stranded DNA. *Phys. Rev. X* **2021**, *11*, 031037. [CrossRef]
31. Best, R.B.; Paci, E.; Hummer, G.; Dudko, O.K. Pulling direction as a reaction coordinate for the mechanical unfolding of single molecules. *J. Phys. Chem. B* **2008**, *112*, 5968–5976. [CrossRef]
32. Dudko, O.K.; Hummer, G.; Szabo, A. Theory, analysis, and interpretation of single-molecule force spectroscopy experiments. *Proc. Natl. Acad. Sci. USA* **2008**, *105*, 15755–15760. [CrossRef]
33. Huguet, J.M.; Bizarro, C.V.; Forns, N.; Smith, S.B.; Bustamante, C.; Ritort, F. Single-molecule derivation of salt dependent base-pair free energies in DNA. *Proc. Natl. Acad. Sci. USA* **2010**, *107*, 15431–15436. [CrossRef] [PubMed]
34. Huguet, J.M.; Ribezzi-Crivellari, M.; Bizarro, C.V.; Ritort, F. Derivation of nearest-neighbor DNA parameters in magnesium from single molecule experiments. *Nucleic Acids Res.* **2017**, *45*, 12921–12931. [CrossRef] [PubMed]
35. Rissone, P.; Bizarro, C.V.; Ritort, F. Stem–loop formation drives RNA folding in mechanical unzipping experiments. *Proc. Natl. Acad. Sci. USA* **2022**, *119*, e2025575119. [CrossRef] [PubMed]
36. Liphardt, J.; Onoa, B.; Smith, S.B.; Tinoco, I., Jr.; Bustamante, C. Reversible unfolding of single RNA molecules by mechanical force. *Science* **2001**, *292*, 733–737. [CrossRef]
37. Chen, S.J.; Dill, K.A. RNA folding energy landscapes. *Proc. Natl. Acad. Sci. USA* **2000**, *97*, 646–651. [CrossRef]
38. Woodson, S.A. Compact intermediates in RNA folding. *Annu. Rev. Biophys.* **2010**, *39*, 61–77. [CrossRef]
39. Liphardt, J.; Dumont, S.; Smith, S.B.; Tinoco, I., Jr.; Bustamante, C. Equilibrium information from nonequilibrium measurements in an experimental test of Jarzynski's equality. *Science* **2002**, *296*, 1832–1835. [CrossRef]
40. Alemany, A.; Mossa, A.; Junier, I.; Ritort, F. Experimental free-energy measurements of kinetic molecular states using fluctuation theorems. *Nat. Phys.* **2012**, *8*, 688. [CrossRef]
41. Koch, S.J.; Shundrovsky, A.; Jantzen, B.C.; Wang, M.D. Probing protein-DNA interactions by unzipping a single DNA double helix. *Biophys. J.* **2002**, *83*, 1098–1105. [CrossRef]
42. Camunas-Soler, J.; Alemany, A.; Ritort, F. Experimental measurement of binding energy, selectivity, and allostery using fluctuation theorems. *Science* **2017**, *355*, 412–415. [CrossRef]
43. Suren, T.; Rutz, D.; Mößmer, P.; Merkel, U.; Buchner, J.; Rief, M. Single-molecule force spectroscopy reveals folding steps associated with hormone binding and activation of the glucocorticoid receptor. *Proc. Natl. Acad. Sci. USA* **2018**, *115*, 11688–11693. [CrossRef] [PubMed]
44. Bizarro, C.V.; Alemany, A.; Ritort, F. Non-specific binding of Na+ and Mg2+ to RNA determined by force spectroscopy methods. *Nucleic Acids Res.* **2012**, *40*, 6922–6935. [CrossRef]
45. Calladine, C.R.; Drew, H. *Understanding DNA: The Molecule and How It Works*; Academic Press: Cambridge, MA, USA, 1997.
46. Saenger, W. *Principles of Nucleic Acid Structure*; Springer: Berlin/Heidelberg, Germany, 1984.
47. Yakovchuk, P.; Protozanova, E.; Frank-Kamenetskii, M.D. Base-stacking and base-pairing contributions into thermal stability of the DNA double helix. *Nucleic Acids Res.* **2006**, *34*, 564–574. [CrossRef] [PubMed]
48. DeVoe, H.; Tinoco, I., Jr. The stability of helical polynucleotides: Base contributions. *J. Mol. Biol.* **1962**, *4*, 500–517. [CrossRef]
49. Crothers, D.M.; Zimm, B.H. Theory of the melting transition of synthetic polynucleotides: Evaluation of the stacking free energy. *J. Mol. Biol.* **1964**, *9*, 1–9. [CrossRef]
50. Breslauer, K.J.; Frank, R.; Blöcker, H.; Marky, L.A. Predicting DNA duplex stability from the base sequence. *Proc. Natl. Acad. Sci. USA* **1986**, *83*, 3746–3750. [CrossRef]
51. Goldstein, R.F.; Benight, A.S. How many numbers are required to specify sequence-dependent properties of polynucleotides? *Biopolymers* **1992**, *32*, 1679–1693. [CrossRef]
52. Licinio, P.; Guerra, J.C.O. Irreducible representation for nucleotide sequence physical properties and self-consistency of nearest-neighbor dimer sets. *Biophys. J.* **2007**, *92*, 2000–2006. [CrossRef]
53. Gray, D.M.; Tinoco, I., Jr. A new approach to the study of sequence-dependent properties of polynucleotides. *Biopolymers* **1970**, *9*, 223–244. [CrossRef]
54. Mathews, D.H.; Sabina, J.; Zuker, M.; Turner, D.H. Expanded sequence dependence of thermodynamic parameters improves prediction of RNA secondary structure. *J. Mol. Biol.* **1999**, *288*, 911–940. [CrossRef]

55. Walter, A.E.; Turner, D.H.; Kim, J.; Lyttle, M.H.; Müller, P.; Mathews, D.H.; Zuker, M. Coaxial stacking of helixes enhances binding of oligoribonucleotides and improves predictions of RNA folding. *Proc. Natl. Acad. Sci. USA* **1994**, *91*, 9218–9222. [CrossRef] [PubMed]
56. Freier, S.M.; Kierzek, R.; Jaeger, J.A.; Sugimoto, N.; Caruthers, M.H.; Neilson, T.; Turner, D.H. Improved free-energy parameters for predictions of RNA duplex stability. *Proc. Natl. Acad. Sci. USA* **1986**, *83*, 9373–9377. [CrossRef] [PubMed]
57. Severino, A.; Monge, A.M.; Rissone, P.; Ritort, F. Efficient methods for determining folding free energies in single-molecule pulling experiments. *J. Stat. Mech. Theory Exp.* **2019**, *2019*, 124001. [CrossRef]
58. Woodside, M.T.; Anthony, P.C.; Behnke-Parks, W.M.; Larizadeh, K.; Herschlag, D.; Block, S.M. Direct measurement of the full, sequence-dependent folding landscape of a nucleic acid. *Science* **2006**, *314*, 1001–1004. [CrossRef]
59. Forns, N.; de Lorenzo, S.; Manosas, M.; Hayashi, K.; Huguet, J.M.; Ritort, F. Improving signal/noise resolution in single-molecule experiments using molecular constructs with short handles. *Biophys. J.* **2011**, *100*, 1765–1774. [CrossRef]
60. Bustamante, C.; Marko, J.; Siggia, E.; Smith, S. Entropic elasticity of λ-phage DNA. *Proc. Nati. Acad. Sci. USA* **1991**, *88*, 10009. [CrossRef]
61. Lipfert, J.; Skinner, G.M.; Keegstra, J.M.; Hensgens, T.; Jager, T.; Dulin, D.; Köber, M.; Yu, Z.; Donkers, S.P.; Chou, F.C.; et al. Double-stranded RNA under force and torque: Similarities to and striking differences from double-stranded DNA. *Proc. Natl. Acad. Sci. USA* **2014**, *111*, 15408–15413. [CrossRef]
62. Tan, Z.J.; Chen, S.J. RNA helix stability in mixed Na^+/Mg^{2+} solution. *Biophys. J.* **2007**, *92*, 3615–3632. [CrossRef]
63. Owczarzy, R.; Moreira, B.G.; You, Y.; Behlke, M.A.; Walder, J.A. Predicting stability of DNA duplexes in solutions containing magnesium and monovalent cations. *Biochemistry* **2008**, *47*, 5336–5353. [CrossRef]
64. Lipfert, J.; Doniach, S.; Das, R.; Herschlag, D. Understanding nucleic acid–ion interactions. *Annu. Rev. Biochem.* **2014**, *83*, 813–841. [CrossRef]
65. Markham, N.R.; Zuker, M. DINAMelt web server for nucleic acid melting prediction. *Nucleic Acids Res.* **2005**, *33*, W577–W581. [CrossRef] [PubMed]
66. Markham, N.; Zuker, M.; Keith, J. UNAFold: Software for nucleic acid folding and hybridization. *Bioinformatics* **2008**, *2*, 3–31.
67. Montanari, A.; Mézard, M. Hairpin formation and elongation of biomolecules. *Phys. Rev. Lett.* **2001**, *86*, 2178. [CrossRef] [PubMed]
68. Dessinges, M.N.; Maier, B.; Zhang, Y.; Peliti, M.; Bensimon, D.; Croquette, V. Stretching single stranded DNA, a model polyelectrolyte. *Phys. Rev. Lett.* **2002**, *89*, 248102. [CrossRef]
69. Viader-Godoy, X.; Manosas, M.; Ritort, F. Sugar-Pucker Force-Induced Transition in Single-Stranded DNA. *Int. J. Mol. Sci.* **2021**, *22*, 4745. [CrossRef]
70. Collin, D.; Ritort, F.; Jarzynski, C.; Smith, S.B.; Tinoco, I.; Bustamante, C. Verification of the Crooks fluctuation theorem and recovery of RNA folding free energies. *Nature* **2005**, *437*, 231–234. [CrossRef]
71. De Lorenzo, S.; Ribezzi-Crivellari, M.; Arias-Gonzalez, J.R.; Smith, S.B.; Ritort, F. A temperature-jump optical trap for single-molecule manipulation. *Biophys. J.* **2015**, *108*, 2854–2864. [CrossRef]
72. Gupta, A.N.; Vincent, A.; Neupane, K.; Yu, H.; Wang, F.; Woodside, M.T. Experimental validation of free-energy-landscape reconstruction from non-equilibrium single-molecule force spectroscopy measurements. *Nat. Phys.* **2011**, *7*, 631–634. [CrossRef]

life

MDPI

Review

Metal Ion-Directed Specific DNA Structures and Their Functions

Toshihiro Ihara *, Yusuke Kitamura and Yousuke Katsuda

Division of Materials Science and Chemistry, Faculty of Advanced Science and Technology, Kumamoto University, 2-39-1, Kurokami, Chuo-ku, Kumamoto 860-8555, Japan; ykita@kumamoto-u.ac.jp (Y.K.); katsuda2243@kumamoto-u.ac.jp (Y.K.)

* Correspondence: toshi@chem.kumamoto-u.ac.jp; Tel.: +81-96-342-3873

Abstract: Various DNA structures, including specific metal ion complexes, have been designed based on the knowledge of canonical base pairing as well as general coordination chemistry. The role of metal ions in these studies is quite broad and diverse. Metal ions can be targets themselves in analytical applications, essential building blocks of certain DNA structures that one wishes to construct, or they can be responsible for signal generation, such as luminescence or redox. Using DNA conjugates with metal chelators, one can more freely design DNA complexes with diverse structures and functions by following the simple HSAB rule. In this short review, the authors summarize a part of their DNA chemistries involving specific metal ion coordination. It consists of three topics: (1) significant stabilization of DNA triple helix by silver ion; (2) metal ion-directed dynamic sequence edition through global conformational change by intramolecular complexation; and (3) reconstruction of luminescent lanthanide complexes on DNA and their analytical applications.

Keywords: DNA conjugate; metal ion; triple helix; silver ion; lanthanide; ATP sensor; aptamer; terpyridine; sequence edition; DNAzyme

Citation: Ihara, T.; Kitamura, Y.; Katsuda, Y. Metal Ion-Directed Specific DNA Structures and Their Functions. *Life* **2022**, *12*, 686. https://doi.org/10.3390/life12050686

Academic Editors: Tigran Chalikian and Jens Völker

Received: 5 April 2022
Accepted: 2 May 2022
Published: 5 May 2022

1. Introduction

Almost 20 years after the completion of the Human Genome Project [1,2], nucleic acid chemistry is once again a research focus, with the emergence of a number of new fields, including epigenetics, RNA interference, noncoding RNA, mRNA vaccines, iPS cells, and nucleic acid medicine. Molecular engineering in nucleic acid chemistry has become more flexible than ever before to meet the demands and new challenges in these emerging research fields, taking advantage of functional nucleic acids such as aptamers [3], ribozyme [4], and DNAzyme [5–7], as well as programmed spontaneous strand exchange reactions such as DNA circuits [8,9].

The basis of any molecular engineering of nucleic acids, after all, is the knowledge and techniques for the formation of canonical and some noncanonical structures of nucleic acids. Under certain conditions, we are now able to logically design a variety of static and dynamic structures of DNA/RNA by predicting the most stable duplex structures that will form in the solutions containing these mixtures. The pioneering work in predicting the thermodynamic stability of duplex structures based on the nearest-neighbor model was undoubtedly revolutionary [10–13]. These advances have contributed to the development of almost all modern hybridization-based techniques widely used for gene expression control, gene editing, and analysis, such as antisense, RNAi, CRISPR/Cas9, and in situ hybridization, among others.

In our previous work, using synthetic DNAs and DNA conjugates, we reported various conjugates consisting of oligo DNAs and functional molecules, e.g., anthracene [14], β-cyclodextrin [15], ferrocene [16], and several metal ion chelators [17]. The DNA conjugates were programmed or designed to change their structures in various ways in response to specific stimuli. Outputs, including photochemical ligations, luminescence, and electrochemical responses resulting from the structural changes, have been used to detect the stimuli themselves, complementary DNA/RNA, or other biomolecules. The merit of nucleic acids as molecular platforms is that pre-designed structures can be precisely constructed in a bottom-up fashion, providing an unparalleled advantage with respect to all biotechnological applications as mentioned above. This short article summarizes some of the works we conducted after one of the authors (T.I.) visited the laboratory of Prof. Breslauer at Rutgers University in 2001–2002, especially for the systems related to complexation with metal ions. In these works, metal ions were used as a critical structural factor and a signal generator in the design of both static and dynamic DNA structures.

2. Stabilization of a Parallel-Motif DNA Triplex by Silver Ion

When designing the ligands for specific sequences in DNA duplexes, triple helix formation is a useful recognition motif, inasmuch as the formation of the base triplet follows the simple rule of complementary Hoogsteen hydrogen bonding, CG.C$^+$ and TA.T, for the parallel motif of the triplex. However, the triplexes containing CG.C$^+$ triplets form only in a weak acidic solution, because the N_3 position of cytosines (pK_a = 4.5) in the third strand must be protonated to fulfill its complementarity [18]. With the aim of achieving sufficient stability under physiological conditions, a large quantity of chemically modified DNA has been developed by taking advantage of the highly advanced techniques of organic synthesis [19].

We reported an effective alternative method for the stabilization of the parallel motif triple helix of DNA using silver ions (Ag^+) [20]. Ono et al. reported that the formation of C–C and T–T mispairings in the duplex is promoted by Ag^+ and Hg^{2+}, respectively [21]. In these duplexes, the ions were placed between the bases to form specific bridges (C–Ag^+–C, T–Hg^{2+}–T). These results led to the idea that it might be possible to stabilize triplex structures containing CG.C$^+$ base triplexes with Ag^+. The silver ion was expected to displace an N_3 proton of a cytosine in the CG.C$^+$ to form a metal ion-mediated base triplet, CG.CAg$^+$, as shown in Figure 1a. This process was expected to stabilize parallel motif triplexes even at neutral pH.

Figure 1b shows the UV melting curves at a pH of 7.0 and pH dependence of the temperatures of triplex–duplex transition. Surprisingly, the addition of an equal amount of Ag^+ (to CG.C$^+$) increased the melting temperature of the triplex by more than 30 °C under neutral conditions [20]. In the absence of Ag^+, the relation of the melting temperature to pH was clearly evident. Meanwhile, in the presence of Ag^+, the correlation disappeared, and a biphasic feature consisting of two temperature-independent regions was observed. A phase diagram of the structure of the Ag^+-mediated nucleobase complex could be drawn based on this characteristic melting temperature–pH property. Mass spectrometry (ESI-TOF MS) clearly showed the quantitative formation of the Ag^+-mediated base triplet, CG.CAg$^+$. The results of modeling studies by DFT (B3LYP/6-31G*//3-21G) suggest that the cytosines on the third strand are forced to be twisted from the plane of Watson–Crick GC pairs in CG.CAg$^+$ triplets, because the coordination distance in N–Ag^+–N would be longer than that of the Hoogsteen hydrogen bonds, N–H^+–N, in CG.C$^+$. The deviation from the typical triplex structure observed in studies using CD is consistent with this non-planarity of CG.CAg$^+$.

The method described here for the stabilization of DNA triplexes is both simple and effective. All that is required is the addition of an equimolar amount of Ag^+ into the solution containing the DNA triplex. The triplexes mediated by Ag^+ were found to be stable even in a weak basic solution and can be applied in various research tasks, including the regulation of DNAzyme activity [22], sensing [23,24], and luminous Ag nanocluster formation [25].

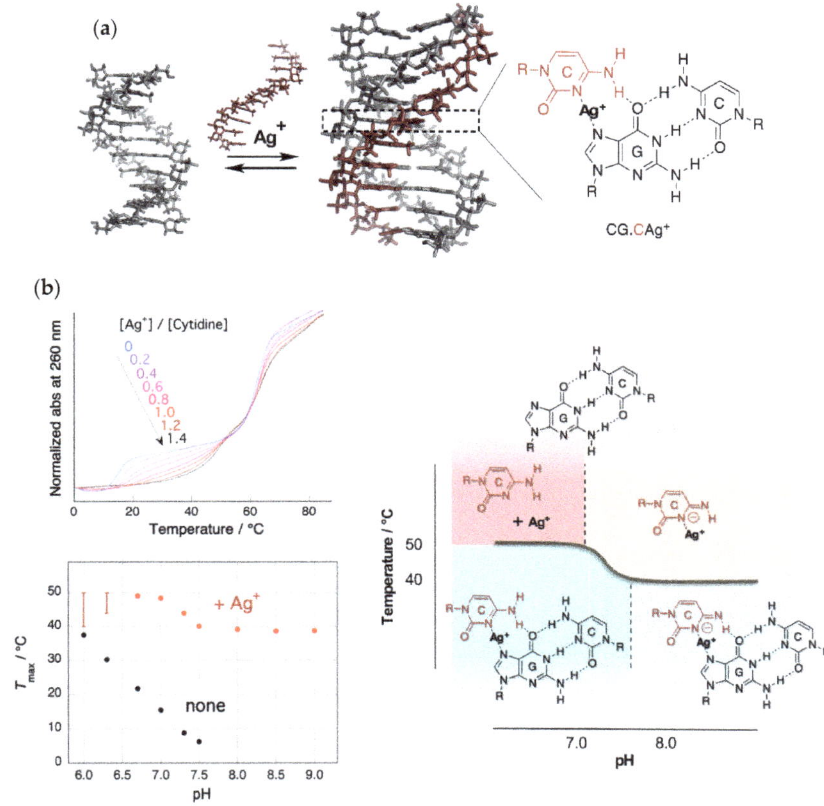

Figure 1. Triplex stabilization by silver ion. (**a**) Structure of the triplex and CG.CAg$^+$ base triplet; (**b**) upper left: UV melting curves in the presence of Ag$^+$ with various feeding ratios. Only the temperature of the triplex–duplex transition increased with the addition of Ag$^+$. Bottom left: pH dependence of the melting temperatures of triplex in the absence and presence of Ag$^+$. The melting temperature in the presence of Ag$^+$ consists of two pH-independent regions. Right: Phase diagram of the structure of Ag$^+$-mediated nucleobase complexes.

3. Metal Ion-Directed Dynamic Splicing of DNA through Global Conformational Change by Intramolecular Complexation

The metal ion-directed global conformational control of DNA was performed as follows. Two terpyridine units were built into the distal sites on the DNA backbone to prepare a conjugate, i.e., **terpy$_2$DNA**. The two terpyridines formed a stable intramolecular 1:2 complex, $[M(terpy)_2]^{2+}$, with divalent transition metal ions, M^{2+}, namely Fe^{2+}, Ni^{2+}, Cu^{2+}, and Zn^{2+}. By the specific formation of an intramolecular metal complex, a part of the sequence of the DNA in between the two terpyridine units was reversibly excluded, and the two flanking external DNA segments were directly connected with each other to form an Ω-shaped structure presenting a new sequence (Figure 2). This can be regarded as a metal ion-directed reversible edition of the DNA sequence or dynamic DNA splicing [26].

Conformational control of **terpy$_2$DNA** was confirmed via UV melting with the complementary tandem sequence of the two external segments. The results show that the duplex structure was significantly stabilized in the presence of an equimolar amount (to **terpy$_2$DNA**) of transition metal ions. In addition, in the presence of the metal ions, the shape of the melting curves changed to be more cooperative, indicating that the two sequences outside the terpyridines were cooperatively dissociated in a narrow temperature range. The dependences of duplex stabilization on the metal ion feeding ratio (r = [M^{2+}]/[**terpy$_2$DNA**]) were different for each of the metal ions. In the case of Fe^{2+} and Ni^{2+}, the duplex remained stable even when additional metal ions were added to $r = 2$ or 3. In contrast, the duplex was destabilized at higher feeding ratios of Cu^{2+} and Zn^{2+}. The stability of the duplex was maintained even in the presence of the excess amounts of Fe^{2+} and Ni^{2+}, because the Ω-shaped conformation of **terpy$_2$DNA** was preserved due to the magnitudes of the two successive binding constants with terpyridine, $K_1 < K_2$. As for Cu^{2+} and Zn^{2+}, the global conformation of **terpy$_2$DNA** was changed from Ω-form to a linear form accompanying the transition of the complex types formed on **terpy$_2$DNA** from $[M(terpy)_2]^{2+}$ (on **terpy$_2$DNA**·M^{2+}) to $2[M(terpy)]^{2+}$ (on **terpy$_2$DNA**·$2M^{2+}$) with increasing amounts of ions due to their binding properties with terpyridine, $K_1 > K_2$. This indicates that the general trend of the complexation of transition metal ions found in the text books of coordination chemistry is still valid on DNA.

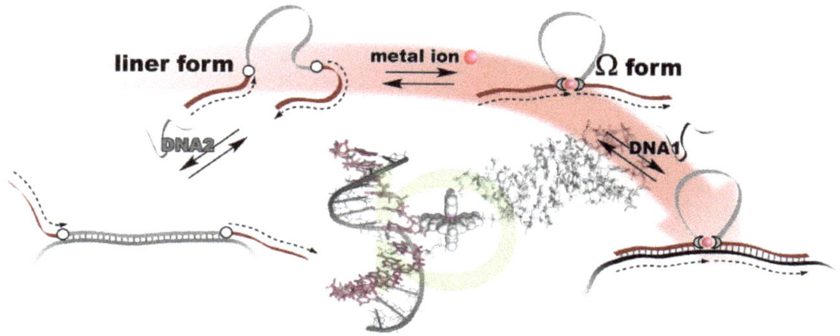

Figure 2. Metal ion-directed reversible edition of the DNA sequence. The sequence of **terpy$_2$DNA** is edited by intramolecular complexation with appropriate metal ions through Ω-shaped global conformational change.

We then applied the metal ion-directed sequence edition based on the Ω-motif to regulate the function of the split DNAzyme with peroxidase-like activity. To activate the split DNAzyme, they need to be reconstituted to form a G-quadruplex structure. As shown in Figure 3a, **terpy$_2$DNA** was used as the tunable template to activate the split DNAzyme. The reaction was monitored by the color change associated with the oxidation of the substrate, 2,2'-azino-bis(3-ethylbenzothiazoline-6-sulfonic acid (ABTS). Figure 3b shows the time course of the reaction profiles. Equivalent concentrations of Fe^{2+} and Ni^{2+} ($r = 1$) restored the activity of the split DNAzyme in the presence of **terpy$_2$DNA** [26]. Cu^{2+} and Zn^{2+} also showed a moderate effect on the restoration of split DNAzyme activity. As we expected, the global conformation of **terpy$_2$DNA** was fixed to a Ω-shape by the intramolecular formation of $[M(terpy)_2]^{2+}$. Subsequently, the new sequence presented on **terpy$_2$DNA** ·M^{2+} worked as an effective template to reconstruct the integrated active form of DNAzyme.

The results demonstrated that the global DNA structure and, furthermore, the activity of DNAzyme were controlled by local metal complexation events that could be rationally designed based on general coordination chemistry. The technique of dynamic DNA splicing proposed in this study would be a compatible technique with the construction of the molecular systems consisting of functional DNA, such as aptamer and DNAzyme. Based on the Ω-motif, one could control the activity of reconstituted functional DNA or RNA, thermodynamics and kinetics of strand exchange, and gene expression.

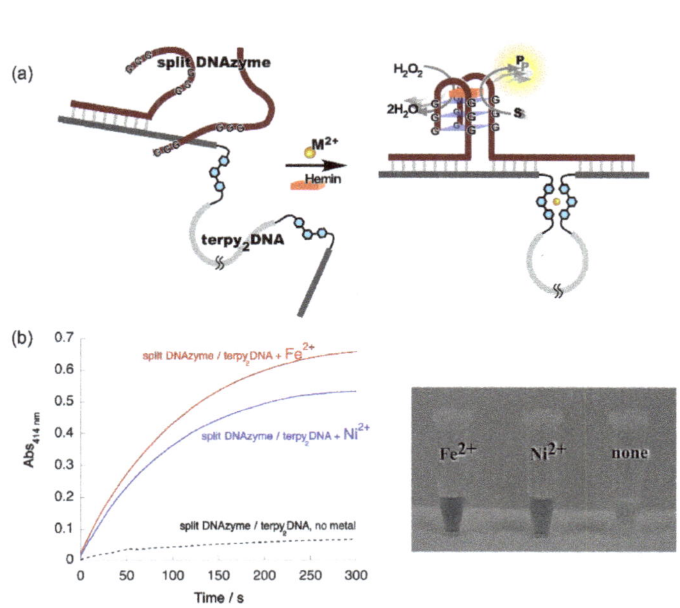

Figure 3. Metal ion-directed regulation of DNAzyme activity. (**a**) Allosteric regulation of split DNAzyme activity by metal ion-directed dynamic sequence edition of the template, **terpy$_2$DNA**. (**b**) Left: Time courses of the ABTS oxidation by the split DNAzyme with **terpy$_2$DNA** in the presence of Fe^{2+} and Ni^{2+}. Red, split DNAzyme/**terpy$_2$DNA** + Fe^{2+}; blue, split DNAzyme/**terpy$_2$DNA** + Ni^{2+}; black, split DNAzyme/**terpy$_2$DNA**, no metal ions. Right: Images of reaction solutions shown in the time courses.

4. Reconstruction of Luminescent Lanthanide Complexes on DNA and Their Analytical Applications

The present study demonstrated a straightforward genetic analysis using DNA-templated cooperative complexation between a luminescent lanthanide ion (Ln^{3+}: Tb^{3+} or Eu^{3+}) and two DNA conjugates. Ethylenediaminetetraacetic acid (EDTA) and 1,10-phenanthrorine (phen) were covalently attached to the end of oligo DNAs to form a pair of the conjugates, i.e., capture and sensitizer probes, respectively. The sequences of these split probes were designed so as to form a tandem duplex with targets (templates) with their auxiliary units facing each other, providing a microenvironment to accommodate Ln^{3+} (Figure 4a) [27]. The results of time-resolved luminescence studies showed that the formation of luminous ternary complexes, EDTA/Ln^{3+}/phen, depends on the sequence of the targets. The intensity of the luminescence is affected by the binding affinities of the probes or the local structural disruption caused by one-base mispairing [28].

This technique was applied to the multicolored allele typing based on single nucleotide polymorphisms (SNPs) in thiopurine S-methyltransferase gene by the concomitant use of the two capture probes, which are complementary to a part of the wild-type (**wt**) and the mutant (**mut**) of the gene. First, the capture probes for **wt** and **mut** were mixed with equimolar amounts of Tb^{3+} and Eu^{3+}, respectively. Both the allele-specific capture probe with Ln^{3+} and the sensitizer probe were then added to three different solutions containing the targets, **wt/wt**, **mut/mut**, and **wt/mut**. The solutions emitted distinctive colors, i.e., green, red, and yellow for **wt/wt**, **mut/mut**, and **wt/mut**, respectively; the colors were identifiable with the naked eye (Figure 4b) [29].

The system was applied as a molecular nanodevice consisting of the lanthanide complex and stem-loop structured oligo DNA. The nanodevice was synthesized by the introduction of EDTA and phen at the 5′- and the 3′-end of the DNA, respectively. This device was named the lanthanide complex molecular beacon (**LCMB**). In the stem-loop structure of **LCMB**, the two auxiliary units were placed in close proximity, providing a microenvironment to accommodate Ln^{3+}. The characteristic emissions of Tb^{3+} and Eu^{3+} were clearly observed in the solution containing the nanodevice and the corresponding Ln^{3+} ("on" state). In contrast, scarce emission was observed in the presence of the DNA complementary to the loop region; the auxiliary units were separated from each other when the duplex was formed ("off" state). The ATP aptamer (**iATP**) was used as an interface for the application of **LCMB** to ATP sensing. The sequence of **LCMB** was designed to be complementary to a part of **iATP** (Figure 5a). With the addition of ATP to the **LCMB**/**iATP** duplex, the fluorescence signal turned on as the result of the restoration of **LCMB** stem-loop structure accompanying the displacement of **iATP** from **LCMB** by ATP. A highly specific response was observed for ATP among NTPs, as shown in Figure 5b [30].

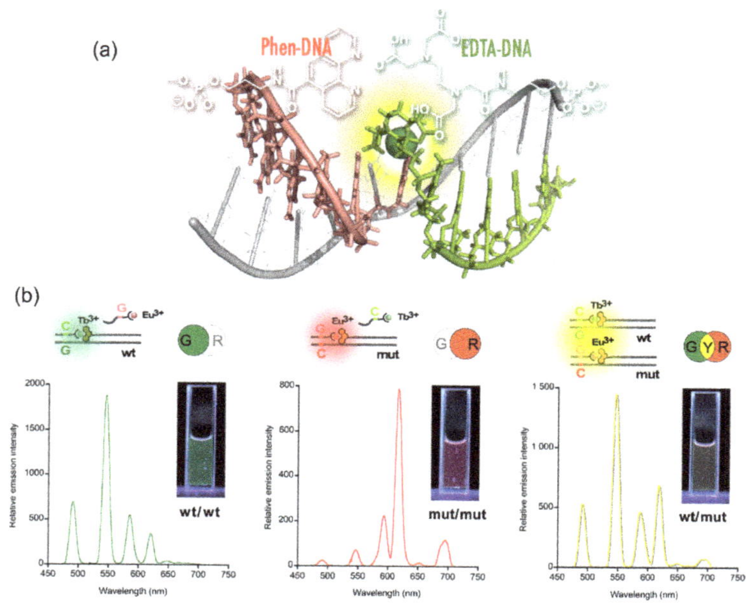

Figure 4. Multicolored allele typing using time-resolved luminescence from lanthanide complexes (Tb^{3+} and Eu^{3+}) cooperatively formed with a pair of split probes. (**a**) The structure of the Ln^{3+} complex formed on tandem duplex of the split probes with target sequence. (**b**) Allele typing of thiopurine S-methyltransferase gene.

Nonenzymatic amplification of the luminescent signal from the Ln complexes on the DNA scaffold was performed through catalytic hairpin assembly (CHA) and hybridization chain reaction (HCR), which are the typical DNA circuits consisting of the autonomous successive strand exchange reactions [31,32]. For HCR, four hairpin DNA conjugates were prepared; two of them carry EDTA on both ends, and phens are attached to both ends of another two hairpin strands DNAs. The sequences of the four hairpin DNA strands were designed so as to provide the long DNA wire as the product with Ln complexes at every junction. The HCR was initiated by a small amount of target DNA, acting as an initiator. Figure 6a shows the scheme of the HCR amplification. The luminescence signal significantly increased with the progress of HCR after target addition. Signal contrast was very high, and the sequence selectivity was preserved in this system [32]. To improve the amplification rate, the system was redesigned to form a cruciform product consisting of four hairpins by catalytic hairpin assembly (CHA) (Figure 6b). The sequences of hairpin monomers were modified so as to hybridize convergently to form a closed cruciform structure. Ln complexes were expected to form at each of the four tips of the cruciform.

The target miRNA *let-7a* was detected using time-resolved luminescence measurement techniques [32]. The CHA system (cruciform formation) was found to be more efficient than that of the earlier version of HCR (DNA wire), probably due to the difference in molecular sizes of the products.

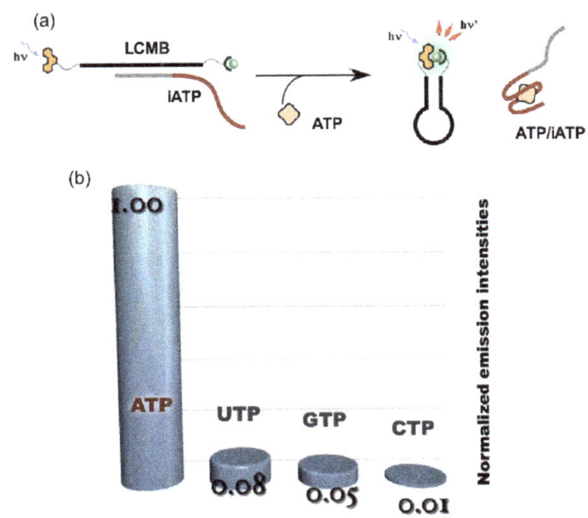

Figure 5. ATP sensing using **LCMB** and **iATP**. (**a**) operating principle of ATP sensing using competitive reaction over **iATP** between ATP and **LCMB**; (**b**) luminescence signal response of ATP sensor to NTPs.

(**a**)

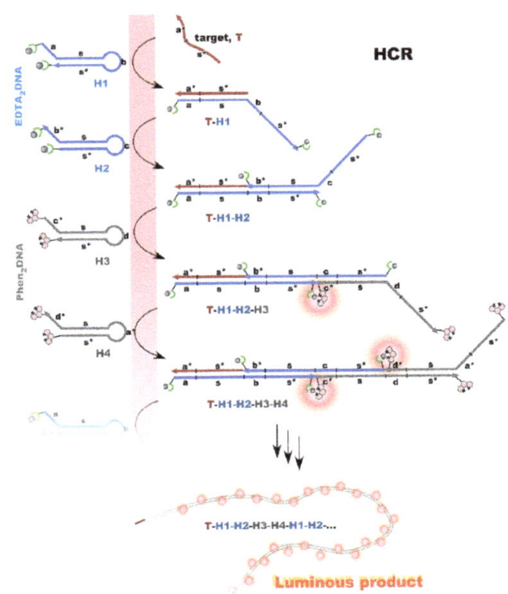

Figure 6. *Cont.*

(b)

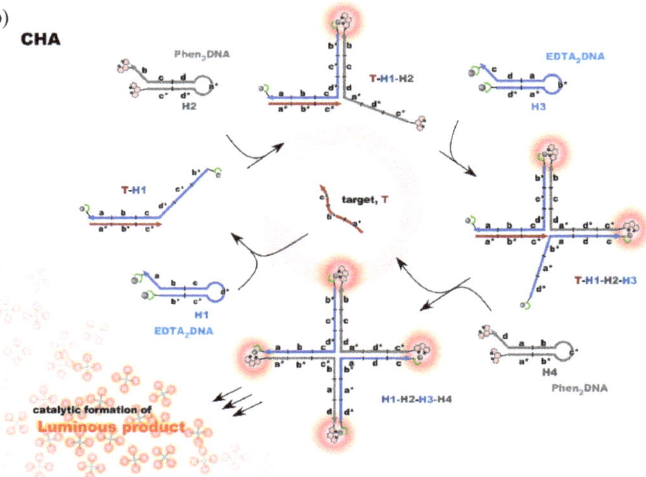

Figure 6. Nonenzymatic signal amplification by DNA circuits: (**a**) Luminous DNA wire was produced by target-initiated HCR; (**b**) Luminous cruciform DNA was produced by CHA.

5. Perspective

In recent years, research on nucleic acids has uncovered new and challenging issues as mentioned above, and nucleic acid conjugates show promise as a molecular tool that can be used to meet those challenges. In addition to the standard complementary nucleic acids, functional nucleic acids, such as aptamers and DNAzyme, as well as nonnatural nucleic acids have been added to the options as nucleic acid components of the conjugates. Furthermore, given the diversity of functional molecules that pair with DNAs, an infinite number of combinations are possible in the design of nucleic acid conjugates. With the emergence of "click chemistry", the in situ synthesis of conjugate molecules is now possible, further expanding the potential of these molecules [33,34]. As demonstrated in the above-referenced studies, it is always critical to accurately predict the structure of nucleic acid conjugates. The fundamentals in the physical chemistry of nucleic acids that Professor Breslauer and his research groups have achieved are pioneering and of universal value. The sum of these works can be considered a milestone in the history of nucleic acid science. We would like to conclude this brief note by sending our best wishes from Japan to Professor Breslauer, on the occasion of his 75th birthday.

Author Contributions: Conceptualization, T.I.; writing—original draft, review, editing, T.I., Y.K. (Yusuke Kitamura) and Y.K. (Yousuke Katsuda). All authors have read and agreed to the published version of the manuscript.

Funding: This work was supported by the Grant-in-Aid for Scientific Research on Innovative Areas (Coordination Programming, Area 2107) (no. 24108734 to T.I.), Challenging Exploratory Research (No. 19K22259 to T.I.), and Scientific Research (B) (no. 24350040, 15H03829, and 20H02769 to T.I.) from the Ministry of Education, Culture, Sports, Science and Technology, Japan.

Acknowledgments: We are grateful to our collaborators for their many contributions to all the works discussed in this paper.

Conflicts of Interest: The authors declare no conflict of interest.

References

1. International Human Genome Sequencing Consortium. Initial sequencing and analysis of the human genome. *Nature* **2001**, *409*, 860–921.
2. Venter, J.C.; Adams, M.D.; Myers, E.W.; Li, P.W.; Mural, R.J.; Sutton, G.G.; Smith, H.O.; Yandell, M.; Evans, C.A.; Holt, R.A.; et al. The sequence of the human genome. *Science* **2001**, *291*, 1304–1351. [CrossRef] [PubMed]
3. Ellington, A.D.; Szostak, J.W. In vitro selection of RNA molecules that bind specific ligands. *Nature* **1990**, *346*, 818–822. [CrossRef]
4. Robertson, D.L.; Joyce, G.F. Selection in vitro of an RNA enzyme that specifically cleaves single-stranded DNA. *Nature* **1990**, *344*, 467–468. [CrossRef] [PubMed]
5. Santoro, S.W.; Joyce, G.F. A general purpose RNA-cleaving DNA enzyme. *Proc. Natl. Acad. Sci. USA* **1997**, *94*, 4262–4266. [CrossRef]
6. Santoro, S.W.; Joyce, G.F.; Sakthivel, K.; Gramatikova, S.; Barbas, C.F. RNA cleavage by a DNA enzyme with expanded chemical functionality. *J. Am. Chem. Soc.* **2000**, *122*, 2433–2439. [CrossRef]
7. Breaker, R.R. Making catalytic DNAs. *Science* **2000**, *290*, 2095–2096. [CrossRef]
8. Seelig, G.; Soloveichik, D.; Zhang, D.Y.; Winfree, E. Enzyme-free nucleic acid logic circuits. *Science* **2006**, *314*, 1585–1588. [CrossRef]
9. Zhang, D.U.; Seelig, G. Dynamic DNA nanotechnology using strand-displacement reactions. *Nat. Chem.* **2011**, *3*, 103–113. [CrossRef]
10. Breslauer, K.J.; Frank, R.; Blöcker, H.; Marky, L.A. Predicting DNA duplex stability from the base sequence. *Proc. Natl. Acad. Sci. USA* **1986**, *83*, 3746–3750. [CrossRef]
11. SantaLucia, J.; Allawi, H.T.; Seneviratne, P.A. Improved nearest-neighbor parameters for predicting DNA duplex stability. *Biochemistry* **1996**, *35*, 3555–3562. [CrossRef] [PubMed]
12. Xia, T.; SantaLucia, J., Jr.; Burkard, M.E.; Kierzek, R.; Schroeder, S.J.; Jiao, X.; Cox, C.; Turner, D.H. Thermodynamic parameters for an expanded nearest-neighbor model for formation of RNA duplexes with Watson–Crick base pairs. *Biochemistry* **1998**, *37*, 14719–14735. [CrossRef] [PubMed]
13. Sugimoto, N.; Nakano, S.; Katoh, M.; Matsumura, A.; Nakamuta, H.; Ohmichi, T.; Yoneyama, M.; Sasaki, M. Thermodynamic parameters to predict stability of RNA/DNA hybrid duplexes. *Biochemistry* **1995**, *34*, 11211–11216. [CrossRef] [PubMed]
14. Ihara, T.; Fujii, T.; Mukae, M.; Kitamura, Y.; Jyo, A. Photochemical ligation of DNA conjugates through anthracene cyclodimer formation and its fidelity to the template sequences. *J. Am. Chem. Soc.* **2004**, *126*, 8880–8881. [CrossRef]
15. Ihara, T.; Uemura, A.; Futamura, A.; Shimizu, M.; Baba, N.; Nishizawa, S.; Teramae, N.; Jyo, A. Cooperative DNA probing using a β-cyclodextrin–DNA conjugate and a nucleobase-specific fluorescent ligand. *J. Am. Chem. Soc.* **2009**, *131*, 1386–1387. [CrossRef] [PubMed]
16. Ihara, T.; Wasano, T.; Nakatake, R.; Arslan, P.; Futamura, A.; Jyo, A. Electrochemical signal modulation in homogeneous solutions using the formation of an inclusion complex between ferrocene and β-cyclodextrin on a DNA scaffold. *Chem. Commun.* **2011**, *47*, 12388–12390. [CrossRef] [PubMed]
17. Ihara, T.; Takeda, Y.; Jyo, A. Metal ion-directed cooperative triple helix formation of glutamic acid–oligonucleotide conjugate. *J. Am. Chem. Soc.* **2001**, *123*, 1772–1773. [CrossRef]
18. Frank-Kamenetskii, M.D.; Mirkin, S.M. Triplex DNA structures. *Annu. Rev. Biochem.* **1995**, *64*, 65–95. [CrossRef]
19. Silver, G.C.; Sun, J.-S.; Nguyen, C.H.; Boutorine, A.S.; Bisagni, E.; Hélène, C. Stable triple-helical DNA Complexes formed by benzopyridoindole- and benzopyridoquinoxaline-oligonucleotide conjugates. *J. Am. Chem. Soc.* **1997**, *119*, 263–268. [CrossRef]
20. Ihara, T.; Ishi, T.; Araki, N.; Wilson, A.W.; Jyo, A. Silver ion unusually stabilizes the structure of parallel-motif DNA triplex. *J. Am. Chem. Soc.* **2009**, *131*, 3826–3827. [CrossRef]
21. Ono, A.; Torigoe, H.; Tanaka, Y.; Okamoto, I. Binding of metal ion by pyrimidine base pairs in DNA duplexes. *Chem. Rev.* **2011**, *40*, 5855–5866. [CrossRef] [PubMed]
22. Urata, S.; Miyahata, T.; Matsuura, H.; Kitamura, Y.; Ihara, T. Alteration of DNAzyme activity by silver ion. *Chem. Lett.* **2014**, *43*, 1020–1022. [CrossRef]
23. Zhu, D.; Zhu, J.; Zhu, Y.; Wang, L.; Jiang, W. Sensitive detection of transcription factors using an Ag^+-stabilized self-assembly triplex DNA molecular switch. *Chem. Commun.* **2014**, *50*, 14987–14990. [CrossRef]
24. Guo, S.; Du, Y.; Yang, X.; Dong, S.; Wang, E. Solid-state label-free integrated aptasensor based on graphene-mesoporous silica–Gold nanoparticle hybrids and silver microspheres. *Anal. Chem.* **2011**, *83*, 8035–8040. [CrossRef] [PubMed]
25. Feng, L.; Huang, Z.; Ren, J.; Qu, X. Toward site-specific, homogeneous and highly stable fluorescent silver nanoclusters fabrication on triplex DNA scaffolds. *Nucleic Acids Res.* **2012**, *40*, e122. [CrossRef]
26. Ihara, T.; Ohura, H.; Shirahama, C.; Furuzono, T.; Shimada, H.; Matsuura, H.; Kitamura, Y. Metal ion-directed dynamic splicing of DNA through global conformational change by intramolecular complexation. *Nat. Commun.* **2015**, *6*, 6640. [CrossRef] [PubMed]
27. Kitamura, Y.; Ihara, T.; Tsujimura, Y.; Tazaki, M.; Jyo, A. DNA-templated cooperative formation of the luminous lanthanide complex and its analytical application to gene detection. *Chem. Lett.* **2005**, *34*, 1606–1607. [CrossRef]
28. Ihara, T.; Kitamura, Y.; Tsujimura, Y.; Jyo, A. DNA analysis based on the local structural disruption to the duplexes carrying a luminous lanthanide complex. *Anal. Sci.* **2011**, *27*, 585–590. [CrossRef]
29. Kitamura, Y.; Ihara, T.; Tsujimura, Y.; Osawa, Y.; Sasahara, D.; Yamamoto, M.; Okada, K.; Tazaki, M.; Jyo, A. Template-directed formation of luminescent lanthanide complexes: Versatile tools for colorimetric identification of single nucleotide polymorphism. *J. Inorg. Biochem.* **2008**, *102*, 1921–1931. [CrossRef]

30. Kitamura, Y.; Yamamoto, S.; Osawa, Y.; Matsuura, H.; Ihara, T. Versatile allosteric molecular devices based on reversible formation of luminous lanthanide complexes. *Chem. Commun.* **2013**, *49*, 285–287. [CrossRef]
31. Kitamura, Y.; Azuma, Y.; Katsuda, Y.; Ihara, T. Catalytic formation of luminescent lanthanide complexes using an entropy-driven DNA circuit. *Chem. Commun.* **2020**, *56*, 3863–3866. [CrossRef] [PubMed]
32. Kitamura, Y.; Nozaki, A.; Ozaki, R.; Katsuda, Y.; Ihara, T. Catalytic formation of luminescent complex clusters based on autonomous strand exchange reaction of DNA. *ACS Appl. Bio Mater.* **2019**, *2*, 2988–2993. [CrossRef] [PubMed]
33. Fantoni, N.Z.; El-Sagheer, A.H.; Brown, T. A hitchhiker's guide to click-chemistry with nucleic acids. *Chem. Rev.* **2021**, *121*, 7122–7154. [CrossRef] [PubMed]
34. Bhardwaj, A.; Kaur, J.; Wuest, M.; Wuest, F. In situ click chemistry generation of cyclooxygenase-2 inhibitors. *Nat. Commun.* **2017**, *8*, 1. [CrossRef] [PubMed]

life

Review

Developing Community Resources for Nucleic Acid Structures

Helen M. Berman [1], Catherine L. Lawson [2] and Bohdan Schneider [3,*]

[1] Department of Chemistry and Chemical Biology, Rutgers, The State University of New Jersey, Piscataway, NJ 08854, USA; berman@rcsb.rutgers.edu
[2] Institute for Quantitative Biomedicine, Rutgers, The State University of New Jersey, Piscataway, NJ 08854, USA; cathy.lawson@rutgers.edu
[3] Institute of Biotechnology of the Czech Academy of Sciences, 252 50 Vestec, Czech Republic
* Correspondence: bohdan.schneider@ibt.cas.cz

Abstract: In this review, we describe the creation of the Nucleic Acid Database (NDB) at Rutgers University and how it became a testbed for the current infrastructure of the RCSB Protein Data Bank. We describe some of the special features of the NDB and how it has been used to enable research. Plans for the next phase as the Nucleic Acid Knowledgebase (NAKB) are summarized.

Keywords: nucleic acid structures; nucleic acid conformation; biological structure database; DNA; RNA; validation standards

Citation: Berman, H.M.; Lawson, C.L.; Schneider, B. Developing Community Resources for Nucleic Acid Structures. *Life* **2022**, *12*, 540. https://doi.org/10.3390/life12040540

Academic Editors: Tigran Chalikian, Jens Völker and Bruce J. Nicholson

Received: 14 March 2022
Accepted: 31 March 2022
Published: 6 April 2022

1. Introduction

The first single crystal structures of nucleic acids were determined in the 1970s, almost twenty years after the model of the DNA double helix based on fiber data was published [1,2]. Short fragments of RNA yielded the first atomic-level views of the double helix and demonstrated conformational flexibility [3–5]. These structures were archived as small molecules in the Cambridge Crystallographic Database (CSD) [6]. The structure of tRNA, determined in 1974 [7–9], showed that RNA can fold into a compact structure and demonstrated the importance of tertiary interactions. As DNA synthesis became possible, structures of the DNA double helix with predefined sequences were determined. The first structures were left-handed Z-form DNA fragments [10], and in 1981, the first single crystal structure of a full turn of B-form DNA was published [11]. The tRNA structures and larger nucleic acid fragments were archived in the Protein Data Bank (PDB [12]). By 1990, there were nearly 100 publicly released nucleic acid structures, thus allowing analyses of sequence-dependent features, hydration patterns, and ligand interactions.

During the late 1970s and 1980s, several faculty members in the Chemistry Department at Rutgers University focused their research on nucleic acids. Ken Breslauer worked on the macroscopic properties of nucleic acids using calorimetric approaches [13–16]; these works, seminal for the understanding of thermodynamics of DNA, have continued to this day [17–20]. Roger Jones developed new methods to synthesize DNA [21]. Jerry Manning developed the counterion condensation theory to understand DNA folding [22], and continued this work in collaboration with the Breslauer group [23]. Wilma Olson performed detailed analyses of the structure of DNA [24]. During that period, Helen Berman carried out nucleic acid crystallography research at the Institute for Cancer Research in Philadelphia and had close interactions with the Rutgers group. In 1989, she joined the Chemistry faculty at Rutgers.

The setting at Rutgers was ideal for collaborative studies using both experimental and computational approaches to investigate nucleic acid structure. It was necessary to have a resource that contained the structural information which resided in the CSD, in the PDB, or in the laboratories of individual researchers to facilitate these efforts. In collaboration with David Beveridge, with whom Berman was collaborating on computational analyses of nucleic acid hydration, Olson and Berman proposed to create the Nucleic Acid Database

(NDB). In the early 1990s, funding was received from the National Science Foundation to establish "A Comprehensive Database of the Three-Dimensional Structures of Nucleic Acids". The goal was to create a searchable database that would integrate information from several sources and make a variety of reports, thus enabling research on nucleic acid structure.

2. Development of the Nucleic Acid Database

The first step in the development of the NDB was to collect and curate the structural data [25]. Coordinates were accessed from the CSD and the PDB. Each structure and experiment were carefully reviewed to create appropriate annotations beyond what was available from each resource. Rather than working directly with the flat files maintained by the PDB, the NDB imported the parsed data files into a relational database management system (DBMS). Sybase [26] was chosen as the DBMS in large part because it was being used by Genbank [27,28]. A query system called NDBquery was put into place. In the early years, distribution was accomplished via FTP and a system called Gopher [29]. By 1995, a web server was set up, which generated a modest amount of activity to access and analyze the 350 structures represented in the NDB. The NDB was actively involved in the development of mmCIF, whose data model is compatible with a relational DBMS. By 1996, mmCIF [30] became the master format for the NDB. The software that was developed and the experience gained using this data representation set the stage for the management of the Protein Data Bank using mmCIF as the master format by the Research Collaboratory for Structural Bioinformatics (RCSB) beginning in 1998.

The NDB also became a driver for the creation of geometrical standards for nucleic acid structures. Careful analysis of high-resolution structures from CSD permitted the calculation of standard reference bond distances and angles for the bases, sugars, and phosphates of nucleic acids [31,32]. Using these values, Parkinson et al. [33] created new parameters that enabled improved refinement of nucleic acid-containing crystal structures against their experimental data. Those standards were widely used. In 1998, the NDB helped organize a conference whose outcome was the standard coordinate frame definition for nucleic acid bases [34]. This standard became widely adopted by researchers studying nucleic acid base morphology.

3. Features of the NDB

In addition to facilitating access to primary data for nucleic acid structures, the NDB provides tables of derived features, such as classifications of base pairing topologies [35], backbone torsion angles, and conformational and base pair classifications [36,37].

The NDB also offers different types of data visualization and presentation. The most important is the NDB Atlas page (Figure 1), which gives summary information about the structure, visualizations of the crystal asymmetric unit, the biological unit, unit cells, and for RNA structures; it provides a view that combines the secondary and tertiary structural features. Links to other resources are also provided.

The functionality of the NDB and its query engine was first and foremost driven by research projects on the nucleic acid structural and computational biologists. Careful attention was given to the quality and uniformity of the metadata so that it would be possible to use Boolean logic to create queries; individual questions could be made into logical constructs joined by logical AND, OR, and NOT. This requirement represented a challenge for building a robust system of precisely defined terms incorporated into a formal computer-readable language; mmCIF was that dictionary.

The NDB website was designed so that the user could select structures with features of interest and then use those structures for further analysis, e.g., through the creation of detailed tabular or graphical reports. Soon after the first functional version of the NDB was available, we started to use its potential to study the geometrical features of nucleic acids. The original NDB reporting capability allowed the user to obtain tabular reports of various properties of the selected nucleic acid structures from basic information about the publication or refinement parameters and graphical reports of selected geometric features such as bond distances (Figure 2) or torsion angles (Figure 3). Once funding for the NDB became limited in the 2000s, it was not possible to maintain these reporting capabilities.

Figure 1. NDB Atlas page of a tRNA structure.

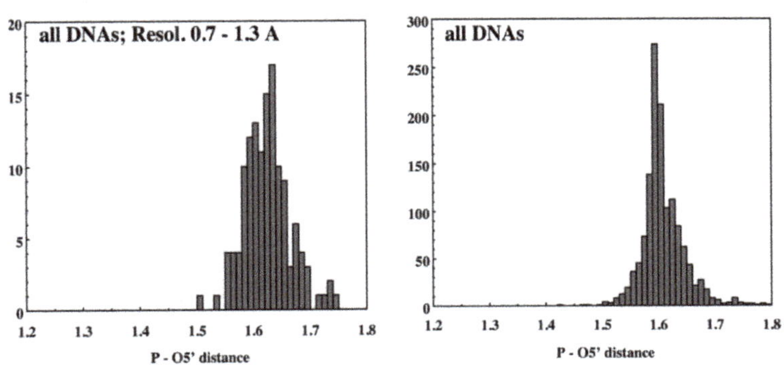

Figure 2. Example NDB report of geometric features of nucleic acids, based on structures available in the 1990s. Histograms show the P–O5′ valence distances in high-resolution DNA, and in all DNA structures.

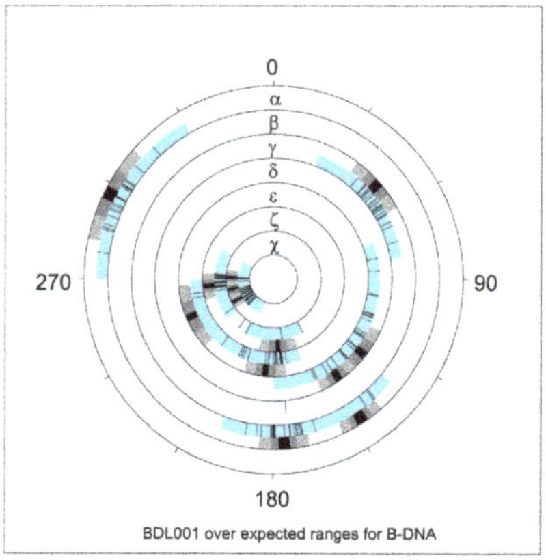

Figure 3. NDB graphical report of torsion angle distribution [38] for the Drew–Dickerson dodecamer, PDB ID 1BNA, NDB ID BDL001 [11]. Blue sectors indicate torsion angle limits for all structures annotated as B–DNA. Overlaid black tick marks are measured torsion values for BDL001. Adjacent black/grey sectors denote average values and spreads of 1 and 2 estimated standard deviations. Note that two averages are indicated for several torsions, e.g., for δ (two distinct sugar puckers) and ε (BI versus BII forms). Values reflect NDB data available in 1996.

4. Research Enabled by the NDB

The NDB has been used by many researchers to analyze the structures of nucleic acids. There are over 1100 citations to the original NDB article. The type of research enabled by the NDB includes DNA conformational analyses [39], DNA structure prediction [40], RNA structure prediction [41], analyses of protein-nucleic acid interactions [42,43], and the creation of new specialty databases [44]. In our research, we have used the NDB to study a variety of aspects of nucleic acids. For example, we surveyed A, B, and Z-form double helical DNA structures and used Fourier averaging to determine hydration patterns, e.g., for DNA nitrogenous bases [45]. Both base and later phosphate studies showed sequence and conformation-dependent water position preferences (Figure 4).

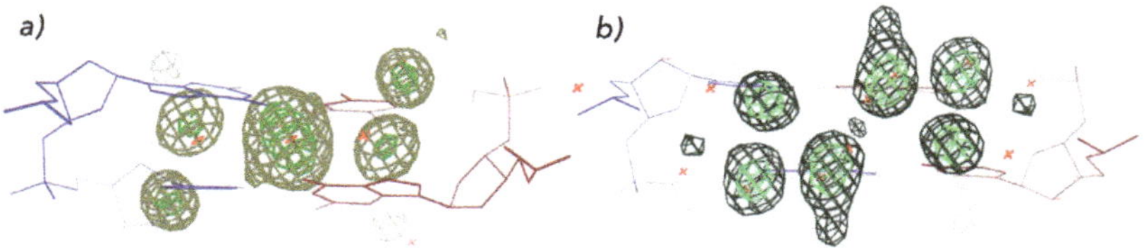

Figure 4. Sequence dependence of DNA hydration. Two distinct hydration patterns are shown in for the A-form major groove for (**a**) 5′–GC–3′ and (**b**) 5′–CG–3′, based on analyses of structures available in the mid–1990s NDB [44]. A more recent analysis of hydration using larger and functionally more relevant dinucleotide fragments is available at watlas.datmos.org/watna (accessed on 30 March 2022).

The growing volume of available crystal structures with ever growing sequence variability also led us to ask whether conformational properties of various DNA and RNA forms could be better characterized. This task posed new challenges to NDB querying and reporting capabilities. Specific subsets of structures were selected based on sequence, function, or structural features using SQL queries; their properties were reported as text or graphs (Figure 5). Ultimately, we were able to sharpen conformational definitions for established subtypes of A-B-Z forms (Figure 6) [46].

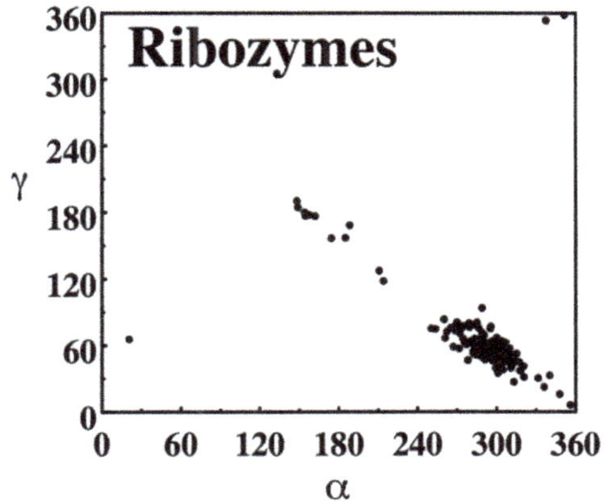

Figure 5. NDB torsion angle scatter plot. The distribution of backbone torsion angles α and γ observed in crystal structures of RNA annotated in the NDB as ribozymes in 1996 is shown. α describes rotations around the P–O5′ phosphodiester bond, γ around C5′–C4′ bond. Plots were created directly on the NDB website as PostScript formatted reports.

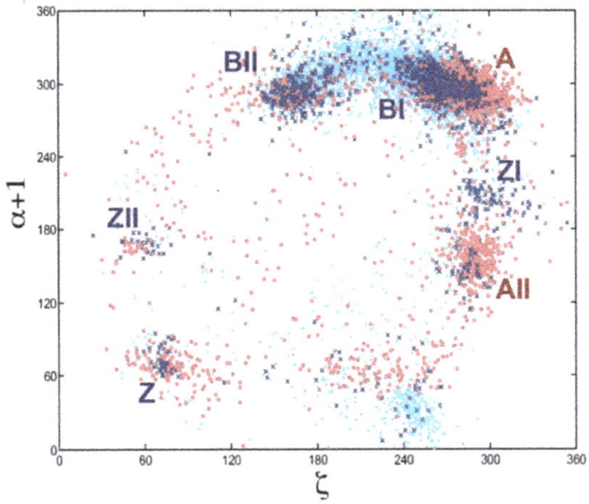

Figure 6. Scatter plot of backbone torsion angles ζ and α + 1. ζ shows the variation in rotation around the O3′–P phosphodiester bond and α + 1 around the P–O5′ bond (labeled α + 1 because this bond belongs to the sequentially following nucleotide). Data for the DNA alone is shown as dark blue crosses, for protein-DNA complexes: light blue dots, and for RNA: red dots. The scattergram shows data for all nucleotide residues in the 1998 NDB. Clusters of some major conformational types are labeled. This analysis revealed that no nucleic acid form can be unequivocally classified by torsion angle pairs; a more sophisticated multidimensional analysis was needed.

The growing number of nucleic acid structures and the appearance of new forms such as quadruplexes and large-folded RNAs demonstrated the plasticity of nucleic acid molecules. It became clear that the conformational space of nucleic acids is extremely complex and that capturing it would require a concerted understanding of base pairing motifs and the backbone structural variability.

Early analyses showed that backbone conformational variability was fundamentally influenced by flexibility around the O3′–P–O5′ phosphodiester bonds that connected adjacent nucleotide residues, described by torsion angles ζ and α [47]. Our multidimensional statistical analysis, therefore, focused on dinucleotide fragments analyzed in torsion space, taking full advantage of the availability of the NDB and PDB.

In the 2000s, research conducted by several groups concentrated on analysis of RNA backbone flexibility culminated in an RNA Consortium consensus set of dinucleotide conformers [48]. The effort was later complemented by an analogous set of DNA conformers [49] and, ultimately, a comprehensive classification system for dinucleotide fragments covering both DNA and RNA [50]. This classification algorithm provides an automated structural ranking of dinucleotide fragments at two levels of detail: fully geometrical classification into dinucleotide conformational classes (NtC) and a more human-accessible structural alphabet (CANA). The assignment of the CANA and NtC classes makes it possible to study the structural propensities of dinucleotide sequences. For example, analysis of DNA in transcription factors and in histone core particle complexes showed important trends of protein interactions with specific bending associated NtC classes (Figure 7) [49].

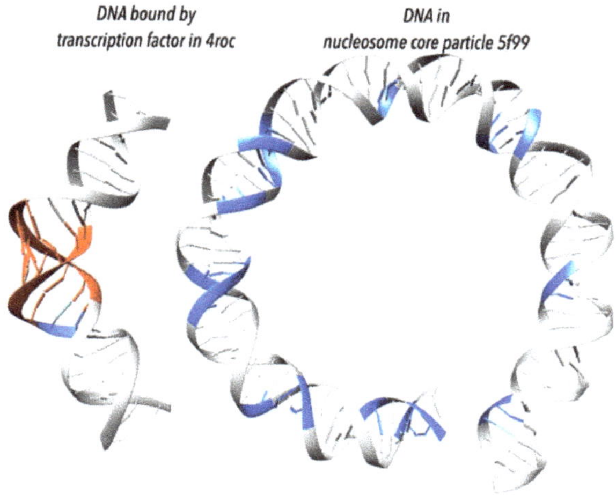

Figure 7. Transcription factors and proteins of the histone core particle bend DNA duplex differently [49]. (**Left**) bending by transcription factors is acquired mostly by local adaptation to the A form (highlighted in red); shown is DNA from complex with TFIIB–Related Factor Brf2 (PDB id 4ROC [51]). (**Right**) bending by the histone core particle is associated with the BII form (highlighted in blue); shown are first 75 base pairs from a histone core particle (PDB id 5F99 [52]); when statistically measured over many structures, the BII form appears in histone-wrapped DNA every tenth step corresponding to one full turn of duplex; the periodicity of the BII form appearance explains the DNA bending.

NtC assignments have also inspired development of a new validation tool linking the global geometry criterion (closeness of fit to the nearest NtC class) and the quality of fit into electron density (Figure 8) [50]. It offers a simple information-rich graphical representation of the overall quality of nucleic acid structure in the form of a 2D graph.

In an additional effort to understand, classify, and validate nucleic acids, we have developed a procedure similar to Ramachandran analysis for proteins, making use of eta (η) and theta (θ) virtual torsion angles (pseudotorsions) [53,54]. Measured (η,θ) pairs define backbone conformations for each central residue within a trinucleotide. Plots are designed to quickly reveal rare conformations that may need extra checking (Figure 9). A web server was recently set up to investigate the utility of this approach for RNA structures determined using cryoEM (ptp.emdataresource.org) (accessed on 30 March 2022).

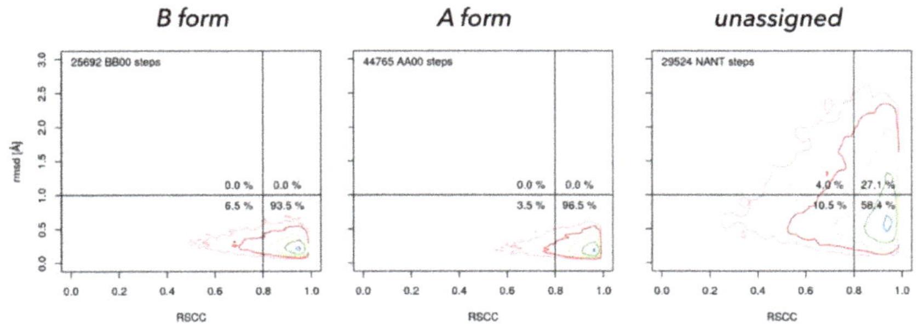

Figure 8. Geometrically well-defined dinucleotides fit well into their electron densities. Real Space Correlation Coefficient (RSCC, horizontal axis) measures how closely the model electron density resembles the experimental density and rmsd (vertical axis) measures how closely the geometry of the model resembles the closest NtC class in the so called golden set [50]. NtC class BB00 (**left**) characterizes the B form, AA00 (**center**) A form in both DNA and RNA, and NANT (**right**) are all unclassified dinucleotides. Geometrically unclassified dinucleotides fit significantly worse to the electron density.

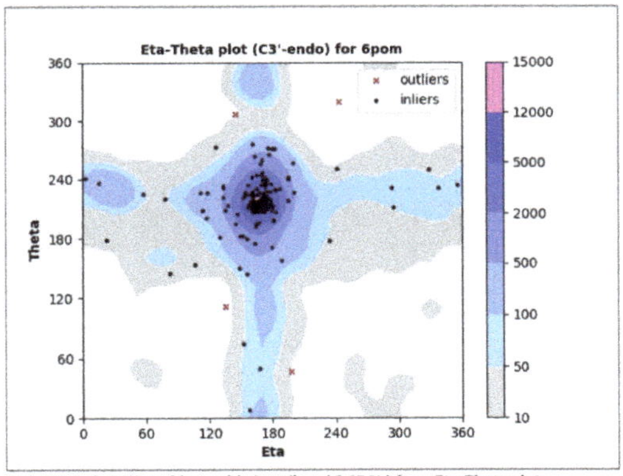

Figure 9. Pseudotorsion plot, a simple coarse-level RNA backbone conformation validation tool. Measured eta and theta (η,θ) pseudotorsion values for each trinucleotide (black dots, red x's) are plotted against a quality-filtered virtual torsion angle distribution derived from a large number of RNA structures (contours), analogous to the Ramachandran plot for proteins.

5. Current State of Nucleic Acid Structural Biology

When the NDB was established in the early 1990s, most of the nucleic acid structures were small fragments with the exception of tRNA. There were a few structures of protein-nucleic acid complexes, limited to virus capsids with viral genomic RNA or DNA and transcription factors bound to duplex DNA. Molecular machines, such as the ribosome, were yet to be determined. In contrast, there are now more than 14,000 nucleic acid-containing structures in the PDB and NDB (Figure 10). A notable trend is the recent increase in the use of electron microscopy (EM) for structure determination. Protein/DNA complexes are the most abundant, followed by protein/RNA, DNA-only, and RNA-only. In addition to the increase in the number of structures, the structures are very diverse, as shown in Figure 11.

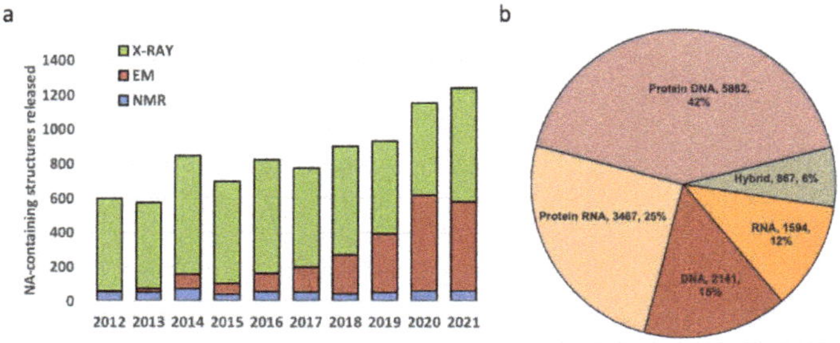

Figure 10. Current statistics for nucleic acid-containing structures. (**a**) New structures released into the PDB, by year and method; (**b**) Distribution of nucleic acid-containing structures. NDB archives and annotates structures determined using X-ray crystallography or NMR. Electron microscopy structures are not included in the NDB, but will be included in the NAKB, the planned successor to NDB.

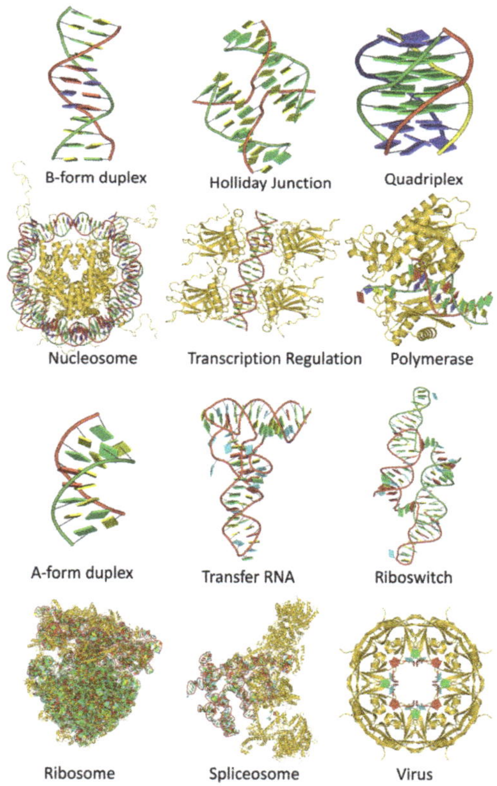

Figure 11. Diversity of structures containing DNA (top rows) and RNA (bottom rows). Nucleic acids are shown with ribbon backbones (random colors) and base blocks (A—red, C—yellow, G—green, T—blue, U—cyan). Proteins are shown as gold ribbons. From top left: B-form duplex DNA (1BNA [11]), Holliday junction (5DSB [55]), parallel-stranded DNA quadruplex (139D [56]), nucleosome core particle (1KX5 [57]), *trp* repressor/operator complex (1TRR [58]), DNA repair enzyme rev1 (6X6Z [59]), A-form duplex RNA (402D [60]), tRNA Asp (6UGG [61]), glutamine II riboswitch (6QN3 [62]), bacterial ribosome (4YBB [63]), spliceosomal E complex (6N7P [64]), Ebola virus matrix protein octamer (7K5L [65]). Images were generated using DSSR and PyMOL [66].

These structures have significantly expanded our knowledge of structure/function relationships and raised the potential of new knowledge from systematic analyses of structure collections. Many different databases and tools have been created to enable specialized analyses of nucleic acid structures. Some have focused on DNA [67], some on RNA [68–72], and some on the interactions between proteins and nucleic acids [73,74]. A systematic long-term analysis of dinucleotides led to a unified RNA + DNA automated classification system [50] available at DNATCO (dnatco.datmos.org) (accessed on 30 March 2022). The NDB (ndbserver.rutgers.edu) (accessed on 30 March 2022) is unique in that all nucleic acid structures and their complexes are contained in a single resource.

6. Going Forward

The NDB is maintained to the extent that new structures and manually curated annotations are added each week, but there is little significant development since its last full funding in 2003. Even so, thousands of users from the Americas, Asia, Europe, and other locations continue to make multiple visits to the NDB website each month. The most heavily visited pages are Advanced Search and DNA and RNA galleries.

In 2018, the collaborative group of scientists managing both the NDB (at Rutgers) and RNAhub services (at Bowling Green State University) proposed to create the Nucleic Acid Knowledge Base (NAKB), with the goal of integrating information already in the NDB with additional sequence, structure, function, and interaction-based annotations for all major classes of NA-containing 3D structures. This new service, which will ultimately replace the NDB, is currently under construction. The NAKB aims to enable users to quickly find and download all structures and metadata relevant to their search topic, whether broad or focused, based on the NDB's internal curation scheme, computationally generated annotations, and/or external database references for DNA, RNA, mixed NA, and for NA-binding enzymatic, regulatory, and structural proteins. All NA-containing structures in the PDB will be indexed, including structures obtained using Electron Microscopy. The NAKB will be updated weekly.

The NDB has employed manual expert curation collected over three decades to identify major NA secondary structure features (duplex, triplex, and quadruplex) and high-level classifications (e.g., ribosomal RNA or telomeric DNA), as well as interactions with ligands (e.g., minor groove binding) and protein classification [37]. Integrated computationally created annotations have included bond distance, angle, and torsion geometries, base and base-pair morphologies, as well as RNA 3D motifs, interactions (base pair types and parameters, base-to-backbone, and base stacking interactions), and RNA equivalence (3D structure similarity) classes [75,76].

New NAKB content will include equivalence class calculations for all nucleic acid molecule types (RNA, DNA, hybrid nucleic acids), enabling more accurate retrieval for closely related NA structures, analogous to the way that UniProt identifier mapping has improved search capabilities for related proteins in PDB [77]. Computationally derived annotations produced by DSSR software [78], including secondary structure features, sugar pucker type, and pseudo-torsion angles, will be added.

New search capabilities will be developed for specific classes of chemical modifications of nucleotides; nucleic acid 3D structure motifs by their common names, for example, G-Quadruplex, R-loop, Holliday Junction, Sarcin–Ricin, Kink-turn; ribosome functional states, e.g., full or single subunit, translational state, and numbers and positions of bound tRNAs; and deeper classification levels for selected proteins such as transcription factors.

The NAKB website will employ a modern web infrastructure with flexible data representation viewable on phones and tablets as well as desktop computers. For each NA-containing 3D structure, an atlas page will provide a summary overview of annotations as well as access to 1D, 2D, and 3D visualizations, external analysis tools, and file downloads. Mappings to external database links will initially include: PDB, Uniprot, RNACentral, Rfam. External analysis tools will include DNATCO and DNAproDB. Some of the reporting functions that were available in the original NDB so that the types of conformational analysis described earlier will be reenabled.

Author Contributions: Conceptualization, H.M.B.; writing—original draft, review, editing, H.M.B., C.L.L. and B.S. All authors have read and agreed to the published version of the manuscript.

Funding: The early funding of the NDB by the National Science Foundation is gratefully acknowledged. Funding for B.S. is from program Inter excellence of Ministry of Education, Youth, and Sports of the Czech Republic, grant number LTAUSA18197, and institutional support to the Institute of Biotechnology of the Czech Academy of Sciences, grant number RVO 86652036. Funding for C.L.L. is from the National Institutes of Health General Medical Sciences, grant numbers R01 GM079429 and R01 GM085238.

Acknowledgments: We are grateful to our collaborators for their many contributions over NDB's three decades of existence.

Conflicts of Interest: The authors declare no conflict of interest.

References

1. Watson, J.D.; Crick, F.H.C. A structure for deoxyribose nucleic acid. *Nature* **1953**, *171*, 737–738. [CrossRef] [PubMed]
2. Franklin, R.E.; Gosling, R.G. Molecular configuration in sodium thymonucleate. *Nature* **1953**, *171*, 740–741. [CrossRef] [PubMed]
3. Sussman, J.L.; Seeman, N.C.; Kim, S.-H.; Berman, H.M. The crystal structure of a naturally occurring dinucleotide phosphate uridylyl 3′,5′-adenosine phosphate. models for RNA chain folding. *J. Mol. Biol.* **1972**, *66*, 403–421. [CrossRef]
4. Seeman, N.C.; Rosenberg, J.M.; Suddath, F.L.; Kim, J.J.P.; Rich, A. RNA double helical fragments at atomic resolution: I. The crystal and molecular structure of sodium adenylyl-3′-5′-uridine hexahydrate. *J. Mol. Biol.* **1976**, *104*, 109–144. [CrossRef]
5. Rosenberg, J.M.; Seeman, N.C.; Day, R.O.; Rich, A. RNA double helical fragments at atomic resolution: II. The structure of sodium guanylyl-3′,5′-cytidine nonhydrate. *J. Mol. Biol.* **1976**, *104*, 145–167. [CrossRef]
6. Allen, F.H.; Bellard, S.; Brice, M.D.; Cartright, B.A.; Doubleday, A.; Higgs, H.; Hummelink, T.; Hummelink-Peters, B.G.; Kennard, O.; Motherwell, W.D.S.; et al. The Cambridge Crystallographic Data Centre: Computer-based search, retrieval, analysis and display of information. *Acta Crystallogr.* **1979**, *B35*, 2331–2339. [CrossRef]
7. Robertus, J.D.; Ladner, J.E.; Finch, J.T.; Rhodes, D.; Brown, R.S.; Clark, B.F.C.; Klug, A. Structure of yeast phenylalanine tRNA at 3 A resolution. *Nature* **1974**, *250*, 546–551. [CrossRef]
8. Suddath, F.; Quigley, G.; McPherson, A.; Sneden, D.; Kim, J.; Kim, S.; Rich, A. Three-dimensional structure of yeast phenylalanine transfer RNA at 3.0 Ångstroms resolution. *Nature* **1974**, *248*, 20–24. [CrossRef]
9. Quigley, G.J.; Seeman, N.C.; Wang, A.H.; Suddath, F.L.; Rich, A. Yeast phenylalanine transfer RNA: Atomic coordinates and torsion angles. *Nucleic Acids Res.* **1975**, *2*, 2329–2341. [CrossRef]
10. Wang, A.H.-J.; Quigley, G.J.; Kolpak, F.J.; Crawford, J.L.; van Boom, J.H.; van der Marel, G.A.; Rich, A. Molecular structure of a left-handed double helical DNA fragment at atomic resolution. *Nature* **1979**, *282*, 680–686. [CrossRef]
11. Drew, H.R.; Wing, R.M.; Takano, T.; Broka, C.; Tanaka, S.; Itakura, K.; Dickerson, R.E. Structure of a B-DNA dodecamer: Conformation and dynamics. *Proc. Natl. Acad. Sci. USA* **1981**, *78*, 2179–2183. [CrossRef] [PubMed]
12. Bernstein, F.C.; Koetzle, T.F.; Williams, G.J.; Meyer, E.F., Jr.; Brice, M.D.; Rodgers, J.R.; Kennard, O.; Shimanouchi, T.; Tasumi, M. The Protein Data Bank: A computer-based archival file for macromolecular structures. *J. Mol. Biol.* **1977**, *112*, 535–542. [CrossRef]
13. Breslauer, K.J. A calorimetric determination of enthalpies and heat capacities of protonation. *J. Chem. Thermodyn.* **1979**, *11*, 527–530. [CrossRef]
14. Breslauer, K.J. Methods for Obtaining Thermodynamic Data on Oligonucleotide Transitions. In *Thermodynamic Data for Biochemistry and Biotechnology*; Hinz, H., Ed.; Springer: New York, NY, USA, 1986; pp. 402–427.
15. Breslauer, K.J. A thermodynamic perspective of DNA Bending. *Curr. Biol.* **1991**, *1*, 416–422. [CrossRef]
16. Breslauer, K.J. Extracting Thermodynamic Data From Equilibrium Melting Curves for Oligonucleotide Order-Disorder Transitions. In *Methods in Molecular Biology, Vol. 26: Protocols for Oligonucleotide Conjugates*; Agrawal, S., Ed.; Humana Press: Totowa, NJ, USA, 1994; Chapter 14; pp. 347–372.
17. Chalikian, T.V.; Breslauer, K.J. Thermodynamic analysis of biomolecules: A volumetric approach. *Curr. Opin. Struct. Biol.* **1998**, *8*, 657–664. [CrossRef]
18. Chalikian, T.V.; Volker, J.; Plum, G.E.; Breslauer, K.J. A more unified picture for the thermodynamics of nucleic acid duplex melting: A characterization by calorimetric and volumetric techniques. *Proc. Natl. Acad. Sci. USA* **1999**, *96*, 7853–7858. [CrossRef]
19. Klump, H.H.; Volker, J.; Breslauer, K.J. Energy mapping of the genetic code and genomic domains: Implications for code evolution and molecular Darwinism. *Q. Rev. Biophys.* **2020**, *53*, e11. [CrossRef]
20. Volker, J.; Klump, H.H.; Breslauer, K.J. DNA metastability and biological regulation: Conformational dynamics of metastable omega-DNA bulge loops. *J. Am. Chem. Soc.* **2007**, *129*, 5272–5280. [CrossRef]
21. Jones, R. Preparation of protected deoxyribonucleosides. In *Oligonucleotide Synthesis, a Practical Approach*; Gait, M.J., Ed.; IRL Press: Washington, DC, USA, 1984; pp. 22–34.

22. Manning, G. The molecular theory of polyelectrolyte solutions with applications to the electrostatic properties of polynucleotides. *Q. Rev. Biophys.* **1978**, *11*, 179–246. [CrossRef]
23. Volker, J.; Klump, H.H.; Manning, G.S.; Breslauer, K.J. Counterion association with native and denatured nucleic acids: An experimental approach. *J. Mol. Biol.* **2001**, *310*, 1011–1025. [CrossRef]
24. Erie, D.; Sinha, N.; Olson, W.; Jones, R.; Breslauer, K. A dumbbell-shaped, double-hairpin structure of DNA: A thermodynamic investigation. *Biochemistry* **1987**, *26*, 7150–7159. [CrossRef] [PubMed]
25. Berman, H.M.; Olson, W.K.; Beveridge, D.L.; Westbrook, J.; Gelbin, A.; Demeny, T.; Hsieh, S.H.; Srinivasan, A.R.; Schneider, B. The nucleic acid database. A comprehensive relational database of three-dimensional structures of nucleic acids. *Biophys. J.* **1992**, *63*, 751–759. [CrossRef]
26. Kitakami, H.; Tateno, Y.; Gojobori, T. Toward unification of taxonomy databases in a distributed computer environment. *Proc. Int. Conf. Intell. Syst. Mol. Biol.* **1994**, *2*, 227–235. [PubMed]
27. Bilofsky, H.S.; Burks, C.; Fickett, J.W.; Goad, W.B.; Lewitter, F.I.; Rindone, W.P.; Swindell, C.D.; Tung, C.S. The GenBank genetic sequence databank. *Nucleic Acids Res.* **1986**, *14*, 1861–1863. [CrossRef]
28. Benson, D.A.; Karsch-Mizrachi, I.; Lipman, D.J.; Ostell, J.; Rapp, B.A.; Wheeler, D.L. GenBank. *Nucleic Acids Res.* **2000**, *28*, 15–18. [CrossRef]
29. Parker, M. Biological data access through Gopher. *Trends Biochem. Sci.* **1993**, *18*, 485–486. [CrossRef]
30. Fitzgerald, P.M.D.; Westbrook, J.D.; Bourne, P.E.; McMahon, B.; Watenpaugh, K.D.; Berman, H.M. 4.5 Macromolecular dictionary (mmCIF). In *International Tables for Crystallography G. Definition and Exchange of Crystallographic Data*; Hall, S.R., McMahon, B., Eds.; Springer: Dordrecht, The Netherlands, 2005; pp. 295–443.
31. Clowney, L.; Jain, S.C.; Srinivasan, A.R.; Westbrook, J.; Olson, W.K.; Berman, H.M. Geometric Parameters in Nucleic Acids: Nitrogenous Bases. *J. Am. Chem. Soc.* **1996**, *118*, 509–518. [CrossRef]
32. Gelbin, A.; Schneider, B.; Clowney, L.; Hsieh, S.-H.; Olson, W.K.; Berman, H.M. Geometric parameters in nucleic acids: Sugar and phosphate constituents. *J. Am. Chem. Soc.* **1996**, *118*, 519–528. [CrossRef]
33. Parkinson, G.; Vojtechovsky, J.; Clowney, L.; Brunger, A.T.; Berman, H.M. New parameters for the refinement of nucleic acid-containing structures. *Acta Cryst. D Biol. Cryst.* **1996**, *52 Pt 1*, 57–64. [CrossRef]
34. Olson, W.K.; Bansal, M.; Burley, S.K.; Dickerson, R.E.; Gerstein, M.; Harvey, S.C.; Heinemann, U.; Lu, X.J.; Neidle, S.; Shakked, Z.; et al. A standard reference frame for the description of nucleic acid base-pair geometry. *J. Mol. Biol.* **2001**, *313*, 229–237. [CrossRef]
35. Leontis, N.B.; Westhof, E. Geometric nomenclature and classification of RNA base pairs. *RNA* **2001**, *7*, 499–512. [CrossRef] [PubMed]
36. Schneider, B.; de la Cruz, J.; Feng, Z.; Chen, L.; Dutta, S.; Persikova, I.; Westbrook, J.; Yang, H.; Young, J.; Zardecki, C.; et al. The Nucleic Acid Database. In *Structural Bioinformatics*, 2nd ed.; Gu, J., Bourne, P.E., Eds.; Wiley-Blackwell: Hoboken, NJ, USA, 2009; pp. 305–319.
37. Coimbatore Narayanan, B.; Westbrook, J.; Ghosh, S.; Petrov, A.I.; Sweeney, B.; Zirbel, C.L.; Leontis, N.B.; Berman, H.M. The Nucleic Acid Database: New features and capabilities. *Nucleic Acids Res.* **2014**, *42*, D114–D122. [CrossRef]
38. Srinivasan, A.R.; Olson, W.K. Yeast tRNAPhe conformation wheels: A novel probe of the monoclinic and orthorhombic models. *Nucleic Acids Res.* **1980**, *8*, 2307–2329. [CrossRef]
39. Dans, P.D.; Perez, A.; Faustino, I.; Lavery, R.; Orozco, M. Exploring polymorphisms in B-DNA helical conformations. *Nucleic Acids Res.* **2012**, *40*, 10668–10678. [CrossRef]
40. Gupta, A.; Kulkarni, M.; Mukherjee, A. Accurate prediction of B-form/A-form DNA conformation propensity from primary sequence: A machine learning and free energy handshake. *Patterns* **2021**, *2*, 100329. [CrossRef] [PubMed]
41. Bayrak, C.S.; Kim, N.; Schlick, T. Using sequence signatures and kink-turn motifs in knowledge-based statistical potentials for RNA structure prediction. *Nucleic Acids Res.* **2017**, *45*, 5414–5422. [CrossRef]
42. Corsi, F.; Lavery, R.; Laine, E.; Carbone, A. Multiple protein-DNA interfaces unravelled by evolutionary information, physico-chemical and geometrical properties. *PLoS Comput. Biol.* **2020**, *16*, e1007624. [CrossRef]
43. Srivastava, A.; Ahmad, S.; Gromiha, M.M. Deciphering RNA-Recognition Patterns of Intrinsically Disordered Proteins. *Int. J. Mol. Sci.* **2018**, *19*, 1595. [CrossRef] [PubMed]
44. Sagendorf, J.M.; Markarian, N.; Berman, H.M.; Rohs, R. DNAproDB: An expanded database and web-based tool for structural analysis of DNA-protein complexes. *Nucleic Acids Res.* **2020**, *48*, D277–D287. [CrossRef]
45. Schneider, B.; Berman, H.M. Hydration of the DNA bases is local. *Biophys. J.* **1995**, *69*, 2661–2669. [CrossRef]
46. Schneider, B.; Neidle, S.; Berman, H.M. Conformations of the sugar-phosphate backbone in helical DNA crystal structures. *Biopolymers* **1997**, *42*, 113–124. [CrossRef]
47. Kim, S.-H.; Berman, H.M.; Seeman, N.C.; Newton, M.D. Seven basic conformations of nucleic acid structural units. *Acta Crystallogr. Sect. B* **1973**, *29*, 703–710. [CrossRef]
48. Richardson, J.S.; Schneider, B.; Murray, L.W.; Kapral, G.J.; Immormino, R.M.; Headd, J.J.; Richardson, D.C.; Ham, D.; Hershkovits, E.; Williams, L.D.; et al. RNA backbone: Consensus all-angle conformers and modular string nomenclature (an RNA Ontology Consortium contribution). *RNA* **2008**, *14*, 465–481. [CrossRef] [PubMed]
49. Schneider, B.; Bozikova, P.; Cech, P.; Svozil, D.; Cerny, J. A DNA Structural Alphabet Distinguishes Structural Features of DNA Bound to Regulatory Proteins and in the Nucleosome Core Particle. *Genes* **2017**, *8*, 278. [CrossRef]

50. Cerny, J.; Bozikova, P.; Svoboda, J.; Schneider, B. A unified dinucleotide alphabet describing both RNA and DNA structures. *Nucleic Acids Res.* **2020**, *48*, 6367–6381. [CrossRef]
51. Gouge, J.; Satia, K.; Guthertz, N.; Widya, M.; Thompson, A.J.; Cousin, P.; Dergai, O.; Hernandez, N.; Vannini, A. Redox Signaling by the RNA Polymerase III TFIIB-Related Factor Brf2. *Cell* **2015**, *163*, 1375–1387. [CrossRef]
52. Frouws, T.D.; Duda, S.C.; Richmond, T.J. X-ray structure of the MMTV-A nucleosome core. *Proc. Natl. Acad. Sci. USA* **2016**, *113*, 1214–1219. [CrossRef]
53. Olson, W.K. Configurational statistics of polynucleotide chains. An updated virtual bond model to treat effects of base stacking. *Macromolecules* **1980**, *13*, 721–728. [CrossRef]
54. Wadley, L.M.; Keating, K.S.; Duarte, C.M.; Pyle, A.M. Evaluating and learning from RNA pseudotorsional space: Quantitative validation of a reduced representation for RNA structure. *J. Mol. Biol.* **2007**, *372*, 942–957. [CrossRef]
55. Vander Zanden, C.M.; Rowe, R.K.; Broad, A.J.; Robertson, A.B.; Ho, P.S. Effect of Hydroxymethylcytosine on the Structure and Stability of Holliday Junctions. *Biochemistry* **2016**, *55*, 5781–5789. [CrossRef]
56. Wang, Y.; Patel, D.J. Solution structure of a parallel-stranded G-quadruplex DNA. *J. Mol. Biol.* **1993**, *234*, 1171–1183. [CrossRef] [PubMed]
57. Davey, C.A.; Sargent, D.F.; Luger, K.; Maeder, A.W.; Richmond, T.J. Solvent mediated interactions in the structure of the nucleosome core particle at 1.9 a resolution. *J. Mol. Biol.* **2002**, *319*, 1097–1113. [CrossRef]
58. Lawson, C.L.; Carey, J. Tandem binding in crystals of a trp repressor/operator half-site complex. *Nature* **1993**, *366*, 178–182. [CrossRef]
59. Weaver, T.M.; Cortez, L.M.; Khoang, T.H.; Washington, M.T.; Agarwal, P.K.; Freudenthal, B.D. Visualizing Rev1 catalyze protein-template DNA synthesis. *Proc. Natl. Acad. Sci. USA* **2020**, *117*, 25494–25504. [CrossRef]
60. Jang, S.B.; Hung, L.W.; Chi, Y.I.; Holbrook, E.L.; Carter, R.J.; Holbrook, S.R. Structure of an RNA internal loop consisting of tandem C-A+ base pairs. *Biochemistry* **1998**, *37*, 11726–11731. [CrossRef] [PubMed]
61. Chan, C.W.; Badong, D.; Rajan, R.; Mondragon, A. Crystal structures of an unmodified bacterial tRNA reveal intrinsic structural flexibility and plasticity as general properties of unbound tRNAs. *RNA* **2020**, *26*, 278–289. [CrossRef]
62. Huang, L.; Wang, J.; Watkins, A.M.; Das, R.; Lilley, D.M.J. Structure and ligand binding of the glutamine-II riboswitch. *Nucleic Acids Res.* **2019**, *47*, 7666–7675. [CrossRef] [PubMed]
63. Noeske, J.; Wasserman, M.R.; Terry, D.S.; Altman, R.B.; Blanchard, S.C.; Cate, J.H. High-resolution structure of the Escherichia coli ribosome. *Nat. Struct. Mol. Biol.* **2015**, *22*, 336–341. [CrossRef]
64. Li, X.; Liu, S.; Zhang, L.; Issaian, A.; Hill, R.C.; Espinosa, S.; Shi, S.; Cui, Y.; Kappel, K.; Das, R.; et al. A unified mechanism for intron and exon definition and back-splicing. *Nature* **2019**, *573*, 375–380. [CrossRef]
65. Landeras-Bueno, S.; Wasserman, H.; Oliveira, G.; VanAernum, Z.L.; Busch, F.; Salie, Z.L.; Wysocki, V.H.; Andersen, K.; Saphire, E.O. Cellular mRNA triggers structural transformation of Ebola virus matrix protein VP40 to its essential regulatory form. *Cell Rep.* **2021**, *35*, 108986. [CrossRef]
66. Lu, X.J. DSSR-enabled innovative schematics of 3D nucleic acid structures with PyMOL. *Nucleic Acids Res.* **2020**, *48*, e74. [CrossRef]
67. Schneider, B.; Bozikova, P.; Necasova, I.; Cech, P.; Svozil, D.; Cerny, J. A DNA structural alphabet provides new insight into DNA flexibility. *Acta Cryst. D Struct. Biol.* **2018**, *74 Pt 1*, 52–64. [CrossRef] [PubMed]
68. Appasamy, S.D.; Hamdani, H.Y.; Ramlan, E.I.; Firdaus-Raih, M. InterRNA: A database of base interactions in RNA structures. *Nucleic Acids Res.* **2016**, *44*, D266–D471. [CrossRef] [PubMed]
69. Boccaletto, P.; Machnicka, M.A.; Purta, E.; Piatkowski, P.; Baginski, B.; Wirecki, T.K.; de Crecy-Lagard, V.; Ross, R.; Limbach, P.A.; Kotter, A.; et al. MODOMICS: A database of RNA modification pathways. 2017 update. *Nucleic Acids Res.* **2018**, *46*, D303–D307. [CrossRef] [PubMed]
70. Chojnowski, G.; Walen, T.; Bujnicki, J.M. RNA Bricks—A database of RNA 3D motifs and their interactions. *Nucleic Acids Res.* **2014**, *42*, D123–D131. [CrossRef]
71. Zok, T.; Antczak, M.; Zurkowski, M.; Popenda, M.; Blazewicz, J.; Adamiak, R.W.; Szachniuk, M. RNApdbee 2.0: Multifunctional tool for RNA structure annotation. *Nucleic Acids Res.* **2018**, *46*, W30–W35. [CrossRef]
72. The RNAcentral Consortium; Petrov, A.I.; Kay, S.J.E.; Kalvari, I.; Howe, K.L.; Gray, K.A.; Bruford, E.A.; Kersey, P.J.; Cochrane, G.; Finn, R.D. RNAcentral: A comprehensive database of non-coding RNA sequences. *Nucleic Acids Res.* **2017**, *45*, D128–D134.
73. Bernier, C.R.; Petrov, A.S.; Waterbury, C.C.; Jett, J.; Li, F.; Freil, L.E.; Xiong, X.; Wang, L.; Migliozzi, B.L.; Hershkovits, E.; et al. RiboVision suite for visualization and analysis of ribosomes. *Faraday Discuss.* **2014**, *169*, 195–207. [CrossRef]
74. Sagendorf, J.M.; Berman, H.M.; Rohs, R. DNAproDB: An interactive tool for structural analysis of DNA-protein complexes. *Nucleic Acids Res.* **2017**, *45*, W89–W97. [CrossRef]
75. Petrov, A.I.; Zirbel, C.L.; Leontis, N.B. Automated classification of RNA 3D motifs and the RNA 3D Motif Atlas. *RNA* **2013**, *19*, 1327–1340. [CrossRef]
76. Zirbel, C.; Leontis, N. Nonredundant 3D Structure Datasets for RNA Knowledge Extraction and Benchmarking. In *RNA 3D Structure Analysis and Prediction*; Leontis, N., Westhof, E., Eds.; Springer: Berlin/Heidelberg, Germany, 2012; Volume 27, pp. 281–298.

77. Dana, J.M.; Gutmanas, A.; Tyagi, N.; Qi, G.; O'Donovan, C.; Martin, M.; Velankar, S. SIFTS: Updated Structure Integration with Function, Taxonomy and Sequences resource allows 40-fold increase in coverage of structure-based annotations for proteins. *Nucleic Acids Res.* **2019**, *47*, D482–D489. [CrossRef] [PubMed]
78. Lu, X.J.; Bussemaker, H.J.; Olson, W.K. DSSR: An integrated software tool for dissecting the spatial structure of RNA. *Nucleic Acids Res.* **2015**, *43*, e142. [CrossRef] [PubMed]

MDPI
St. Alban-Anlage 66
4052 Basel
Switzerland
Tel. +41 61 683 77 34
Fax +41 61 302 89 18
www.mdpi.com

Life Editorial Office
E-mail: life@mdpi.com
www.mdpi.com/journal/life